Understanding Molecular Simulation

From Algorithms to Applications

Computational Science

From Theory to Applications

Series Editors

Daan Frenkel
FOM Institute for Atomic and Molecular Physics
Amsterdam, The Netherlands

Michael Klein
Laboratory for Research on the Structure of Matter
University of Pennsylvania, USA

Michele Parrinello
CSCS Swiss Center for Scientific Computing / ETH Zurich
Switzerland

Berend Smit
Department of Chemical Engineering
University of Amsterdam
Amsterdam, The Netherlands

Bringing together theories and techniques from a variety of fields and providing details on applying these theories in a multitude of applications, this series provides timely and thorough coverage of a broad range of topical and interdisciplinary areas. The books identify similarities across disciplines thereby helping researches take their work in new directions.

www.academicpress.com / computationalscience

Volume 1 in the
COMPUTATIONAL SCIENCE SERIES

Understanding Molecular Simulation

From Algorithms to Applications

Daan Frenkel
FOM Institute for Atomic and Molecular Physics,
Amsterdam, The Netherlands

Department of Chemical Engineering,
Faculty of Sciences
University of Amsterdam
Amsterdam, The Netherlands

Berend Smit
Department of Chemical Engineering
Faculty of Sciences
University of Amsterdam
Amsterdam, The Netherlands

ACADEMIC PRESS

An Imprint of Elsevier

San Diego San Francisco New York
Boston London Sydney Tokyo

This book is printed on acid-free paper. ♾

Academic Press
An Imprint of Elsevier
525 B Street, Suite 1900, San Diego, California 92101-4495, USA
http://www.academicpress.com

Academic Press
84 Theobald's Road, London WC1X 8RR, UK
http://www.academicpress.com

Library of Congress Catalog Number: 2001091477

ISBN-13: 978-0-12-267351-1
ISBN-10: 0-12-267351-4

Cover illustration: a 2,5-dimethyloctane molecule inside a pore of a TON type zeolite (figure by David Dubbeldam)

Typeset by the authors

PRINTED IN THE UNITED STATES OF AMERICA
12 13 14 15 16 17 16 15 14 13

Contents

Preface to the Second Edition

Why did we write a second edition? A minor revision of the first edition would have been adequate to correct the (admittedly many) typographical mistakes. However, many of the nice comments that we received from students and colleagues alike, ended with a remark of the type: "unfortunately, you don't discuss topic *x*". And indeed, we feel that, after only five years, the simulation world has changed so much that the title of the book was no longer covered by the contents.

The first edition was written in 1995 and since then several new techniques have appeared or matured. Most (but not all) of the major changes in the second edition deal with these new developments. In particular, we have included a section on:

- Transition path sampling and diffusive barrier crossing to simulate rare events
- Dissipative particle dynamic as a course-grained simulation technique
- Novel schemes to compute the long-ranged forces
- Discussion on Hamiltonian and non-Hamiltonian dynamics in the context of constant-temperature and constant-pressure Molecular Dynamics simulations
- Multiple-time-step algorithms as an alternative for constraints
- Defects in solids
- The pruned-enriched Rosenbluth sampling, recoil growth, and concerted rotations for complex molecules
- Parallel tempering for glassy Hamiltonians

We have updated some of the examples to include also recent work. Several new Examples have been added to illustrate recent applications.

We have taught several courses on Molecular Simulation, based on the first edition of this book. As part of these courses, Dr Thijs Vlugt prepared many *Questions*, *Exercises*, and *Case Studies*, most of which have been included in the present edition. Some additional exercises can be found on the Web. We are very grateful to Thijs Vlugt for the permission to reproduce this material.

Many of the advanced Molecular Dynamics techniques described in this book are derived using the Lagrangian or Hamilton formulations of classical mechanics. However, many chemistry and chemical engineering students are not familiar with these formalisms. While a full description of classical mechanics is clearly beyond the scope of the present book, we have added an Appendix that summarizes the necessary essentials of Lagrangian and Hamiltonian mechanics.

Special thanks are due to Giovanni Ciccotti, Rob Groot, Gavin Crooks, Thijs Vlugt, and Peter Bolhuis for their comments on parts of the text. In addition, we thank everyone who pointed out mistakes and typos, in particular Drs. J.B. Freund, R. Akkermans, and D. Moroni.

Preface

This book is not a computer simulation cookbook. Our aim is to explain the physics that is behind the "recipes" of molecular simulation. Of course, we also give the recipes themselves, because otherwise the book would be too abstract to be of much practical use. The scope of this book is necessarily limited: we do not aim to discuss all aspects of computer simulation. Rather, we intend to give a unified presentation of those computational tools that are currently used to study the equilibrium properties and, in particular, the phase behavior of molecular and supramolecular substances. Moreover, we intentionally restrict the discussion to simulations of *classical* many-body systems, even though some of the techniques mentioned can be applied to quantum systems as well. And, within the context of classical many-body systems, we restrict our discussion to the modeling of systems at, or near, equilibrium.

The book is aimed at readers who are active in computer simulation or are planning to become so. Computer simulators are continuously confronted with questions concerning the choice of technique, because a bewildering variety of computational tools is available. We believe that, to make a rational choice, a good understanding of the physics behind each technique is essential. Our aim is to provide the reader with this background.

We should state at the outset that we consider some techniques to be more useful than others, and therefore our presentation is biased. In fact, we believe that the reader is well served by the fact that we do not present all techniques as equivalent. However, whenever we express our personal preference, we try to back it up with arguments based in physics, applied mathematics, or simply experience. In fact, we mix our presentation with practical examples that serve a twofold purpose: first, to show how a given technique works in practice, and second, to give the reader a flavor of the kind of phenomena that can be studied by numerical simulation.

The reader will also notice that two topics are discussed in great detail, namely simulation techniques to study first-order phase transitions, and various aspects of the configurational bias Monte Carlo method. The reason why we devote so much space to these topics is not that we consider them to be more important than other subjects that get less coverage, but rather because we feel that, at present, the discussion of both topics in the literature is rather fragmented.

The present introduction is written for the nonexpert. We have done so on purpose. The community of people who perform computer simulations is rapidly expanding as computer experiments become a general research tool. Many of the new simulators will use computer simulation as a tool and will not be primarily interested in techniques. Yet, we hope to convince those readers who consider a computer simulation program a *black box*, that the inside of the black box is interesting and, more importantly, that a better understanding of the working of a simulation program may greatly improve the efficiency with which the black box is used.

In addition to the theoretical framework, we discuss some of the practical tricks and rules of thumb that have become "common" knowledge in the simulation community and are routinely used in a simulation. Often, it is difficult to trace back the original motivation behind these rules. As a result, some "tricks" can be very useful in one case yet result in inefficient programs in others. In this book, we discuss the rationale behind the various tricks, in order to place them in a proper context. In the main text of the book we describe the theoretical framework of the various techniques. To illustrate how these ideas are used in practice we provide Algorithms, Case Studies and Examples.

Algorithms

The description of an algorithm forms an essential part of this book. Such a description, however, does not provide much information on how to implement the algorithm efficiently. Of course, details about the implementation of an algorithm can be obtained from a listing of the complete program. However, even in a well-structured program, the code contains many lines that, although necessary to obtain a working program, tend to obscure the essentials of the algorithm that they express. As a compromise solution, we provide a pseudo-code for each algorithm. These pseudo-codes contain only those aspects of the implementation directly related to the particular algorithm under discussion. This implies that some aspects that are essential for using this pseudo-code in an actual program have to be added. For example, the pseudo-codes consider only the x directions; similar lines have to be added for the y and z direction if the code is going to be used in a simulation. Furthermore, we have omitted the initialization of most variables.

Case Studies

In the Case Studies , the algorithms discussed in the main text are combined in a complete program. These programs are used to illustrate some elementary aspects of simulations. Some Case Studies focus on the problems that can occur in a simulation or on the errors that are sometimes made. The complete listing of the FORTRAN codes that we have used for the Case Studies is accessible to the reader through the Internet.[1]

Examples

In the Examples, we demonstrate how the techniques discussed in the main text are used in an application. We have tried to refer as much as possible to research topics of current interest. In this way, the reader may get some feeling for the type of systems that can be studied with simulations. In addition, we have tried to illustrate in these examples how simulations can contribute to the solution of "real" experimental or theoretical problems.

Many of the topics that we discuss in this book have appeared previously in the open literature. However, the Examples and Case Studies were prepared specifically for this book. In writing this material, we could not resist including a few computational tricks that, to our knowledge, have not been reported in the literature.

In computer science it is generally assumed that any source code over 200 lines contains at least one error. The source codes of the Case Studies contain over 25,000 lines of code. Assuming we are no worse than the average programmer this implies that we have made at least 125 errors in the source code. If you spot these errors and send them to us, we will try to correct them (we can not promise this!). It also implies that, before you use part of the code yourself, you should convince yourself that the code is doing what you expect it to do.

In the light of the previous paragraph, we must add the following disclaimer:

> We make no warranties, express or implied, that the programs contained in this work are free of error, or that they will meet your requirements for any particular application. They should not be relied on for solving problems whose incorrect solution could result in injury, damage, or loss of property. The authors and publishers disclaim all liability for direct or consequential damages resulting from your use of the programs.

Although this book and the included programs are copyrighted, we authorize the readers of this book to use parts of the programs for their own use, provided that proper acknowledgment is made.

Finally, we gratefully acknowledge the help and collaboration of many of our colleagues. In fact, many dozens of our colleagues collaborated with us on topics described in the text. Rather than listing them all here, we mention their names at the appropriate place in the text. Yet, we do wish to

[1] `http://molsim.chem.uva.nl/frenkel_smit`

express our gratitude for their input. Moreover, Daan Frenkel should like to acknowledge numerous stimulating discussions with colleagues at the FOM Institute for Atomic and Molecular Physics in Amsterdam and at the van 't Hoff Laboratory of Utrecht University, while Berend Smit gratefully acknowledges discussions with colleagues at the University of Amsterdam and Shell. In addition, several colleagues helped us directly with the preparation of the manuscript, by reading the text or part thereof. They are Giovanni Ciccotti, Mike Deem, Simon de Leeuw, Toine Schlijper, Stefano Ruffo, Maria-Jose Ruiz, Guy Verbist and Thijs Vlugt. In addition, we thank Klaas Esselink and Sami Karaborni for the cover figure. We thank them all for their efforts. In addition we thank the many readers who have drawn our attention to errors and omissions in the first print. But we stress that the responsibility for the remainder of errors in the text is ours alone.

List of Symbols

A	dynamical variable	(2.2.6)
$\text{acc}(o \to n)$	acceptance probability of a move from o to n	
$\mathbf{b}$	trial position or orientation	
c	concentration	
c_i	concentration of species i	
C_V	specific heat at constant volume	(4.4.3)
d	dimensionality	
D	diffusion coefficient	
E	total energy	
f	number of degrees of freedom	
f_i	fugacity component i	(9.1.9)
$\mathbf{f}_i$	force on particle i	
F	Helmholtz free energy	(2.1.15)
$g(r)$	radial distribution function	
G	Gibbs free energy	(5.4.9)
$h = 2\pi\hbar$	Planck's constant	
$\mathcal{H}(\mathbf{p}, \mathbf{r})$	Hamiltonian	
$\mathbf{j}_i$	flux of species i	
$\mathbf{k}$	wave vector	
k_B	Boltzmann's constant	
K	kinetic energy	
$K(o \to n)$	flow of configurations from o to n	
L	box length	
$\mathcal{L}(\mathbf{q}, \dot{\mathbf{q}})$	Lagrangian	(A.1.2)
ℓ	total number of (pseudo-)atoms in a molecule (chain length)	
n	new configuration or conformation	
m	mass	
M	total number of Monte Carlo samples	
N	number of particles	
$\mathcal{N}(o)$	prob. density to find a system in configuration o	
$\mathcal{N}_{N,V,T}$	prob. density for canonical ensemble	(5.2.2)
$\mathcal{N}_{N,P,T}$	prob. density for isobaric-isothermal ensemble	(5.4.8)
$\mathcal{N}_{\mu,V,T}$	prob. density for grand-canonical ensemble	(5.6.6)
o	old configuration or conformation	
$\mathcal{O}(x^n)$	terms of order x^n or smaller	

$\mathbf{p}$	momentum of a particle	
P	pressure	
$\mathcal{P}$	total linear momentum	
q_i	electric charge on particle i	
$\mathbf{q}$	generalized coordinates	
Q	mass associated with time scaling coordinate s	(6.1.3)
$Q(N,V,T)$	canonical partition function	(5.2.1)
$Q(N,P,T)$	isothermal-isobaric partition function	(5.4.7)
$Q(\mu,V,T)$	grand-canonical partition function	(5.6.5)
$\mathbf{r}_i$	Cartesian coordinate of particle i	
r_c	cut-off radius of the potential	
Ranf	random number uniform in [0, 1]	
s	time-scaling coordinate in Nose scheme	(6.1.3)
$\mathbf{s}_i$	scaled coordinate of particle i	(5.4.2)
S	entropy	
t	time	
T	temperature	
$\mathcal{U}(o)$	potential energy of configuration o	
u_i	potential energy per particle	
$u(r)$	pair potential	
$\mathbf{v}_i$	Cartesian velocity of particle i	
V	volume	
vir	virial	(3.4.2)
w_i	Rosenbluth factor of (pseudo-)atom i	
$W(o)$	total Rosenbluth factor configuration o	
$\mathcal{W}(o)$	normalized total Rosenbluth factor configuration o	
Z	configurational part of the partition function	
$\alpha(o \to n)$	probability of generating conf. n starting from o	(3.1.14)
β	reciprocal temperature $(1/k_BT)$	
Δt	Molecular Dynamics time step	
Γ	coordinate in phase space	
ϵ	characteristic energy in pair potential	
η	shear viscosity	(4.4.12)
μ	chemical potential	
λ_T	thermal conductivity	(4.4.14)
Λ	thermal de Broglie wavelength	(5.4.1)
$\pi(o \to n)$	transition probability from o to n	(3.1.13)
ρ	number density	
σ	characteristic distance in pair potential	
σ_e	electrical conductivity	(4.4.16)
$\sigma_{\alpha\beta}$	$\alpha\beta$ component of the stress tensor	(4.4.13)
σ_A^2	variance in dynamical variable A	
ξ	thermodynamic friction coefficient	(6.1.25)
ξ_i	fugacity coefficient of component i	(9.1.15)
$\Omega(E)$	quantum: degeneracy of energy level E	(2.1.1)
	classical: phase space subvolume with energy E	(2.2.9)

ω	orientation of a molecule
$\langle \cdots \rangle$	ensemble average
$\langle \cdots \rangle_{\text{sub}}$	average under condition indicated by sub

Super- and subscripts

$*$	reduced units (default, usually omitted)
$\mathbf{r}_\alpha$	α component of vector $\mathbf{r}$
$\mathbf{r}_i$	vector $\mathbf{r}$ associated with particle i
f^{ex}	excess part of quantity f
f^{id}	ideal gas part of quantity f
$\hat{\mathbf{u}}$	unit vector $\mathbf{u}$

Symbol List: Algorithms

`b(j)`	trial orientation/position j
`beta`	reciprocal temperature ($1/k_B T$)
`box`	simulation box length
`delx`	maximum displacement
`dt`	time step in an MD simulation
`ell`	chain length
`eni`	energy of atom i
`enn`	energy of the new configuration
`eno`	energy of the old configuration
`etot`	total energy of the system
`f`	force
`k`	total number of trial orientations/position
`kv`	bond vibration energy constant
`l`	bond length
`n`	selected trial position
`ncycle`	total number of MC cycles
`nhis`	number of bins in a histogram
`NINT`	nearest integer
`npart`	total number of particles
`nsamp`	number of MC cycles of MD steps between two samples
`o`	particle number of the old configuration
`p`	pressure
`phi`	bond-bending angle
`pi`	$\pi = 3.14159$
`r2`	distance squared between two atoms
`ranf()`	random number $\in [0, 1]$
`rho`	density
`rc2`	cutoff radius squared (of the potential)
`switch`	= 0 initialization; = 1 sample; = 2 print result

`t`	time in a MD simulation
`temp`	temperature
`tempa`	instantaneous temperature, from kinetic energy
`theta`	torsional angle
`tmax`	maximum simulation time
`tors`	torsion energy
`v(i)`	velocity of atom `i`
`vmax`	maximum displacement volume
`vol`	volume simulation box
`w`	Rosenbluth factor (new or old)
`wn, wo`	Rosenbluth factor n(ew)/o(ld) configuration
`x(i)`	position of atom `i`
`xm(i)`	position of atom `i` at previous time step
`xn`	new configuration of a particle
`xn(i)`	positions of atoms that have been grown
`xo`	old configuration of a particle
`xt(j)`	`j`th trial position for a given atom
`ubb`	bond-bending energy
`utors`	torsion energy
•	vector dot product
×	vector cross product
$\|\cdots\|$	length of the vector
a.le.b	test: true if a is less than or equal to b
a.lt.b	test: true if a is less than b
a.ge.b	test: true if a is greater than or equal to b
a.gt.b	test: true if a is greater than b
a.and.b	test: true if both a and b are true
a.or.b	test: true if a or b are true
+	continuation symbol
*	multiplication
**	to the power
/	division
`sqrt`	square root

Chapter 1

Introduction

(Pre)history of Computer Simulation

It usually takes decades rather than years before a fundamentally new invention finds widespread application. For computer simulation, the story is rather different. Computer simulation started as a tool to exploit the electronic computing machines that had been developed during and after the Second World War. These machines had been built to perform the very heavy computation involved in the development of nuclear weapons and code breaking. In the early 1950s, electronic computers became partly available for nonmilitary use and this was the beginning of the discipline of computer simulation. W. W. Wood [1] recalls: "When the Los Alamos MANIAC became operational in March 1952, Metropolis was interested in having as broad a spectrum of problems as possible tried on the machine, in order to evaluate its logical structure and demonstrate the capabilities of the machine."

The strange thing about computer simulation is that it is also a discovery, albeit a delayed discovery that grew slowly after the introduction of the technique. In fact, *discovery* is probably not the right word, because it does not refer to a new insight into the working of the natural world but into our description of nature. Working with computers has provided us with a new metaphor for the laws of nature: they carry as much (and as little) information as algorithms. For any nontrivial algorithm (i.e., loosely speaking, one that cannot be solved analytically), you cannot predict the outcome of a computation simply by looking at the program, although it often is possible to make precise statements about the general nature (e.g., the symmetry) of the result of the computation. Similarly, the basic laws of nature as we know them have the unpleasant feature that they are expressed in terms of equations we cannot solve exactly, except in a few very special cases. If we wish to study the motion of more than two interacting bodies, even the relatively simple laws of Newtonian mechanics become essentially unsolvable. That is to say, they cannot be solved analytically, using only pencil and the back of the proverbial envelope. However, using a computer, we can get the answer to any desired accuracy. Most of materials science deals with the properties of systems of many atoms or molecules. *Many* almost always means more than two; usually, very much more. So if we wish to compute the properties of a liquid (to take a particularly nasty example), there is no hope of finding the answer exactly using only pencil and paper.

Before computer simulation appeared on the scene, there was only one way to predict the properties of a molecular substance, namely by making use of a theory that provided an approximate description of that material. Such approximations are inevitable precisely because there are very few systems for which the equilibrium properties can be computed exactly (examples are the ideal gas, the harmonic crystal, and a number of lattice models, such as the two-dimensional Ising model for ferromagnets). As a result, most properties of real materials were predicted on the basis of approximate theories (examples are the van der Waals equation for dense gases, the Debye-Hückel theory for electrolytes, and the Boltzmann equation to describe the transport properties of dilute gases). Given sufficient information about the intermolecular interactions, these theories will provide us with an estimate of the properties of interest. Unfortunately, our knowledge of the intermolecular interactions of all but the simplest molecules is also quite limited. This leads to a problem if we wish to test the validity of a particular theory by comparing directly to experiment. If we find that theory and experiment disagree, it may mean that our theory is wrong, or that we have an incorrect estimate of the intermolecular interactions, or both.

Clearly, it would be very nice if we could obtain essentially exact results for a given model system without having to rely on approximate theories. Computer simulations allow us to do precisely that. On the one hand, we can now compare the calculated properties of a model system with those of an experimental system: if the two disagree, our model is inadequate; that is, we have to improve on our estimate of the intermolecular interactions. On the other hand, we can compare the result of a simulation of a given model system with the predictions of an approximate analytical theory applied to the same model. If we now find that theory and simulation disagree, we know that the *theory* is flawed. So, in this case, the computer simulation plays the role of the experiment designed to test the theory. This method of screening theories before we apply them to the real world is called a *computer experiment*. This application of computer simulation is of tremendous importance. It has led to the revision of some very respectable theories, some of them dating back to Boltzmann. And it has changed the way in which we construct new theories. Nowadays it is becoming increasingly rare that a theory is applied to the real world before being tested by computer simula-

$\rho k_B T$	P
1	1.03 ± 0.04
2	1.99 ± 0.03
3	2.98 ± 0.05
4	4.04 ± 0.03
5	5.01 ± 0.04

Table 1.1: Simulated equation of state of an ideal gas

tion. The simulation then serves a twofold purpose: it gives the theoretician a feeling for the physics of the problem, and it generates some "exact" results that can be used to test the quality of the theory to be constructed. Computer experiments have become standard practice, to the extent that they now provide the first (and often the last) test of a new theoretical result.

But note that the computer as such offers us no understanding, only numbers. And, as in a real experiment, these numbers have statistical errors. So what we get out of a simulation is never directly a theoretical relation. As in a real experiment, we still have to extract the useful information. To take a not very realistic example, suppose we were to use the computer to measure the pressure of an ideal gas as a function of density. This example is unrealistic because the volume dependence of the ideal-gas pressure has, in fact, been well known since the work of Boyle and Gay-Lussac. The Boyle-Gay-Lussac law states that the product of volume and pressure of an ideal gas is constant. Now suppose we were to measure this product by computer simulation. We might, for instance, find the set of experimental results in Table 1.1. The data suggest that P equals $\rho k_B T$, but no more than that. It is left to us to infer the conclusions.

The early history of computer simulation (see, e.g., ref. [2]) illustrates this role of computer simulation. Some areas of physics appeared to have little need for simulation because very good analytical theories were available (e.g., to predict the properties of dilute gases or of nearly harmonic crystalline solids). However, in other areas, few if any exact theoretical results were known, and progress was much hindered by the lack of unambiguous tests to assess the quality of approximate theories. A case in point was the theory of dense liquids. Before the advent of computer simulations, the only way to model liquids was by mechanical simulation [3–5] of large assemblies of macroscopic spheres (e.g., ball bearings). Then the main problem becomes how to arrange these balls in the same way as atoms in a liquid. Much work on this topic was done by the famous British scientist J. D. Bernal, who built and analyzed such mechanical models for liquids. Actually, it would be fair to say that the really tedious work of analyzing the resulting three-dimensional structures was done by his research students, such as the unfortunate Miss Wilkinson whose research assignment was to identify all distinct local packing geometries of plastic foam spheres: she found that there were at least 197. It is instructive to see how Bernal built some of his models. The following quote from the 1962 Bakerian lecture describes Bernal's attempt to build a ball-and-spoke model of a liquid [5]:

> ...I took a number of rubber balls and stuck them together with rods of a selection of different lengths ranging from 2.75 to 4 inch. I tried to do this in the first place as casually as possible, working in my own office, being interrupted every five minutes or so and not remembering what I had done before the interruption. However,....

Subsequent models were made, for instance, by pouring thousands of steel balls from ball bearings into a balloon. It should be stressed that these mechanical models for liquids were in some respects quite realistic. However, the analysis of the structures generated by mechanical simulation was very laborious and, in the end, had to be performed by computer anyway.

In view of the preceding, it is hardly surprising that, when electronic computers were, for the first time, made available for unclassified research, numerical simulation of dense liquids was one of the first problems to be tackled. In fact, the first simulation of a liquid was carried out by Metropolis, Rosenbluth, Rosenbluth, Teller, and Teller on the MANIAC computer at Los Alamos [6], using (or, more properly, introducing) the Metropolis Monte Carlo (MC) method. The name *Monte Carlo simulation* had been coined earlier by Metropolis and Ulam (see Ref. [7]), because the method makes heavy use of computer-generated random numbers. Almost at the same time, Fermi, Pasta, and Ulam [8] performed their famous numerical study of the dynamics of an anharmonic, one-dimensional crystal. The first proper Molecular Dynamics (MD) simulations were reported in 1956 by Alder and Wainwright [9] at Livermore, who studied the dynamics of an assembly of hard spheres. The first MD simulation of a model for a "real" material was reported in 1959 (and published in 1960) by the group led by Vineyard at Brookhaven [10], who simulated radiation damage in crystalline Cu (for a historical account, see [11]). The first MD simulation of a real liquid (argon) was reported in 1964 by Rahman at Argonne [12]. After that, computers were increasingly becoming available to scientists outside the US government labs, and the practice of simulation started spreading to other continents [13–16]. Much of the methodology of computer simulations has been developed since then, although it is fair to say that the basic algorithms for MC and MD have hardly changed since the 1950s.

The most common application of computer simulations is to predict the properties of materials. The need for such simulations may not be immediately obvious. After all it is much easier to measure the freezing point of water than to extract it from a computer simulation. The point is, of course, that it is easy to measure the freezing point of water at 1 atmosphere but often very difficult and therefore expensive to measure the properties of real materials at very high pressures or temperatures. The computer does not care: it does not go up in smoke when you ask it to simulate a system at 10,000 K. In addition, we can use computer simulation to predict the properties of materials that have not yet been made. And finally, computer simulations are increasingly used in data analysis. For instance, a very efficient technique for obtaining structural information about macromolecules from 2D-NMR is to feed the experimental data into a Molecular Dynamics simulation and let the computer find the structure that is both energetically favorable and compatible with the available NMR data.

Initially, such simulations were received with a certain amount of skepticism, and understandably so. Simulation did not fit into the existing idea that whatever was not experiment had to be theory. In fact, many scientists much preferred to keep things the way they were: theory for the theoreticians and experiments for the experimentalists and no computers to confuse the issue. However, this position became untenable, as is demonstrated by the following autobiographical quote of George Vineyard [11], who was the first to study the dynamics of radiation damage by numerical simulation:

> ...In the summer of 1957 at the Gordon Conference on Chemistry and Physics of Metals, I gave a talk on radiation damage in metals .... After the talk there was a lively discussion .... Somewhere the idea came up that a computer might be used to follow in more detail what actually goes on in radiation damage cascades. We got into quite an argument, some maintaining that it wasn't possible to do this on a computer, others that it wasn't necessary. John Fisher insisted that the job could be done well enough by hand, and was then goaded into promising to demonstrate. He went off to his room to work. Next morning he asked for a little more time, promising to send me the results soon after he got home. After about two weeks, not having heard from him, I called and he admitted that he had given up. This stimulated me to think further about how to get a high-speed computer into the game in place of John Fisher. ...

Finally, computer simulation can be used as a purely exploratory tool. This sounds strange. One would be inclined to say that one cannot "discover" anything by simulation because you can never get out what you have not put in. Computer discoveries, in this respect, are not unlike mathematical discoveries. In fact, before computers were actually available this kind of numerical charting of unknown territory was never considered.

The best way to explain it is to give an explicit example. In the mid-1950s, one of the burning questions in statistical mechanics was this: can crystals form in a system of spherical particles that have a harsh short-range repulsion, but no mutual attraction whatsoever? In a very famous computer simulation, Alder and Wainwright [17] and Wood and Jacobson [18] showed that such a system does indeed have a first-order freezing transition. This is now accepted wisdom, but at the time it was greeted with skepticism. For instance, at a meeting in New Jersey in 1957, a group of some 15 very distinguished scientists (among whom were 2 Nobel laureates) discussed the issue. When a vote was taken as to whether hard spheres can form a stable crystal, it appeared that half the audience simply could not believe this result. However, the work of the past 30 years has shown that harsh repulsive forces really determine the structural properties of a simple liquid and that attractive forces are in a sense of secondary importance.

Suggested Reading

As stated at the outset, the present book does not cover all aspects of computer simulation. Readers who are interested in aspects of computer simulation not covered in this book are referred to one of the folowing books

- Allen and Tildesley, *Computer Simulation of Liquids* [19]
- Haile, *Molecular Dynamics Simulations: Elementary Methods* [20]
- Landau and Binder, *A Guide to Monte Carlo Simulations in Statistical Physics* [21]
- Rapaport, *The Art of Molecular Dynamics Simulation* [22]
- Newman and Barkema, *Monte Carlo Methods in Statistical Physics* [23]

Also of interest in this context are the books by Hockney and Eastwood [24], Hoover [25,26], Vesely [27], and Heermann [28] and the book by Evans and Morriss [29] for the theory and simulation of transport phenomena. The latter book is out of print and has been made available in electronic form.[1]

A general discussion of Monte Carlo sampling (with examples) can be found in Koonin's *Computational Physics* [30]. As the title indicates, this is a textbook on computational physics in general, as is the book by Gould and Tobochnik [31]. In contrast, the book by Kalos and Whitlock [32] focuses specifically on the Monte Carlo method. A good discussion of (quasi) random-number generators can be found in *Numerical Recipes* [33], while Ref. [32] gives a detailed discussion of tests for random-number generators. A discussion of Monte Carlo simulations with emphasis on techniques relevant for atomic and molecular systems may be found in two articles by Valleau and Whittington in *Modern Theoretical Chemistry* [34,35]. The books by Binder [36,37] and Mouritsen [38] emphasize the application of MC simulations to discrete systems, phase transitions and critical phenomena. In addition, there exist several very useful proceedings of summer schools [39–42] on computer simulation.

[1] See `http://rsc.anu.edu.au/~evans/evansmorrissbook.htm`

Part I

Basics

Chapter 2

Statistical Mechanics

The topic of this book is computer simulation. Computer simulation allows us to study properties of many-particle systems. However, not all properties can be directly measured in a simulation. Conversely, most of the quantities that can be measured in a simulation do not correspond to properties that are measured in real experiments. To give a specific example: in a Molecular Dynamics simulation of liquid water, we could measure the instantaneous positions and velocities of all molecules in the liquid. However, this kind of information cannot be compared to experimental data, because no real experiment provides us with such detailed information. Rather, a typical experiment measures an average property, averaged over a large number of particles and, usually, also averaged over the time of the measurement. If we wish to use computer simulation as the numerical counterpart of experiments, we must know what kind of averages we should aim to compute. In order to explain this, we need to introduce the language of statistical mechanics. This we shall do here. We provide the reader with a quick (and slightly dirty) derivation of the basic expressions of statistical mechanics. The aim of these derivations is only to show that there is nothing mysterious about concepts such as phase space, temperature and entropy and many of the other statistical mechanical objects that will appear time and again in the remainder of this book.

2.1 Entropy and Temperature

Most of the computer simulations that we discuss are based on the assumption that classical mechanics can be used to describe the motions of atoms and molecules. This assumption leads to a great simplification in almost all calculations, and it is therefore most fortunate that it is justified in many cases of practical interest. Surprisingly, it turns out to be easier to derive the basic laws of statistical mechanics using the language of quantum mechanics. We will follow this route of least resistance. In fact, for our derivation, we need only little quantum mechanics. Specifically, we need the fact that a quantum mechanical system can be found in different states. For the time being, we limit ourselves to quantum states that are eigenvectors of the Hamiltonian $\mathcal{H}$ of the system (i.e., energy eigenstates). For any such state $|i>$, we have that $\mathcal{H}|i> = E_i|i>$, where E_i is the energy of state $|i>$. Most examples discussed in quantum mechanics textbooks concern systems with only a few degrees of freedom (e.g., the one-dimensional harmonic oscillator or a particle in a box). For such systems, the degeneracy of energy levels will be small. However, for the systems that are of interest to statistical mechanics (i.e., systems with $\mathcal{O}(10^{23})$ particles), the degeneracy of energy levels is astronomically large. In what follows, we denote by $\Omega(E, V, N)$ the number of eigenstates with energy E of a system of N particles in a volume V. We now express the basic assumption of statistical mechanics as follows: a system with fixed N, V, and E is equally likely to be found in any of its $\Omega(E)$ eigenstates. Much of statistical mechanics follows from this simple (but highly nontrivial) assumption.

To see this, let us first consider a system with total energy E that consists of two weakly interacting subsystems. In this context, *weakly interacting* means that the subsystems can exchange energy but that we can write the total energy of the system as the sum of the energies E_1 and E_2 of the subsystems. There are many ways in which we can distribute the total energy over the two subsystems such that $E_1 + E_2 = E$. For a given choice of E_1, the total number of degenerate states of the system is $\Omega_1(E_1) \times \Omega_2(E_2)$. Note that the total number of states is not the sum but the product of the number of states in the individual systems. In what follows, it is convenient to have a measure of the degeneracy of the subsystems that is additive. A logical choice is to take the (natural) logarithm of the degeneracy. Hence:

$$\ln \Omega(E_1, E - E_1) = \ln \Omega_1(E_1) + \ln \Omega_2(E - E_1). \tag{2.1.1}$$

We assume that subsystems 1 and 2 can exchange energy. What is the most likely distribution of the energy? We know that *every* energy state of the total system is equally likely. But the number of eigenstates that correspond to a given distribution of the energy over the subsystems depends very strongly on the value of E_1. We wish to know the most likely value of E_1, that is, the one that maximizes $\ln \Omega(E_1, E - E_1)$. The condition for this maximum is that

$$\left(\frac{\partial \ln \Omega(E_1, E - E_1)}{\partial E_1}\right)_{N,V,E} = 0 \tag{2.1.2}$$

or, in other words,

$$\left(\frac{\partial \ln \Omega_1(E_1)}{\partial E_1}\right)_{N_1,V_1} = \left(\frac{\partial \ln \Omega_2(E_2)}{\partial E_2}\right)_{N_2,V_2} \tag{2.1.3}$$

We introduce the shorthand notation

$$\beta(E, V, N) \equiv \left(\frac{\partial \ln \Omega(E, V, N)}{\partial E} \right)_{N,V}. \tag{2.1.4}$$

With this definition, we can write equation (2.1.3) as

$$\beta(E_1, V_1, N_1) = \beta(E_2, V_2, N_2). \tag{2.1.5}$$

Clearly, if initially we put all energy in system 1 (say), there will be energy transfer from system 1 to system 2 until equation (2.1.3) is satisfied. From that moment on, no net energy flows from one subsystem to the other, and we say that the two subsystems are in (thermal) equilibrium. When this equilibrium is reached, $\ln \Omega$ of the total system is at a maximum. This suggests that $\ln \Omega$ is somehow related to the thermodynamic entropy S of the system. After all, the second law of thermodynamics states that the entropy of a system N, V, and E is at its maximum when the system is in thermal equilibrium. There are many ways in which the relation between $\ln \Omega$ and entropy can be established. Here we take the simplest route; we simply define the entropy to be equal to $\ln \Omega$. In fact, for (unfortunate) historical reasons, entropy is not simply equal to $\ln \Omega$; rather we have

$$S(N, V, E) \equiv k_B \ln \Omega(N, V, E), \tag{2.1.6}$$

where k_B is Boltzmann's constant, which in S.I. units has the value 1.38066 10^{-23} J/K. With this identification, we see that our assumption that all degenerate eigenstates of a quantum system are equally likely immediately implies that, in thermal equilibrium, the entropy of a composite system is at a maximum. It would be a bit premature to refer to this statement as the second law of thermodynamics, as we have not yet demonstrated that the present definition of entropy is, indeed, equivalent to the thermodynamic definition. We simply take an advance on this result.

The next thing to note is that thermal equilibrium between subsystems 1 and 2 implies that $\beta_1 = \beta_2$. In everyday life, we have another way to express the same thing: we say that two bodies brought into thermal contact are in equilibrium if their temperatures are the same. This suggests that β must be related to the absolute temperature. The thermodynamic definition of temperature is

$$1/T = \left(\frac{\partial S}{\partial E} \right)_{V,N}. \tag{2.1.7}$$

If we use the same definition here, we find that

$$\beta = 1/(k_B T). \tag{2.1.8}$$

Now that we have defined temperature, we can consider what happens if we have a system (denoted by A) that is in thermal equilibrium with a large heat bath (B). The total system is closed; that is, the total energy $E = E_B + E_A$ is fixed (we assume that the system and the bath are weakly coupled, so that we may ignore their interaction energy). Now suppose that the system A is prepared in one specific quantum state i with energy E_i. The bath then has an energy $E_B = E - E_i$ and the degeneracy of the bath is given by $\Omega_B(E - E_i)$. Clearly, the degeneracy of the bath determines the probability P_i to find system A in state i:

$$P_i = \frac{\Omega_B(E - E_i)}{\sum_j \Omega_B(E - E_j)}. \tag{2.1.9}$$

To compute $\Omega_B(E - E_i)$, we expand $\ln \Omega_B(E - E_i)$ around $E_i = 0$:

$$\ln \Omega_B(E - E_i) = \ln \Omega_B(E) - E_i \frac{\partial \ln \Omega_B(E)}{\partial E} + \mathcal{O}(1/E) \tag{2.1.10}$$

or, using equations (2.1.6) and (2.1.7),

$$\ln \Omega_B(E - E_i) = \ln \Omega_B(E) - E_i/k_B T + \mathcal{O}(1/E). \tag{2.1.11}$$

If we insert this result in equation (2.1.9), we get

$$P_i = \frac{\exp(-E_i/k_B T)}{\sum_j \exp(-E_j/k_B T)}. \tag{2.1.12}$$

This is the well-known Boltzmann distribution for a system at temperature T. Knowledge of the energy distribution allows us to compute the average energy $\langle E \rangle$ of the system at the given temperature T:

$$\begin{aligned} \langle E \rangle &= \sum_i E_i P_i \qquad (2.1.13) \\ &= \frac{\sum_i E_i \exp(-E_i/k_B T)}{\sum_j \exp(-E_j/k_B T)} \\ &= -\frac{\partial \ln \sum_i \exp(-E_i/k_B T)}{\partial 1/k_B T} \\ &= -\frac{\partial \ln Q}{\partial 1/k_B T}, \qquad (2.1.14) \end{aligned}$$

where, in the last line, we have defined the partition function Q. If we compare equation (2.1.13) with the thermodynamic relation

$$E = \frac{\partial F/T}{\partial 1/T},$$

where F is the Helmholtz free energy, we see that F is related to the partition function Q:

$$F = -k_B T \ln Q = -k_B T \ln \left(\sum_i \exp(-E_i/k_B T) \right). \tag{2.1.15}$$

Strictly speaking, F is fixed only up to a constant. Or, what amounts to the same thing, the reference point of the energy can be chosen arbitrarily. In what follows, we can use equation (2.1.15) without loss of generality. The relation between the Helmholtz free energy and the partition function is often more convenient to use than the relation between $\ln \Omega$ and the entropy. As a consequence, equation (2.1.15) is the workhorse of equilibrium statistical mechanics.

2.2 Classical Statistical Mechanics

Thus far, we have formulated statistical mechanics in purely quantum mechanical terms. The entropy is related to the density of states of a system with energy E, volume V, and number of particles N. Similarly, the Helmholtz free energy is related to the partition function Q, a sum over all quantum states i of the Boltzmann factor $\exp(-E_i/k_B T)$. To be specific, let us consider the average value of some observable A. We know the probability that a system at temperature T will be found in an energy eigenstate with energy E_i and we can therefore compute the thermal average of A as

$$\langle A \rangle = \frac{\sum_i \exp(-E_i/k_B T) < i|A|i >}{\sum_j \exp(-E_j/k_B T)}, \qquad (2.2.1)$$

where $< i|A|i >$ denotes the expectation value of the operator A in quantum state i. This equation suggests how we should go about computing thermal averages: first we solve the Schrödinger equation for the (many-body) system of interest, and next we compute the expectation value of the operator A for all those quantum states that have a nonnegligible statistical weight. Unfortunately, this approach is doomed for all but the simplest systems. First of all, we cannot hope to solve the Schrödinger equation for an arbitrary many-body system. And second, even if we could, the number of quantum states that contribute to the average in equation (2.2.1) would be so astronomically large ($\mathcal{O}(10^{10^{25}})$) that a numerical evaluation of all expectation values would be unfeasible. Fortunately, equation (2.2.1) can be simplified to a more workable expression in the classical limit. To this end, we first rewrite equation (2.2.1) in a form that is independent of the specific basis set. We note that $\exp(-E_i/k_B T) = < i|\exp(-\mathcal{H}/k_B T)|i >$, where $\mathcal{H}$ is the Hamiltonian of the system. Using this relation, we can write

$$\begin{aligned} \langle A \rangle &= \frac{\sum_i < i|\exp(-\mathcal{H}/k_B T)A|i >}{\sum_j < j|\exp(-\mathcal{H}/k_B T)|j >} \\ &= \frac{\mathrm{Tr}\exp(-\mathcal{H}/k_B T)A}{\mathrm{Tr}\exp(-\mathcal{H}/k_B T)}, \end{aligned} \qquad (2.2.2)$$

where Tr denotes the trace of the operator. As the value of the trace of an operator does not depend on the choice of the basis set, we can compute thermal averages using any basis set we like. Preferably, we use simple basis sets, such as the set of eigenfunctions of the position or the momentum operator. Next, we use the fact that the Hamiltonian $\mathcal{H}$ is the sum of a kinetic part $\mathcal{K}$ and a potential part $\mathcal{U}$. The kinetic energy operator is a quadratic function of the momenta of all particles. As a consequence, momentum eigenstates are also eigenfunctions of the kinetic energy operator. Similarly, the potential energy operator is a function of the particle coordinates. Matrix elements of $\mathcal{U}$ therefore are most conveniently computed in a basis set of position eigenfunctions. However, $\mathcal{H} = \mathcal{K} + \mathcal{U}$ itself is not diagonal in either basis set nor is $\exp[-\beta(\mathcal{K} + \mathcal{U})]$. However, if we could replace $\exp(-\beta\mathcal{H})$ by $\exp(-\beta\mathcal{K})\exp(-\beta\mathcal{U})$, then we could simplify equation (2.2.2) considerably. In general, we cannot make this replacement because

$$\exp(-\beta\mathcal{K})\exp(-\beta\mathcal{U}) = \exp\{-\beta[\mathcal{K} + \mathcal{U} + \mathcal{O}([\mathcal{K},\mathcal{U}])]\},$$

where $[\mathcal{K},\mathcal{U}]$ is the commutator of the kinetic and potential energy operators while $\mathcal{O}([\mathcal{K},\mathcal{U}])$ is meant to note all terms containing commutators and higher-order commutators of $\mathcal{K}$ and $\mathcal{U}$. It is easy to verify that the commutator $[\mathcal{K},\mathcal{U}]$ is of order $\hbar$ ($\hbar = h/(2\pi)$, where h is Planck's constant). Hence, in the limit $\hbar \to 0$, we may ignore the terms of order $\mathcal{O}([\mathcal{K},\mathcal{U}])$. In that case, we can write

$$\mathrm{Tr}\exp(-\beta\mathcal{H}) \approx \mathrm{Tr}\exp(-\beta\mathcal{U})\exp(-\beta\mathcal{K}). \tag{2.2.3}$$

If we use the notation $|r>$ for eigenvectors of the position operator and $|k>$ for eigenvectors of the momentum operator, we can express equation (2.2.3) as

$$\mathrm{Tr}\exp(-\beta\mathcal{H}) = \sum_{r,k} < r|e^{-\beta\mathcal{U}}|r >< r|k >< k|e^{-\beta\mathcal{K}}|k >< k|r > . \tag{2.2.4}$$

All matrix elements can be evaluated directly:

$$< r|\exp(-\beta\mathcal{U})|r >= \exp\left[-\beta\mathcal{U}(\mathbf{r}^N)\right],$$

where $\mathcal{U}(\mathbf{r}^N)$ on the right-hand side is no longer an operator but a function of the coordinates of all N particles. Similarly,

$$< k|\exp(-\beta\mathcal{K})|k >= \exp\left[-\beta\sum_{i=1}^{N} p_i^2/(2m_i)\right],$$

where $p_i = \hbar k_i$, and

$$< r|k >< k|r >= 1/V^N,$$

where V is the volume of the system and N the number of particles. Finally, we can replace the sum over states by an integration over all coordinates and momenta. The final result is

$$\begin{aligned} \operatorname{Tr}\exp(-\beta\mathcal{H}) &\approx \frac{1}{h^{dN}N!}\int d\mathbf{p}^N d\mathbf{r}^N \exp\left\{-\beta\left[\sum_i p_i^2/(2m_i) + \mathcal{U}(\mathbf{r}^N)\right]\right\} \\ &\equiv Q_{\text{classical}}, \end{aligned} \tag{2.2.5}$$

where d is the dimensionality of the system and the last line defines the classical partition function. The factor $1/N!$ has been inserted afterward to take the indistinguishability of identical particles into account. Every N-particle quantum state corresponds to a volume h^{dN} in classical phase space, but not all such volumes correspond to distinct quantum states. In particular, all points in phase space that only differ in the labeling of the particles correspond to the same quantum state (for more details, see, e.g., [43]).

Similarly, we can derive the classical limit for $\operatorname{Tr}\exp(-\beta\mathcal{H})A$, and finally, we can write the classical expression for the thermal average of the observable A as

$$\langle A\rangle = \frac{\int d\mathbf{p}^N d\mathbf{r}^N \exp\left\{-\beta\left[\sum_i p_i^2/(2m_i) + \mathcal{U}(\mathbf{r}^N)\right]\right\} A(\mathbf{p}^N, \mathbf{q}^N)}{\int d\mathbf{p}^N d\mathbf{r}^N \exp\left\{-\beta\left[\sum_j p_j^2/(2m_j) + \mathcal{U}(\mathbf{r}^N)\right]\right\}}. \tag{2.2.6}$$

Equations (2.2.5) and (2.2.6) are the starting point for virtually all classical simulations of many-body systems.

2.2.1 Ergodicity

Thus far, we have discussed the average behavior of many-body systems in a purely static sense: we introduced only the assumption that every quantum state of a many-body system with energy E is equally likely to be occupied. Such an average over all possible quantum states of a system is called an *ensemble* average. However, this is not the way we usually think about the average behavior of a system. In most experiments we perform a series of measurements during a certain time interval and then determine the average of these measurements. In fact, the idea behind Molecular Dynamics simulations is precisely that we can study the average behavior of a many-particle system simply by computing the natural time evolution of that system numerically and averaging the quantity of interest over a sufficiently long time. To take a specific example, let us consider a fluid consisting of atoms. Suppose that we wish to compute the average density of the fluid at a distance r from a given atom i, $\rho_i(r)$. Clearly, the instantaneous density depends on the coordinates $\mathbf{r}_j$ of all particles j in the system. As time progresses, the atomic coordinates will change (according to Newton's equations of motion), and hence the density around atom i will change.

Provided that we have specified the initial coordinates and momenta of all atoms $(\mathbf{r}^N(0), \mathbf{p}^N(0))$ we know, at least in principle, the time evolution of $\rho_i(r; \mathbf{r}^N(0), \mathbf{p}^N(0), t)$. In a Molecular Dynamics simulation, we measure the time-averaged density $\overline{\rho_i(r)}$ of a system of N atoms, in a volume V, at a constant total energy E:

$$\overline{\rho_i(r)} = \lim_{t\to\infty} \frac{1}{t} \int_0^t dt' \, \rho_i(r; t'). \tag{2.2.7}$$

Note that, in writing down this equation, we have implicitly assumed that, for t sufficiently long, the time average does not depend on the initial conditions. This is, in fact, a subtle assumption that is not true in general (see, e.g., [44]). However, we shall disregard subtleties and simply assume that, once we have specified N, V, and E, time averages do not depend on the initial coordinates and momenta. If that is so, then we would not change our result for $\overline{\rho_i(r)}$ if we average over many different initial conditions; that is, we consider the hypothetical situation where we run a large number of Molecular Dynamics simulations at the same values for N, V, and E, but with different initial coordinates and momenta,

$$\overline{\rho_i(r)} = \frac{\sum_{\text{initial conditions}} \left(\lim_{t\to\infty} \frac{1}{t} \int_0^t dt' \, \rho_i(r; \mathbf{r}^N(0), \mathbf{p}^N(0), t') \right)}{\text{number of initial conditions}}. \tag{2.2.8}$$

We now consider the limiting case where we average over all initial conditions compatible with the imposed values of N, V, and E. In that case, we can replace the sum over initial conditions by an integral:

$$\frac{\sum_{\text{initial conditions}} f(\mathbf{r}^N(0), \mathbf{p}^N(0))}{\text{number of initial conditions}} \to \frac{\int_E d\mathbf{r}^N d\mathbf{p}^N \, f(\mathbf{r}^N(0), \mathbf{p}^N(0))}{\Omega(N, V, E)}, \tag{2.2.9}$$

where f denotes an arbitrary function of the initial coordinates $\mathbf{r}^N(0), \mathbf{p}^N(0)$, while $\Omega(N, V, E) = \int_E d\mathbf{r}^N d\mathbf{p}^N$ (we have ignored a constant factor[1]). The subscript E on the integral indicates that the integration is restricted to a shell of constant energy E. Such a "phase space" average, corresponds to the classical limit of the *ensemble* average discussed in the previous sections.[2] We

[1] If we consider a quantum mechanical system, then $\Omega(N, V, E)$ is simply the number of quantum states of that system, for given N, V, and E. In the classical limit, the number of quantum states of a d-dimensional system of N distinguishable, structureless particles is given by $\Omega(N, V, E) = (\int d\mathbf{p}^N d\mathbf{r}^N)/h^{dN}$. For N *indistinguishable* particles, we should divide the latter expression by a factor N!.

[2] Here we consider the classical equivalent of the so-called microcanonical ensemble, i.e., the ensemble of systems with fixed N, V, and E. The classical expression for phase space integrals in the microcanonical ensemble can be derived from the quantum mechanical expression involving a sum over quantum states in much the same way that we used to derive the classical constant N, V, T ("canonical") ensemble from the corresponding quantum mechanical expression.

denote an ensemble average by $\langle\cdots\rangle$ to distinguish it from a time average, denoted by a bar. If we switch the order of the time averaging and the averaging over initial conditions, we find

$$\overline{\rho_i(r)} = \lim_{t\to\infty} \frac{1}{t} \int dt' \, \langle \rho_i(r; \mathbf{r}^N(0), \mathbf{p}^N(0), t') \rangle_{NVE}. \tag{2.2.10}$$

However, the ensemble average in this equation does not depend on the time t'. This is so, because there is a one-to-one correspondence between the initial phase space coordinates of a system and those that specify the state of the system at a later time t' (see e.g., [44, 45]). Hence, averaging over all initial phase space coordinates is equivalent to averaging over the time-evolved phase space coordinates. For this reason, we can leave out the time averaging in equation (2.2.10), and we find

$$\overline{\rho_i(r)} = \langle \rho_i(r) \rangle_{NVE}. \tag{2.2.11}$$

This equation states that, if we wish to compute the average of a function of the coordinates and momenta of a many-particle system, we can *either* compute that quantity by time averaging (the "MD" approach) *or* by ensemble averaging (the "MC" approach). It should be stressed that the preceding paragraphs are meant only to make equation (2.2.11) *plausible*, not as a proof. In fact, that would have been quite impossible because equation (2.2.11) is not true in general. However, in what follows, we shall simply assume that the "ergodic hypothesis", as equation (2.2.11) is usually referred to, applies to the systems that we study in computer simulations. The reader, however, should be aware that many examples of systems are not ergodic *in practice*, such as glasses and metastable phases, or even *in principle*, such as nearly harmonic solids.

2.3 Questions and Exercises

Question 1 (Number of Configurations)

1. *Consider a system A consisting of subsystems A_1 and A_2, for which $\Omega_1 = 10^{20}$ and $\Omega_2 = 10^{22}$. What is the number of configurations available to the combined system? Also, compute the entropies S, S_1, and S_2.*
2. *By what factor does the number of available configurations increase when $10\ m^3$ of air at 1.0 atm and 300 K is allowed to expand by 0.001% at constant temperature?*
3. *By what factor does the number of available configurations increase when 150 kJ is added to a system containing 2.0 mol of particles at constant volume and $T = 300$ K?*

4. *A sample consisting of five molecules has a total energy* 5ϵ*. Each molecule is able to occupy states of energy* ϵj*, with* $j = 0, 1, 2, \cdots, \infty$*. Draw up a table with columns by the energy of the states and write beneath them all configurations that are consistent with the total energy. Identify the type of configuration that is most probable.*

Question 2 (Thermodynamic Variables in the Canonical Ensemble) *Starting with an expression for the Helmholtz free energy* (F) *as a function of* N, V, T

$$F = \frac{-\ln\left[Q\left(N, V, T\right)\right]}{\beta}$$

one can derive all thermodynamic properties. Show this by deriving equations for U*,* p*, and* S*.*

Question 3 (Ideal Gas (Part 1)) *The canonical partition function of an ideal gas consisting of monoatomic particles is equal to*

$$Q\left(N, V, T\right) = \frac{1}{h^{3N} N!} \int d\Gamma \exp\left[-\beta H\right] = \frac{V^N}{\lambda^{3N} N!}$$

in which $\lambda = h/\sqrt{2\pi m/\beta}$ *and* $d\Gamma = dq_1 \cdots dq_N dp_1 \cdots dp_N$.

Derive expressions for the following thermodynamic properties:

- $F\left(N, V, T\right)$ *(hint:* $\ln\left(N!\right) \approx N \ln\left(N\right) - N$*)*
- $p\left(N, V, T\right)$ *(which leads to the ideal gas law !!!)*
- $\mu\left(N, V, T\right)$ *(which leads to* $\mu = \mu_0 + RT \ln \rho$*)*
- $U\left(N, V, T\right)$ *and* $S\left(N, V, T\right)$
- C_v *(heat capacity at constant volume)*
- C_p *(heat capacity at constant pressure)*

Question 4 (Ising Model) *Consider a system of* N *spins arranged on a lattice. In the presence of a magnetic field,* H*, the energy of the system is*

$$U = -\sum_{i=1}^{N} H\mu s_i - J \sum_{i>j} s_i s_j$$

in which J *is called the coupling constant* ($J > 0$) *and* $s_i = \pm 1$*. The second summation is a summation over all pairs* ($D \times N$ *for a periodic system,* D *is the dimensionality of the system). This system is called the Ising model.*

1. *Show that for positive* J*, and* $H = 0$*, the lowest energy of the Ising model is equal to*

$$U_0 = -DNJ$$

in which D *is the dimensionality of the system.*

2. *Show that the free energy per spin of a 1D Ising model with zero field is equal to*

$$\frac{F(\beta, N)}{N} = -\frac{\ln(2\cosh(\beta J))}{\beta}$$

when $N \to \infty$. *The function* $\cosh(x)$ *is defined as*

$$\cosh(x) = \frac{\exp[-x] + \exp[x]}{2}. \tag{2.3.1}$$

3. *Derive equations for the energy and heat capacity of this system.*

Question 5 (The Photon Gas) *An electromagnetic field in thermal equilibrium can be described as a phonon gas. From the quantum theory of the electromagnetic field, it is found that the total energy of the system (*U*) can be written as the sum of photon energies:*

$$U = \sum_{j=1}^{N} n_j \omega_j \hbar = \sum_{j=1}^{N} n_j \epsilon_j$$

in which ϵ_j *is the characteristic energy of a photon with frequency* ω, j, $n_j = 0, 1, 2, \cdots, \infty$ *is the so-called occupancy number of mode* j, *and* N *is the number of field modes (here we take* N *to be finite).*

1. *Show that the canonical partition function of the system can be written as*

$$Q = \prod_{j=1}^{N} \frac{1}{1 - \exp[-\beta \epsilon_j]}. \tag{2.3.2}$$

Hint: you will have to use the following identity for $|x| < 1$*:*

$$\sum_{i=0}^{i=\infty} x^i = \frac{1}{1-x}. \tag{2.3.3}$$

For the product of partition functions of two independent systems A *and* B *we can write*

$$Q_A \times Q_B = Q_{AB} \tag{2.3.4}$$

when $A \cap B = \oslash$ *and* $A \cup B = AB$.

2. *Show that the average occupancy number of state* j, $\langle n_j \rangle$, *is equal to*

$$\langle n_j \rangle = \frac{\partial \ln Q}{\partial(-\beta \epsilon_j)} = \frac{1}{\exp[\beta \epsilon_j] - 1}. \tag{2.3.5}$$

3. *Describe the behavior of* $\langle n_j \rangle$ *when* $T \to \infty$ *and when* $T \to 0$.

Question 6 (Ideal Gas (Part 2)) *An ideal gas is placed in a constant gravitational field. The potential energy of* N *gas molecules at height* z *is* Mgz*, where* $M = mN$ *is the total mass of* N *molecules. The temperature in the system is uniform and the system infinitely large. We assume that the system is locally in equilibrium, so we are allowed to use a local partition function.*

1. *Show that the grand-canonical partition function of a system in volume* V *at height* z *is equal to*

$$Q(\mu, V, T, z) = \sum_{N=0}^{\infty} \frac{\exp[\beta\mu N]}{h^{3N}N!} \int d\Gamma \exp[-\beta(H_0 + Mgz)] \tag{2.3.6}$$

   in which H_0 *is the Hamiltonian of the system at* $z = 0$.

2. *Explain that a change in* z *is equivalent to a change in chemical potential,* μ. *Use this to show that the pressure of the gas at height* z *is equal to*

$$p(z) = p(z = 0) \times \exp[-\beta mgz]. \tag{2.3.7}$$

   (Hint: you will need the formula for the chemical potential of an ideal gas.)

Exercise 1 (Distribution of Particles)

Consider an ideal gas of N particles in a volume V at constant energy E. Let us divide the volume in p identical compartments. Every compartment contains n_i molecules such that

$$N = \sum_{i=1}^{i=p} n_i. \tag{2.3.8}$$

An interesting quantity is the distribution of molecules over the p compartments. Because the energy is constant, every possible eigenstate of the system will be equally likely. This means that in principle it is possible that one of the compartments is empty.

1. On the book's website you can find a program that calculates the distribution of molecules among the p compartments. Run the program for different numbers of compartments (p) and total number of gas molecules (N). Note that the code has to be completed first (see the file *distribution.f*). The output of the program is the probability of finding x particles in a particular compartment as a function of x. This is printed in the file *output.dat*.
2. What is the probability that one of the compartments is empty?
3. Consider the case $p = 2$ and N even. The probability of finding $N/2 + n_1$ molecules in compartment 1 and $N/2 - n_1$ molecules in compartment 2 is given by

$$P(n_1) = \frac{N!}{(N/2 - n_1)!(N/2 + n_1)!2^N}. \tag{2.3.9}$$

Compare your numerical results with the analytical expression for different values of N. Show that this distribution is a Gaussian for small n_1/N. Hint: For $x > 10$, it might be useful to use Stirling's approximation:

$$x! \approx (2\pi)^{\frac{1}{2}} x^{x+\frac{1}{2}} \exp[-x]. \tag{2.3.10}$$

Exercise 2 (Boltzmann Distribution)
Consider a system of N energy levels with energies $0, \epsilon, 2\epsilon, \cdots, (N-1)\epsilon$ and $\epsilon > 0$.

1. Calculate, using the given program, the occupancy of each level for different values of the temperature. What happens at high temperatures?
2. Change the program in such a way that the degeneracy of energy level i equals $i + 1$. What do you see?
3. Modify the program in such a way that the occupation of the energy levels as well as the partition function (q) is calculated for a hetero nuclear linear rotor with moment of inertia I. Compare your result with the approximate result

$$q = \frac{2I}{\beta \hbar^2} \tag{2.3.11}$$

for different temperatures. Note that the energy levels of a linear rotor are

$$U = J(J+1)\frac{\hbar^2}{2I} \tag{2.3.12}$$

with $J = 0, 1, 2, \cdots, \infty$. The degeneracy of level J equals $2J + 1$.

Exercise 3 (Coupled Harmonic Oscillators)
Consider a system of N harmonic oscillators with a total energy U. A single harmonic oscillator has energy levels $0, \epsilon, 2\epsilon, \cdots, \infty$ ($\epsilon > 0$). All harmonic oscillators in the system can exchange energy.

1. Invent a computational scheme to update the system at constant total energy (U). Compare your scheme with the scheme that is incorporated in the computer code that you can find on the book's website (see the file *harmonic.f*).
2. Make a plot of the energy distribution of the first oscillator as a function of the number of oscillators for a constant value of U/N (*output.dat*). Which distribution is recovered when N becomes large? What is the function of the other $N - 1$ harmonic oscillators? Explain.
3. Compare this distribution with the canonical distribution of a single oscillator at the same average energy (use the option NVT).
4. How does this exercise relate to the derivation of the Boltzmann distribution for a system at temperature T?

Exercise 4 (Random Walk on a 1D Lattice)
Consider the random walk of a single particle on a line. The particle performs jumps of fixed length 1. Assuming that the probability for forward or backward jumps is equal, the mean-squared displacement of a particle after N jumps is equal to N. The probability that, after N jumps, the net distance covered by the particle equals n is given by

$$\ln\left[P\left(n, N\right)\right] \approx \frac{1}{2}\ln\left(\frac{2}{\pi N}\right) - \frac{n^2}{2N}.$$

1. Derive this equation using Stirling's approximation for $\ln x!$.
2. Compare your numerical result for the root mean-squared displacement with the theoretical prediction (the computed function $P\left(n, N\right)$, see the file *output.dat*). What is the diffusivity of this system?
3. Modify the program in such a way that the probability to jump in the forward direction equals 0.8. What happens?

Exercise 5 (Random Walk on a 2D Lattice)
Consider the random walk of N particles on a $M \times M$ lattice. Two particles cannot occupy the same lattice site. On this lattice, periodic boundaries are used. This means that when a particle leaves the lattices it returns on the opposite side of the lattice; i.e., the coordinates are given modulo M.

1. What is the fraction of occupied sites (θ) of the lattice as a function of M and N?
2. Make a plot of the diffusivity D as a function of θ for $M = 32$. For low values of θ, the diffusivity can be approximated by

   $$D \approx D_0 \left(1 - \theta\right).$$

   Why is this equation reasonable at low densities? Why does it break down at higher densities?
3. Modify the program in such a way that the probability to jump in one direction is larger than the probability to jump in the other direction. Explain the results.
4. Modify the program in such a way that periodic boundary conditions are used in one direction and reflecting boundary conditions in the other. What happens?

Chapter 3

Monte Carlo Simulations

In the present chapter, we describe the basic principles of the Monte Carlo method. In particular, we focus on simulations of systems of a fixed number of particles (N) in a given volume (V) at a temperature (T).

3.1 The Monte Carlo Method

In the previous chapter, we introduced some of the basic concepts of (classical) statistical mechanics. Our next aim is to indicate where the Monte Carlo method comes in. We start from the classical expression for the partition function Q, equation (2.2.5):

$$Q = c \int d\mathbf{p}^N d\mathbf{r}^N \exp[-\mathcal{H}(\mathbf{r}^N \mathbf{p}^N)/k_B T], \tag{3.1.1}$$

where $\mathbf{r}^N$ stands for the coordinates of all N particles, and $\mathbf{p}^N$ for the corresponding momenta. The function $\mathcal{H}(\mathbf{q}^N, \mathbf{p}^N)$ is the Hamiltonian of the system. It expresses the total energy of an isolated system as a function of the coordinates and momenta of the constituent particles: $\mathcal{H} = \mathcal{K} + \mathcal{U}$, where $\mathcal{K}$ is the kinetic energy of the system and $\mathcal{U}$ is the potential energy. Finally, c is a constant of proportionality, chosen such that the sum over quantum states in equation (2.1.15) approaches the classical partition function in the limit $\hbar \to 0$. For instance, for a system of N identical atoms, $c = 1/(h^{3N} N!)$. The classical equation corresponding to equation (2.2.1) is

$$\langle A \rangle = \frac{\int d\mathbf{p}^N d\mathbf{r}^N \; A(\mathbf{p}^N, \mathbf{r}^N) \exp[-\beta \mathcal{H}(\mathbf{p}^N, \mathbf{r}^N)]}{\int d\mathbf{p}^N d\mathbf{r}^N \; \exp[-\beta \mathcal{H}(\mathbf{p}^N, \mathbf{r}^N)]}, \tag{3.1.2}$$

where $\beta = 1/k_B T$. In this equation, the observable A has been expressed as a function of coordinates and momenta. As $\mathcal{K}$ is a quadratic function of the momenta the integration over momenta can be carried out analytically. Hence, averages of functions that depend on momenta only are usually easy to evaluate.[1] The difficult problem is the computation of averages of functions $A(\mathbf{r}^N)$. Only in a few exceptional cases can the multidimensional integral over particle coordinates be computed analytically; in all other cases numerical techniques must be used.

Having thus defined the nature of the numerical problem that we must solve, let us next look at possible solutions. It might appear that the most straightforward approach would be to evaluate $\langle A \rangle$ in equation (3.1.2) by numerical quadrature, for instance using Simpson's rule. It is easy to see, however, that such a method is completely useless even if the number of independent coordinates DN (D is the dimensionality of the system) is still very small $\mathcal{O}(100)$. Suppose that we plan to carry out the quadrature by evaluating the integrand on a mesh of points in the DN-dimensional configuration space. Let us assume that we take m equidistant points along each coordinate axis. The total number of points at which the integrand must be evaluated is then equal to m^{DN}. For all but the smallest systems this number becomes astronomically large, even for small values of m. For instance, if we take 100 particles in three dimensions, and $m = 5$, then we would have to evaluate the integrand at 10^{210} points! Computations of such magnitude cannot be performed in the known universe. And this is fortunate, because the answer that would be obtained would have been subject to a large statistical error. After all, numerical quadratures work best on functions that are smooth over distances corresponding to the mesh size. But for most intermolecular potentials, the Boltzmann factor in equation (3.1.2) is a rapidly varying function of the particle coordinates. Hence an accurate quadrature requires a small mesh spacing (i.e., a large value of m). Moreover, when evaluating the integrand for a dense liquid (say), we would find that for the overwhelming majority of points this Boltzmann factor is vanishingly small. For instance, for a fluid of 100 hard spheres at the freezing point, the Boltzmann factor would be nonzero for 1 out of every 10^{260} configurations!

The preceding example clearly demonstrates that better numerical techniques are needed to compute thermal averages. One such a technique is the Monte Carlo method or, more precisely, the Monte Carlo importance-sampling algorithm introduced in 1953 by Metropolis *et al.* [6]. The application of this method to the numerical simulation of dense molecular systems is the subject of the present chapter.

3.1.1 Importance Sampling

Before discussing importance sampling, let us first look at the simplest Monte Carlo technique, that is, random sampling. Suppose we wish to evaluate

[1] This is not the case when hard constraints are used, see section 11.2.1.

numerically a one-dimensional integral I:

$$I = \int_a^b dx\, f(x). \tag{3.1.3}$$

Instead of using a conventional quadrature where the integrand is evaluated at predetermined values of the abscissa, we could do something else. Note that equation (3.1.3) can be rewritten as

$$I = (b - a)\, \langle f(x) \rangle\,, \tag{3.1.4}$$

where $\langle f(x) \rangle$ denotes the unweighted average of $f(x)$ over the interval $[a, b]$. In brute force Monte Carlo, this average is determined by evaluating $f(x)$ at a large number (say, L) of x values randomly distributed over the interval $[a, b]$. It is clear that, as $L \to \infty$, this procedure should yield the correct value for I. However, as with the conventional quadrature procedure, this method is of little use to evaluate averages such as in equation (3.1.2) because most of the computing is spent on points where the Boltzmann factor is negligible. Clearly, it would be much preferable to sample many points in the region where the Boltzmann factor is large and few elsewhere. This is the basic idea behind importance sampling.

How should we distribute our sampling through configuration space? To see this, let us first consider a simple, one-dimensional example. Suppose we wish to compute the definite integral in equation (3.1.3) by Monte Carlo sampling, but with the sampling points distributed nonuniformly over the interval $[a, b]$ (for convenience we assume $a = 0$ and $b = 1$), according to some nonnegative probability density $w(x)$. Clearly, we can rewrite equation (3.1.3) as

$$I = \int_0^1 dx\, w(x) \frac{f(x)}{w(x)}. \tag{3.1.5}$$

Let us assume that we know that $w(x)$ is the derivative of another (nonnegative, nondecreasing) function $u(x)$, with $u(0) = 0$ and $u(1) = 1$ (these boundary conditions imply that $w(x)$ is normalized). Then I can be written as

$$I = \int_0^1 du\, \frac{f[x(u)]}{w[x(u)]}. \tag{3.1.6}$$

In equation (3.1.6) we have written $x(u)$ to indicate that, if we consider u as the integration variable, then x must be expressed as a function of u. The next step is to generate L random values of u uniformly distributed in the interval $[0, 1]$. We then obtain the following estimate for I:

$$I \approx \frac{1}{L} \sum_{i=1}^{L} \frac{f[x(u_i)]}{w[x(u_i)]}, \tag{3.1.7}$$

What have we gained by rewriting I in this way? The answer depends crucially on our choice for $w(x)$. To see this, let us estimate σ_I^2, the variance in I_L, where I_L denotes the estimate for I obtained from equation (3.1.7) with L random sample points:

$$\sigma_I^2 = \frac{1}{L^2}\sum_{i=1}^{L}\sum_{j=1}^{L}\left\langle\left(\frac{f[x(u_i)]}{w[x(u_i)]} - \langle f/w\rangle\right)\left(\frac{f[x(u_j)]}{w[x(u_j)]} - \langle f/w\rangle\right)\right\rangle, \tag{3.1.8}$$

where the angular brackets denote the true average, that is, the one that would be obtained in the limit $l \to \infty$. As different samples i and j are assumed to be totally independent, all cross terms in equation (3.1.8) vanish, and we are left with

$$\begin{aligned}\sigma_I^2 &= \frac{1}{L^2}\sum_{i=1}^{L}\left\langle\left(\frac{f[x(u_i)]}{w[x(u_i)]} - \langle f/w\rangle\right)^2\right\rangle \\ &= \frac{1}{L}\left[\langle (f/w)^2\rangle - \langle f/w\rangle^2\right].\end{aligned} \tag{3.1.9}$$

Equation (3.1.9) shows that the variance in I still goes as $1/L$, but the magnitude of this variance can be reduced greatly by choosing $w(x)$ such that $f(x)/w(x)$ is a smooth function of x. Ideally, we should have $f(x)/w(x)$ constant, in which case the variance would vanish altogether. In contrast, if $w(x)$ is constant, as is the case for the brute force Monte Carlo sampling, then the relative error in I can become very large. For instance, if we are sampling in a (multidimensional) configuration space of volume Ω, of which only a small fraction f is accessible (for instance, $f = 10^{-260}$, see previous section), then the relative error that results in a brute force MC sampling will be of order $1/(Lf)$. As the integrand in equation (3.1.2) is nonzero only for those configurations where the Boltzmann factor is nonzero, it would clearly be advisable to carry out a nonuniform Monte Carlo sampling of configuration space, such that the weight function w is approximately proportional to the Boltzmann factor. Unfortunately, the simple importance sampling scheme described previously cannot be used to sample multidimensional integrals over configuration space, such as equation (3.1.2). The reason is simply that we do not know how to construct a transformation such as the one from equation (3.1.5) to equation (3.1.6) that would enable us to generate points in configuration space with a probability density proportional to the Boltzmann factor. In fact, a necessary (but not nearly sufficient) condition for the solution to the latter problem is that we must be able to compute analytically the partition function of the system under study. If we could do that for the systems of interest to us, there would be hardly any need for computer simulation.

3.1.2 The Metropolis Method

The closing lines of the previous section suggest that it is in general not possible to evaluate an integral, such as $\int d\mathbf{r}^N \exp[-\beta\mathcal{U}(\mathbf{r}^N)]$, by direct Monte Carlo sampling. However, in many cases, we are not interested in the configurational part of the partition function itself but in averages of the type

$$\langle A\rangle = \frac{\int d\mathbf{r}^N \exp[-\beta\mathcal{U}(\mathbf{r}^N)]A(\mathbf{r}^N)}{\int d\mathbf{r}^N \exp[-\beta\mathcal{U}(\mathbf{r}^N)]}. \tag{3.1.10}$$

Hence, we wish to know the *ratio* of two integrals. What Metropolis *et al.* [6] showed is that it is possible to devise an efficient Monte Carlo scheme to sample such a ratio.[2] To understand the Metropolis method, let us first look more closely at the structure of equation (3.1.10). In what follows we denote the configurational part of the partition function by Z:

$$Z \equiv \int d\mathbf{r}^N \exp[-\beta\mathcal{U}(\mathbf{r}^N)]. \tag{3.1.11}$$

Note that the ratio $\exp(-\beta\mathcal{U})/Z$ in equation (3.1.10) is the probability density of finding the system in a configuration around $\mathbf{r}^N$. Let us denote this probability density by

$$\mathcal{N}(\mathbf{r}^N) \equiv \frac{\exp[-\beta\mathcal{U}(\mathbf{r}^N)]}{Z}.$$

Clearly, $\mathcal{N}(\mathbf{r}^N)$ is nonnegative.

Suppose now that we are somehow able to randomly generate points in configuration space according to this probability distribution $\mathcal{N}(\mathbf{r}^N)$. This means that, on average, the number of points n_i generated per unit volume around a point $\mathbf{r}^N$ is equal to $L\mathcal{N}(\mathbf{r}^N)$, where L is the total number of points that we have generated. In other words,

$$\langle A\rangle \approx \frac{1}{L}\sum_{i=1}^{L} n_i A(\mathbf{r}_i^N). \tag{3.1.12}$$

By now the reader is almost certainly confused about the difference, if any, between equation (3.1.12) and equation (3.1.7) of section 3.1.1. The difference is that in the case of equation (3.1.7) we know *a priori* the probability of sampling a point in a (hyper)volume $d\mathbf{r}^N$ around $\mathbf{r}^N$. In other words we know both $\exp[-\beta\mathcal{U}(\mathbf{r}^N)]$ *and* Z. In contrast, in equation (3.1.12) we know only $\exp[-\beta\mathcal{U}(\mathbf{r}^N)]$; that is, we know only the relative but not the absolute probability of visiting different points in configuration space. This may sound

[2] An interesting account of the early history of the Metropolis method may be found in refs. [1,46].

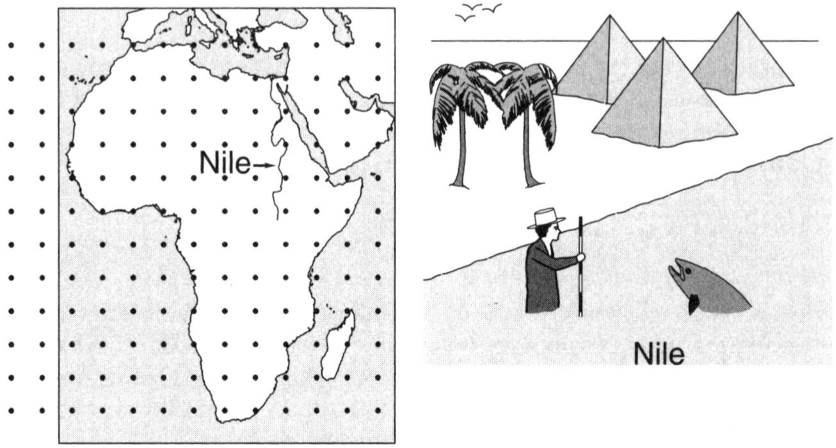

Figure 3.1: Measuring the depth of the Nile: a comparison of conventional quadrature (left), with the Metropolis scheme (right).

rather abstract: let us therefore try to clarify the difference with the help of a simple example (see Figure 3.1). In this figure, we compare two ways of measuring the depth of the river Nile, by conventional quadrature (left) and by Metropolis sampling; that is, the construction of an importance-weighted random walk (right). In the conventional quadrature scheme, the value of the integrand is measured at a predetermined set of points. As the choice of these points does not depend on the value of the integrand, many points may be located in regions where the integrand vanishes. In contrast, in the Metropolis scheme, a random walk is constructed through that region of space where the integrand is nonnegligible (i.e., through the Nile itself). In this random walk, a trial move is rejected if it takes you out of the water and is accepted otherwise. After *every* trial move (accepted or not), the depth of the water is measured. The (unweighted) average of all these measurements yields an estimate of the average depth of the Nile. This, then, is the essence of the Metropolis method. In principle, the conventional quadrature scheme would also give results for the *total* area of the Nile. In the importance sampling scheme, however, information on the total area cannot be obtained directly, since this quantity is similar to Z.

Let us next consider how to generate points in configuration space with a relative probability proportional to the Boltzmann factor. The general approach is first to prepare the system in a configuration $\mathbf{r}^N$, which we denote by o (old), that has a nonvanishing Boltzmann factor $\exp[-\beta \mathcal{U}(o)]$. This configuration, for example, may correspond to a regular crystalline lattice with no hard-core overlaps. Next, we generate a new trial configuration $\mathbf{r}'^N$, which we denote by n (new), by adding a small random displacement Δ to o. The Boltzmann factor of this trial configuration is $\exp[-\beta\mathcal{U}(n)]$. We must now decide whether we will accept or reject the trial configuration. Many rules for making this decision satisfy the constraint that on average the probability of finding the system in a configuration n is proportional to $\mathcal{N}(n)$. Here we discuss only the Metropolis scheme, because it is simple and generally applicable.

Let us now "derive" the Metropolis scheme to determine the transition probability $\pi(o \to n)$ to go from configuration o to n. It is convenient to start with a thought experiment (actually a thought simulation). We carry out a very large number (say M) Monte Carlo simulations in parallel, where M is much larger than the total number of accessible configurations. We denote the number of points in any configuration o by $m(o)$. We wish that, on average, $m(o)$ is proportional to $\mathcal{N}(o)$. The matrix elements $\pi(o \to n)$ must satisfy one obvious condition: they do not destroy such an equilibrium distribution once it is reached. This means that, in equilibrium, the average number of accepted trial moves that result in the system leaving state o must be exactly equal to the number of accepted trial moves from all other states n to state o. It is convenient to impose a much stronger condition; namely, that in equilibrium the average number of accepted moves from o to any other state n is exactly canceled by the number of reverse moves. This detailed balance condition implies the following:

$$\mathcal{N}(o)\pi(o \to n) = \mathcal{N}(n)\pi(n \to o). \qquad (3.1.13)$$

Many possible forms of the transition matrix $\pi(o \to n)$ satisfy equation (3.1.13). Let us look how $\pi(o \to n)$ is constructed in practice. We recall that a Monte Carlo move consists of two stages. First, we perform a trial move from state o to state n. We denote the transition matrix that determines the probability of performing a trial move from o to n by $\alpha(o \to n)$, where α is usually referred to as the underlying matrix of the Markov chain [47]. The next stage is the decision to either accept or reject this trial move. Let us denote the probability of accepting a trial move from o to n by $\text{acc}(o \to n)$. Clearly,

$$\pi(o \to n) = \alpha(o \to n) \times \text{acc}(o \to n). \qquad (3.1.14)$$

In the original Metropolis scheme, α is chosen to be a symmetric matrix ($\alpha(o \to n) = \alpha(n \to o)$). However, in later sections we shall see several examples where α is *not* symmetric. If α is symmetric, we can rewrite equation (3.1.13) in terms of the $\text{acc}(o \to n)$:

$$\mathcal{N}(o) \times \text{acc}(o \to n) = \mathcal{N}(n) \times \text{acc}(n \to o). \qquad (3.1.15)$$

From equation (3.1.15) follows

$$\frac{\text{acc}(o \to n)}{\text{acc}(n \to o)} = \frac{\mathcal{N}(n)}{\mathcal{N}(o)} = \exp\{-\beta[\mathcal{U}(n) - \mathcal{U}(o)]\}. \qquad (3.1.16)$$

Again, many choices for $\text{acc}(o \to n)$ satisfy this condition (and the obvious condition that the probability $\text{acc}(o \to n)$ cannot exceed 1). The choice of Metropolis *et al.* is

$$\begin{aligned} \text{acc}(o \to n) &= \mathcal{N}(n)/\mathcal{N}(o) \quad \text{if } \mathcal{N}(n) < \mathcal{N}(o) \\ &= 1 \quad \text{if } \mathcal{N}(n) \geq \mathcal{N}(o). \end{aligned} \tag{3.1.17}$$

Other choices for $\text{acc}(o \to n)$ are possible (for a discussion, see for instance [19]), but the original choice of Metropolis *et al.* appears to result in a more efficient sampling of configuration space than most other strategies that have been proposed.

In summary, then, in the Metropolis scheme, the transition probability for going from state o to state n is given by

$$\begin{aligned} \pi(o \to n) &= \alpha(o \to n) & \mathcal{N}(n) \geq \mathcal{N}(o) \\ &= \alpha(o \to n)[\mathcal{N}(n)/\mathcal{N}(o)] & \mathcal{N}(n) < \mathcal{N}(o) \\ \pi(o \to o) &= 1 - \textstyle\sum_{n \neq o} \pi(o \to n). \end{aligned} \tag{3.1.18}$$

Note that we still have not specified the matrix α, except for the fact that it must be symmetric. This reflects considerable freedom in the choice of our trial moves. We will come back to this point in subsequent sections.

One thing that we have not yet explained is how to decide whether a trial move is to be accepted or rejected. The usual procedure is as follows. Suppose that we have generated a trial move from state o to state n, with $\mathcal{U}(n) > \mathcal{U}(o)$. According to equation (3.1.16) this trial move should be accepted with a probability

$$\text{acc}(o \to n) = \exp\{-\beta[\mathcal{U}(n) - \mathcal{U}(o)]\} < 1.$$

In order to decide whether to accept or reject the trial move, we generate a random number, denoted by Ranf, from a uniform distribution in the interval $[0, 1]$. Clearly, the probability that Ranf is less than $\text{acc}(o \to n)$ is equal to $\text{acc}(o \to n)$. We now accept the trial move if Ranf $< \text{acc}(o \to n)$ and reject it otherwise. This rule guarantees that the probability to accept a trial move from o to n is indeed equal to $\text{acc}(o \to n)$. Obviously, it is very important that our random-number generator does indeed generate numbers uniformly in the interval $[0, 1]$. Otherwise the Monte Carlo sampling will be biased. The quality of random-number generators should never be taken for granted. A good discussion of random-number generators can be found in *Numerical Recipes* [33] and in *Monte Carlo Methods* by Kalos and Whitlock [32].

Thus far, we have not mentioned another condition that $\pi(o \to n)$ should satisfy, namely that it is *ergodic* (i.e., every accessible point in configuration space can be reached in a finite number of Monte Carlo steps from any other point). Although some simple MC schemes are guaranteed to be ergodic, these are often not the most efficient schemes. Conversely, many efficient Monte Carlo schemes have either not been proven to be ergodic or, worse, been proven to be nonergodic. The solution is usually to mix the efficient, nonergodic scheme with an occasional trial move of the less-efficient but ergodic scheme. The method as a whole will then be ergodic (at least, in principle).

At this point, we should stress that, in the present book, we focus on Monte Carlo methods to model phenomena that do not depend on time. In the literature one can also find dynamic Monte Carlo schemes. In such dynamic algorithms, Monte Carlo methods are used to generate a numerical solution of the master equation that is supposed to describe the time evolution of the system under study. These dynamic techniques fall outside the scope of this book. The reader interested in dynamic MC schemes is referred to the relevant literature, for example Ref. [48] and references therein.

3.2 A Basic Monte Carlo Algorithm

It is difficult to talk about Monte Carlo or Molecular Dynamics programs in abstract terms. The best way to explain how such programs work is to write them down. This will be done in the present section.

Most Monte Carlo or Molecular Dynamics programs are only a few hundred to several thousand lines long. This is very short compared to, for instance, a typical quantum-chemistry code. For this reason, it is not uncommon that a simulator will write many different programs that are tailor-made for specific applications. The result is that there is no such thing as a standard Monte Carlo or Molecular Dynamics program. However, the cores of most MD/MC programs are, if not identical, at least very similar. Next, we shall construct such a core. It will be very rudimentary, and efficiency has been traded for clarity. But it should demonstrate how the Monte Carlo method works.

3.2.1 The Algorithm

The prime purpose of the kind of Monte Carlo or Molecular Dynamics program that we shall be discussing is to compute equilibrium properties of classical many-body systems. From now on, we shall refer to such programs simply as MC or MD programs, although it should be remembered that there exist many other applications of the Monte Carlo method (and, to a lesser extent, of the Molecular Dynamics method). Let us now look at a simple Monte Carlo program.

In the previous section, the Metropolis method was introduced as a Markov process in which a random walk is constructed in such a way that the probability of visiting a particular point $\mathbf{r}^N$ is proportional to the Boltzmann factor $\exp[-\beta\mathcal{U}(\mathbf{r}^N)]$. There are many ways to construct such a random walk. In the approach introduced by Metropolis *et al.* [6], the following scheme is proposed:

1. Select a particle at random, and calculate its energy $\mathcal{U}(\mathbf{r}^N)$.
2. Give the particle a random displacement, $r' = r + \Delta$, and calculate its new energy $\mathcal{U}(\mathbf{r}'^N)$.
3. Accept the move from $\mathbf{r}^N$ to $\mathbf{r}'^N$ with probability

$$\text{acc}(o \to n) = \min\left(1, \exp\{-\beta[\mathcal{U}(\mathbf{r}'^N) - \mathcal{U}(\mathbf{r}^N)]\}\right). \tag{3.2.1}$$

An implementation of this basic Metropolis scheme is shown in Algorithms 1 and 2.

3.2.2 Technical Details

In this section, we discuss a number of computational tricks that are of great practical importance for the design of an efficient simulation program. It should be stressed that most of these tricks, although undoubtedly very useful, are not unique and have no deep physical significance. But this does not imply that the use of such computational tools is free of risks or subtleties. Ideally, schemes to save computer time should not affect the results of a simulation in a systematic way. Yet, in some cases, time-saving tricks do have a measurable effect on the outcome of a simulation. This is particularly true for the different procedures used to avoid explicit calculation of intermolecular interactions between particles that are far apart. Fortunately, once this is recognized, it is usually possible to estimate the undesirable side effect of the time-saving scheme and correct for it.

Boundary Conditions

Monte Carlo and Molecular Dynamics simulations of atomic or molecular systems aim to provide information about the properties of a macroscopic sample. Yet, the number of degrees of freedom that can be conveniently handled in present-day computers ranges from a few hundred to a few million. Most simulations probe the structural and thermodynamical properties of a system of a few hundred to a few thousand particles. Clearly, this number is still far removed from the thermodynamic limit. To be more precise, for such small systems it cannot be safely assumed that the choice of the boundary conditions (e.g., free or hard or periodic) has a negligible effect on the properties of the system. In fact, in a three-dimensional N-particle system

Algorithm 1 (Basic Metropolis Algorithm)

```
   PROGRAM mc                               basic Metropolis algorithm

   do icycl=1,ncycl                         perform ncycl MC cycles
     call mcmove                            displace a particle
     if (mod(icycl,nsamp).eq.0)
+      call sample                          sample averages
   enddo
   end
```

Comments to this algorithm:

1. *Subroutine* `mcmove` *attempts to displace a randomly selected particle (see Algorithm 2).*
2. *Subroutine* `sample` *samples quantities every* `nsamp`*th cycle.*

Algorithm 2 (Attempt to Displace a Particle)

```
   SUBROUTINE mcmove                        attempts to displace a particle

   o=int(ranf()*npart)+1                    select a particle at random
   call ener(x(o),eno)                      energy old configuration
   xn=x(o)+(ranf()-0.5)*delx                give particle random displacement
   call ener(xn,enn)                        energy new configuration
   if (ranf().lt.exp(-beta                  acceptance rule (3.2.1)
+     *(enn-eno)) x(o)=xn                   accepted: replace x(o) by xn
   return
   end
```

Comments to this algorithm:

1. *Subroutine* `ener` *calculates the energy of a particle at the given position.*
2. *Note that, if a configuration is rejected, the old configuration is retained.*
3. *The* `ranf()` *is a random number uniform in* $[0, 1]$.

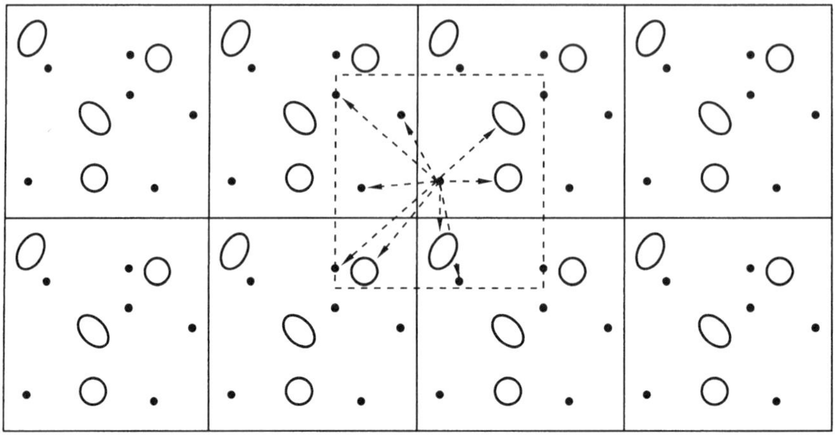

Figure 3.2: Schematic representation of periodic boundary conditions.

with free boundaries, the fraction of all molecules that is at the surface is proportional to $N^{-1/3}$. For instance, in a simple cubic crystal of 1000 atoms, some 49% of all atoms are at the surface, and for 10^6 atoms this fraction has decreased to only 6%.

In order to simulate bulk phases it is essential to choose boundary conditions that mimic the presence of an infinite bulk surrounding our N-particle model system. This is usually achieved by employing periodic boundary conditions. The volume containing the N particles is treated as the primitive cell of an infinite periodic lattice of identical cells (see Figure 3.2). A given particle (i, say) now interacts with all other particles in this infinite periodic system, that is, all other particles in the same periodic cell and all particles (including its own periodic image) in all other cells. For instance, if we assume that all intermolecular interactions are pairwise additive, then the total potential energy of the N particles in any one periodic box is

$$\mathcal{U}_{\text{tot}} = \frac{1}{2} \sum_{i,j,\mathbf{n}}{}' u(|\mathbf{r}_{ij} + \mathbf{n}L|),$$

where L is the diameter of the periodic box (assumed cubic, for convenience) and $\mathbf{n}$ is an arbitrary vector of three integer numbers, while the prime over the sum indicates that the term with $i = j$ is to be excluded when $\mathbf{n} = \mathbf{0}$. In this very general form, periodic boundary conditions are not particularly useful, because to simulate bulk behavior, we had to rewrite the potential energy as an infinite sum rather than a finite one.[3] In practice, however, we

[3]In fact, in the first MC simulation of three-dimensional Lennard-Jones particles, Wood and Parker [49] discuss the use of such infinite sums in relation to the now conventional approach discussed here.

are often dealing with short-range interactions. In that case it is usually permissible to truncate all intermolecular interactions beyond a certain cutoff distance r_c. How this is done in practice is discussed next.

Although the use of periodic boundary conditions proves to be a surprisingly effective method for simulating homogeneous bulk systems, one should always be aware that the use of such boundary conditions may lead to spurious correlations not present in a truly macroscopic bulk system. In particular, one consequence of the periodicity of the model system is that only those fluctuations are allowed that have a wavelength compatible with the periodic lattice: the longest wavelength that still fits in the periodic box is the one for which $\lambda = L$. If long wavelength fluctuations are expected to be important (as, for instance, in the vicinity of a continuous phase transition), then one should expect problems with the use of periodic boundary conditions. Another unphysical effect that is a manifestation of the use of periodic boundary conditions is that the radial distribution function $g(r)$ of a dense atomic fluid is found to be not exactly isotropic [50].

Finally, it is useful to point out one common misconception about periodic boundary conditions, namely, the idea that the boundary of the periodic box itself has any special significance. It has none. The origin of the periodic lattice of primitive cells may be chosen anywhere, and this choice will not affect any property of the model system under study. In contrast, what *is* fixed is the shape of the periodic cell and its orientation.

Truncation of Interactions

Let us now consider the case that we perform a simulation of a system with *short-range* interactions. In this context, short-ranged means that the total potential energy of a given particle i is dominated by interactions with neighboring particles that are closer than some cutoff distance r_c. The error that results when we ignore interactions with particles at larger distances can be made arbitrarily small by choosing r_c sufficiently large. If we use periodic boundary conditions, the case that r_c is less than $L/2$ (half the diameter of the periodic box) is of special interest because in that case we need to consider the interaction of a given particle i only with the nearest periodic image of any other particles j (see the dotted box in Figure 3.2). If the intermolecular potential is not rigorously zero for $r \geq r_c$, truncation of the intermolecular interactions at r_c will result in a systematic error in $\mathcal{U}^{tot}$. If the intermolecular interactions decay rapidly, one may correct for the systematic error by adding a tail contribution to $\mathcal{U}^{tot}$:

$$\mathcal{U}^{tot} = \sum_{i<j} u_c(r_{ij}) + \frac{N\rho}{2} \int_{r_c}^{\infty} dr \, u(r) 4\pi r^2, \qquad (3.2.2)$$

where u_c stands for the truncated potential energy function and ρ is the average number density. In writing down this expression, it is implicitly assumed that the radial distribution function $g(r) = 1$ for $r > r_c$. Clearly, the nearest periodic image convention can be applied only if the tail correction is small. From equation (3.2.2) it can be seen that the tail correction to the potential energy is infinite unless the potential energy function $u(r)$ decays more rapidly than r^{-3} (in three dimensions). This condition is satisfied if the long-range interaction between molecules is dominated by dispersion forces. However, for the very important case of Coulomb and dipolar interactions, the tail correction diverges and hence the nearest-image convention cannot be used for such systems. In such cases, the interactions with all periodic images should be taken into account explicitly. Ways to do this are described in Chapter 12.1.

Several factors make truncation of the potential a tricky business. First of all, it should be realized that, although the absolute value of the potential energy function decreases with interparticle separation r, for sufficiently large r, the number of neighbors is a rapidly increasing function of r. In fact, the number of particles at a distance r of a given atom increases asymptotically as r^{d-1}, where d is the dimensionality of the system. As an example, let us compute the effect of truncating the pair potential for a simple example — the three-dimensional Lennard-Jones fluid. The pair potential for this rather popular model system is given by

$$u^{lj}(r) = 4\epsilon \left[\left(\frac{\sigma}{r} \right)^{12} - \left(\frac{\sigma}{r} \right)^{6} \right]. \tag{3.2.3}$$

The average potential energy (in three dimensions) of any given atom i is given by

$$u_i = (1/2) \int_0^\infty dr\, 4\pi r^2 \rho(r) u(r),$$

where $\rho(r)$ denotes the average number density at a distance r from a given atom i. The factor $(1/2)$ has been included to correct for double counting of intermolecular interactions. If we truncate the potential at a distance r_c, we ignore the tail contribution u^{tail}:

$$u^{tail} \equiv (1/2) \int_{r_c}^\infty dr\, 4\pi r^2 \rho(r) u(r), \tag{3.2.4}$$

where we have dropped the subscript i, because all atoms in the system are identical. To simplify the calculation of u^{tail}, we assume that for $r \geq r_c$, the density $\rho(r)$ is equal to the average number density ρ. If $u(r)$ is the Lennard-

Jones potential, we find for u^{tail}

$$\begin{aligned} u^{tail} &= \frac{1}{2}4\pi\rho \int_{r_c}^{\infty} dr\, r^2 u(r) \\ &= \frac{1}{2}16\pi\rho\epsilon \int_{r_c}^{\infty} dr\, r^2 \left[\left(\frac{\sigma}{r}\right)^{12} - \left(\frac{\sigma}{r}\right)^{6} \right] \\ &= \frac{8}{3}\pi\rho\epsilon\sigma^3 \left[\frac{1}{3}\left(\frac{\sigma}{r_c}\right)^{9} - \left(\frac{\sigma}{r_c}\right)^{3} \right]. \end{aligned} \tag{3.2.5}$$

For a cutoff distance $r_c = 2.5\ \sigma$ the potential has decayed to a value that is about 1/60th of the well depth. This seems to be a very small value, but in fact the tail correction is usually nonnegligible. For instance, at a density $\rho\sigma^3 = 1$, we find $u^{tail} = -0.535\epsilon$. This number is certainly not negligible compared to the total potential energy per atom (almost 10% at a typical liquid density); hence although we can truncate the potential at 2.5 σ, we cannot ignore the effect of this truncation.

There are several ways to truncate potentials in a simulation. Although the methods are designed to yield similar results, it should be realized that they yield results that may differ significantly, in particular in the vicinity of critical points [51–53] (see Figure 3.3). Often used methods to truncate the potential are

1. Simple truncation
2. Truncation and shift
3. Minimum image convention.

Simple Truncation The simplest method to truncate potentials is to ignore all interaction beyond r_c, the potential that is simulated is

$$u^{trunc}(r) = \begin{cases} u^{lj}(r) & r \leq r_c \\ 0 & r > r_c \end{cases} . \tag{3.2.6}$$

As already explained, this may result in an appreciable error in our estimate of the potential energy of the true Lennard-Jones potential (3.2.3). Moreover, as the potential changes discontinuously at r_c, a truncated potential is not particularly suitable for a Molecular Dynamics simulation. It can, however, be used in Monte Carlo simulations. In that case, one should be aware that there is an "impulsive" contribution to the pressure due to the discontinuous change of the potential at r_c. That contribution can by no means be ignored.

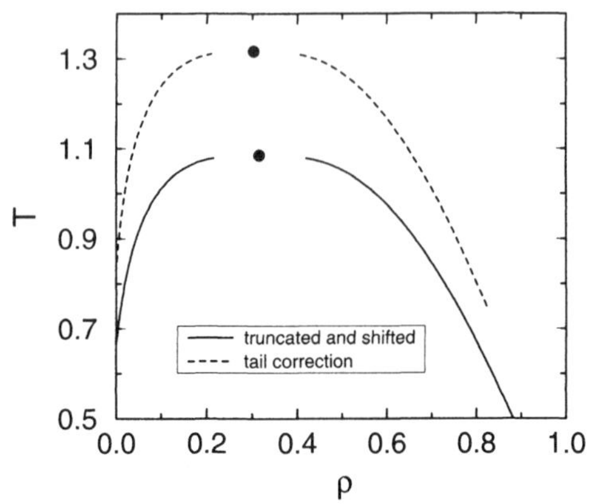

Figure 3.3: Vapor-liquid coexistence curves of various three-dimensional Lennard-Jones fluids: effect of the truncation of the potential on the location of the critical point (large black dots). The upper curve gives the phase envelope for the full Lennard-Jones potential (i.e., a truncated potential with tail correction); the lower curve gives the envelope for the potential that is used in most Molecular Dynamics simulations (truncated and shifted potential with $r_c = 2.5\sigma$), data from [53].

For instance, for the three-dimensional Lennard-Jones system,

$$\begin{aligned} \Delta P^{imp} &= \frac{8\pi}{3}\rho^2 g(r_c)\epsilon\sigma^3 \left[\left(\frac{\sigma}{r_c}\right)^9 - \left(\frac{\sigma}{r_c}\right)^3\right] \\ &\approx \frac{8\pi}{3}\rho^2 \epsilon\sigma^3 \left[\left(\frac{\sigma}{r_c}\right)^9 - \left(\frac{\sigma}{r_c}\right)^3\right]. \end{aligned} \tag{3.2.7}$$

It is rare, however, to see this impulsive correction to the pressure applied in simulations of systems with truncated potentials. Usually, it is simply assumed that we can correct for the truncation of the intermolecular potential by adding the correction given by equation (3.2.5) to the potential energy. The corresponding correction to the pressure is

$$\begin{aligned} \Delta P^{tail} &= (1/2)4\pi\rho^2 \int_{r_c}^{\infty} dr\, r^2 \mathbf{r}\cdot\mathbf{f}(r) \\ &= \frac{16}{3}\pi\rho^2\epsilon\sigma^3\left[\frac{2}{3}\left(\frac{\sigma}{r_c}\right)^9 - \left(\frac{\sigma}{r_c}\right)^3\right]. \end{aligned} \tag{3.2.8}$$

But, as is immediately obvious from a comparison of equations (3.2.7) and (3.2.8), the impulsive correction to the pressure is not equivalent to the tail correction. Rather, the impulsive pressure is the contribution that must be included if one wishes to compute the true pressure of a system with a truncated potential, whereas the tail correction should be included to obtain an estimate of the pressure in a system with untruncated interactions.

Truncated and Shifted In Molecular Dynamics simulations, it is common to use another procedure: the potential is truncated and shifted, such that the potential vanishes at the cutoff radius:

$$u^{tr-sh}(r) = \begin{cases} u^{lj}(r) - u^{lj}(r_c) & r \leq r_c \\ 0 & r > r_c \end{cases} . \tag{3.2.9}$$

In this case, there are no discontinuities in the intermolecular potential and hence no impulsive corrections to the pressure. The advantage of using such a truncated and shifted potential is that the intermolecular forces are always finite.[4] This is important because impulsive forces cannot be handled in those Molecular Dynamics algorithms to integrate the equations of motion that are based on a Taylor expansion of the particle positions. Of course, the potential energy and pressure of a system with a truncated and shifted potential differ from the corresponding properties of both the models with untruncated and with truncated but unshifted pair potentials. But, as before, we can approximately correct for the effect of the modification of the intermolecular potential on both the potential energy and the pressure. For the pressure, the tail correction is the same as in equation (3.2.8). For the potential energy, we must add to the long-range correction (3.2.5) a contribution equal to the average number of particles that are within a distance r_c from a given particle, multiplied by half the value of the (untruncated) pair potential at r_c. The factor one-half is included to correct for overcounting of the intermolecular interactions. One should be extremely careful when applying truncated and shifted potentials in models with anisotropic interactions. In that case, truncation should not be carried out at a fixed value of the distance between the molecular centers of mass but at a point where the pair potential has a fixed value, because otherwise the potential cannot be shifted to 0 at the point where it is truncated. For Monte Carlo simulations, this is not serious, but for Molecular Dynamics simulations this would be quite disastrous, as the system would no longer conserve energy, unless the impulsive forces due to the truncating and shifting are taken into account explicitly.

Minimum Image Convention Sometimes the minimum image convention is used. The truncation is in this case not at a spherical cutoff; instead the

[4]The first derivative of the force is discontinuous at the cutoff radius; some authors remove this discontinuity as well (for more details, see [19]).

interaction with the (nearest image) of all the particles in the simulation box is calculated. As a consequence, the potential is not a constant on the surface of a cube around a given particle. Hence, for the same reasons as mentioned in the previous paragraph, the simple minimum image convention should never be used in Molecular Dynamics simulations.

In the preceding, we described some details on how the energy should be calculated. The implementation of a simple, order-N^2, algorithm to compute the energy will be discussed in section 4.2.2 in the context of a Molecular Dynamics simulation (see Algorithm 5). More advanced schemes to simulate large systems efficiently are described in Appendix F.

Initialization

To start the simulation, we should assign initial positions to all particles in the system. As the equilibrium properties of the system do not (or, at least, should not) depend on the choice of initial conditions, all reasonable initial conditions are in principle acceptable. If we wish to simulate the solid state of a particular model system, it is logical to prepare the system in the crystal structure of interest. In contrast, if we are interested in the fluid phase, we simply prepare the system in any convenient crystal structure. This crystal subsequently melts, because at the temperature and density of a typical liquid-state point, the solid state is not thermodynamically stable. Actually, one should be careful here, because the crystal structure may be metastable, even if it is not absolutely stable. For this reason, it is unwise to use a crystal structure as the starting configuration of a liquid close to the freezing curve. In such cases, it is better to use the final (liquid) configuration of a system at a higher temperature or lower density, where the solid is unstable and has melted spontaneously. In any event, it is usually preferable to use the final (well-equilibrated) configuration of an earlier simulation at a nearby state point as the starting configuration for a new run and adjust the temperature and density to the desired values.

The equilibrium properties of a system should not depend on the initial conditions. If such a dependence nevertheless is observed in a simulation, there are two possibilities. The first is that our results reflect the fact that the system that we simulate really behaves nonergodically. This is the case, for instance, in glassy materials or low-temperature, substitutionally disordered alloys. The second (and much more likely) explanation is the system we simulate is ergodic but our sampling of configuration space is inadequate; in other words, we have not yet reached equilibrium.

Reduced Units

In simulations it is often convenient to express quantities such as temperature, density, pressure, and the like in reduced units. This means that we choose a convenient unit of energy, length and mass and then express all other quantities in terms of these basic units. In the example of a Lennard-Jones system, we use a pair potential that is of the form $u(r) = \epsilon f(r/\sigma)$ (see equation (3.2.3)). A natural (though not unique) choice for our basic units is the following:

- Unit of length, σ
- Unit of energy, ϵ
- Unit of mass, m (the mass of the atoms in the system)

and from these basic units, all other units follow. For instance, our unit of time is

$$\sigma\sqrt{m/\epsilon}$$

and the unit of temperature is

$$\epsilon/k_B.$$

In terms of these reduced units, denoted with superscript *, the reduced pair potential $u^* \equiv u/\epsilon$ is a dimensionless function of the reduced distance $r^* \equiv r/\sigma$. For instance, the reduced form for the Lennard-Jones potential is

$$u^{*lj}(r^*) = 4\left[\left(\frac{1}{r^*}\right)^{12} - \left(\frac{1}{r^*}\right)^{6}\right]. \tag{3.2.10}$$

With these conventions we can define the following reduced units: the potential energy $U^* = U\epsilon^{-1}$, the pressure $P^* = P\sigma^3\epsilon^{-1}$, the density $\rho^* = \rho\sigma^3$, and the temperature $T^* = k_B T\epsilon^{-1}$.

One may wonder why it is convenient to introduce reduced units. The most important reason is that (infinitely) many combinations of ρ, T, ϵ, and σ all correspond to the same state in reduced units. This is the law of corresponding states: the same simulation of a Lennard-Jones model can be used to study Ar at 60 K and a density of 840 kg/m^3 and Xe at 112 K and a density of 1617 kg/m^3. In reduced units, both simulations correspond to the state point $\rho^* = 0.5$, $T^* = 0.5$. If we had not used reduced units, we might have easily missed the equivalence of these two simulations. Another, practical, reason for using reduced units is the following: when we work with real (SI) units, we find that the absolute numerical values of the quantities that we are computing (e.g., the average energy of a particle or its acceleration) are either much less or much larger than 1. If we multiply several such quantities using standard floating-point multiplication, we face a distinct risk that, at some stage, we will obtain a result that creates an overflow or underflow. Conversely, in reduced units, almost all quantities of interest are of order 1 (say, between 10^{-3} and 10^3). Hence, if we suddenly find a very large (or very small) number in our simulations (say, 10^{42}), then there is a good chance that we have made an error somewhere. In other words, reduced

Quantity	Reduced units		Real units
temperature	$T^* = 1$	$\leftrightarrow$	$T = 119.8$ K
density	$\rho^* = 1.0$	$\leftrightarrow$	$\rho = 1680$ kg/m^3
time	$\Delta t^* = 0.005$	$\leftrightarrow$	$\Delta t = 1.09 \times 10^{-14}$ s
pressure	$P^* = 1$	$\leftrightarrow$	$P = 41.9$ MPa

Table 3.1: Translation of reduced units to real units for Lennard-Jones argon ($\epsilon/k_B = 119.8$ K, $\sigma = 3.405 \times 10^{-10}$m, $M = 0.03994$ kg/mol)

units make it easier to spot errors. Simulation results that are obtained in reduced units can always be translated back into real units. For instance, if we wish to compare the results of a simulation on a Lennard-Jones model at $T^* = 1$ and $P^* = 1$ with experimental data for argon ($\epsilon/k_B = 119.8$ K, $\sigma = 3.405 \times 10^{-10}$ m, $M = 0.03994$ kg/mol), then we can use the translation given in Table 3.1 to convert our simulation parameters to real SI units.[5]

3.2.3 Detailed Balance versus Balance

Throughout this book we use the condition of detailed balance as a test of the validity of a Monte Carlo scheme. However, as stated before, the detailed-balance condition is sufficient, but not necessary. Manousiouthakis and Deem [54] have shown that the weaker "balance condition" is a necessary and sufficient requirement. A consequence of this proof is that one has more freedom in developing Monte Carlo moves. For example, in the simple Monte Carlo scheme shown in Algorithm 2 we select a particle at random and give it a random displacement. During the next trial move, the *a priori* probability to select the *same* particle is the same. Thus the reverse trial move has the same *a priori* probability and detailed balance is satisfied. An alternative scheme is to attempt moving all particles sequentially, i.e., first an attempt to move particle one, followed by an attempt to move particle two, etc. In this sequential scheme, the probability that a single-particle move is followed by its reverse is zero. Hence, this scheme clearly violates detailed balance. However, Manousiouthakis and Deem have shown that such a sequential updating scheme does obey balance and does therefore (usually — see Ref. [54]) result in correct MC sampling.

We stress that the detailed-balance condition remains an important guiding principle in developing novel Monte Carlo schemes. Moreover, most algorithms that do not satisfy detailed balance are simply wrong. This is true in particular for "composite" algorithms that combine different trial moves. Therefore, we suggest that it is good practice to impose detailed balance

[5]In what follows we will always use reduced units, unless explicitly indicated otherwise. We, therefore, omit the superscript * to denote reduced units.

when writing a code. Of course, if subsequently it turns out that the performance of a working program can be improved considerably by using a "balance-only" algorithm, then it is worth implementing it. At present, we are not aware of examples in the literature where a "balance-only" algorithm is shown to be much faster than its "detailed-balance" counterpart.

3.3 Trial Moves

Now that we have specified the general structure of the Metropolis algorithm, we should consider its implementation. We shall not go into the problem of selecting intermolecular potentials for the model system under study. Rather, we shall simply assume that we have an atomic or molecular model system in a suitable starting configuration and that we have specified all intermolecular interactions. We must now set up the underlying Markov chain, that is, the matrix α. In more down to earth terms: we must decide how we are going to generate trial moves. We should distinguish between trial moves that involve only the molecular centers of mass and those that change the orientation or possibly even the conformation of a molecule.

3.3.1 Translational Moves

We start our discussion with trial moves of the molecular centers of mass. A perfectly acceptable method for creating a trial displacement is to add random numbers between $-\Delta/2$ and $+\Delta/2$ to the x, y, and z coordinates of the molecular center of mass:

$$\begin{aligned} x_i' &\rightarrow x_i + \Delta\,(\text{Ranf} - 0.5) \\ y_i' &\rightarrow y_i + \Delta\,(\text{Ranf} - 0.5) \\ z_i' &\rightarrow z_i + \Delta\,(\text{Ranf} - 0.5), \end{aligned} \qquad (3.3.1)$$

where Ranf are random numbers uniformly distributed between 0 and 1. Clearly, the reverse trial move is equally probable (hence, α is symmetric).[6] We are now faced with two questions: how large should we choose Δ? and should we attempt to move all particles simultaneously or one at a time? In the latter case we should pick the molecule that is to be moved at random to ensure that the underlying Markov chain remains symmetric. All

[6] Although almost all published MC simulations on atomic and molecular systems generate trial displacements in a cube centered around the original center of mass position, this is by no means the only possibility. Sometimes, it is more convenient to generate trial moves in a spherical volume, and it is not even necessary that the distribution of trial moves in such a volume be uniform, as long as it has inversion symmetry. For an example of a case where another sampling scheme is preferable, see ref. [55].

other things being equal, we should choose the most efficient sampling procedure. But, to this end, we must first define what we mean by *efficient sampling*. In very vague terms, sampling is efficient if it gives you good value for money. Good value in a simulation corresponds to high statistical accuracy, and "money" is simply *money*: the money that buys your computer time and even your own time. For the sake of the argument, we assume the average scientific programmer is poorly paid. In that case we have to worry only about your computer budget.[7] Then we could use the following definition of an optimal sampling scheme: a Monte Carlo sampling scheme can be considered optimal if it yields the lowest statistical error in the quantity to be computed for a given expenditure of computing budget. Usually, computing budget is equivalent to CPU time.

From this definition it is clear that, in principle, a sampling scheme may be optimal for one quantity but not for another. Actually, the preceding definition is all but useless in practice (as are most definitions). For instance, it is just not worth the effort to measure the error estimate in the pressure for a number of different Monte Carlo sampling schemes in a series of runs of fixed length. However, it is reasonable to assume that the mean-square error in the observables is inversely proportional to the number of uncorrelated configurations visited in a given amount of CPU time. And the number of independent configurations visited is a measure for the distance covered in configuration space. This suggests a more manageable, albeit rather ad hoc, criterion to estimate the efficiency of a Monte Carlo sampling scheme: the sum of the squares of all accepted trial displacements divided by computing time. This quantity should be distinguished from the mean-squared displacement per unit of computing time, because the latter quantity goes to 0 in the absence of diffusion (e.g., in a solid or a glass), whereas the former does not.

Using this criterion it is easy to show that for simulations of condensed phases it is usually advisable to perform random displacements of one particle at a time (as we shall see later, the situation is different for correlated displacements). To see why random single-particle moves are preferred, consider a system of N spherical particles, interacting through a potential energy function $\mathcal{U}(\mathbf{r}^N)$. Typically, we expect that a trial move will be rejected if the potential energy of the system changes by much more than $k_B T$. At the same time, we try to make the Monte Carlo trial steps as large as is possible without having a very low acceptance. A displacement that would, on average, give rise to an increase of the potential energy by $k_B T$ would still have a reasonable acceptance. In the case of a single-particle trial move, we

[7] Still, we should stress that it is not worthwhile to spend a lot of time developing a fancy computational scheme that will be only marginally better than existing, simpler schemes, unless your program will run very often and speed is crucial.

then have

$$\begin{aligned}\langle \Delta \mathcal{U} \rangle &= \left\langle \frac{\partial \mathcal{U}}{\partial r_i^\alpha} \right\rangle \overline{\Delta r_i^\alpha} + \frac{1}{2} \left\langle \frac{\partial^2 \mathcal{U}}{\partial r_i^\alpha \partial r_i^\beta} \right\rangle \overline{\Delta r_i^\alpha \Delta r_i^\beta} + \cdots \\ &= 0 + f(\mathcal{U})\, \overline{\Delta r_i^2} + \mathcal{O}(\Delta^4), \qquad (3.3.2)\end{aligned}$$

where the angle brackets denote averaging over the ensemble and the horizontal bar denotes averaging over random trial moves. The second derivative of $\mathcal{U}$ has been absorbed into the function $f(\mathcal{U})$, the precise form of which does not concern us here. If we now equate $\langle \Delta \mathcal{U} \rangle$ on the left-hand side of equation (3.3.2) to $k_B T$, we find the following expression for $\overline{\Delta r_i^2}$:

$$\overline{\Delta r_i^2} \approx k_B T / f(\mathcal{U}). \qquad (3.3.3)$$

If we attempt to move N particles, one at a time, most of the computation involved is spent on the evaluation of the change in potential energy. Assuming that we use a neighbor list or a similar time-saving device (see Appendix F), the total time spent on evaluating the potential energy change is proportional to nN, where n is the average number of interaction partners per molecule. The sum of the mean-squared displacements will be proportional to $N\overline{\Delta r^2} \sim N k_B T / f(\mathcal{U})$. Hence, the mean-squared displacement per unit of CPU time will be proportional to $k_B T / (n\, f(\mathcal{U}))$. Now suppose that we try to move all particles at once. The cost in CPU time will still be proportional to nN. But, using the same reasoning as in equations (3.3.2) and (3.3.3), we estimate that the sum of the mean-squared displacements is smaller by a factor $1/N$. Hence the total efficiency will be down by this same factor. This simple argument explains why most simulators use single-particle, rather than collective trial moves. It is important to note that we have assumed that a collective MC trial move consists of N independent trial displacements of the particles. As will be discussed in section 14.2, efficient collective MC moves *can* be constructed if the trial displacements of the individual particles are not chosen independently.

Next, consider the choice of the parameter Δ which determines the size of the trial move. How large should Δ be? If it is very large, it is likely that the resulting configuration will have a high energy and the trial move will probably be rejected. If it is very small, the change in potential energy is probably small and most moves will be accepted. In the literature, one often finds the mysterious statement that an acceptance of approximately 50% should be optimal. This statement is not necessarily true. The optimum acceptance ratio is the one that leads to the most efficient sampling of configuration space. If we express efficiency as mean-squared displacement per CPU time, it is easy to see that different Monte Carlo codes will have different optimal acceptance ratios. The reason is that it makes a crucial difference whether the amount of computing required to test whether a trial

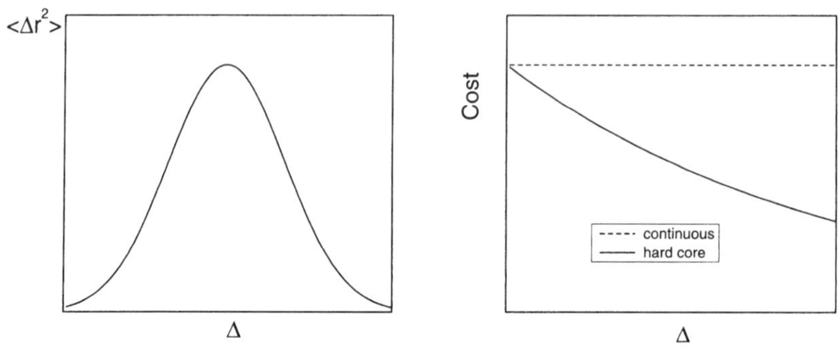

Figure 3.4: (left) Typical dependence of the mean-squared displacement of a particle on the average size Δ of the trial move. (right) Typical dependence of the computational cost of a trial move on the step-size Δ. For continuous potentials, the cost is constant, while for hard-core potentials it decreases rapidly with the size of the trial move.

move is accepted depends on the magnitude of the move (see Figure 3.4). In the conventional Metropolis scheme, all continuous interactions have to be computed before a move can be accepted or rejected. Hence, for continuous potentials, the amount of computation does not depend on the size of a trial move. In contrast, for simulations of molecules with hard repulsive cores, a move can be rejected as soon as overlap with any neighbor is detected. In that case, a rejected move is cheaper than an accepted one, and hence the average computing time per trial move goes down as the step size is increased. As a result, the optimal acceptance ratio for hard-core systems is appreciably lower than for systems with continuous interactions. Exactly how much depends on the nature of the program, in particular on whether it is a scalar or a vector code (in the latter case, hard-core systems are treated much like continuous systems), on how the information about neighbor lists is stored, and even on the computational "cost" of random numbers and exponentiation. The consensus seems to be that for hard-core systems the optimum acceptance ratio is closer to 20 than to 50%, but this is just another rule of thumb that should be checked.[8]

A distinct disadvantage of the efficiency criterion discussed previously is that it does not allow us to detect whether the sampling of configuration space is ergodic. To take a specific example, suppose that our system consists of a number of particles that are trapped in different potential energy min-

[8]In section 14.3.1, we show how, even in the case of continuous potentials, it is possible to reject trial moves before all interactions have been evaluated. With such a sampling scheme, the distinction between the sampling of hard-core and continuous potentials all but disappears.

ima. Clearly, we can sample the vicinity of these minima quite well and still have totally inadequate sampling of the whole of the configuration space. A criterion that would detect such nonergodicity has been proposed by Mountain and Thirumalai [56]. These authors consider the difference between the variance of the time average of the (potential) energy of all particles. Let us denote the time average of the energy of particle j in time interval t by $e_j(t)$:

$$e_j(t) = \frac{1}{t}\int_0^t dt'\, e_j(t').$$

And the average single-particle energy for this interval is

$$\bar{e}(t) \equiv \frac{1}{N}\sum_{j=1}^{N} e_j(t).$$

The variance of interest is

$$\sigma_E^2(t) \equiv \frac{1}{N}\sum_{j=1}^{N} \left[e_j(t) - \bar{e}(t)\right]^2 .$$

If all particles sample the whole of configuration space, $\sigma_E^2(t)$ will approach zero as $t \to \infty$:

$$\sigma_E^2(t)/\sigma_E^2(0) \to \tau_E/t,$$

where τ_E is a measure for the characteristic time to obtain uncorrelated samples. However, if the system is nonergodic, as in a (spin) glass, σ_E will not decay to 0. The work of Mountain and Thirumalai suggests that a good method for optimizing the efficiency of a Monte Carlo scheme is to minimize the product of τ_E and the computer time per trial move. Using this scheme, Mountain and Thirumalai concluded that, even for the Lennard-Jones system, a trial move acceptance of 50% is far from optimal. They found that an acceptance probability of 20% was twice as efficient.

Of course, a scheme based on the energy fluctuations of a particle is not very useful to monitor the rate of convergence of simulations of hard-core systems. But the essence of the method is not that one measures the energy but any quantity that is sensitive to the local environment of a particle. For instance, a robust criterion would look at the convergence of the time-averaged Voronoi signature of a particle. Different environments yield different signatures. Only if every particle samples all environments will the variance of Voronoi signatures decay to 0.

Of course, in some situations an efficiency criterion based on ergodicity is not useful. By construction, it cannot be used to optimize simulations of glasses. But also when studying interfaces (e.g., solid-liquid or liquid vapor) the ergodicity criterion would suggest that every particle should have ample time to explore both coexisting phases. This is clearly unnecessary: ice can be in equilibrium with water, even though the time of equilibration is far too short to allow complete exchange of the molecules in the two phases.

3.3.2 Orientational Moves

If we are simulating molecules rather than atoms we must also generate trial moves that change the molecular orientation. As we discussed already, it almost requires an effort for generating translational trial moves with a distribution that does not satisfy the symmetry requirement of the underlying Markov chain. For rotational moves, the situation is very different. It is only too easy to introduce a systematic bias in the orientational distribution function of the molecules by using a nonsymmetrical orientational sampling scheme. Several different strategies to generate rotational displacements are discussed in [19]. Here we only mention one possible approach.

Rigid, Linear Molecules

Consider a system consisting of N linear molecules. We specify the orientation of the ith molecule by a unit vector $\hat{\mathbf{u}}_i$. One possible procedure to change $\hat{\mathbf{u}}_i$ by a small, random amount is the following. First, we generate a unit vector $\hat{\mathbf{v}}$ with a random orientation. This is quite easy to achieve (see Algorithm 42). Next we multiply this random unit vector $\hat{\mathbf{v}}$ by a scale factor γ. The magnitude of γ determines the magnitude of the trial rotation. We now add $\gamma\hat{\mathbf{v}}$ to $\hat{\mathbf{u}}_i$. Let us denote the resulting sum vector by $\mathbf{t}$: $\mathbf{t} = \gamma\hat{\mathbf{v}} + \hat{\mathbf{u}}_i$. Note that $\mathbf{t}$ is not a unit vector. Finally, we normalize $\mathbf{t}$, and the result is our trial orientation vector $\hat{\mathbf{u}}_i'$. We still have to fix γ, which determines the acceptance probability for the orientational trial move. The optimum value of γ is determined by essentially the same criteria as for translational moves. We have not yet indicated whether the translational and orientational trial moves should be performed simultaneously. Both procedures are acceptable. However, if rotation and translation correspond to separate moves, then the selection of the type of move should be probabilistic rather than deterministic.

Rigid, Nonlinear Molecules

Only slightly more complex is the case of a nonlinear, rigid molecule. It is conventional to describe the orientation of nonlinear molecules in terms of the Eulerian angles (ϕ, θ, ψ). However, for most simulations, use of these angles is less convenient because all rotation operations should then be expressed in terms of trigonometric functions, and these are computationally expensive. It is usually better to express the orientation of such a molecule in terms of quaternion parameters (for a discussion of quaternions in the context of computer simulation, see [19]). The rotation of a rigid body can be specified by a quaternion of unit norm **Q**. Such a quaternion may be thought of as a unit vector in four-dimensional space:

$$\mathbf{Q} \equiv (q_0, q_1, q_2, q_3) \quad \text{with } q_0^2 + q_1^2 + q_2^2 + q_3^2 = 1. \tag{3.3.4}$$

There is a one-to-one correspondence between the quaternion components q_α and the Eulerian angles,

$$\begin{aligned} q_0 &= \cos\frac{\theta}{2}\cos\left(\frac{\phi+\psi}{2}\right) \\ q_1 &= \sin\frac{\theta}{2}\cos\left(\frac{\phi-\psi}{2}\right) \\ q_2 &= \sin\frac{\theta}{2}\sin\left(\frac{\phi-\psi}{2}\right) \\ q_3 &= \cos\frac{\theta}{2}\sin\left(\frac{\phi+\psi}{2}\right), \end{aligned} \tag{3.3.5}$$

and the rotation matrix R, which describes the rotation of the molecule-fixed vector in the laboratory frame, is given by (see, e.g., [57])

$$\mathsf{R} = \begin{pmatrix} q_0^2 + q_1^2 - q_2^2 - q_3^2 & 2(q_1 q_2 - q_0 q_3) & 2(q_1 q_3 + q_0 q_2) \\ 2(q_1 q_2 + q_0 q_3) & q_0^2 - q_1^2 + q_2^2 - q_3^2 & 2(q_2 q_3 - q_0 q_1) \\ 2(q_1 q_3 - q_0 q_2) & 2(q_2 q_3 + q_0 q_1) & q_0^2 - q_1^2 - q_2^2 + q_3^2 \end{pmatrix}. \tag{3.3.6}$$

To generate trial rotations of nonlinear, rigid bodies, we must rotate the vector (q_0, q_1, q_2, q_3) on the four-dimensional (4D) unit sphere. The procedure just described for the rotation of a 3D unit vector is easily generalized to 4D. An efficient method for generating random vectors uniformly on the 4D unit sphere has been suggested by Vesely [57].

Nonrigid Molecules

If the molecules under consideration are not rigid then we must also consider Monte Carlo trial moves that change the internal degrees of freedom of a molecule. In practice, it makes an important difference whether we have frozen out some of the internal degrees of freedom of a molecule by imposing rigid constraints on, say, bond lengths and possibly even some bond angles. If not, the situation is relatively simple: we can carry out normal trial moves on the Cartesian coordinates of the individual atoms in the molecule (in addition to center-of-mass moves). If some of the atoms are strongly bound, it is advisable to carry out small trial moves on those particles (no rule forbids the use of trial moves of different size for different atoms, as long as the moves for one particular atom are always sampled from the same distribution).

However, when the bonds between different atoms become very stiff, this procedure does not sample conformational changes of the molecule efficiently. In Molecular Dynamics simulations it is common practice to replace very stiff intramolecular interactions with rigid constraints (see Chapter 15). For Monte Carlo simulations this is also possible. In fact, elegant techniques have been developed for this purpose [58]. However, the corresponding MD techniques [59] are so much easier to use, in particular for large molecules, that we cannot recommend the use of the Monte Carlo technique for any but the smallest flexible molecules with internal constraints.

To understand why Monte Carlo simulations of flexible molecules with a number of stiff (or even rigid) bonds (or bond angles) can become complicated, let us return to the original expression (3.1.2) for a thermal average of a function $A(\mathbf{r}^N)$:

$$\langle A \rangle = \frac{\int d\mathbf{p}^N d\mathbf{r}^N \; A(\mathbf{r}^N) \exp[-\beta \mathcal{H}(\mathbf{p}^N, \mathbf{r}^N)]}{\int d\mathbf{p}^N d\mathbf{r}^N \; \exp[-\beta \mathcal{H}(\mathbf{p}^N, \mathbf{r}^N)]}.$$

If we are dealing with flexible molecules, it is convenient to perform Monte Carlo sampling not on the Cartesian coordinates $\mathbf{r}^N$ but on the generalized coordinates $\mathbf{q}^N$, where q may be, for instance, a bond length or an internal angle. We must now express the Hamiltonian in equation (3.1.2) in terms of these generalized coordinates and their conjugate momenta. This is done most conveniently by first considering the Lagrangian $\mathcal{L} = \mathcal{K} - \mathcal{U}$, where $\mathcal{K}$ is the kinetic energy of the system ($\mathcal{K} = \sum (1/2) m \dot{r}^2$) and $\mathcal{U}$ the potential energy. When we transform from Cartesian coordinates $\mathbf{r}$ to generalized coordinates $\mathbf{q}$, $\mathcal{L}$ changes to

$$\begin{aligned} \mathcal{L} &= \sum_{i=1}^{N} \frac{1}{2} m_i \frac{\partial \mathbf{r}_i}{\partial q_\alpha} \frac{\partial \mathbf{r}_i}{\partial q_\beta} \dot{q}_\alpha \dot{q}_\beta - \mathcal{U}(\mathbf{q}^N) \\ &\equiv \frac{1}{2} \dot{\mathbf{q}} \cdot \mathsf{G} \cdot \dot{\mathbf{q}} - \mathcal{U}(\mathbf{q}^N). \end{aligned} \tag{3.3.7}$$

In the second line of equation (3.3.7) we have defined the matrix G. The momenta conjugate to $\mathbf{q}^N$ are easily derived using

$$p^\alpha \equiv \frac{\partial \mathcal{L}}{\partial \dot{q}_\alpha}.$$

This yields $p^\alpha = \mathsf{G}_{\alpha\beta} \dot{q}_\beta$. We can now write down the Hamiltonian $\mathcal{H}$ in terms of the generalized coordinates and conjugate momenta:

$$\mathcal{H}(\mathbf{p}, \mathbf{q}) = \frac{1}{2} \mathbf{p} \cdot \mathsf{G}^{-1} \cdot \mathbf{p} + \mathcal{U}(\mathbf{q}^N). \tag{3.3.8}$$

If we now insert this form of the Hamiltonian into equation (3.1.2), and carry out the (Gaussian) integration over the momenta, we find that

$$\begin{aligned}\langle A \rangle &= \frac{\int d\mathbf{q}^N \exp[-\beta\mathcal{U}(\mathbf{q}^N)]A(\mathbf{q}^N)\int d\mathbf{p}^N \exp(-\beta\mathbf{p}\cdot G^{-1}\cdot\mathbf{p}/2)}{\int d\mathbf{q}^N d\mathbf{p}^N \exp(-\beta\mathcal{H})} \\ &= \frac{\int d\mathbf{q}^N \exp[-\beta\mathcal{U}(\mathbf{q}^N)]A(\mathbf{q}^N)|G|^{\frac{1}{2}}}{\int d\mathbf{q}^N d\mathbf{p}^N \exp(-\beta\mathcal{H})}. \end{aligned} \tag{3.3.9}$$

The problem with equation (3.3.9) is the term $|G|^{\frac{1}{2}}$. Although the determinant $|G|$ can be computed fairly easily for small flexible molecules, its evaluation can become quite an unpleasant task in the case of larger molecules.

Thus far we have considered the effect of introducing generalized coordinates only on the form of the expression for thermal averages. If we are considering a situation where some of the generalized coordinates are actually constrained to have a fixed value, then the picture changes again, because such hard constraints are imposed at the level of the Lagrangian equations of motion. Hard constraints therefore lead to a different form for the Hamiltonian in equation (3.3.8) and to another determinant in equation (3.3.9). Again, all this can be taken into account in the Monte Carlo sampling (see [58]). An example of such a Monte Carlo scheme is the concerted rotation algorithm that has been developed by Theodorou and co-workers [60] to simulate polymer melts and glasses (see section 13.4.4). The idea of this algorithm is to select a set of adjacent skeletal bonds in a chain (up to seven bonds). These bonds are given a collective rotation while the rest of the chain is unaffected. By comparison, Molecular Dynamics simulations of flexible molecules with hard constraints have the advantage that these constraints enter directly into the equations of motion (see [59]). The distinction between Molecular Dynamics and Monte Carlo, however, is more apparent than real, since it is possible to use MD techniques to generate collective Monte Carlo moves (see section 14.2). In Chapter 13, we shall discuss other Monte Carlo sampling schemes that are particularly suited for flexible molecules.

3.4 Applications

In this section we give several case studies using the basic NVT Monte Carlo algorithm.

Case Study 1 (Equation of State of the Lennard-Jones Fluid)
One of the more important applications of molecular simulation is to compute the phase diagram of a given model system. In fact, in Chapter 8 several numerical techniques that have been developed specifically to study

phase transitions will be discussed. It may not be immediately obvious to the reader, however, that there is any need for the sophisticated numerical schemes presented in Chapter 8. In this Case Study, we illustrate some of the problems that occur when we use standard Monte Carlo simulation to determine a phase diagram. As an example, we focus on the vapor-liquid curve of the Lennard-Jones fluid. Of course, as was already mentioned in section 3.2.2, the phase behavior is quite sensitive to the detailed form of the intermolecular potential that is used. In this Case Study, we approximate the full Lennard-Jones potential as follows:

$$u(r) = \begin{cases} u^{lj}(r) & r \leq r_c \\ 0 & r > r_c, \end{cases}$$

where the cutoff radius r_c is set to half the box length. The contribution of the particles beyond this cutoff is estimated with the usual tail corrections; that is, for the energy

$$u^{tail} = \frac{8}{3}\pi\rho \left[\frac{1}{3}\left(\frac{1}{r_c}\right)^9 - \left(\frac{1}{r_c}\right)^3 \right]$$

and for the pressure

$$p^{tail} = \frac{16}{3}\pi\rho^2 \left[\frac{2}{3}\left(\frac{1}{r_c}\right)^9 - \left(\frac{1}{r_c}\right)^3 \right].$$

The equation of state of the Lennard-Jones fluid has been investigated by many groups using Molecular Dynamics or Monte Carlo simulations starting with the work of Wood and Parker [49]. A systematic study of the equation of state of the Lennard-Jones fluid was reported by Verlet [13]. Subsequently, many more studies have been published. In 1979, the data available at that time were compiled by Nicolas *et al.* [61] into an accurate equation of state. This equation has been refitted by Johnson *et al.* [62] in the light of more recent data. In the present study we compare our numerical results with the equation of state by Johnson *et al.*

We performed several simulations using Algorithms 1 and 2. During the simulations we determined the energy per particle and the pressure. The pressure was calculated using the virial

$$P = \frac{\rho}{\beta} + \frac{\text{vir}}{V}, \tag{3.4.1}$$

where the virial is defined by

$$\text{vir} = \frac{1}{3} \sum_i \sum_{j>i} \mathbf{f}(\mathbf{r}_{ij}) \cdot \mathbf{r}_{ij}, \tag{3.4.2}$$

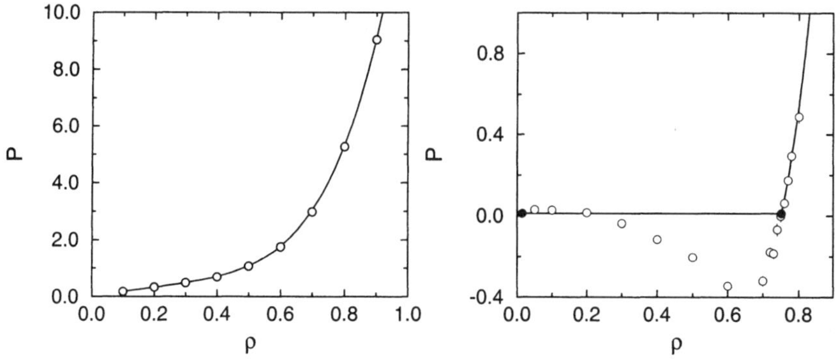

Figure 3.5: Equation of state of the Lennard-Jones fluid. (left) Isotherm at $T = 2.0$. (right) Isotherm below the critical temperature ($T = 0.9$); the horizontal line is the saturated vapor pressure and the filled circles indicate the densities of the coexisting vapor and liquid phases. The solid curve represents the equation of state of Johnson *et al.* [62] and the circles are the results of the simulations ($N = 500$). The errors are smaller than the symbol size.

where $\mathbf{f}(\mathbf{r}_{ij})$ is the intermolecular force. Figure 3.5 (left) compares the pressure as obtained from a simulation above the critical temperature with the equation of state of Johnson *et al.* [62]. The agreement is excellent (as is to be expected).

Figure 3.5 (right) shows a typical isotherm below the critical temperature. If we cool the system below the critical temperature, we should expect to observe vapor-liquid coexistence. However, conventional Monte Carlo or Molecular Dynamics simulations of small model systems are not suited to study the coexistence between two phases. Using the Johnson equation of state, we predict how the pressure of a macroscopic Lennard-Jones system would behave in the two-phase region (see Figure 3.5). For densities inside the coexistence region the pressure is expected to be constant and equal to the saturated vapor pressure. If we now perform a Monte Carlo simulation of a finite system (500 LJ particles), we find that the computed pressure is not at all constant in the coexistence region (see Figure 3.5). In fact we observe that, over a wide density range, the simulated system is metastable and may even have a negative pressure. The reason is that, in a finite system, a relatively important free-energy cost is associated with the creation of a liquid-vapor interface. So much so that, for sufficiently small systems, it is favorable for the system not to phase separate at all [63]. Clearly these problems will be most severe for small systems and in cases where the interfacial free energy is large. For this reason, standard NVT-simulations are not recommended to determine the vapor-liquid coexistence curve or, for that matter, any strong first-order phase transition.

To determine the liquid-vapor coexistence curve we should determine the equation of state for a large number of state points outside the coexistence region. These data can then be fitted to an analytical equation of state. With this equation of state we can determine the vapor-liquid curve (this is exactly the procedure used by Nicolas *et al.* [61] and Johnson *et al.* [62]).

Of course, if we simulate a system consisting of a very large number of particles, it is possible to simulate a liquid phase in coexistence with its vapor. However, such simulations are quite time consuming, because it takes a long time to equilibrate a two-phase system.

Case Study 2 (Importance of Detailed Balance)
For a Monte Carlo simulation to sample points in configuration space according to their correct Boltzmann weight, it is sufficient, but not necessary, to impose the detailed-balance condition on the sampling algorithm. Of course, as the condition of detailed balance is stronger than strictly necessary, it is not excluded that correct sampling schemes exist that violate detailed balance. However, unless one can actually prove that a non-detailed-balance scheme yields the correct distribution, the use of such schemes is strongly to be discouraged. Even seemingly reasonable schemes may give rise to serious, systematic errors.

Here we give an example of such a scheme. Consider an ordinary N,V,T move; a new position is generated by giving a randomly selected particle, say i, a random displacement:

$$x_n(i) = x_o(i) + \Delta_x(\text{Ranf} - 0.5),$$

where Δ_x is twice the maximum displacement. We now make a small error and generate a new position using

$$x_n(i) = x_o(i) + \Delta_x(\text{Ranf} - 0.0) \qquad \text{wrong!}$$

We give the particles only a *positive* displacement. With such a move detailed balance is violated, since the reverse move — putting the particle back at x_o — is not possible.

For the Lennard-Jones fluid we can use the program of Case Study 1 to compare the two sampling schemes. The results of these simulations are shown in Figure 3.6. Note that, at first sight, the results of the incorrect scheme look reasonable; in fact, at low densities the results of the two schemes do not show significant differences. But at high densities the wrong scheme overestimates the pressure. It is important to note that the incorrect scheme leads to a systematic error that does not disappear when we perform longer simulations.

This example illustrates that one can generate numerical results that *look* reasonable, even with an incorrect sampling scheme. For this reason, it is

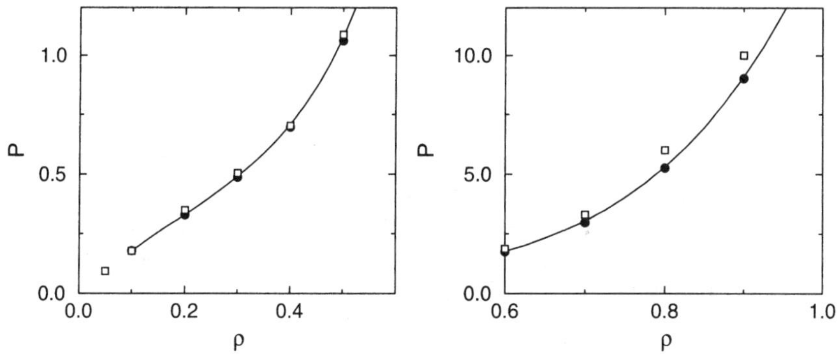

Figure 3.6: Equation of state of the Lennard-Jones fluid (T = 2.0); comparison of a displacement scheme that obeys detailed balance (circles) and one that does not (squares). Both simulations have been performed with 500 particles. The solid curve is the equation of state of Johnson *et al.* [62]. The figure at the left corresponds to the low-pressure regime. The high-pressure regime is shown in the right-hand figure.

important always to compare the results obtained with a new Monte Carlo program with known numerical results or, better still, with exact results that may be known in some limiting case (dilute vapor, dense solid, etc.).

In the present example, the error due to the neglect of detailed balance is quite obvious. In many cases, the effects are less clear. The most common source of non-detailed-balance sampling schemes is the following: in many programs, we can choose from a repertoire of trial moves (e.g., translation, rotation, volume changes). It is recommended that these trial moves are not carried out in fixed order, because then the reverse sequence is impossible and detailed balance is no longer satisfied.[9]

In practice one often does not know *a priori* the optimal maximum displacement in a Monte Carlo simulation. A practical solution is to adjust during the simulation the maximum displacement in such a way that the optimum acceptance probability is obtained. The ideal situation is to determine this optimum during the equilibration. However, if one would keep adjusting the maximum step-size during a production run, then one would violate detailed balance [65]. For example, if from one move to the next, the maximum displacement is decreased, then the *a priori* probability for a particle to return to its previous position could be zero. Hence, if one would change the maximum displacement after every Monte Carlo step serious errors are to be expected. Of course, if one changes the maximum displacement only a few

[9]It has been shown [64] that in this case the detailed-balance condition is indeed sufficient but not necessary to maintain equilibrium.

times during the simulation, then the error will be negligible. Yet, it is better to stay on the safe side and never change the maximum displacement during the projection run.

Case Study 3 (Why Count the Old Configuration Again?)
A somewhat counterintuitive feature of the Metropolis sampling scheme is that, if a trial move is rejected, we should once again count the contributions of the old configuration to the average that we are computing (see acceptance rule (3.1.18)). The aim of this Case Study is to show that this recounting is really essential. In the Metropolis scheme the acceptance rule for a move from o to n is

$$\begin{aligned} \mathrm{acc}(o \to n) &= \exp\{-\beta[\mathcal{U}(n) - \mathcal{U}(o)]\} & \mathcal{U}(n) \geq \mathcal{U}(o) \\ &= 1 & \mathcal{U}(n) < \mathcal{U}(o). \end{aligned}$$

These acceptance rules lead to a transition probability

$$\begin{aligned} \pi(o \to n) &= \exp\{-\beta[\mathcal{U}(n) - \mathcal{U}(o)]\} & \mathcal{U}(n) \geq \mathcal{U}(o) \\ &= 1 & \mathcal{U}(n) < \mathcal{U}(o). \end{aligned}$$

Note that this transition probability must be normalized:

$$\sum_n \pi(o \to n) = 1.$$

From this normalization it follows that the probability that we accept the old configuration again is by definition

$$\pi(o \to o) = 1 - \sum_{n \neq o} \pi(o \to n).$$

This last equation implies that we should count the contribution of the old configuration again.

It is instructive to use the Lennard-Jones program from Case Study 1 to investigate numerically the error that is made when we only include accepted configurations in our averaging. In essence, this means that in Algorithm 2 we continue attempting to displace the selected particle until a trial move has been accepted.[10] In Figure 3.7 we compare the results of the correct scheme with those obtained by the scheme in which we continue to displace a particle until a move is accepted. Again the results look reasonable, but the figure shows that large, systematic errors are being made.

[10]It is easy to see that this approach leads to the wrong answer if we try to compute the average energy of a two-level system with energy levels E_0 and E_1. If we include only accepted trial moves in our averaging, we would find that $\langle E \rangle = (E_0 + E_1)/2$, independent of temperature.

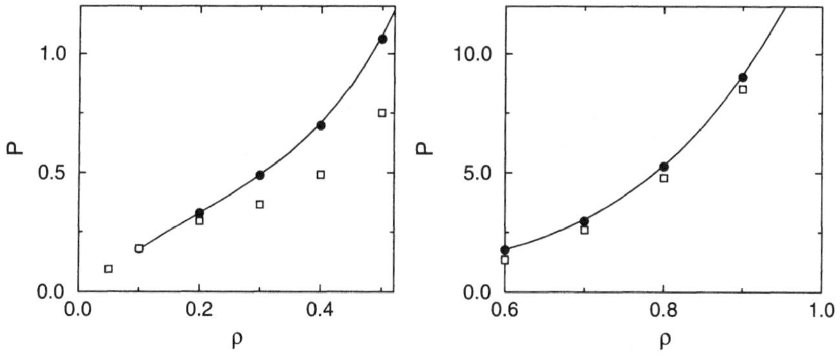

Figure 3.7: Equation of state of the Lennard-Jones fluid ($T = 2.0$); comparison of a scheme in which particles are displaced until a move is accepted (squares) with the conventional scheme (circles). Both simulations have been performed with 108 particles. The solid curve is the equation of state of Johnson *et al.* [62]. The left figure is at low pressure and the right one at high pressure.

One of the important disadvantages of the Monte Carlo scheme is that it does not reproduce the natural dynamics of the particles in the system. However, sometimes this limitation of the method can be made to work to our advantage. In Example 1 we show how the equilibration of a Monte Carlo simulation can be speeded up by many orders of magnitude through the use of unphysical trial moves.

Example 1 (Mixture of Hard Disks)

In a Molecular Dynamics simulation of, for instance, a binary ($A - B$) mixture of hard disks (see Figure 3.8), the efficiency with which configuration space is sampled is greatly reduced by the fact that concentration fluctuations decay very slowly (typically the relaxation time $\tau \sim D_{AB}/\lambda^2$, where D_{AB} is the mutual diffusion coefficient and λ is the wavelength of the concentration fluctuation). This implies that very long runs are needed to ensure equilibration of the local composition of the mixture. In solids, equilibration may not take place at all (even on time scales accessible in nature). In contrast, in a Monte Carlo simulation, it is permissible to carry out trial moves that swap the identities of two particles of species A and B. Such moves, even if they have only a moderate rate of acceptance (a few percent will do), greatly speed up the sampling of concentration fluctuations.

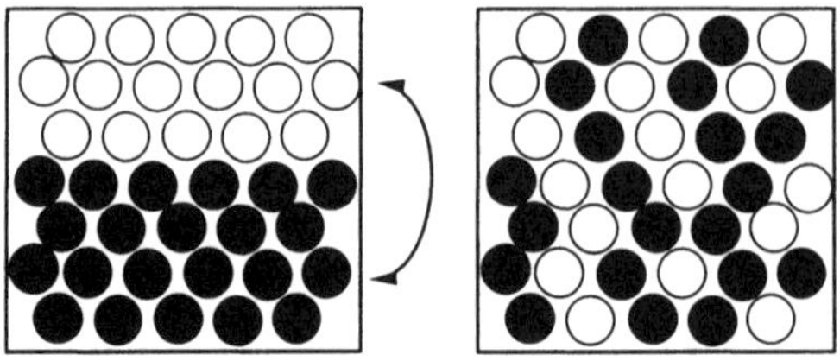

Figure 3.8: A mixture of hard disks, where the identities of two particles are swapped.

3.5 Questions and Exercises

Question 7 (Reduced Units) *Typical sets of Lennard-Jones parameters for argon and krypton are* $\sigma_{Ar} = 3.41$ Å, $\epsilon_{Ar}/k_B = 119.8$ K *and* $\sigma_{Kr} = 3.38$ Å, $\epsilon_{Kr}/k_B = 164.0$ K *[19].*

1. *At the reduced temperature* $T^* = 2.0$, *what is the temperature in kelvin of argon and krypton?*
2. *A typical time step for MD is* $\Delta t^* = 0.001$. *What is this in SI units for argon and krypton?*
3. *If we simulate argon at* $T = 278$ K *and density* $\rho = 2000$ kg/m^3 *with a Lennard-Jones potential, for which conditions of krypton can we use the same data? If we assume ideal gas behavior, compute the pressure in reduced and normal units.*
4. *List the main reasons to use reduced units.*

Question 8 (Heat Capacity) *Heat capacity can also be calculated from fluctuations in the total energy in the canonical ensemble:*

$$C_v = \frac{\langle U^2 \rangle - \langle U \rangle^2}{k_B T^2}. \qquad (3.5.1)$$

1. *Derive this equation.*
2. *In a MC NVT simulation, one does not calculate fluctuations in the total energy but in the potential energy. Is it then still possible to calculate the heat capacity? Explain.*
3. *Heat capacity can be also calculated from differentiating the total energy of a system with respect to temperature. Discuss the advantages or disadvantages of this approach.*

Question 9 (A New Potential) *On the planet Krypton, the pair potential between two Gaia atoms is given by the Lennard-Jones 10-5 potential*

$$U(r) = 5\epsilon \left[\left(\frac{\sigma}{r}\right)^{10} - \left(\frac{\sigma}{r}\right)^{5} \right].$$

Kryptonians are notoriously lazy and it is therefore up to you to derive the tail corrections for the energy, pressure, and chemical potential. If we use this potential in an MD simulation in the truncated and shifted form we still have a discontinuity in the force. Why? If you compare this potential with the Lennard-Jones potential, will there be any difference in efficiency of the simulation? (Hint: there are two effects!)

Exercise 6 (Calculation of π)
Consider a circle of diameter d surrounded by a square of length l ($l \geq d$). Random coordinates are generated within the square. The value of π can be calculated from the fraction of points that fall within the circle.

1. How can π be calculated from the fraction of points that fall in the circle? Remark: the "exact" value of π can be computed numerically using $\pi = 4 \times \arctan(1)$.
2. Complete the small Monte Carlo program to calculate π using this method.
3. How does the accuracy of the result depend on the ratio l/d and the number of generated coordinates? Derive a formula to calculate the relative standard deviation of the estimate of π.
4. Why is this not a very efficient method for computing π accurately?

Exercise 7 (The Photon Gas)
The average occupancy number of state j of the photon gas, $\langle n_j \rangle$, can be calculated analytically; see equation (2.3.5). It is possible to estimate this quantity using a Monte Carlo scheme. In this exercise, we will use the following procedure to calculate $\langle n_j \rangle$:

(i) Start with an arbitrary n_j.

(ii) Decide at random to perform a trial move to increase or decrease n_j by 1.

(iii) Accept the trial move with probability

$$\text{acc}(o \to n) = \min\left(1, \exp\left[-\beta\left(U(n) - U(o)\right)\right]\right).$$

Of course, n_j cannot become negative!

1. Does this scheme obey detailed balance when $n_j = 0$?

2. Is the algorithm still correct when trial moves are performed that change n_j with a random integer from the interval $[-5, 5]$? What happens when only trial moves are performed that change n_j with either -3 or $+3$?
3. Assume that $N = 1$ and $\epsilon_j = \epsilon$. Write a small Monte Carlo program to calculate $\langle n_j \rangle$ as a function of $\beta\epsilon$. Compare your result with the analytical solution.
4. Modify the program in such a way that the averages are not updated when a trail move is rejected. Why does this lead to erroneous results? At which values of β does this error become more pronounced?
5. Modify the program in such a way that the distribution of n_j is calculated as well. Compare this distribution with the analytical expression.

Exercise 8 (Monte Carlo Simulation of a Lennard-Jones System)
In this exercise, we study a 3D Lennard-Jones system. See also Case Study 1.

1. In the code that you can find on the book's website, the pressure of the system is not calculated. Modify the code in such a way that the average pressure can be calculated. You will only have to make some changes in the subroutine *ener.f*.
2. Perform a simulation at $T = 2.0$ and at various densities. Up to what density is the ideal gas law
$$\beta p = \rho \tag{3.5.2}$$
a good approximation?
3. The program produces a sequence of snapshots of the state of the system. Try to visualize these snapshots using, for example, the program *MOLMOL*.
4. For the heat capacity at constant volume one can derive
$$C_v = \frac{\langle U^2 \rangle - \langle U \rangle^2}{k_B T^2}$$
in which U is the total energy of the system. Derive a formula for the dimensionless heat capacity. Modify the program (only in *mc_nvt.f*) in such a way that C_v is calculated.
5. Instead of performing trial moves in which one particle at a time is displaced, one can make trial moves in which all particles are displaced. Compare the maximum displacements of these moves when 50% of all displacements are accepted.
6. Instead of using a uniformly distributed displacement, one can also use a Gaussian displacement. Does this increase the efficiency of the simulation?

Exercise 9 (Scaling as a Monte Carlo Move)
Consider a system in which the energy is a function of one variable (x) only,

$$\exp\left[-\beta U(x)\right] = \theta(x)\,\theta(1-x),$$

in which $\theta(x)$ is the Heaviside step function: $\theta(x<0)=0$ and $\theta(x>0)=1$. We wish to calculate the distribution of x in the canonical ensemble. We will consider two possible algorithms (we will use $\delta > 0$):

(i) Generate a random change in x between $[-\delta, \delta]$. Accept or reject the new x according to its energy.

(ii) Generate a random number ϕ between $[1, 1+\delta]$. With a probability of 0.5, invert the value ϕ thus obtained. The new value of x is obtained by multiplying x with ϕ.

1. Derive the correct acceptance/rejection rules for both schemes.
2. Complete the computer code to calculate the probability density of x. The program writes this distribution to *distri.dat*.
3. What happens when the acceptance rule of method (i) is used in the algorithm of method (ii)?

Chapter 4

Molecular Dynamics Simulations

Molecular Dynamics simulation is a technique for computing the equilibrium and transport properties of a classical many-body system. In this context, the word *classical* means that the nuclear motion of the constituent particles obeys the laws of classical mechanics. This is an excellent approximation for a wide range of materials. Only when we consider the translational or rotational motion of light atoms or molecules (He, H_2, D_2) or vibrational motion with a frequency ν such that $h\nu > k_B T$ should we worry about quantum effects.

Of course, our discussion of this vast subject is necessarily incomplete. Other aspects of the Molecular Dynamics techniques can be found in [19, 39–41].

4.1 Molecular Dynamics: The Idea

Molecular Dynamics simulations are in many respects very similar to real experiments. When we perform a real experiment, we proceed as follows. We prepare a sample of the material that we wish to study. We connect this sample to a measuring instrument (e.g., a thermometer, manometer, or viscosimeter), and we measure the property of interest during a certain time interval. If our measurements are subject to statistical noise (as most measurements are), then the longer we average, the more accurate our measurement becomes. In a Molecular Dynamics simulation, we follow exactly the same approach. First, we prepare a sample: we select a model system consisting of N particles and we solve Newton's equations of motion for this system until the properties of the system no longer change with time (we equilibrate the system). After equilibration, we perform the actual measurement. In fact, some of the most common mistakes that can be made when performing a computer experiment are very similar to the mistakes that can be made in real experiments (e.g., the sample is not prepared correctly, the measurement is too short, the system undergoes an irreversible change during the experiment, or we do not measure what we think).

To measure an observable quantity in a Molecular Dynamics simulation, we must first of all be able to express this observable as a function of the positions and momenta of the particles in the system. For instance, a convenient definition of the temperature in a (classical) many-body system makes use of the equipartition of energy over all degrees of freedom that enter quadratically in the Hamiltonian of the system. In particular for the average kinetic energy per degree of freedom, we have

$$\left\langle \frac{1}{2} m v_\alpha^2 \right\rangle = \frac{1}{2} k_B T. \tag{4.1.1}$$

In a simulation, we use this equation as an operational definition of the temperature. In practice, we would measure the total kinetic energy of the system and divide this by the number of degrees of freedom N_f (= $3N - 3$ for a system of N particles with fixed total momentum[1]). As the total kinetic energy of a system fluctuates, so does the instantaneous temperature:

$$T(t) = \sum_{i=1}^{N} \frac{m_i v_i^2(t)}{k_B N_f}. \tag{4.1.2}$$

The relative fluctuations in the temperature will be of order $1/\sqrt{N_f}$. As N_f is typically on the order of 10^2–10^3, the statistical fluctuations in the temperature are on the order of 5–10%. To get an accurate estimate of the temperature, one should average over many fluctuations.

4.2 Molecular Dynamics: A Program

The best introduction to Molecular Dynamics simulations is to consider a simple program. The program we consider is kept as simple as possible to illustrate a number of important features of Molecular Dynamics simulations.

The program is constructed as follows:

1. We read in the parameters that specify the conditions of the run (e.g., initial temperature, number of particles, density, time step).

[1] Actually, if we define the temperature of a microcanonical ensemble through $(k_B T)^{-1} = (\partial \ln \Omega / \partial E)$, then we find that, for a d-dimensional system of N atoms with fixed total momentum, $k_B T$ is equal to $2E/(d(N-1)-2)$.

Algorithm 3 (A Simple Molecular Dynamics Program)

```
program md                          simple MD program

call init                           initialization
t=0
do while (t.lt.tmax)                MD loop
   call force(f,en)                 determine the forces
   call integrate(f,en)             integrate equations of motion
   t=t+delt
   call sample                      sample averages
enddo
stop
end
```

Comment to this algorithm:

1. *Subroutines* `init`, `force`, `integrate`, *and* `sample` *will be described in Algorithms 4, 5, and 6, respectively. Subroutine* `sample` *is used to calculate averages like pressure or temperature.*

2. We initialize the system (i.e., we select initial positions and velocities).
3. We compute the forces on all particles.
4. We integrate Newton's equations of motion. This step and the previous one make up the core of the simulation. They are repeated until we have computed the time evolution of the system for the desired length of time.
5. After completion of the central loop, we compute and print the averages of measured quantities, and stop.

Algorithm 3 is a short pseudo-algorithm that carries out a Molecular Dynamics simulation for a simple atomic system. We discuss the different operations in the program in more detail.

4.2.1 Initialization

To start the simulation, we should assign initial positions and velocities to all particles in the system. The particle positions should be chosen compatible with the structure that we are aiming to simulate. In any event, the particles should not be positioned at positions that result in an appreciable overlap of the atomic or molecular cores. Often this is achieved by initially placing

Algorithm 4 (Initialization of a Molecular Dynamics Program)

```
subroutine init                    initialization of MD program
sumv=0
sumv2=0
do i=1,npart
  x(i)=lattice_pos(i)              place the particles on a lattice
  v(i)=(ranf()-0.5)                give random velocities
  sumv=sumv+v(i)                   velocity center of mass
  sumv2=sumv2+v(i)**2              kinetic energy
enddo
sumv=sumv/npart                    velocity center of mass
sumv2=sumv2/npart                  mean-squared velocity
fs=sqrt(3*temp/sumv2)              scale factor of the velocities
do i=1,npart                       set desired kinetic energy and set
  v(i)=(v(i)-sumv)*fs              velocity center of mass to zero
  xm(i)=x(i)-v(i)*dt               position previous time step
enddo
return
end
```

Comments to this algorithm:

1. *Function* `lattice_pos` *gives the coordinates of lattice position* `i` *and* `ranf()` *gives a uniformly distributed random number. We do not use a Maxwell-Boltzmann distribution for the velocities; on equilibration it will become a Maxwell-Boltzmann distribution.*
2. *In computing the number of degrees of freedom, we assume a three-dimensional system (in fact, we approximate* N_f *by* 3N*).*

the particles on a cubic lattice, as described in section 3.2.2 in the context of Monte Carlo simulations.

In the present case (Algorithm 4), we have chosen to start our run from a simple cubic lattice. Assume that the values of the density and initial temperature are chosen such that the simple cubic lattice is mechanically unstable and melts rapidly. First, we put each particle on its lattice site and then we attribute to each velocity component of every particle a value that is drawn from a uniform distribution in the interval $[-0.5, 0.5]$. This initial velocity distribution is Maxwellian neither in shape nor even in width. Subsequently, we shift all velocities, such that the total momentum is zero and we scale the resulting velocities to adjust the mean kinetic energy to the de-

N dimensions $\sum_{\alpha=1}^{N} \frac{1}{2} m V_\alpha^2 = \frac{N}{2} k_B T.$

sired value. We know that, in thermal equilibrium, the following relation should hold:

$$\langle v_\alpha^2 \rangle = k_B T/m, \tag{4.2.1}$$

where v_α is the α component of the velocity of a given particle. We can use this relation to define an instantaneous temperature at time t $T(t)$:

$$k_B T(t) \equiv \sum_{i=1}^{N} \frac{m v_{\alpha,i}^2(t)}{N_f}. \tag{4.2.2}$$

Clearly, we can adjust the instantaneous temperature $T(t)$ to match the desired temperature T by scaling all velocities with a factor $(T/T(t))^{1/2}$. This initial setting of the temperature is not particularly critical, as the temperature will change anyway during equilibration.

As will appear later, we do not really use the velocities themselves in our algorithm to solve Newton's equations of motion. Rather, we use the positions of all particles at the present (`x`) and previous (`xm`) time steps, combined with our knowledge of the force (`f`) acting on the particles, to predict the positions at the next time step. When we start the simulation, we must bootstrap this procedure by generating approximate previous positions. Without much consideration for any law of mechanics but the conservation of linear momentum, we approximate `x` for a particle in a direction by `xm(i) = x(i) - v(i)*dt`. Of course, we could make a better estimate of the true previous position of each particle. But as we are only bootstrapping the simulation, we do not worry about such subtleties.

4.2.2 The Force Calculation

What comes next is the most time-consuming part of almost all Molecular Dynamics simulations: the calculation of the force acting on every particle. If we consider a model system with pairwise additive interactions (as we do in the present case), we have to consider the contribution to the force on particle i due to all its neighbors. If we consider only the interaction between a particle and the nearest image of another particle, this implies that, for a system of N particles, we must evaluate $N \times (N-1)/2$ pair distances.

This implies that, if we use no tricks, the time needed for the evaluation of the forces scales as N^2. There exist efficient techniques to speed up the evaluation of both short-range and long-range forces in such a way that the computing time scales as N, rather than N^2. In Appendix F, we describe some of the more common techniques to speed up the simulations. Although the examples in this Appendix apply to Monte Carlo simulations, the same techniques can also be used in a Molecular Dynamics simulation. However, in the present, simple example (see Algorithm 5) we will not attempt to make

Algorithm 5 (Calculation of the Forces)

```
subroutine force(f,en)                    determine the force
en=0                                      and energy
do i=1,npart
   f(i)=0                                 set forces to zero
enddo
do i=1,npart-1
  do j=i+1,npart                          loop over all pairs
     xr=x(i)-x(j)
     xr=xr-box*nint(xr/box)               periodic boundary conditions
     r2=xr**2
     if (r2.lt.rc2) then                  test cutoff
       r2i=1/r2
       r6i=r2i**3
       ff=48*r2i*r6i*(r6i-0.5)            Lennard-Jones potential
       f(i)=f(i)+ff*xr                    update force
       f(j)=f(j)-ff*xr
       en=en+4*r6i*(r6i-1)-ecut           update energy
     endif
  enddo
enddo
return
end
```

Comments to this algorithm:

1. *For efficiency reasons the factors 4 and 48 are usually taken out of the force loop and taken into account at the end of the calculation for the energy.*
2. *The term* `ecut` *is the value of the potential at* $r = r_c$*; for the Lennard-Jones potential, we have*

$$\texttt{ecut} = 4\left(\frac{1}{r_c^{12}} - \frac{1}{r_c^{6}}\right).$$

the program particularly efficient and we shall, in fact, consider all possible pairs of particles explicitly.

We first compute the current distance in the x, y, and z directions between each pair of particles `i` and `j`. These distances are indicated by `xr`. As in the Monte Carlo case, we use periodic boundary conditions (see section 3.2.2). In the present example, we use a cutoff at a distance r_c in the explicit calculation of intermolecular interactions, where r_c is chosen to be less than half the diameter of the periodic box. In that case we can always limit the evaluation of intermolecular interactions between i and j to the interaction between i and the nearest periodic image of j.

In the present case, the diameter of the periodic box is denoted by box. If we use simple cubic periodic boundary conditions, the distance in any direction between i and the nearest image of j should always be less (in absolute value) than box/2. A compact way to compute the distance between i and the nearest periodic image of j uses the nearest integer function (nint(x) in FORTRAN). The nint function simply rounds a real number to the nearest integer.[2] Starting with the x-distance (say) between i and any periodic image of j, xr, we compute the x-distance between i and the nearest image of j as xr=xr-box*nint(xr/box). Having thus computed all Cartesian components of $\mathbf{r}_{ij}$, the vector distance between i and the nearest image of j, we compute r_{ij}^2 (denoted by r2 in the program). Next we test if r_{ij}^2 is less than r_c^2, the square of the cutoff radius. If not, we immediately skip to the next value of j. It perhaps is worth emphasizing that we do not compute $|r_{ij}|$ itself, because this would be both unnecessary and expensive (as it would involve the evaluation of a square root).

If a given pair of particles is close enough to interact, we must compute the force between these particles, and the contribution to the potential energy. Suppose that we wish to compute the x-component of the force

$$\begin{aligned} f_x(r) &= -\frac{\partial u(r)}{\partial x} \\ &= -\left(\frac{x}{r}\right)\left(\frac{\partial u(r)}{\partial r}\right). \end{aligned}$$

For a Lennard-Jones system (in reduced units),

$$f_x(r) = \frac{48x}{r^2}\left(\frac{1}{r^{12}} - 0.5\frac{1}{r^6}\right).$$

4.2.3 Integrating the Equations of Motion

Now that we have computed all forces between the particles, we can integrate Newton's equations of motion. Algorithms have been designed to do this. Some of these will be discussed in a bit more detail. In the program (Algorithm 6), we have used the so-called Verlet algorithm. This algorithm is not only one of the simplest, but also usually the best.

To derive it, we start with a Taylor expansion of the coordinate of a particle, around time t,

$$r(t+\Delta t) = r(t) + v(t)\Delta t + \frac{f(t)}{2m}\Delta t^2 + \frac{\Delta t^3}{3!}\dddot{r} + \mathcal{O}(\Delta t^4),$$

[2] Unfortunately, many FORTRAN compilers yield very slow nint functions. It is often cheaper to write your own code to replace the nint library routine.

Algorithm 6 (Integrating the Equations of Motion)

Verlet Algorithm

```
subroutine integrate(f,en)               integrate equations of motion
sumv=0
sumv2=0
do i=1,npart                             MD loop
   xx=2*x(i)-xm(i)+delt**2*f(i)          Verlet algorithm (4.2.3)
   vi=(xx-xm(i))/(2*delt)                velocity (4.2.4)
   sumv=sumv+vi                          velocity center of mass
   sumv2=sumv2+vi**2                     total kinetic energy
   xm(i)=x(i)                            update positions previous time
   x(i)=xx                               update positions current time
enddo
temp=sumv2/(3*npart)                     instantaneous temperature
etot=(en+0.5*sumv2)/npart                total energy per particle
return
end
```

Comments to this algorithm:

1. *The total energy* `etot` *should remain approximately constant during the simulation. A drift of this quantity may signal programming errors. It therefore is important to monitor this quantity. Similarly, the velocity of the center of mass* `sumv` *should remain zero.*
2. *In this subroutine we use the Verlet algorithm (4.2.3) to integrate the equations of motion. The velocities are calculated using equation (4.2.4).*

similarly,

$$r(t-\Delta t) = r(t) - v(t)\Delta t + \frac{f(t)}{2m}\Delta t^2 - \frac{\Delta t^3}{3!}\dddot{r} + \mathcal{O}(\Delta t^4).$$

Summing these two equations, we obtain

$$r(t+\Delta t) + r(t-\Delta t) = 2r(t) + \frac{f(t)}{m}\Delta t^2 + \mathcal{O}(\Delta t^4)$$

or

$$r(t+\Delta t) \approx 2r(t) - r(t-\Delta t) + \frac{f(t)}{m}\Delta t^2. \quad (4.2.3)$$

The estimate of the new position contains an error that is of order Δt^4, where Δt is the time step in our Molecular Dynamics scheme. Note that the

Verlet algorithm does not use the velocity to compute the new position. One, however, can derive the velocity from knowledge of the trajectory, using

$$r(t+\Delta t) - r(t-\Delta t) = 2v(t)\Delta t + \mathcal{O}(\Delta t^3)$$

or

$$v(t) = \frac{r(t+\Delta t) - r(t-\Delta t)}{2\Delta t} + \mathcal{O}(\Delta t^2). \tag{4.2.4}$$

This expression for the velocity is only accurate to order Δt^2. However, it is possible to obtain more accurate estimates of the velocity (and thereby of the kinetic energy) using a Verlet-like algorithm (i.e., an algorithm that yields trajectories identical to that given by equation (4.2.3)). In our program, we use the velocities only to compute the kinetic energy and, thereby, the instantaneous temperature.

Now that we have computed the new positions, we may discard the positions at time $t-\Delta t$. The current positions become the old positions and the new positions become the current positions.

After each time step, we compute the current temperature (`temp`), the current potential energy (`en`) calculated in the force loop, and the total energy (`etot`). Note that the total energy should be conserved.

This completes the introduction to the Molecular Dynamics method. The reader should now be able to write a basic Molecular Dynamics program for liquids or solids consisting of spherical particles. In what follows, we shall do two things. First of all, we discuss, in a bit more detail, the methods available to integrate the equations of motion. Next, we discuss measurements in Molecular Dynamics simulations. Important extensions of the Molecular Dynamics technique are discussed in Chapter 6.

4.3 Equations of Motion

It is obvious that a good Molecular Dynamics program requires a good algorithm to integrate Newton's equations of motion. In this sense, the choice of algorithm is crucial. However, although it is easy to recognize a *bad* algorithm, it is not immediately obvious what criteria a *good* algorithm should satisfy. Let us look at the different points to consider.

Although, at first sight, speed seems important, it is usually not very relevant because the fraction of time spent on integrating the equations of motion (as opposed to computing the interactions) is small, at least for atomic and simple molecular systems.

Accuracy for large time steps is more important, because the longer the time step that we can use, the fewer evaluations of the forces are needed per unit of simulation time. Hence, this would suggest that it is advantageous to use a sophisticated algorithm that allows use of a long time step.

Algorithms that allow the use of a large time step achieve this by storing information on increasingly higher-order derivatives of the particle coordinates. As a consequence, they tend to require more memory storage. For a typical simulation, this usually is not a serious drawback because, unless one considers very large systems, the amount of memory needed to store these derivatives is small compared to the total amount available even on a normal workstation.

Energy conservation is an important criterion, but actually we should distinguish two kinds of energy conservation, namely, short time and long time. The sophisticated higher-order algorithms tend to have very good energy conservation for short times (i.e., during a few time steps). However, they often have the undesirable feature that the overall energy drifts for long times. In contrast, Verlet-style algorithms tend to have only moderate short-term energy conservation but little long-term drift.

It would seem to be most important to have an algorithm that accurately predicts the trajectory of all particles for both short and long times. In fact, no such algorithm exists. For essentially all systems that we study by MD simulations, we are in the regime where the trajectory of the system through phase space (i.e., the 6N-dimensional space spanned by all particle coordinates and momenta) depends sensitively on the initial conditions. This means that two trajectories that are initially very close will diverge exponentially as time progresses. We can consider the integration error caused by the algorithm as the source for the initial small difference between the "true" trajectory of the system and the trajectory generated in our simulation. We should expect that any integration error, no matter how small, will always cause our simulated trajectory to diverge exponentially from the true trajectory compatible with the same initial conditions. This so-called Lyapunov instability (see section 4.3.4) would seem to be a devastating blow to the whole idea of Molecular Dynamics simulations but we have good reasons to assume that even this problem need not be serious.

Clearly, this statement requires some clarification. First of all, one should realize that the aim of an MD simulation is *not* to predict precisely what will happen to a system that has been prepared in a precisely known initial condition: we are always interested in statistical predictions. We wish to predict the average behavior of a system that was prepared in an initial state about which we know something (e.g., the total energy) but by no means everything. In this respect, MD simulations differ fundamentally from numerical schemes for predicting the trajectory of satellites through space: in the latter case, we really wish to predict the true trajectory. We cannot afford to launch an ensemble of satellites and make statistical predictions about their destination. However, in MD simulations, statistical predictions are good enough. Still, this would not justify the use of inaccurate trajectories unless the trajectories obtained numerically, in some sense, are close to true trajectories.

This latter statement is generally believed to be true, although, to our

Lyapunov instability:
$$\Delta X(t) = \Delta X(0) \exp(\lambda t).$$

knowledge, it has not been proven for any class of systems that is of interest for MD simulations. However, considerable numerical evidence (see, e.g., [66]) suggests that there exist so-called shadow orbits. A shadow orbit is a true trajectory of a many-body system that closely follows the numerical trajectory for a time that is long compared to the time it takes the Lyapunov instability to develop. Hence, the results of our simulation are representative of a true trajectory in phase space, even though we cannot tell *a priori* which. Surprisingly (and fortunately), it appears that shadow orbits are better behaved (i.e., track the numerical trajectories better) for systems in which small differences in the initial conditions lead to an exponential divergence of trajectories than for the, seemingly, simpler systems that show no such divergence [66]. Despite this reassuring evidence (see also section 4.3.5 and the article by Gillilan and Wilson [67]), it should be emphasized that it is just evidence and not proof. Hence, our trust in Molecular Dynamics simulation as a tool to study the time evolution of many-body systems is based largely on belief. To conclude this discussion, let us say that there is clearly still a corpse in the closet. We believe this corpse will not haunt us, and we quickly close the closet. For more details, the reader is referred to [27, 67, 68].

Why

Newton's equations of motion are time reversible, and so should be our algorithms. In fact, many algorithms (for instance the predictor-corrector schemes, see Appendix E, and many of the schemes used to deal with constraints) are *not* time reversible. That is, future and past phase space coordinates do not play a symmetric role in such algorithms. As a consequence, if one were to reverse the momenta of all particles at a given instant, the system would not trace back its trajectory in phase space, even if the simulation would be carried out with infinite numerical precision. Only in the limit of an infinitely short time step will such algorithms become reversible. However, what is more important, many seemingly reasonable algorithms differ in another crucial respect from Hamilton's equation of motion: true Hamiltonian dynamics leaves the magnitude of any volume element in phase space unchanged, but many numerical schemes, in particular those that are not time reversible, do not reproduce this area-preserving property. This may sound like a very esoteric objection to an algorithm, but it is not. Again, without attempting to achieve a rigorous formulation of the problem, let us simply note that all trajectories that correspond to a particular energy E are contained in a (hyper) volume Ω in phase space. If we let Hamilton's equation of motion act on all points in this volume (i.e., we let the volume evolve in time), then we end up with exactly the same volume. However, a non-area-preserving algorithm will map the volume Ω on another (usually larger) volume Ω'. After sufficiently long times, we expect that the non-area-preserving algorithm will have greatly expanded the volume of our system in phase space. This is not compatible with energy conservation. Hence, it is plausible that nonreversible algorithms will have serious long-term energy drift problems. Reversible, area-preserving algorithms will not change the magnitude of the volume in phase space. This property is not sufficient to guarantee the absence of long-term energy drift, but it is at least compatible with it. It is possible to check whether an algorithm is area preserving by computing the Jacobian associated with the transformation of old to new phase space coordinates.

Finally, it should be noted that even when we integrate a time-reversible algorithm, we shall find that the numerical implementation is hardly ever truly time reversible. This is so, because we work on a computer with finite machine precision using floating-point arithmetic that results in rounding errors (on the order of the machine precision).

In the remainder of this section, we shall discuss some of these points in more detail. Before we do so, let us first consider how the Verlet algorithm scores on these points. First of all, the Verlet algorithm is fast. But we had argued that this is relatively unimportant. Second, it is not particularly accurate for long time steps. Hence, we should expect to compute the forces on all particles rather frequently. Third, it requires about as little memory as is at all possible. This is useful when we simulate very large systems, but in general it is not a crucial advantage. Fourth, its short-term energy conservation is fair (in particular in the versions that use a more accurate expression for the velocities) but, more important, it exhibits little long-term energy drift. This is related to the fact that the Verlet algorithm is time reversible and area preserving. In fact, although the Verlet algorithm does not conserve the total energy of this system exactly, strong evidence indicates that it does conserve a pseudo-Hamiltonian approaching the true Hamiltonian in the limit of infinitely short time steps (see section 4.3.3). The accuracy of the trajectories generated with the Verlet algorithm is not impressive. But then, it would hardly help to use a better algorithm. Such an algorithm may postpone the unavoidable exponential growth of the error in the trajectory by a few hundred time steps (see section 4.3.4), but no algorithm is good enough that it will keep the trajectories close to the true trajectories for a time comparable to the duration of a typical Molecular Dynamics run.[3]

4.3.1 Other Algorithms

Let us now briefly look at some alternatives to the Verlet algorithm. The most naive algorithm is based simply on a truncated Taylor expansion of the particle coordinates:

$$r(t+\Delta t) = r(t) + v(t)\Delta t + \frac{f(t)}{2m}\Delta t^2 + \cdots .$$

[3]Error-free integration of the equations of motion is possible for certain discrete models, such as lattice-gas cellular automata. But these models do not follow Newton's equation of motion.

If we truncate this expansion beyond the term in Δt^2, we obtain the so-called Euler algorithm. Although it looks similar to the Verlet algorithm, it is much worse on virtually all counts. In particular, it is not reversible or area preserving and suffers from a (catastrophic) energy drift. The Euler algorithm therefore is not recommended.

Several algorithms are equivalent to the Verlet scheme. The simplest among these is the so-called Leap Frog algorithm [24]. This algorithm evaluates the velocities at half-integer time steps and uses these velocities to compute the new positions. To derive the Leap Frog algorithm from the Verlet scheme, we start by defining the velocities at half-integer time steps as follows:

$$v(t - \Delta t/2) \equiv \frac{r(t) - r(t - \Delta t)}{\Delta t}$$

and

$$v(t + \Delta t/2) \equiv \frac{r(t + \Delta t) - r(t)}{\Delta t}.$$

From the latter equation we immediately obtain an expression for the new positions, based on the old positions and velocities:

$$r(t + \Delta t) = r(t) + \Delta t v(t + \Delta t/2). \tag{4.3.1}$$

From the Verlet algorithm, we get the following expression for the update of the velocities:

$$v(t + \Delta t/2) = v(t - \Delta t/2) + \Delta t \frac{f(t)}{m}. \tag{4.3.2}$$

As the Leap Frog algorithm is derived from the Verlet algorithm, it gives rise to identical trajectories. Note, however, that the velocities are not defined at the same time as the positions. As a consequence, kinetic and potential energy are also not defined at the same time, and hence we cannot directly compute the total energy in the Leap Frog scheme.

It is, however, possible to cast the Verlet algorithm in a form that uses positions and velocities computed at equal times. This velocity Verlet algorithm [69] looks like a Taylor expansion for the coordinates:

$$r(t + \Delta t) = r(t) + v(t)\Delta t + \frac{f(t)}{2m}\Delta t^2. \tag{4.3.3}$$

However, the update of the velocities is different from the Euler scheme:

$$v(t + \Delta t) = v(t) + \frac{f(t + \Delta t) + f(t)}{2m}\Delta t. \tag{4.3.4}$$

Note that, in this algorithm, we can compute the new velocities only after we have computed the new positions and, from these, the new forces. It is not immediately obvious that this scheme, indeed, is equivalent to the original Verlet algorithm. To show this, we note that

$$r(t+2\Delta t) = r(t+\Delta t) + v(t+\Delta t)\Delta t + \frac{f(t+\Delta t)}{2m}\Delta t^2$$

and equation (4.3.3) can be written as

$$r(t) = r(t+\Delta t) - v(t)\Delta t - \frac{f(t)}{2m}\Delta t^2.$$

By addition we get

$$r(t+2\Delta t) + r(t) = 2r(t+\Delta t) + [v(t+\Delta t) - v(t)]\Delta t + \frac{f(t+\Delta t) - f(t)}{2m}\Delta t^2.$$

Substitution of equation (4.3.4) yields

$$r(t+2\Delta t) + r(t) = 2r(t+\Delta t) + \frac{f(t+\Delta t)}{m}\Delta t^2,$$

which, indeed, is the coordinate version of the Verlet algorithm.

Let us end the discussion of Verlet-like algorithms by mentioning two schemes that yield the same trajectories as the Verlet algorithm, but provide better estimates of the velocity. The first is the so-called Beeman algorithm. It looks quite different from the Verlet algorithm:

$$r(t+\Delta t) = r(t) + v(t)\Delta t + \frac{4f(t) - f(t-\Delta t)}{6m}\Delta t^2 \tag{4.3.5}$$

$$v(t+\Delta t) = v(t) + \frac{2f(t+\Delta t) + 5f(t) - f(t-\Delta t)}{6m}\Delta t. \tag{4.3.6}$$

However, by eliminating $v(t)$ from equation (4.3.5), using equation (4.3.6), it is easy to show that the positions satisfy the Verlet algorithm. However, the velocities are more accurate than in the original Verlet algorithm. As a consequence, the total energy conservation looks somewhat better. A disadvantage of the Beeman algorithm is that the expression for the velocities does not have time-reversal symmetry. A very simple solution to this problem is to use the so-called velocity-corrected Verlet algorithm for which the error both in the positions and in the velocities is of order $\mathcal{O}(\Delta t^4)$.

The velocity-corrected Verlet algorithm is derived as follows. First write down a Taylor expansion for $r(t+2\Delta t)$, $r(t+\Delta t)$, $r(t-\Delta t)$ and $r(t-2\Delta t)$:

$$\begin{aligned}
r(t+2\Delta t) &= r(t) + 2v(t)\Delta t + \dot{v}(t)(2\Delta t)^2/2! + \ddot{v}(2\Delta t)^3/3! + \cdots \\
r(t+\Delta t) &= r(t) + v(t)\Delta t + \dot{v}(t)\Delta t^2/2! + \ddot{v}\Delta t^3/3! + \cdots \\
r(t-\Delta t) &= r(t) - v(t)\Delta t + \dot{v}(t)\Delta t^2/2! - \ddot{v}\Delta t^3/3! + \cdots \\
r(t-2\Delta t) &= r(t) - 2v(t)\Delta t + \dot{v}(t)(2\Delta t)^2/2! - \ddot{v}(2\Delta t)^3/3! + \cdots .
\end{aligned}$$

By combining these equations, we can write

$$12v(t)\Delta t = 8[r(t+\Delta t) - r(t-\Delta t)] - [r(t+2\Delta t) - r(t-2\Delta t)] + \mathcal{O}(\Delta t^4)$$

or, equivalently,

$$v(t) = \frac{v(t+\Delta t/2) + v(t-\Delta t/2)}{2} + \frac{\Delta t}{12}[\dot{v}(t-\Delta t) - \dot{v}(t+\Delta t)] + \mathcal{O}(\Delta t^4). \quad (4.3.7)$$

Note that this velocity can be computed only after the next time step (i.e., we must know the positions and forces at $t + \Delta t$ to compute $v(t)$).

4.3.2 Higher-Order Schemes

For most Molecular Dynamics applications, Verlet-like algorithms are perfectly adequate. However, sometimes it is convenient to employ a higher-order algorithm (i.e., an algorithm that employs information about higher-order derivatives of the particle coordinates). Such an algorithm makes it possible to use a longer time step without loss of (short-term) accuracy or, alternatively, to achieve higher accuracy for a given time step. But, as mentioned before, higher-order algorithms require more storage and are, more often than not, neither reversible nor area preserving. This is true in particular of the so-called predictor-corrector algorithms, the most popular class of higher-order algorithms used in Molecular Dynamics simulations. For the sake of completeness, the predictor-corrector scheme is described in Appendix E.1. We refer the reader who wishes to know more about the relative merits of algorithms for Molecular Dynamics simulations to the excellent review by Berendsen and van Gunsteren [70].

4.3.3 Liouville Formulation of Time-Reversible Algorithms

Thus far we have considered algorithms for integrating Newton's equations of motion from the point of view of applied mathematics. However, recently Tuckerman *et al.* [71] have shown how to systematically derive time-reversible, area-preserving MD algorithms from the Liouville formulation of classical mechanics. The same approach has been developed independently by Sexton and Weingarten [72] in the context of hybrid Monte Carlo simulations (see section 14.2). As the Liouville formulation provides considerable insight into what makes an algorithm a good algorithm, we briefly review the Liouville approach.

Let us consider an arbitrary function f that depends on all the coordinates and momenta of the N particles in a classical many-body system. The term $f(\mathbf{p}^N(t), \mathbf{r}^N(t))$ depends on the time t implicitly, that is, through the dependence of $(\mathbf{p}^N, \mathbf{r}^N)$ on t. The time derivative of f is $\dot{f}$:

$$\begin{aligned} \dot{f} &= \dot{\mathbf{r}}\frac{\partial f}{\partial \mathbf{r}} + \dot{\mathbf{p}}\frac{\partial f}{\partial \mathbf{p}} \\ &\equiv iLf, \end{aligned} \tag{4.3.8}$$

where we have used the shorthand notation $\mathbf{r}$ for $\mathbf{r}^N$ and $\mathbf{p}$ for $\mathbf{p}^N$. The last line of equation (4.3.8) defines the *Liouville operator*

$$iL = \dot{\mathbf{r}}\frac{\partial}{\partial \mathbf{r}} + \dot{\mathbf{p}}\frac{\partial}{\partial \mathbf{p}}. \tag{4.3.9}$$

We can formally integrate equation (4.3.8) to obtain

$$f\left[\mathbf{p}^N(t), \mathbf{r}^N(t)\right] = \exp(iLt) f\left[\mathbf{p}^N(0), \mathbf{r}^N(0)\right]. \tag{4.3.10}$$

In all cases of practical interest, we cannot do much with this formal solution, because evaluating the right-hand side is still equivalent to the exact integration of the classical equations of motion. However, in a few simple cases the formal solution is known explicitly. In particular, suppose that our Liouville operator contained only the first term on the right-hand side of equation (4.3.9). We denote this part of iL by iL_r:

$$iL_r \equiv \dot{\mathbf{r}}(0)\frac{\partial}{\partial \mathbf{r}}, \tag{4.3.11}$$

where $\dot{\mathbf{r}}(0)$ is the value of $\dot{\mathbf{r}}$ at time $t = 0$. If we insert iL_r in equation (4.3.10) and use a Taylor expansion of the exponential on the right-hand side, we get

$$\begin{aligned} f(t) &= f(0) + iL_r t f(0) + \frac{(iL_r t)^2}{2!} f(0) + \cdots \\ &= \exp\left(\dot{\mathbf{r}}(0)t\frac{\partial}{\partial \mathbf{r}}\right) f(0) \\ &= \sum_{n=0}^{\infty} \frac{(\dot{\mathbf{r}}(0)t)^n}{n!} \frac{\partial^n}{\partial \mathbf{r}^n} f(0) \\ &= f\left[\mathbf{p}^N(0), (\mathbf{r} + \dot{\mathbf{r}}(0)t)^N\right]. \end{aligned} \tag{4.3.12}$$

Hence, the effect of $\exp(iL_r t)$ is a simple shift of coordinates. Similarly, the effect of $\exp(iL_p t)$, with iL_p defined as

$$iL_p \equiv \dot{\mathbf{p}}(0)\frac{\partial}{\partial \mathbf{p}}, \tag{4.3.13}$$

is a simple shift of momenta. The total Liouville operator, iL, is equal to iL_r + iL_p. Unfortunately, we cannot replace $\exp(iLt)$ by $\exp(iL_r t) \times \exp(iL_p t)$, because iL_r and iL_p are noncommuting operators. For noncommuting operators A and B, we have

$$\exp(A+B) \neq \exp(A)\exp(B). \tag{4.3.14}$$

However, we do have the following *Trotter identity*:

$$e^{(A+B)} = \lim_{P\to\infty} \left(e^{A/2P} e^{B/P} e^{A/2P}\right)^P. \tag{4.3.15}$$

In the limit $P \to \infty$, this relation is formally correct, but of limited practical value. However, for large but finite P, we have

$$e^{(A+B)} = \left(e^{A/2P} e^{B/P} e^{A/2P}\right)^P e^{\mathcal{O}(1/P^2)}. \tag{4.3.16}$$

Now let us apply this expression to the formal solution of the Liouville equation. To this end, we make the identification

$$\frac{A}{P} \equiv \frac{iL_p t}{P} \equiv \Delta t \dot{\mathbf{p}}(0) \frac{\partial}{\partial \mathbf{p}}$$

$\frac{1}{P} = \frac{\Delta t}{t}.$

and

$$\frac{B}{P} \equiv \frac{iL_r t}{P} \equiv \Delta t \dot{\mathbf{r}}(0) \frac{\partial}{\partial \mathbf{r}},$$

where $\Delta t = t/P$. The idea is now to replace the formal solution of the Liouville equation by the discretized version, equation (4.3.16). In this scheme, one time step corresponds to applying the operator

$$e^{iL_p \Delta t/2} e^{iL_r \Delta t} e^{iL_p \Delta t/2}$$

once. Let us see what the effect is of this operator on the coordinates and momenta of the particles. First, we apply $\exp(iL_p \Delta t/2)$ to f and obtain

$$e^{iL_p \Delta t/2} f\left[\mathbf{p}^N(0), \mathbf{r}^N(0)\right] = f\left\{\left[\mathbf{p}(0) + \frac{\Delta t}{2}\dot{\mathbf{p}}(0)\right]^N, \mathbf{r}^N(0)\right\}.$$

Next, we apply $\exp(iL_r \Delta t)$ to the result of the previous step

$$\begin{aligned} e^{iL_r \Delta t} f &\left\{\left[\mathbf{p}(0) + \frac{\Delta t}{2}\dot{\mathbf{p}}(0)\right]^N, \mathbf{r}^N(0)\right\} \\ &= f\left\{\left[\mathbf{p}(0) + \frac{\Delta t}{2}\dot{\mathbf{p}}(0)\right]^N, [\mathbf{r}(0) + \Delta t \dot{\mathbf{r}}(\Delta t/2)]^N\right\}, \end{aligned}$$

and finally we apply $\exp(iL_p\Delta t/2)$ once more, to obtain

$$f\left\{\left[\mathbf{p}(0)+\frac{\Delta t}{2}\dot{\mathbf{p}}(0)+\frac{\Delta t}{2}\dot{\mathbf{p}}(\Delta t)\right]^N, [\mathbf{r}(0)+\Delta t\dot{\mathbf{r}}(\Delta t/2)]^N\right\}.$$

Note that every step in the preceding sequence corresponds to a simple shift operation in either $\mathbf{r}^N$ or $\mathbf{p}^N$. It is of particular importance to note that the shift in $\mathbf{r}$ is a function of $\mathbf{p}$ only (because $\dot{\mathbf{r}} = \mathbf{p}/m$), while the shift in $\mathbf{p}$ is a function of $\mathbf{r}$ only (because $\dot{\mathbf{p}} = \mathbf{F}(\mathbf{r}^N)$). The Jacobian of the transformation from $\{\mathbf{p}^N(0), \mathbf{r}^N(0)\}$ to $\{\mathbf{p}^N(\Delta t), \mathbf{r}^N(\Delta t)\}$ is simply the product of the Jacobians of the three elementary transformations. But, as each of these Jacobians is equal to 1, the overall Jacobian is also equal to 1. In other words, the algorithm is area preserving.

If we now consider the overall effect of this sequence of operations on the positions and momenta, we find the following:

$$\mathbf{p}(0) \rightarrow \mathbf{p}(0)+\frac{\Delta t}{2}(\mathbf{F}(0)+\mathbf{F}(\Delta t)) \tag{4.3.17}$$

$$\begin{aligned}\mathbf{r}(0) &\rightarrow \mathbf{r}(0)+\Delta t\dot{\mathbf{r}}(\Delta t/2)\\ &= \mathbf{r}(0)+\Delta t\dot{\mathbf{r}}(0)+\frac{\Delta^2 t}{2m}\mathbf{F}(0).\end{aligned} \tag{4.3.18}$$

But these are precisely the equations of the Verlet algorithm (in the velocity form). Hence, we have shown that the Verlet algorithm is area preserving. That it is reversible follows directly from the fact that past and future coordinates enter symmetrically in the algorithm.

Finally, let us try to understand the absence of long-term energy drift in the Verlet algorithm. When we use the Verlet algorithm, we replace the true Liouville operator $\exp(iLt)$ by $\exp(iL_r\Delta t/2)\exp(iL_P\Delta t)\exp(iL_r\Delta t/2)$. In doing so, we make an error. If all (nth-order) commutators of L_p and L_r exist (i.e., if the Hamiltonian is an infinitely differentiable function of $\mathbf{p}^N$ and $\mathbf{r}^N$) then, at least in principle, we can evaluate the error that is involved in this replacement:

$$\exp(iL_r\Delta t/2)\exp(iL_P\Delta t)\exp(iL_r\Delta t/2) = \exp(iL\Delta t+\epsilon), \tag{4.3.19}$$

where ϵ is an operator that can be expressed in terms of the commutators of L_p and L_r:

$$\epsilon = \sum_{n=1}^{\infty}(\Delta t)^{2n+1}c_{2n+1}, \tag{4.3.20}$$

where c_m denotes a combination of mth-order commutators. For instance, the leading term is

$$-(\Delta t)^3\left(\frac{1}{24}[iL_r,[iL_r,iL_p]]+\frac{1}{12}[iL_p,[iL_r,iL_p]]\right).$$

Now the interesting thing to note is that, if the expansion in equation (4.3.20) converges, then we can define a pseudo-Liouville operator

$$iL_{pseudo} \equiv iL + \epsilon/\Delta t.$$

This pseudo-Liouville operator corresponds to a pseudo-Hamiltonian, and the remarkable thing is that this pseudo-Hamiltonian (H_{pseudo}) is rigorously conserved by Verlet style (or generalized multi-time-step) algorithms [73–75]. The difference between the conserved pseudo-Hamiltonian and the true Hamiltonian of the system is of order $(\Delta t)^{2n}$ (where n depends on the order of the algorithm). Clearly, by choosing Δt small (and, if necessary, n large), we can make the difference between the true and the pseudo-Hamiltonian as small as we like. As the true Hamiltonian is forced to remain close to a conserved quantity, we can now understand why there is no long-term drift in the energy with Verlet-style algorithms. In some cases, we can explicitly compute the commutators (for instance, for a harmonic system) and can verify that the pseudo-Hamiltonian is indeed conserved [68]. And, even if we cannot compute the complete series of commutators, the leading term will give us an improved estimate of the pseudo-Hamiltonian. Toxvaerd [68] has verified that even for a realistic many-body system, such an approximate pseudo-Hamiltonian is very nearly a constant of motion.

The Liouville formalism allows us to derive the Verlet algorithm as a special case of the Trotter expansion of the time-evolution operator. It should be realized that the decomposition of iL as a sum of iL_r and iL_p is arbitrary. Other decompositions are possible and may lead to algorithms that are more convenient.

4.3.4 Lyapunov Instability

To end this discussion of algorithms, we wish to illustrate the extreme sensitivity of the trajectories to small differences in initial conditions. Let us consider the position ($\mathbf{r}^N$) of one of the N particles at time t. This position is a function of the initial positions and momenta at $t = 0$:

$$\mathbf{r}(t) = f\left[\mathbf{r}^N(0), \mathbf{p}^N(0); t\right].$$

Let us now consider the position at time t that would result if we perturbed the initial conditions (say, some of the momenta) by a small amount ϵ. In that case, we would obtain a different value for r at time t:

$$\mathbf{r}'(t) = f\left[\mathbf{r}^N(0), \mathbf{p}^N(0) + \epsilon; t\right].$$

We denote the difference between $\mathbf{r}(t)$ and $\mathbf{r}'(t)$ by $\Delta\mathbf{r}(t)$. For sufficiently short times, $\Delta\mathbf{r}(t)$ is linear in ϵ. However, the coefficient of the linear dependence diverges exponentially; that is,

$$|\Delta\mathbf{r}(t)| \sim \epsilon \exp(\lambda t). \tag{4.3.21}$$

This so-called Lyapunov instability of the trajectories is responsible for our inability to accurately predict a trajectory for all but the shortest simulations. The exponent λ is called the Lyapunov exponent (more precisely, the largest Lyapunov exponent; there are more such exponents, 6N in fact, but the largest dominates the long-time exponential divergence of initially close trajectories). Suppose that we wish to maintain a certain bound Δ_{max} on $|\Delta\mathbf{r}(t)|$, in the interval $0 < t < t_{max}$. How large an initial error (ϵ) can we afford? From equation (4.3.21), we deduce

$$\epsilon \sim \Delta_{max} \exp(-\lambda t_{max}).$$

Hence, the acceptable error in our initial conditions decreases exponentially with t_{max}, the length of the run. To illustrate that this effect is real, we show the result of two almost identical simulations: the second differs from the first in that the x components of the velocities of 2 particles (out of 1000) have been changed by $+10^{-10}$ and -10^{-10} (in reduced units). We monitor the sum of the squares of the differences of the positions of all particles:

$$\sum_{i=1}^{N} |\mathbf{r}_i(t) - \mathbf{r}'_i(t)|^2 .$$

As can be seen in Figure 4.1, this measure of the distance does indeed grow exponentially with time.

After 1000 time steps, the two systems that were initially very close have become very nearly uncorrelated. It should be stressed that this run was performed using perfectly normal parameters (density, temperature, time step). The only unrealistic thing about this simulation is that it is extremely short. Most Molecular Dynamics simulations do require many thousands of time steps.

4.3.5 One More Way to Look at the Verlet Algorithm...

In Molecular Dynamics simulations, the Newtonian equations of motion are integrated approximately. An alternative route would be to *first* write down a time-discretized version of the action. (See Appendix A, and *then* find the set of coordinates (i.e., the discretized trajectory) that minimizes this action. This approach is discussed in some detail in a paper by Gillilan and Wilson [67].) Let us start with the continuous-time version of the action

$$S = \int_{t_b}^{t_e} dt \left[\frac{1}{2} m \left(\frac{dx(t)}{dt} \right)^2 - U(x) \right]$$

and discretize it as follows:

$$S_{discr} = \sum_{i=i_b}^{i_e-1} \Delta t \left[\frac{1}{2} m \left(\frac{x_{i+1} - x_i}{\Delta t} \right)^2 - U(x_i) \right],$$

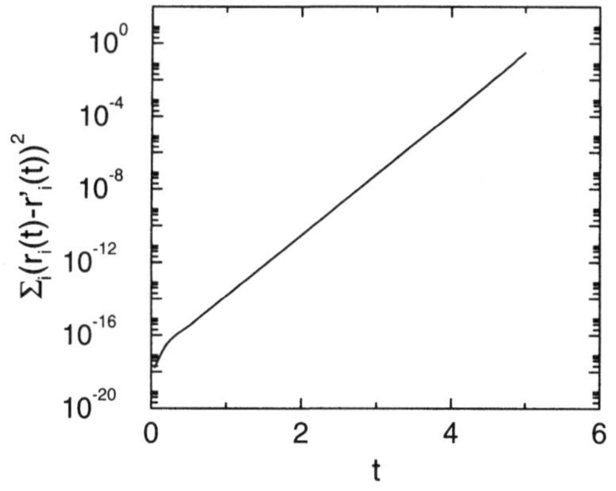

Figure 4.1: Illustration of the Lyapunov instability in a simulation of a Lennard-Jones system. The figure shows the time dependence of the sum of squared distances between two trajectories that were initially very close (see text). The total length of the run in reduced units was 5, which corresponds to 1000 time steps. Note that, within this relatively short time, the two trajectories become essentially uncorrelated.

where $t_b = i_b \Delta t$ and $t_e = i_e \Delta t$. As in the continuous case, we can determine the set of values of the coordinates x_i for which S_{discr} is stationary. At stationarity, the derivative of S_{discr} with respect to all x_i vanishes. It is easy to verify that this implies that

$$m\left(\frac{2x_i - x_{i+1} - x_{i-1}}{\Delta t}\right) - \Delta t \frac{\partial \mathcal{U}(x_i)}{\partial x_i} = 0$$

or

$$x_{i+1} = 2x_i - x_{i-1} - \frac{\Delta t^2}{m}\left(\frac{\partial \mathcal{U}(x_i)}{\partial x_i}\right),$$

which is, of course, the Verlet algorithm. This illustrates that the trajectories generated by the Verlet algorithms have an interesting "shadow" property (see ref. [67] and section 4.3): a "Verlet trajectory" that connects point x_{i_b} and x_{i_e} in a time interval $t_e - t_b$ will tend to lie close to the true trajectory that connects these two points. However, this true trajectory is not at all the one that has the same initial velocity as the Verlet trajectory. That is,

$$\left(\frac{dx(t_b)}{dt}\right)_{\text{true}} \neq \left(\frac{x_{i_b+1} - x_{i_b-1}}{2\Delta t}\right)_{\text{Verlet}}.$$

Nevertheless, as discussed in section 4.3, the Verlet algorithm is a good algorithm in the sense that it follows from a minimization principle that forces it to approximate a true dynamical trajectory of the system under consideration.

This attractive feature of algorithms that can be derived from a discretized action has inspired Elber and co-workers to construct a novel class of MD algorithms that are designed to yield reasonable long-time dynamics with very large time steps [76,77]. In fact, Elber and co-workers do not base their approach on the discretization of the classical action but on the so-called Onsager-Machlup action [78]. The reason for selecting this more general action is that the Onsager-Machlup action is a *minimum* for the true trajectory, while the Lagrangian action is only an extremum. It would carry too far to discuss the practical implementation of the algorithm based on the Onsager-Machlup action. For details, we refer the reader to refs. [76,77].

4.4 Computer Experiments

Now that we have a working Molecular Dynamics program, we wish to use it to "measure" interesting properties of many-body systems. What properties are interesting? First of all, of course, those quantities that can be compared with real experiments. Simplest among these are the thermodynamic properties of the system under consideration, such as the temperature T, the pressure P, and the heat capacity C_V. As mentioned earlier, the temperature is measured by computing the average kinetic energy per degree of freedom. For a system with f degrees of freedom, the temperature T is given by

$$k_B T = \frac{\langle 2\mathcal{K} \rangle}{f}. \tag{4.4.1}$$

There are several different (but equivalent) ways to measure the pressure of a classical N-body system. The most common among these is based on the virial equation for the pressure. For pairwise additive interactions, we can write (see, e.g., [79])

$$P = \rho k_B T + \frac{1}{dV} \left\langle \sum_{i<j} \mathbf{f}(\mathbf{r}_{ij}) \cdot \mathbf{r}_{ij} \right\rangle, \tag{4.4.2}$$

where d is the dimensionality of the system, and $\mathbf{f}(\mathbf{r}_{ij})$ is the force between particles i and j at a distance $\mathbf{r}_{ij}$. Note that this expression for the pressure has been derived for a system at constant N, V, and T, whereas our simulations are performed at constant N, V, and E. In fact, the expression for the pressure in the microcanonical ensemble (constant N, V, E) is not identical to the expression that applies to the canonical (constant N, V, T) ensemble. Lebowitz *et al.* [80] have derived a general procedure to convert averages from one ensemble to another. A more recent (and more accessible) description of these interensemble transformations has been given by Allen and

Tildesley [41]. An example of a relation derived by such a transformation is the expression for the heat capacity at constant volume, as obtained from the fluctuations in the kinetic energy in the microcanonical ensemble:

$$\langle \mathcal{K}^2 \rangle_{NVE} - \langle \mathcal{K} \rangle_{NVE}^2 = \frac{3k_B^2 T^2}{2N} \left(1 - \frac{3k_B}{2C_V} \right). \quad (4.4.3)$$

However, one class of thermodynamic functions cannot be measured directly in a simulation, in the sense that these properties cannot be expressed as a simple average of some function of the coordinates and momenta of all the particles in the system. Examples of such properties are the entropy S, the Helmholtz free energy F, and the Gibbs free energy G. Separate techniques are required to evaluate such thermal quantities in a computer simulation. Methods to calculate these properties are discussed in Chapter 7.

A second class of observable properties are the functions that characterize the local structure of a fluid. Most notable among these is the so-called radial distribution function $g(r)$. The radial distribution function is of interest for two reasons: first of all, neutron and X-ray scattering experiments on simple fluids, and light-scattering experiments on colloidal suspensions, yield information about $g(r)$. Second, $g(r)$ plays a central role in theories of the liquid state. Numerical results for $g(r)$ can be compared with theoretical predictions and thus serve as a criterion to test a particular theory. In a simulation, it is straightforward to measure $g(r)$: it is simply the ratio between the average number density $\rho(r)$ at a distance r from any given atom (for simplicity we assume that all atoms are identical) and the density at a distance r from an atom in an ideal gas at the same overall density. In Algorithm 7 an implementation to compute the radial distribution function is described. By construction, $g(r) = 1$ in an ideal gas. Any deviation of $g(r)$ from unity reflects correlations between the particles due to the intermolecular interactions.

Both the thermodynamic properties and the structural properties mentioned previously do not depend on the time evolution of the system: they are static equilibrium averages. Such averages can be obtained by Molecular Dynamics simulations or equally well by Monte Carlo simulations. However, in addition to the static equilibrium properties, we can also measure dynamic equilibrium properties in a Molecular Dynamics simulation (but not with a Monte Carlo simulation). At first sight, a dynamic equilibrium property appears to be a contradiction: in equilibrium all properties are independent of time; hence any time dependence in the macroscopic properties of a system seems to be related to nonequilibrium behavior. This is true, but it turns out that the time-dependent behavior of a system that is only weakly perturbed is completely described by the dynamic equilibrium properties of the system. Later, we shall provide a simple introduction to the quantities that play a central role in the theory of time-dependent

Algorithm 7 (The Radial Distribution Function)

```
subroutine gr(switch)                    radial distribution function
                                         switch = 0 initialization,
                                         = 1 sample, and = 2 results
if (switch.eq.0) then                    initialization
  ngr=0
  delg=box/(2*nhis)                      bin size
  do i=0,nhis                            nhis total number of bins
     g(i)=0
  enddo
else if (switch.eq.1) then               sample
  ngr=ngr+1
  do i=1,npart-1
    do j=i+1,npart                       loop over all pairs
      xr=x(i)-x(j)
      xr=xr-box*nint(xr/box)             periodic boundary conditions
      r=sqrt(xr**2)
      if (r.lt.box/2) then               only within half the box length
         ig=int(r/delg)
         g(ig)=g(ig)+2                   contribution for particle i and j
      endif
    enddo
  enddo
else if (switch.eq.2) then               determine g(r)
  do i=1,nhis
    r=delg*(i+0.5)                       distance r
    vb=((i+1)**3-i**3)*delg**3           volume between bin i+1 and i
    nid=(4/3)*pi*vb*rho                  number of ideal gas part. in vb
    g(i)=g(i)/(ngr*npart*nid)            normalize g(r)
  enddo
endif
return
end
```

Comments to this algorithm:

1. *For efficiency reasons the sampling part of this algorithm is usually combined with the force calculation (for example, Algorithm 5).*
2. *The factor* pi = *3.14159....*

processes near equilibrium, in particular the so-called time-correlation functions. However, we shall not start with a general description of nonequilibrium processes. Rather we start with a discussion of a simple specific example that allows us to introduce most of the necessary concepts.

4.4.1 Diffusion

Diffusion is the process whereby an initially nonuniform concentration profile (e.g., an ink drop in water) is smoothed in the absence of flow (no stirring). Diffusion is caused by the molecular motion of the particles in the fluid. The macroscopic law that describes diffusion is known as Fick's law, which states that the flux $\mathbf{j}$ of the diffusing species is proportional to the negative gradient in the concentration of that species:

$$\mathbf{j} = -D\boldsymbol{\nabla}c, \tag{4.4.4}$$

where D, the constant of proportionality, is referred to as the *diffusion coefficient*. In what follows, we shall be discussing a particularly simple form of diffusion, namely, the case where the molecules of the diffusing species are identical to the other molecules but for a label that does not affect the interaction of the labeled molecules with the others. For instance, this label could be a particular polarization of the nuclear spin of the diffusing species or a modified isotopic composition. Diffusion of a labeled species among otherwise identical solvent molecules is called *self-diffusion*.

Let us now compute the concentration profile of the tagged species, under the assumption that, at time $t = 0$, the tagged species was concentrated at the origin of our coordinate frame. To compute the time evolution of the concentration profile, we must combine Fick's law with an equation that expresses conservation of the total amount of labeled material:

$$\frac{\partial c(r,t)}{\partial t} + \boldsymbol{\nabla} \cdot \mathbf{j}(r,t) = 0. \tag{4.4.5}$$

Combining equation (4.4.5) with equation (4.4.4), we obtain

$$\frac{\partial c(r,t)}{\partial t} - D\nabla^2 c(r,t) = 0. \tag{4.4.6}$$

We can solve equation (4.4.6) with the boundary condition

$$c(r,0) = \delta(r)$$

($\delta(r)$ is the Dirac delta function) to yield

$$c(r,t) = \frac{1}{(4\pi Dt)^{d/2}} \exp\left(-\frac{r^2}{4Dt}\right).$$

As before, d denotes the dimensionality of the system. In fact, for what follows we do not need $c(r,t)$ itself, but just the time dependence of its second moment:

$$\langle r^2(t)\rangle \equiv \int \mathbf{dr}\, c(r,t) r^2,$$

where we have used the fact that we have imposed

$$\int \mathbf{dr}\, c(r,t) = 1.$$

We can directly obtain an equation for the time evolution of $\langle r^2(t)\rangle$ by multiplying equation (4.4.6) by r^2 and integrating over all space. This yields

$$\frac{\partial}{\partial t}\int \mathbf{dr}\, r^2 c(r,t) = D\int \mathbf{dr}\, r^2 \nabla^2 c(r,t). \tag{4.4.7}$$

The left-hand side of this equation is simply equal to

$$\frac{\partial \langle r^2(t)\rangle}{\partial t}.$$

Applying partial integration to the right-hand side, we obtain

$$\begin{aligned}
\frac{\partial \langle r^2(t)\rangle}{\partial t} &= D\int \mathbf{dr}\, r^2\nabla^2 c(r,t) \\
&= D\int \mathbf{dr}\, \boldsymbol{\nabla}\cdot(r^2\boldsymbol{\nabla} c(r,t)) - D\int \mathbf{dr}\, \boldsymbol{\nabla} r^2\cdot\boldsymbol{\nabla} c(r,t) \\
&= D\int \mathbf{dS}\,(r^2\boldsymbol{\nabla} c(r,t)) - 2D\int \mathbf{dr}\,\mathbf{r}\cdot\boldsymbol{\nabla} c(r,t) \\
&= 0 - 2D\int \mathbf{dr}\,(\boldsymbol{\nabla}\cdot \mathbf{r}c(r,t)) + 2D\int \mathbf{dr}\,(\boldsymbol{\nabla}\cdot\mathbf{r})c(r,t) \\
&= 0 + 2dD\int \mathbf{dr}\, c(r,t) \\
&= 2dD.
\end{aligned} \tag{4.4.8}$$

Equation (4.4.8) relates the diffusion coefficient D to the width of the concentration profile. This relation was first derived by Einstein. It should be realized that, whereas D is a macroscopic transport coefficient, $\langle r^2(t)\rangle$ has a microscopic interpretation: it is the mean-squared distance over which the labeled molecules have moved in a time interval t. This immediately suggests how to measure D in a computer simulation. For every particle i, we measure the distance traveled in time t, $\Delta\mathbf{r}_i(t)$, and we plot the mean square of these distances as a function of the time t:

$$\langle \Delta r(t)^2\rangle = \frac{1}{N}\sum_{i=1}^{N}\Delta\mathbf{r}_i(t)^2.$$

This plot would look like the one that will be shown later in Figure 4.6. We should be more specific about what we mean by the displacement of a particle in a system with periodic boundary conditions. The displacement that we are interested in is simply the time integral of the velocity of the tagged particle:

$$\Delta\mathbf{r}(t) = \int_0^t dt'\, \mathbf{v}(t').$$

In fact, there is a relation that expresses the diffusion coefficient directly in terms of the particle velocities. We start with the relation

$$2D = \lim_{t\to\infty} \frac{\partial \langle x^2(t) \rangle}{\partial t}, \tag{4.4.9}$$

where, for convenience, we consider only one Cartesian component of the mean-squared displacement. If we write $x(t)$ as the time integral of the x component of the tagged-particle velocity, we get

$$\begin{aligned} \langle x^2(t) \rangle &= \left\langle \left(\int_0^t dt'\, v_x(t') \right)^2 \right\rangle \\ &= \int_0^t \int_0^t dt'dt''\, \langle v_x(t')v_x(t'') \rangle \\ &= 2 \int_0^t \int_0^{t'} dt'dt''\, \langle v_x(t')v_x(t'') \rangle\,. \end{aligned} \tag{4.4.10}$$

The quantity $\langle v_x(t')v_x(t'') \rangle$ is called the velocity autocorrelation function. It measures the correlation between the velocity of a particle at times t' and t''. The velocity autocorrelation function (VACF) is an equilibrium property of the system, because it describes correlations between velocities at different times along an equilibrium trajectory. As equilibrium properties are invariant under a change of the time origin, the VACF depends only on the difference of t' and t''. Hence,

$$\langle v_x(t')v_x(t'') \rangle = \langle v_x(t'-t'')v_x(0) \rangle\,.$$

Inserting equation (4.4.10) in equation (4.4.9), we obtain

$$\begin{aligned} 2D &= \lim_{t\to\infty} 2 \int_0^t dt''\, \langle v_x(t-t'')v_x(0) \rangle \\ D &= \int_0^\infty d\tau\, \langle v_x(\tau)v_x(0) \rangle\,. \end{aligned} \tag{4.4.11}$$

In the last line of equation (4.4.11) we introduced the coordinate $\tau \equiv t - t''$. Hence, we see that we can relate the diffusion coefficient D to the integral of the velocity autocorrelation function. Such a relation between a transport coefficient and an integral over a time-correlation function is called a *Green-Kubo relation* (see Appendix C for some details). Green-Kubo relations have been derived for many other transport coefficients, such as the shear viscosity η,

$$\eta = \frac{1}{Vk_BT}\int_0^\infty dt\ \langle\sigma^{xy}(0)\sigma^{xy}(t)\rangle \tag{4.4.12}$$

with

$$\sigma^{xy} = \sum_{i=1}^{N}\left(m_i v_i^x v_i^y + \frac{1}{2}\sum_{j\neq i} x_{ij} f_y(r_{ij})\right); \tag{4.4.13}$$

the thermal conductivity λ_T,

$$\lambda_T = \frac{1}{Vk_BT^2}\int_0^\infty dt\ \langle j_z^e(0) j_z^e(t)\rangle \tag{4.4.14}$$

with

$$j_z^e = \frac{d}{dt}\sum_{i=1}^{N} z_i \frac{1}{2}\left(m_i v_i^2 + \sum_{j\neq i} v(r_{ij})\right); \tag{4.4.15}$$

and electrical conductivity σ_e

$$\sigma_e = \frac{1}{Vk_BT}\int_0^\infty dt\ \left\langle j_x^{el}(0) j_x^{el}(t)\right\rangle \tag{4.4.16}$$

with

$$j_x^{el} = \sum_{i=1}^{N} q_i v_i^x. \tag{4.4.17}$$

For details, see, for example, [79]. Time-correlation functions can easily be measured in a Molecular Dynamics simulation. It should be emphasized that for classical systems, the Green-Kubo relation for D and the Einstein relation are strictly equivalent. There may be practical reasons to prefer one approach over the other, but the distinction is never fundamental. In Algorithm 8 an implementation of the calculation of the mean-squared displacement and velocity autocorrelation function is described.

4.4.2 Order-n Algorithm to Measure Correlations

The calculation of transport coefficients from the integral of a time-correlation function, or from a (generalized) Einstein relation, may require a lot of memory and CPU time, in particular if fluctuations decay slowly. As an example, we consider again the calculation of the velocity autocorrelation function and the measurement of the diffusion coefficient. In a dense

Algorithm 8 (Diffusion) Using (4.4.11). velocity

```
 subroutine dif(switch,nsamp)              diffusion; switch = 0 init.
                                           = 1 sample, and = 2 results
 if (switch.eq.0) then                     Initialization
    ntel=0                                 time counter
    dtime=dt*nsamp                         time between two samples
    do i=1,tmax                            tmax total number of time step
       ntime(i)=0                          number of samples for time i
       vacf(i)=0
       r2t(i)=0
    enddo
 else if (switch.eq.1) then                sample
    ntel=ntel+1
    if (mod(ntel,it0).eq.0) then           decide to take a new t = 0
       t0 = t0 + 1                         update number of t = 0
       tt0=mod(t0-1,t0max)+1               see note 1
       time0(tt0)=ntel                     store the time of t = 0
       do i=,npart
          x0(i,tt0)=x(i)                   store position for given t = 0
          vx0(i,tt0)=vx(i)                 store velocity for given t = 0
       enddo
    endif
    do t=1,min(t0,t0max)                   update vacf and r2, for t = 0
      delt=ntel-time0(t)+1                 actual time minus t = 0
      if (delt.lt.tmax) then
       ntime(delt)=ntime(delt)+1
       do i=1,npart
          vacf(delt)=vacf(delt)+           update velocity autocorr.
+            vx(i)*vx0(i,t)
          r2t(delt)=r2t(delt)+             update mean-squared displ.
+             (x(i)-x0(i,t))**2
       enddo
      endif
    enddo
  else if (switch.eq.2) then               determine results
    do i=1,tmax
      time=dtime*(i+0.5)                   time
      vacf(i)=vacf(i)                      volume velocity autocorr.
+              /(npart*ntime(i))
      r2t(i)=r2t(i)                        mean-squared displacement
+             /(npart*ntime(i))
    enddo
 endif
 return
 end
```

Comments to this algorithm:

1. *We define a new* `t = 0` *after each* `it0` *times this subroutine has been called. For each* `t = 0`*, we store the current positions and velocities. The term* `t0max` *is the maximum number of* `t = 0` *we can store. If we sample more, the first* `t = 0` *will be removed and replaced by a new one. This limits the maximum time we collect data to* `t0max*it0`*; this number should not be smaller than* `tmax`*, the total number of time steps we want to sample.*
2. *Because* `nsamp` *gives the frequency at which this subroutine is called, the time between two calls is* `nsamp*delt`*, where* `delt` *is the time step.*

medium, the velocity autocorrelation function changes rapidly on typically microscopic time scales. It therefore is important to have an even shorter time interval between successive samples of the velocity. Yet, when probing the long-time decay of the velocity autocorrelation function, it is not necessary to sample with the same frequency. The conventional schemes for measuring correlation functions do not allow for such adjustable sampling frequencies. Here, we describe an algorithm that allows us to measure fast and slow decay simultaneously at minimal numerical cost. This scheme can be used to measure the correlation function itself, but in the example that we discuss, we show how it can be used to compute the transport coefficient.

Let us denote by Δt the time interval between successive measurements of the velocity of the particles in the system. We can define block sums of the velocity of a given particle as follows:

$$\mathbf{v}^{(i)}(j) \equiv \sum_{l=(j-1)n+1}^{jn} \mathbf{v}^{(i-1)}(l) \tag{4.4.18}$$

with

$$\mathbf{v}^{(0)}(l) \equiv \mathbf{v}(l), \tag{4.4.19}$$

where $\mathbf{v}(l)$ is the velocity of a particle at time l. Equation (4.4.18) is a recursive relation between block sums of level i and $i-1$. The variable n determines the number of terms in the summation. For example, $\mathbf{v}^{(3)}(j)$ can

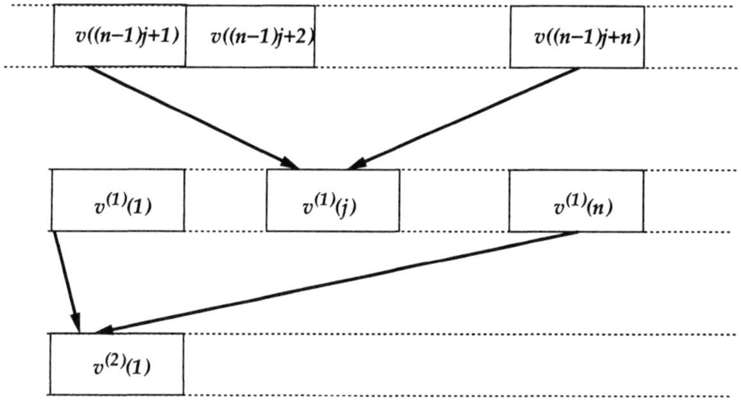

Figure 4.2: Coarse graining the velocities.

be written as

$$\begin{aligned}
\mathbf{v}^{(3)}(j) &= \sum_{l_1=(j-1)n+1}^{jn} \mathbf{v}^{(2)}(l_1) \\
&= \sum_{[l_1=(j-1)n+1]}^{jn} \; \sum_{[l_2=(l_1-1)n+1]}^{l_1 n} \; \sum_{[l_3=(l_2-1)n+1]}^{l_2 n} \mathbf{v}(l_3) \\
&= \sum_{l=(j-1)n^3+1}^{n^3 j} \mathbf{v}(l) \\
&\approx \frac{1}{\Delta t}\int_{l=(j-1)n^3+1}^{n^3 j} dt\, \mathbf{v}(t) = \frac{\mathbf{r}(n^3 j) - \mathbf{r}(n^3(j-1)+1)}{\Delta t}.
\end{aligned}$$

Clearly, the block sum of the velocity is related to the displacement of the particle in a time interval $n^i \Delta t$. In Figure 4.2 the blocking operation is illustrated. From the preceding block sums, it is straightforward to compute the velocity autocorrelation function with a resolution that decreases with increasing time. At each level of blocking, we need to store $n \times N$ block sums, where N is the number of particles (in practice, it will be more convenient to store the *block-averaged* velocities).

The total storage per particle for a simulation of length $t = n^i \Delta t$ is $i \times n$. This should be compared to the conventional approach where, to study correlations over the same time interval, the storage per particle would be n^i. In the conventional calculation of correlation functions, the number of floating-point operations scales as t^2 (or $t \ln t$, if the fast Fourier technique is used). In contrast, in the present scheme the number of operations scales as t.

At each time step we have to update $\mathbf{v}^{(0)}(t)$ and correlate it with all n entries in the $\mathbf{v}^{(0)}$-array. The next block sum has to be updated and correlated once every n time steps, the third every every n^2 steps, etc. This yields, for the total number of operations,

$$\frac{t}{\Delta t} \times n \left(1 + \frac{1}{n} + \frac{1}{n^2} + \cdots + \frac{1}{n^i}\right) < \frac{t}{\Delta t} n \frac{n}{n-1}.$$

Using this approach, we can quickly and efficiently compute a wide variety of correlation functions, both temporal and spatial. However, it should be stressed that each blocking operation leads to more coarse graining. Hence, any high-frequency modulation of long-time behavior of such correlation functions will be washed out.

Interestingly enough, even though the velocity autocorrelation function itself is approximate at long times, we can still compute the integral of the velocity autocorrelation function (i.e., the diffusion coefficient), with no loss in numerical accuracy. Next, we discuss in some detail this technique for computing the diffusion coefficient.

Let us define

$$\Delta \bar{\mathbf{r}}^{(i)}(j) \equiv \sum_{l=0}^{j} \mathbf{v}^{(i)}(l)\Delta t = \mathbf{r}(n^i) - \mathbf{r}(0). \tag{4.4.20}$$

The square of the displacement of the particle in a time interval $n^i \Delta t$ can be written as

$$(\Delta \bar{\mathbf{r}}^2)^{(i)}(j) = \left[\mathbf{r}(n^i) - \mathbf{r}(0)\right]^2 = \Delta \bar{\mathbf{r}}^{(i)}(j) \cdot \Delta \bar{\mathbf{r}}^{(i)}(j). \tag{4.4.21}$$

To compute the diffusion coefficient, we should follow the time dependence of the mean-squared displacement. As a first step, we must determine $\Delta \bar{\mathbf{r}}^{(i)}(j)$ for all i and all j. In fact, to improve the statistics, we wish to use every sample point as a new time origin. To achieve this, we again create arrays of length n. However, these arrays do not contain the same block sums as before, but partial block sums (see Algorithm 9).

1. At every time interval Δt, the lowest-order blocking operation is performed through the following steps:

   (a) We first consider the situation that all lowest-order accumulators have already been filled at least once (this is true if $t > n\Delta t$). The value of the current velocity $v(t)$ is added to

   $$\mathbf{v}_{\text{sum}}(1, j) = \mathbf{v}_{\text{sum}}(1, j+1) + \mathbf{v}(t)$$

   for `j = 1,n-1`, and

   $$\mathbf{v}_{\text{sum}}(1, j) = \mathbf{v}(t)$$

   for `j = n`.

Algorithm 9 (Diffusion: Order-n Algorithm)

```
subroutine dif(switch,nsamp)          diffusion
                                      switch = 0 initialization,
                                      = 1 sample, and = 2 results
if (switch.eq.0) then                 initialization
  ntel=0                              time counter for this subroutine
  dtime=dt*nsamp                      time between two samples
  do ib=1,ibmax                       ibmax max. number of blocks
   ibl(ib)=0                          length of current block
   do j=1,n                           n number of steps in a block
    tel(ib,j)=0                       counter number of averages
    delr2(ib,j)=0                     running average mean-sq. displ.
    do i=1,npart
      vxsum(ib,j,i)=0                 coarse-grained velocity particle i
    enddo
   enddo
  enddo
else if (switch.eq.2) then            print results
  do ib=1,max(ibmax,iblm)
    do j=2,min(ibl(ib),n)
     time=dtime*j*n**(ib-1)           time
     r2=delr2(ib,j)*dtime**2          mean-squared displacement
            /tel(ib,j)
    enddo
  enddo
   ...(continue)....
```

(b) These operations yield

$$\mathbf{v}_{\text{sum}}(1, l) = \sum_{j=t-n+l}^{j=t} \mathbf{v}(j).$$

The equation allow us to update the accumulators for the mean-squared displacement (4.4.21) for $l = 1, 2, \ldots, n$:

$$(\Delta \bar{\mathbf{r}}^2)^{(0)}(l) = (\Delta \bar{\mathbf{r}}^2)^{(0)}(l) + \mathbf{v}_{\text{sum}}^2(1, l)\Delta t^2.$$

2. If the current time step is a multiple of n, we perform the first blocking operation, if it is a multiple of n^2 the second, etc. Performing blocking operation i involves the following steps:

```
      ...(continue)....
   else if (switch.eq.1) then               sample
    ntel=ntel+1
    iblm=MaxBlock(ntel,n)                   maximum number of possible
                                            blocking operations
    do ib=1,iblm
     if (mod(ntel,n**(ib-1))                test if ntel is a multiple of n^ib
+             .eq.0) then
      ibl(ib)=ibl(ib)+1                     increase current block length
      inm=max(ibl(ib),n)                    set maximum block length to n
      do i=1,npart
        if(ib.eq.1) then
          delx=vx(i)                        0th block: ordinary velocity
        else
          delx=vxsum(ib-1,1,i)              previous block velocity
        endif
        do in=1,inm
          if (inm.ne.n) then                test block length equal to n
            inp=in
          else
            inp=in+1
          endif
          if (in.lt.inm) then
           vxsum(ib,in,i)=
+            vxsum(ib,inp,i)+delx           eqns. (4.4.22) or (4.4.25)
          else
           vxsum(ib,in,i)=delx              eqns. (4.4.23) or (4.4.26)
          endif
        enddo
        do in=1,inm
         tel(ib,in)=tel(ib,in)+1            counter number of updates
         delr2(ib,in)=delr2(ib,in)          update equation (4.4.24)
+          +vxsum(ib,inm-in+1,i)**2
        enddo
      enddo
     endif
    enddo
   endif
   return
   end
```

Comment to this algorithm:

1. `MaxBlock(ntel,n)` *gives the maximum number of blocking operations that can be performed on the current time step* `ntel`.

(a) As before, we first consider the situation that all ith-order accumulators have already been filled at least once (i.e., $t > n^i \Delta t$). Using the $i-1$th block sum ($\mathbf{v}_{\text{sum}}(i-1,1)$), we update

$$\mathbf{v}_{\text{sum}}(i,j) = \mathbf{v}_{\text{sum}}(i,j+1) + \mathbf{v}_{\text{sum}}(i-1,1) \tag{4.4.22}$$

for `j = 1,n-1`, and

$$\mathbf{v}_{\text{sum}}(i,j) = \mathbf{v}_{\text{sum}}(i-1,1) \tag{4.4.23}$$

for `j = n`.

(b) These operations yield

$$\mathbf{v}_{\text{sum}}(i,l) = \sum_{j=n-l+1}^{j=n} \mathbf{v}_{\text{sum}}(i-1,j).$$

The equations allows us to update the accumulators for the mean-squared displacement, equation (4.4.21), for $l = 1, 2, \ldots, n$:

$$(\Delta \mathbf{r}^2)^{(i)}(l) = (\Delta \mathbf{r}^2)^{(i)}(l) + \mathbf{v}^2_{\text{sum}}(i,l)\Delta t^2. \tag{4.4.24}$$

3. Finally, we must consider how to handle arrays that have not yet been completely filled. Consider the situation that only `nmax` of the n locations of the array that contains the ith-level sums have been initialized. In that case, we should proceed as follows:

   (a) Update the current block length: `nmax = nmax+1` (`nmax` $< n$).

   (b) For `j = 1,nmax-1`

   $$\mathbf{v}_{\text{sum}}(i,j) = \mathbf{v}_{\text{sum}}(i,j) + \mathbf{v}_{\text{sum}}(i-1,1). \tag{4.4.25}$$

   (c) For `j = nmax`

   $$\mathbf{v}_{\text{sum}}(i,j) = \mathbf{v}_{\text{sum}}(i-1,1). \tag{4.4.26}$$

   The update of equation (4.4.21) remains the same.

In Case Study 6, a detailed comparison is made between the present algorithm and the conventional algorithm for the diffusion of the Lennard-Jones fluid.

4.5 Some Applications

Let us illustrate the results of the previous sections with an example. Like in the section on Monte Carlo simulations we choose the Lennard-Jones fluid as our model system. We use a truncated and shifted potential (see also section 3.2.2):

$$u^{tr-sh}(r) = \begin{cases} u^{lj}(r) - u^{lj}(r_c) & r \leq r_c \\ 0 & r > r_c \end{cases},$$

where $u^{lj}(r)$ is the Lennard-Jones potential and for these simulations $r_c = 2.5\sigma$ is used.

Case Study 4 (Static Properties of the Lennard-Jones Fluid)
Let us start a simulation with 108 particles on a simple cubic lattice. We give the system an initial temperature $T = 0.728$ and density $\rho = 0.8442$, which is close to the triple (gas-liquid-solid) point of the Lennard-Jones fluid [81–83].

In Figure 4.3 the evolution of the total energy, kinetic energy, and potential energy is shown. It is important to note that the total energy remains constant and does not show a (slow) drift during the entire simulation. The kinetic and potential energies do change initially (the equilibration period) but during the end of the simulation they oscillate around their equilibrium value. This figure shows that, for the calculation of the average potential energy or kinetic energy, we need approx. 1000 time steps to equilibrate the simulation. The figure also shows significant fluctuations in the potential energy, some of which may take several (100) time steps before they disappear.

Appendix D shows in detail how to calculate statistical error in the data of a simulation. In this example, we use the method of Flyvbjerg and Petersen [84]. The following operations on the set of data points are performed: we start by calculating the standard deviation of all the data points, then we group two consecutive data points and determine again the standard deviation of the new, blocked, data set. This new data set contains half the number of data points of the original set. The procedure is repeated until there are not enough data points to compute a standard deviation; the number of times we perform this operation is called M. What do we learn from this?

First of all, let us assume that the time between two samples is so large that the data points are uncorrelated. If the data are uncorrelated the standard deviation (as calculated according to the formula in Appendix D, i.e., correcting for the fact we have fewer data points) should be invariant to this blocking operation and we should get a standard deviation that is independent of M. In a simulation, however, the time between two data points is usually too short to obtain a statistically independent sample; as a consequence consecutive data points would be (highly) correlated. If we would calculate a standard deviation using these data, this standard deviation will be too optimistic. The effect of the block operation will be that after grouping two consecutive data points, the correlation between the two new data points will be less. This, however, will increase the standard deviation; the data will have more noise since consecutive data points no longer resemble

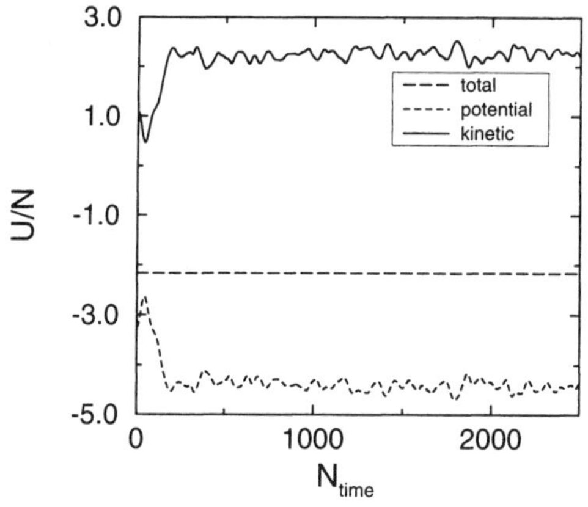

Figure 4.3: Total, potential, and kinetic energy per particle U/N as a function of the number of time steps N_{time}.

each other that closely. This decrease of accuracy as a function of the number of blocking operations will continue until we have grouped so many data points that two consecutive points are really uncorrelated. This is exactly the standard deviation we want to determine. It is important to note that we have to ensure that the standard deviations we are looking at are significant; therefore, we have to determine the standard deviation of the error at the same time.

The results of this error calculation for the potential energy are shown in Figure 4.4, as expected, for a low value of M; the error increases until a plateau is reached. For high values of M, we have only a few data points, which results in a large standard deviation in the error. The advantage of this method is that we have a means of finding out whether we have simulated enough; if we do not find such a plateau, the simulation must have been too short. In addition we find a reliable estimate of the standard deviation. The figure also shows the effect of increasing the total length of the simulation by a factor of 4; the statistical error in the potential energy has indeed decreased by a factor of 2.

In this way we obtained the following results. For the potential energy $U = -4.4190 \pm 0.0012$ and for the kinetic energy $K = 2.2564 \pm 0.0012$, the latter corresponds to an average temperature of $T = 1.5043 \pm 0.0008$. For the pressure, we have obtained 5.16 ± 0.02.

In Figure 4.5, the radial distribution function is shown. To determine this function we used Algorithm 7. This distribution function shows the characteristics of a dense liquid. We can use the radial distribution function to calculate the energy and pressure. The potential energy per particle can be

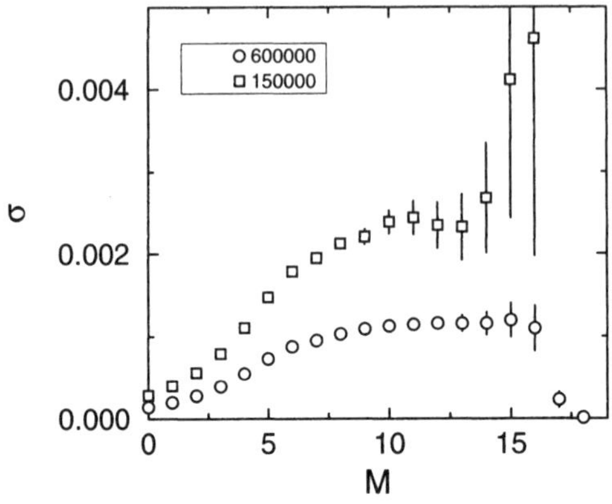

Figure 4.4: The standard deviation σ in the potential energy as a function of the number of block operations M for a simulation of 150,000 and 600,000 time steps. This variance is calculated using equation (D.3.4).

calculated from

$$
\begin{aligned}
U/N &= \frac{1}{2}\rho \int_0^\infty \mathbf{dr}\, u(r)g(r) \\
&= 2\pi\rho \int_0^\infty dr\, r^2 u(r)g(r)
\end{aligned}
\tag{4.5.1}
$$

and for the pressure from

$$
\begin{aligned}
P &= \rho k_B T - \frac{1}{3}\frac{1}{2}\rho^2 \int_0^\infty \mathbf{dr}\, \frac{du(r)}{dr} r g(r) \\
&= \rho k_B T - \frac{2}{3}\pi\rho^2 \int_0^\infty dr\, \frac{du(r)}{dr} r^3 g(r),
\end{aligned}
\tag{4.5.2}
$$

where $u(r)$ is the pair potential.

Equations (4.5.1) and (4.5.2) can be used to check the consistency of the energy and pressure calculations and the determination of the radial distribution function. In our example, we obtained from the radial distribution function for the potential energy $U/N = -4.419$ and for the pressure $P = 5.181$, which is in good agreement with the direct calculation.

Case Study 5 (Dynamic Properties of the Lennard-Jones Fluid)
As an example of a dynamic property we have determined the diffusion coefficient. As shown in the previous section, the diffusion coefficient can be determined from the mean-squared displacement or from the velocity autocorrelation function. We have determined these properties using Algorithm 8.

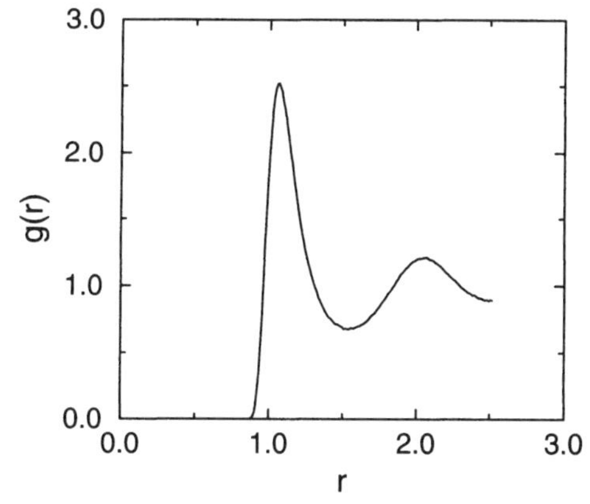

Figure 4.5: Radial distribution function of a Lennard-Jones fluid close to the triple point: $T = 1.5043 \pm 0.0008$ and $\rho = 0.8442$.

In Figure 4.6 the mean-squared displacement is shown as a function of the simulation time. From the mean-squared displacement we can determine the diffusion using equation (4.4.9). This equation, however, is valid only in the limit $t \to \infty$. In practice this means that we have to verify that we have simulated enough that the mean-squared displacement is really proportional to t and not to another power of t.

The velocity autocorrelation function can be used as an independent route to test the calculation of the diffusion coefficient. The diffusion coefficient follows from equation (4.4.11). In this equation we have to integrate to $t \to \infty$. Knowing whether we have simulated sufficiently to perform this integration reliably is equivalent to determining the slope in the mean-squared displacement. A simple trick is to determine the diffusion coefficient as a function of the truncation of the integration; if a plateau has been reached over a sufficient number of integration limits, the calculation is probably reliable.

Case Study 6 (Algorithms to Calculate the Mean-Squared Displacement)
In this case study, a comparison is made between the conventional (Algorithm 8) and the order-n methods (Algorithm 9) to determine the mean-squared displacement. For this comparison we determine the mean-squared displacement of the Lennard-Jones fluid.

In Figure 4.7 the mean-squared displacement as a function of time as computed with the conventional method is compared with that obtained from the order-n scheme. The calculation using the conventional scheme could not be extended to times longer than $t > 10$ without increasing the number of

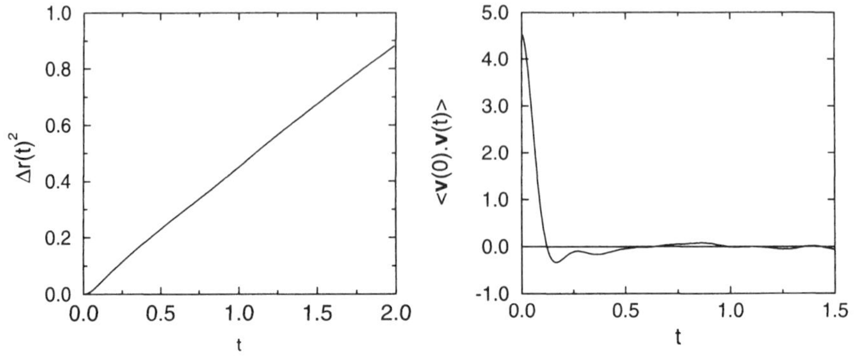

Figure 4.6: (left) Mean-squared displacement $\Delta r(t)^2$ as a function of the simulation time t. Note that for long times, $\Delta r(t)^2$ varies linearly with t. The slope is then given by 2dD, where d is the dimensionality of the system and D the self-diffusion coefficient. (right) Velocity autocorrelation function $\langle \mathbf{v}(0) \cdot \mathbf{v}(t) \rangle$ as a function of the simulation time t.

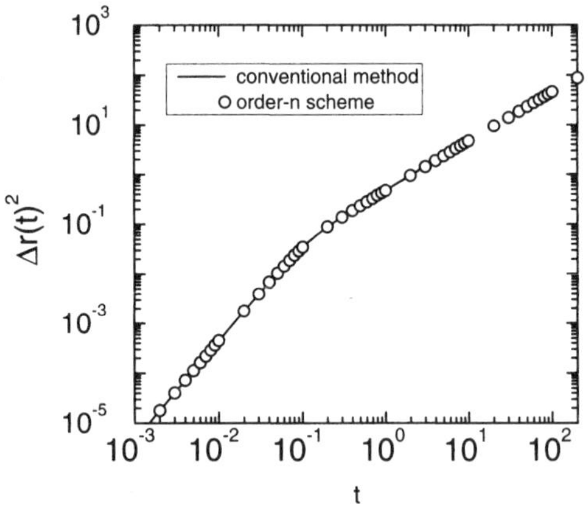

Figure 4.7: Mean-squared displacement as a function of time for the Lennard-Jones fluid ($\rho = 0.844$, $N = 108$, and $T = 1.50$); comparison of the conventional method with the order-n scheme .

time steps between two samples because of lack of memory. With the order-n scheme the calculation could be extended to much longer times with no difficulty. It is interesting to compare the accuracy of the two schemes. In the conventional scheme, the velocities of the particles at the current time step are used to update the mean-squared displacement of all time intervals. In

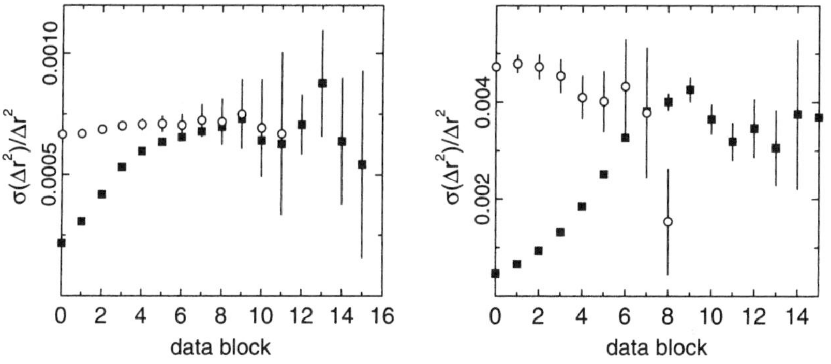

Figure 4.8: Relative error in the mean-squared displacement as a function of the number of data blocks as defined by Flyvbjerg and Petersen. The figures compare the conventional scheme (solid squares) with the order-n method (open circles) to determine the mean-squared displacement. The right figure is for $t = 0.1$ and the left figure for $t = 1.0$.

the order-n scheme the current time step is only used to update the lowest-order array of $\mathbf{v}_{\text{sum}}$ (see Algorithm 9). The block sums of level i are updated only once every n^i time step. Therefore, for a total simulation of M time steps, the number of samples is much less for the order-n scheme; for the conventional scheme, we have M samples for all time steps, whereas the order-n scheme has M/n^i samples for the ith block velocity. Naively, one would think that the conventional scheme therefore is more accurate. In the conventional scheme, however, the successive samples will have much more correlation and therefore are not independent. To investigate the effect of these correlations on the accuracy of the results, we have used the method of Flyvbjerg and Petersen [84] (see Appendix D.3 and Case Study 4). In this method, the standard deviation is calculated as a function of the number of data blocks. If the data are correlated, the standard deviation will increase as a function of the number of blocks until the number of blocks is sufficient that the data in a data block are uncorrelated. If the data are uncorrelated, the standard deviation will be independent of the number of blocks. This limiting value is the standard deviation of interest.

In these simulations the time step was $\Delta t = 0.001$ and the block length was set to $n = 10$. For both methods the total number of time steps was equal. To calculate the mean-squared displacement, we have used 100,000 samples for all times in the conventional scheme. For the order-n scheme, we have used 100,000 samples for $t \in [0, 0.01]$, 10,000 for $t \in [0.01, 0.1]$, 1,000 for for $t \in [0.1, 1]$, etc. This illustrates that the number of samples in the order-n scheme is considerably less than in the conventional scheme. The

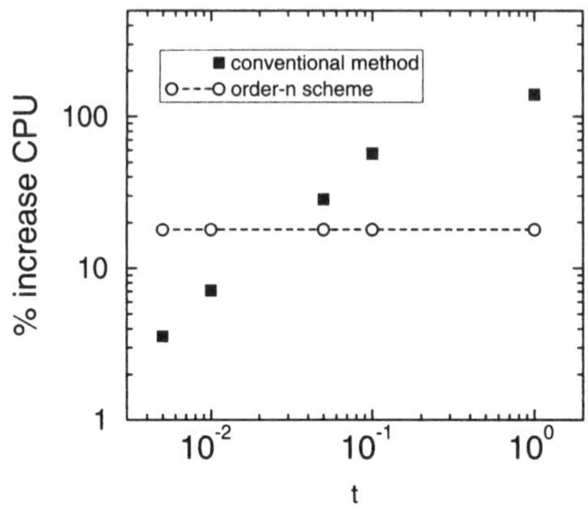

Figure 4.9: Percentage increase of the total CPU time as a function of the total time for which we determine the mean-squared displacement; comparison of the conventional scheme with the order-n scheme for the same system as is considered in Figure 4.7

accuracy of the results, however, turned out to be the same. This is shown in Figure 4.8 for $t = 0.1$ and $t = 1.0$. Since the total number of data blocking operations that can be performed on the data depends on the total number of samples, the number of blocking operations is less for the order-n method. Figure 4.8 shows that for $t = 0.1$ the order-n scheme yields a standard deviation that is effectively constant after three data blocking operations, indicating the samples are independent, whereas the standard deviation using the conventional method shows an increase for the first six to eight data blocking operations. For $t = 1.0$ the order-n method is independent of the number of data blocks, the conventional method only after 10 data blocks. This implies that one has to average over $2^{10} \approx 1000$ successive samples to have two independent data points. In addition, the figure shows that the plateau value of the standard deviation is essentially the same for the two methods, which implies that for this case the two methods are equally accurate.

In Figure 4.9 we compare the CPU requirements of the two algorithms for simulations with a fixed total number of time steps. This figure shows the increase of the total CPU time of the simulation as a function of the total time for which the mean-squared displacement has been calculated. With the order-n scheme the CPU time should be (nearly) independent of the total time for which we determine the mean-squared displacement, which is indeed what we observe. For the conventional scheme, however, the required CPU time increases significantly for longer times. At $t = 1.0$ the order-n scheme gives an increase of the total CPU time of 17%, whereas the conventional scheme shows an increase of 130%.

This example illustrates that the saving in memory as well as in CPU time of the order-n scheme can be significant, especially if we are interested in the mean-squared displacement at long times.

4.6 Questions and Exercises

Question 10 (Integrating the Equations of Motion)

1. *If you do an MD simulation of the Lennard-Jones potential with a time step that is much too large you will find an energy drift. This drift is towards a higher energy. Why?*
2. *Why don't we use Runga-Kutta methods to integrate the equations of motion of particles in MD?*
3. *Which of the following quantities are conserved in the MD simulation of Case Study 4: potential energy, total momentum, position of the center of mass of the system, or total angular momentum?*
4. *Show that the Verlet and velocity Verlet algorithms lead to identical trajectories.*
5. *Derive the Leap-Frog Algorithm by using Taylor expansions for* $v\left(t + \frac{\Delta t}{2}\right)$, $v\left(t - \frac{\Delta t}{2}\right)$, $x(t + \Delta t)$, *and* $x(t)$.

Question 11 (Correlation Functions)

1. *The value of the velocity autocorrelation function (vacf) at* $t = 0$ *is related to an observable quantity. Which one?*
2. *Calculate the limit of the vacf for* $t \to \infty$.
3. *What is the physical significance of vacf* < 0?
4. *When you calculate the mean-squared displacement for particles in a system in which periodic boundary conditions are used and in which particles are placed back in the box, you should be very careful in calculating the displacement. Why?*
5. *What is more difficult to calculate accurately: the self-diffusion coefficient or the viscosity? Explain.*

Exercise 10 (Molecular Dynamics of a Lennard-Jones System)
On the book's website you can find a Molecular Dynamics (MD) program for a Lennard-Jones fluid in the NVE ensemble. Unfortunately, the program does not conserve the total energy because it contains three errors.

1. Find the three errors in the code. Hint: there are two errors in *integrate.f* and one in *force.f*. See the file *system.inc* for documentation about some of the variables used in this code.

2. How is one able to control the temperature in this program? After all, the total energy of the system should be constant (not the temperature).
3. To test the energy drift ΔU of the numerical integration algorithm for a given time step Δt after N integration steps, one usually computes [85]

$$\Delta U(\Delta t) = \frac{1}{N}\sum_{i=1}^{i=N}\left|\frac{U(0)-U(i\Delta t)}{U(0)}\right|.$$

In this equation, $U(x)$ is the total energy (kinetic + potential) of the system at time x. Change the program (only in *mdloop.f*) in such a way that ΔU is computed and make a plot of ΔU as a function of the time step. How does the time step for a given energy drift change with the temperature and density?
4. One of the most time-consuming parts of the program is the calculation of the nearest image of two particles. In the present program, this calculation is performed using an *if-then-else-endif* construction. This works only when the distance between two particles is smaller than 1.5 and larger than -1.5 times the size of the periodic box. A way to overcome this problem is to use a function that calculates the nearest integer `nint`

$$x = x - box * \mathtt{nint}(x * ibox),$$

in which $ibox = 1.0/box$. Which expression is faster? (Hint: You only have to make some modifications in *force.f*.) Which expression will be faster on a vector computer like a Cray C90? Because the `nint` function is usually slow, you can write your own `nint` function. For example, when $x < -998$, we can use

$$\mathtt{nint}(x) = \mathrm{int}(x + 999.5) - 999. \tag{4.6.1}$$

What happens with the speed of the program when you replace the standard `nint` function? Do you have an explanation for this? [4]
5. In equation (4.6.1), $ibox$ is used instead of $1/box$. Why?
6. An important quantity of a liquid or gas is the so-called self-diffusivity D. There are two methods to calculate D:

   (a) by integrating the velocity autocorrelation function:

$$\begin{aligned} D &= \frac{1}{3}\int_0^\infty \left\langle \mathbf{v}(t)\cdot\mathbf{v}\left(t+t'\right)\right\rangle dt' \\ &= \frac{\int_0^\infty \sum_{i=1}^{i=N}\left\langle \mathbf{v}(i,t)\cdot\mathbf{v}\left(i,t+t'\right)\right\rangle dt'}{3N} \end{aligned} \tag{4.6.2}$$

[4] The result will strongly depend on the computer/compiler that is used.

in which N is the number of particles and $\mathbf{v}(i,t)$ is the velocity of particle i at time t. One should choose t in such a way that independent time origins are taken, i.e., $t = ia\Delta t$, $i = 1, 2, \cdots, \infty$ and $\langle \mathbf{v}(t) \cdot \mathbf{v}(t + a\Delta t) \rangle \approx \mathbf{0}$.

(b) by calculating the mean-squared displacement:

$$D = \lim_{t' \to \infty} \frac{\left\langle \left| \mathbf{x}\left(t + t'\right) - \mathbf{x}(t) \right|^2 \right\rangle}{6t'}. \tag{4.6.3}$$

One should be very careful with calculation of the mean-squared displacement when periodic boundary conditions are used. Why?

Modify the program in such a way that the self-diffusivity can be calculated using both methods. Only modifications in subroutine *sample_diff.f* are needed. Why is it sufficient to use only independent time origins for the calculation of the means-squared displacement and the velocity autocorrelation function? What is the unit of D in SI units? How can one transform D into dimensionless units?

7. For Lennard-Jones liquids, Naghizadeh and Rice [86] report the following equation for self-diffusivity (dimensionless units, $T^* < 1.0$ and $p^* < 3.0$):

$$^{10}\log(D^*) = 0.05 + 0.07p^* - \frac{1.04 + 0.1p^*}{T^*}. \tag{4.6.4}$$

Try to verify this equation with simulations. How can one translate D^* to a diffusivity in SI units?

8. Instead of calculating the average energy $\langle U \rangle$ directly, one can use the radial distribution function $g(r)$. Derive an expression for $\langle U \rangle$ using $g(r)$. Compare this calculation with a direct calculation of the average energy. A similar method can be used to compute the average pressure.

9. In the current version of the code, the equations of motion are integrated by the Verlet algorithm. Make a plot of the energy drift ΔU for the following integration algorithms:

- Verlet
- Velocity Verlet
- Euler (never use this algorithm in real simulations).

Part II

Ensembles

Chapter 5

Monte Carlo Simulations in Various Ensembles

In a conventional Molecular Dynamics simulation, the total energy E and the total linear momentum $\mathbf{P}$ are constants of motion. Hence, Molecular Dynamics simulations measure (time) averages in an ensemble that is very similar to the microcanonical (see [87]); namely, the constant-NVE-$\mathbf{P}$ ensemble. In contrast, a conventional Monte Carlo simulation probes the canonical (i.e., constant-NVT) ensemble. The fact that these ensembles are different leads to observable differences in the statistical averages computed in Molecular Dynamics and Monte Carlo simulations. Most of these differences disappear in the thermodynamic limit and are already relatively small for systems of a few hundred particles. However, the choice of ensemble does make a difference when computing the mean-squared value of fluctuations in thermodynamic quantities. Fortunately, techniques exist to relate fluctuations in different ensembles [80]. Moreover, nowadays it is common practice to carry out Molecular Dynamics simulations in ensembles other than the microcanonical. In particular, it is possible to do Molecular Dynamics at constant pressure, at constant stress, and at constant temperature (see Chapter 6). The choice of ensembles for Monte Carlo simulations is even wider: isobaric-isothermal, constant-stress-isothermal, grand canonical (i.e., constant μVT), and even microcanonical [88–93]. A more recent addition to this list is a Monte Carlo method that employs the Gibbs ensemble technique [94], which was developed to study phase coexistence in moderately dense (multicomponent) fluids. The Gibbs ensemble method is discussed in detail in Chapter 8.

As explained in section 3.1 the principal idea of importance sampling is to use a Monte Carlo procedure to generate a random walk in those regions of phase space that have an important contribution to the ensemble averages. The acceptance rules are chosen such that these configurations occur with a frequency prescribed by the desired probability distribution. In section 3.1 it is shown that such a procedure indeed yields the correct distribution of configurations. Essential in the demonstration that our Monte Carlo scheme samples the desired distribution is the condition of detailed balance. To be more precise, detailed balance, in fact, is too strong a condition, but if detailed balance is obeyed we are *guaranteed* to have a correct sampling scheme. It may very well be possible that a scheme that does not obey detailed balance still samples the correct distribution. In a Monte Carlo scheme errors are easily introduced, so one should be extremely careful. We will give some examples where we can show that detailed balance is not obeyed and the results show systematic errors. We have found that we could demonstrate that detailed balance was not obeyed in all cases where we observed strange results.

5.1 General Approach

In the following sections, we will use the following procedure to demonstrate the validity of our Monte Carlo algorithms:

1. Decide which distribution we want to sample. This distribution, denoted $\mathcal{N}$; will depend on the details of the ensemble.
2. Impose the condition of detailed balance,

$$K(o \to n) = K(n \to o), \tag{5.1.1}$$

   where $K(o \to n)$ is the flow of configuration o to n. This flow is given by the product of the probability of being in configuration o, the probability of generating configuration n, and the probability of accepting this move,

$$K(o \to n) = \mathcal{N}(o) \times \alpha(o \to n) \times \text{acc}(o \to n). \tag{5.1.2}$$

3. Determine the probabilities of generating a particular configuration.
4. Derive the condition that needs to be fulfilled by the acceptance rules.

5.2 Canonical Ensemble

It is instructive to apply the preceding recipe to the ordinary Metropolis scheme. In the canonical ensemble, the number of particles, temperature, and volume are constant (see Figure 5.1). The partition function is

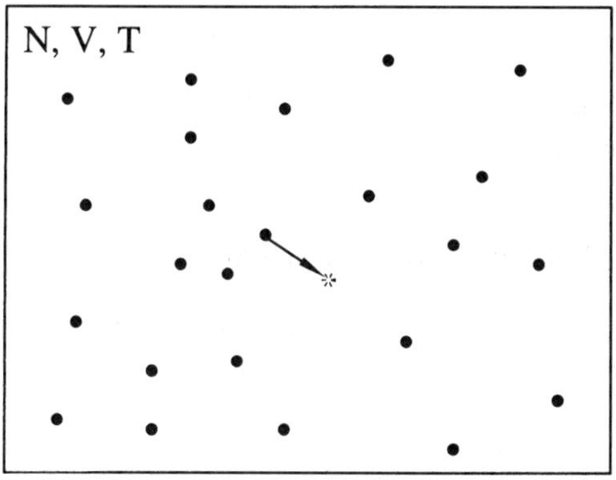

Figure 5.1: Canonical ensemble. The number of particles, volume, and temperature are constant. Shown is a Monte Carlo move in which a particle is displaced.

$$Q(N, V, T) \equiv \frac{1}{\Lambda^{3N} N!} \int d\mathbf{r}^N \exp[-\beta \mathcal{U}(\mathbf{r}^N)], \tag{5.2.1}$$

where $\Lambda = \sqrt{h^2/(2\pi m k_B T)}$ is the thermal de Broglie wavelength. From the partition function it follows that the probability of finding configuration $\mathbf{r}^N$ is given by

$$\mathcal{N}(\mathbf{r}^N) \propto \exp[-\beta \mathcal{U}(\mathbf{r}^N)]. \tag{5.2.2}$$

Equations (5.2.1) and (5.2.2) are the basic equations for a simulation in the canonical ensemble.

5.2.1 Monte Carlo Simulations

In the canonical ensemble, we have to sample distribution (5.2.2). This can be done using the following scheme:

1. Select a particle at random and calculate the energy of this configuration $\mathcal{U}(o)$.
2. Give this particle a random displacement (see Figure 5.1),

   $$\mathbf{r}(o) \rightarrow \mathbf{r}(o) + \Delta(\text{Ranf} - 0.5),$$

   where $\Delta/2$ is the maximum displacement. The value of Δ should be chosen such that the sampling scheme is optimal (see section 3.3). The new configuration is denoted n and its energy $\mathcal{U}(n)$.
3. The move is accepted with a probability (see equation (3.1.17))

   $$\text{acc}(o \rightarrow n) = \min(1, \exp\{-\beta[\mathcal{U}(n) - \mathcal{U}(o)]\}). \tag{5.2.3}$$

If rejected, the old configuration is kept.

An implementation of this basic Metropolis scheme is shown in Section 3.2 (Algorithms 1 and 2).

5.2.2 Justification of the Algorithm

The probability of generating a particular configuration is constant and independent of the conformation of the system

$$\alpha(o \rightarrow n) = \alpha(n \rightarrow o) = \alpha.$$

Substitution of this equation in the condition of detailed balance (5.1.1) and substitution of the desired distribution (5.2.2) gives as condition for the acceptance rules

$$\frac{\text{acc}(o \rightarrow n)}{\text{acc}(n \rightarrow o)} = \exp\{-\beta[\mathcal{U}(n) - \mathcal{U}(o)]\}. \tag{5.2.4}$$

It is straightforward to demonstrate that acceptance rule (5.2.3) obeys this condition.

5.3 Microcanonical Monte Carlo

Most experimental observations are performed at constant N,P,T; sometimes at constant μ, V, T; and occasionally at constant N,V,T. Experiments at constant N, V, E are very rare, to say the least. Under what circumstances, then, would anyone wish to perform Monte Carlo simulations at constant N, V, and E? We suppose that, if you are interested in the simulation of dense liquids or solids, the answer would be "hardly ever". Still there are situations where a microcanonical Monte Carlo method, first suggested by Creutz [93], may be of use. In particular, you might be worried that a poor random-number generator may introduce a bias in the sampling of the Boltzmann distribution or in the unlikely case that the exponentiation of the Boltzmann factor $\exp\{-\beta[U(n) - U(o)]\}$ may account for a nonnegligible fraction of the computing time.

The microcanonical Monte Carlo method uses no random numbers to determine the acceptance of a move. Rather, it uses the following procedure. We start with the system in a configuration $\mathbf{q}^N$. Denote the potential energy for this state by $U(\mathbf{q}^N)$. We now fix the total energy of the system at a value $E > U$. To this end, we introduce an additional degree of freedom that carries the remainder of the energy of the system: $E_D = E - U$. E_D must always be nonnegative. Now we start our Monte Carlo run.

1. After each trial move we compute the change in potential energy of the system,
$$\Delta \mathcal{U} = \mathcal{U}(\mathbf{q}'^N) - \mathcal{U}(\mathbf{q}^N).$$

2. If $\Delta \mathcal{U} < 0$, we accept the move and increase the energy carried by the demon by $|\Delta \mathcal{U}|$. If $\Delta \mathcal{U} > 0$, we test if the demon carries enough energy to make up the difference. Otherwise, we reject the trial move.

Note that no random numbers were used in this decision. Using elementary statistical mechanics it is easy to see that, after equilibration, the probability density to find the demon with an energy E_D is given by the Boltzmann distribution:
$$\mathcal{N}(E_D) = (k_B T)^{-1} \exp(-E_D/k_B T).$$

Hence, the demon acts as a thermometer. Note that this method does not really simulate the microcanonical ensemble. What is kept (almost) constant is the total potential energy. We can, however, mimic the real N, V, E ensemble by introducing a demon for every quadratic term in the kinetic energy. We then apply the same rules as before, randomly selecting a demon to pay or accept the potential energy change for every trial move.

Microcanonical Monte Carlo is rarely, if ever, used to simulate molecular systems.

5.4 Isobaric-Isothermal Ensemble

The isobaric-isothermal (constant-NPT) ensemble is widely used in Monte Carlo simulations. This is not surprising because most real experiments are also carried out under conditions of controlled pressure and temperature. Moreover, constant-NPT simulations can be used to measure the equation of state of a model system even if the virial expression for the pressure cannot be readily evaluated. This may be the case, for instance, for certain models of nonspherical hard-core molecules, but also for the increasingly important class of models where the (nonpairwise additive) potential energy function is computed numerically for each new configuration. Finally, it is often convenient to use constant-NPT Monte Carlo to simulate systems in the vicinity of a first-order phase transition, because at constant pressure the system is free (given enough time, of course) to transform completely into the state of lowest (Gibbs) free energy, whereas in a constant-NVT simulation the system may be kept at a density where it would like to phase separate into two bulk phases of different density but is prevented from doing so by finite-size effects.

Monte Carlo simulations at constant pressure were first described by Wood [88] in the context of a simulation study of two-dimensional hard disks. Although the method introduced by Wood is very elegant, it is not readily applicable to systems with arbitrary continuous potentials. McDonald [89] was the first to apply constant-NPT simulations to a system with continuous intermolecular forces (a Lennard-Jones mixture), and the constant-pressure method of McDonald is now being used almost universally and that is discussed next.

5.4.1 Statistical Mechanical Basis

We will derive the basic equations of constant-pressure Monte Carlo in a way that may appear unnecessarily complicated. However, this derivation has the advantage that the same framework can be used to introduce some of the other non-NVT Monte Carlo methods to be discussed later. For the sake of convenience we shall initially assume that we are dealing with a system of N identical atoms. The partition function for this system is given by

$$Q(N,V,T) = \frac{1}{\Lambda^{3N} N!} \int_0^L \cdots \int_0^L d\mathbf{r}^N \exp[-\beta \mathcal{U}(\mathbf{r}^N)]. \tag{5.4.1}$$

It is convenient to rewrite equation (5.4.1) in a slightly different way. We have assumed that the system is contained in a cubic box with diameter $L = V^{1/3}$. We now define scaled coordinates $\mathbf{s}^N$ by

$$\mathbf{r}_i = L\mathbf{s}_i \qquad \text{for} \quad i = 1, 2, \cdots, N. \tag{5.4.2}$$

If we now insert these scaled coordinates in equation (5.4.1), we obtain

$$Q(N,V,T) = \frac{V^N}{\Lambda^{3N} N!} \int_0^1 \cdots \int_0^1 d\mathbf{s}^N \exp[-\beta \mathcal{U}(\mathbf{s}^N; L)]. \tag{5.4.3}$$

In equation (5.4.3), we have written $\mathcal{U}(\mathbf{s}^N; L)$ to indicate that $\mathcal{U}$ depends on the real rather than the scaled distances between the particles. The expression for the Helmholtz free energy of the system is

$$\begin{aligned} F(N,V,T) &= -k_B T \ln Q \\ &= -k_B T \ln\left(\frac{V^N}{\Lambda^{3N} N!}\right) - k_B T \ln \int d\mathbf{s}^N \exp[-\beta \mathcal{U}(\mathbf{s}^N; L)] \\ &= F^{id}(N,V,T) + F^{ex}(N,V,T). \end{aligned} \tag{5.4.4}$$

In the last line of this equation we have identified the two contributions to the Helmholtz free energy on the previous line as the ideal gas expression plus an excess part. Let us now assume that the system is separated by a piston[1] from an ideal gas reservoir (see Figure 5.2). The total volume of the

[1] Actually, there is no need to assume a real piston. The systems with volume V and $V_0 - V$ may both be isolated systems subject to their individual (periodic) boundary conditions. The only constraint that we impose is that the sum of the volumes of the two systems equals V_0.

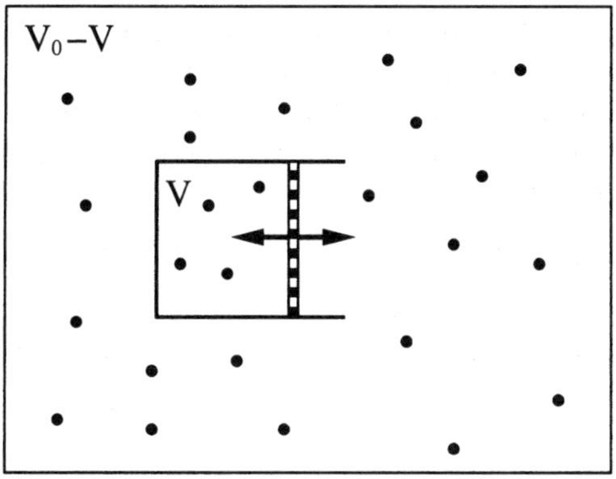

Figure 5.2: Ideal gas (m particles, volume $V_0 - V$) can exchange volume with an N-particle system (volume V).

system plus reservoir is fixed at a value V_0. The total number of particles is M. Hence, the volume accessible to the $M - N$ ideal gas molecules is $V_0 - V$. The partition function of the total system is simply the product of the partition functions of the constituent subsystems:

$$Q(N, M, V, V_0, T) = \frac{V^N (V_0 - V)^{M-N}}{\Lambda^{3M} N!(M-N)!} \int d\mathbf{s}^{M-N} \int d\mathbf{s}^N \exp[-\beta \mathcal{U}(\mathbf{s}^N; L)]. \tag{5.4.5}$$

Note that the integral over the $\mathbf{s}^{M-N}$ scaled coordinates of the ideal gas yields simply 1. For the sake of compactness, we have assumed that the thermal wavelength of the ideal gas molecules is also equal to Λ. The total free energy of this combined system is $F^{tot} = -k_B T \ln Q(N, M, V, V_0, T)$. Now let us assume that the piston between the two subsystems is free to move, so that the volume V of the N-particle subsystem can fluctuate. Of course, the most probable value of V will be the one that minimizes the free energy of the combined system. The probability density $\mathcal{N}(V)$ that the N-particle subsystem has a volume V is given by [2]

$$\mathcal{N}(V) = \frac{V^N (V_0 - V)^{M-N} \int d\mathbf{s}^N \exp[-\beta \mathcal{U}(\mathbf{s}^N; L)]}{\int_0^{V_0} dV' \, V'^N (V_0 - V')^{M-N} \int d\mathbf{s}^N \exp[-\beta \mathcal{U}(\mathbf{s}^N; L')]}. \tag{5.4.6}$$

We now consider the limit that the size of the reservoir tends to infinity

[2] Actually, this step is hard to justify. The reason is that there is no natural "metric" for the volume integration. Unlike the degeneracy of energy levels or the number of particles in a system, we cannot *count* volume. This problem has been addressed by several authors [95,96]. Attard [95] approaches the problem from an information-theory point of view and concludes that the integration variable should be ln V, rather than V. In contrast, Koper and Reiss [96] aim to reduce the problem to one of counting the number of quantum states compatible with a give volume. They end up with an expression that is almost identical to the one discussed here.

($V_0 \to \infty$, $M \to \infty$, $(M-N)/V_0 \to \rho$). In that limit, a small volume change of the small system does not change the pressure P of the large system. In other words, the large system works as a manostat for the small system. In that case, we can simplify equations (5.4.5) and (5.4.6). Note that in the limit $V/V_0 \to 0$, we can write

$$(V_0 - V)^{M-N} = V_0^{M-N}[1 - (V/V_0)]^{M-N} \to V_0^{M-N} \exp(-(M-N)V/V_0).$$

Note that for $M - N \to \infty$, $\exp(-(M-N)V/V_0) \to \exp(-\rho V)$. But, as the reservoir contains an ideal gas, ρ can be written as βP. With these substitutions, the combined partition function (5.4.5) can be written as

$$Q(N,P,T) \equiv \frac{\beta P}{\Lambda^{3N} N!} \int dV V^N \exp(-\beta PV) \int d\mathbf{s}^N \exp[-\beta \mathcal{U}(\mathbf{s}^N; L)], \quad (5.4.7)$$

where we have included a factor βP to make $Q(N,P,T)$ dimensionless (this choice is not obvious —see footnote 2). This gives, for equation (5.4.6),

$$\mathcal{N}_{N,P,T}(V) = \frac{V^N \exp(-\beta PV) \int d\mathbf{s}^N \exp[-\beta \mathcal{U}(\mathbf{s}^N; L)]}{\int_0^{V_0} dV' \, V'^N \exp(-\beta PV') \int d\mathbf{s}^N \exp[-\beta \mathcal{U}(\mathbf{s}^N; L')]}. \quad (5.4.8)$$

In the same limit, the difference in free energy between the combined system and the ideal gas system in the absence of the N-particle subsystem is the well-known Gibbs free energy G:

$$G(N,P,T) = -k_B T \ln Q(N,P,T). \quad (5.4.9)$$

Equation (5.4.8) is the starting point for constant-NPT Monte Carlo simulations. The idea is that the probability density to find the small system in a particular configuration of the N atoms (as specified by $\mathbf{s}^N$) at a given volume V is given by

$$\begin{aligned} \mathcal{N}(V; \mathbf{s}^N) &\propto V^N \exp(-\beta PV) \exp[-\beta \mathcal{U}(\mathbf{s}^N; L)] \\ &= \exp\{-\beta[\mathcal{U}(\mathbf{s}^N, V) + PV - N\beta^{-1} \ln V]\}. \quad (5.4.10) \end{aligned}$$

We can now carry out Metropolis sampling on the reduced coordinates $\mathbf{s}^N$ and the volume V.

In the constant-NPT Monte Carlo method, V is simply treated as an additional coordinate, and trial moves in V must satisfy the same rules as trial moves in $\mathbf{s}$ (in particular, we should maintain the symmetry of the underlying Markov chain). Let us assume that our trial moves consist of an attempted change of the volume from V to $V' = V + \Delta V$, where ΔV is a random number uniformly distributed over the interval $[-\Delta V_{max}, +\Delta V_{max}]$. In the Metropolis scheme such a random, volume-changing move will be accepted with the probability

$$\begin{aligned} \text{acc}(o \to n) &= \min\left(1, \exp\{-\beta[\mathcal{U}(\mathbf{s}^N, V') - \mathcal{U}(\mathbf{s}^N, V) \right. \\ &\quad \left. + P(V' - V) - N\beta^{-1} \ln(V'/V)]\}\right). \quad (5.4.11) \end{aligned}$$

Instead of attempting random changes in the volume itself, one might construct trial moves in the box length L [89] or in the logarithm of the volume [97]. Such trial moves are equally legitimate, as long as the symmetry of the underlying Markov chain is maintained. However, such alternative schemes result in a slightly different form for equation (5.4.11). The partition function (5.4.7) can be rewritten as

$$Q(N,P,T) = \frac{\beta P}{\Lambda^{3N} N!} \int d(\ln V) V^{N+1} \exp(-\beta P V) \int d\mathbf{s}^N \exp[-\beta \mathcal{U}(\mathbf{s}^N; L)]. \tag{5.4.12}$$

This equation shows that, if we perform a random walk in $\ln V$, the probability of finding volume V is given by

$$\mathcal{N}(V; \mathbf{s}^N) \propto V^{N+1} \exp(-\beta P V) \exp[-\beta \mathcal{U}(\mathbf{s}^N; L)]. \tag{5.4.13}$$

This distribution can be sampled with the following acceptance rule:

$$\begin{aligned} \text{acc}(o \to n) &= \min\left(1, \exp\{-\beta[\mathcal{U}(\mathbf{s}^N, V') - \mathcal{U}(\mathbf{s}^N, V) \right. \\ &\quad \left. + P(V' - V) - (N+1)\beta^{-1} \ln(V'/V)]\}\right). \end{aligned} \tag{5.4.14}$$

5.4.2 Monte Carlo Simulations

The frequency with which trial moves in the volume should be attempted is dependent on the efficiency with which volume space is sampled. If, as before, we use as our criterion of efficiency

$$\frac{\text{sum of squares of accepted volume changes}}{t_{CPU}},$$

then it is obvious that the frequency with which we attempt moves depends on their cost. In general, a volume trial move will require that we recompute all intermolecular interactions. It therefore is comparable in cost to carrying out N trial moves on the molecular positions. In such cases it is common practice to perform one volume trial move for every cycle of positional trial moves. Note that, to guarantee the symmetry of the underlying Markov chain, volume moves should not be attempted periodically after a fixed number (say N) positional trial moves. Rather, at every step there should be a probability $1/N$ to attempt a volume move instead of a particle move. The criteria determining the optimal acceptance of volume moves are no different than those for particle moves.

In one class of potential energy functions, volume trial moves are very cheap, namely, those for which the total interaction energy can be written as a sum of powers of the interatomic distances,

$$\begin{aligned} U_n &= \sum_{i<j} \epsilon(\sigma/r_{ij})^n \\ &= \sum_{i<j} \epsilon[\sigma/(Ls_{ij})]^n, \end{aligned} \quad (5.4.15)$$

or, possibly, a linear combination of such sums (the famous Lennard-Jones potential is an example of the latter category). Note that U_n in equation (5.4.15) changes in a trivial way if the volume is modified such that the linear dimensions of the system change for L to L':

$$U_n(L') = \left(\frac{L}{L'}\right)^n U_n(L). \quad (5.4.16)$$

Clearly, in this case, computing the probability of acceptance of a volume-changing trial move is extremely cheap. Hence such trial moves may be attempted with high frequency, for example, as frequent as particle moves. One should be very careful when using the scaling property (5.4.16) if at the same time one uses a cutoff (say r_c) on the range of the potential. Use of equation (5.4.16) implicitly assumes that the cutoff radius r_c scales with L, such that $r_c' = r_c(L'/L)$. The corresponding tail correction to the potential (and the virial) should also be recomputed to take into account both the different cutoff radius and the different density of the system.

Algorithms 2, 10, and 11 show the basic structure of a simulation in the NPT ensemble.

Finally, it is always useful to compute the virial pressure during a constant pressure simulation. On average, the virial pressure should always be equal to the applied pressure. This is easy to prove as follows. First of all, note that the virial pressure $P_v(V)$ of an N-particle system at volume V is equal to

$$P_v(V) = -\left(\frac{\partial F}{\partial V}\right)_{NT}. \quad (5.4.17)$$

In an isothermal-isobaric ensemble, the probability-density $\mathcal{P}(V)$ of finding the system with volume V is equal to $\exp[-\beta(F(V) + PV)]/Q(NPT)$, where

$$Q(NPT) \equiv \beta P \int dV \, \exp[-\beta(F(V) + PV)].$$

Algorithm 10 (Basic NPT-Ensemble Simulation)

```
   PROGRAM mc_npt                       basic NPT ensemble simulation

   do icycl=1,ncycl                     perform ncycl MC cycles
     ran=ranf()*(npart+1)+1
     if (ran.le.npart) then
       call mcmove                      perform particle displacement
     else
       call mcvol                       perform volume change
     endif
     if (mod(icycl,nsamp).eq.0)
  +    call sample                      sample averages
   enddo
   end
```

Comments to this algorithm:

1. *This algorithm ensures that, after each MC step, detailed balance is obeyed and that per cycle we perform (on average)* `npart` *attempts to displace particles and one attempt to change the volume.*
2. *Subroutine* `mcmove` *attempts to displace a randomly selected particle (Algorithm 2), and subroutine* `mcvol` *attempts to change the volume (Algorithm 11), and subroutine* `sample` *updates ensemble averages every* `nsamp`*th cycle.*

Let us now compute the average value of the virial pressure:

$$\begin{aligned}
\langle P_v \rangle &= -\frac{\beta P}{Q(NPT)} \int dV \, (\partial F(V)/\partial V) \exp[-\beta(F(V)+PV)] \\
&= \frac{\beta P}{Q(NPT)} \int dV \beta^{-1} \, (\partial \exp[-\beta F(V)]/\partial V) \exp(-\beta PV) \\
&= \frac{\beta P}{Q(NPT)} \int dV P \exp[-\beta(F(V)+PV)] \\
&= P.
\end{aligned} \tag{5.4.18}$$

The third line in this equation follows from partial integration.

Thus far we have limited our discussion of Monte Carlo at constant pressure to pure, atomic systems. Extension of the technique to mixtures is straightforward. The method is also easily applicable to molecular systems. However, in the latter case, it is crucial to note that only the center-of-mass

Algorithm 11 (Attempt to Change the Volume)

```
SUBROUTINE mcvol                          attempts to change
                                          the volume
call toterg(box,eno)                      total energy old conf.
vo=box**3                                 determine old volume
lnvn=log(vo)+(ranf()-0.5)*vmax            perform random walk in ln V
vn=exp(lnvn)
boxn=vn**(1/3)                            new box length
do i=1,npart
  x(i)=x(i)*boxn/box                      rescale center of mass
enddo
call toterg(boxn,enn)                     total energy new conf.
arg=-beta*((enn-eno)+p*(vn-vo)
+ -(npart+1)*log(vn/vo)/beta)             appropriate weight function!
if (ranf().gt.exp(arg)) then              acceptance rule (5.2.3)
  do i=1,npart                            REJECTED
    x(i)=x(i)*box/boxn                    restore the old positions
  enddo
endif
return
end
```

Comments to this algorithm:

1. *A random walk in* ln V *is performed using acceptance rule (5.4.14).*
2. *The subroutine* `toterg` *calculates the total energy. Usually the energy of the old configuration is known; therefore this subroutine is called only once.*

positions of the molecules should be scaled in a volume move, never the relative positions of the constituent atoms in the molecule. This has one practical consequence, namely, that the simple scaling relation (5.4.16) can never be used in molecular systems with site-site interactions. The reason is that, even if the center-of-mass separations between molecules scale as a simple power of the system size, the site-site separations do not.

5.4.3 Applications

Case Study 7 (Equation of State of the Lennard-Jones Fluid)
Simulations at constant pressure can be used also to determine the equation of state of a pure component. In such a simulation the density is determined

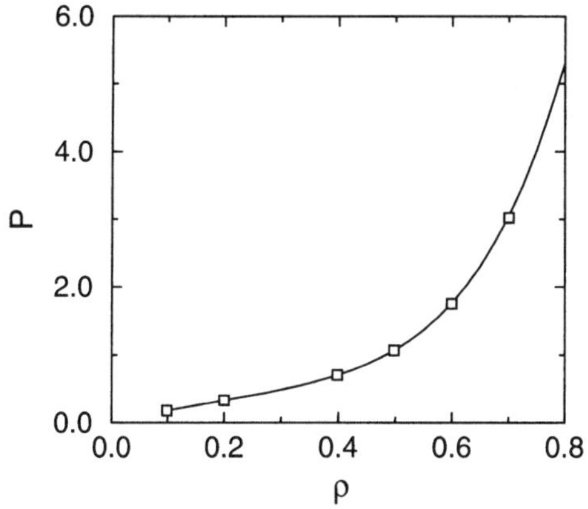

Figure 5.3: Equation of state of the Lennard-Jones fluid as obtained from N,P,T simulations; isotherms at $T = 2.0$. The solid line is the equation of state of Johnson *et al.* [62] and the squares are the results from the simulations ($N = 108$).

as a function of the applied pressure and temperature. Figure 5.3 shows that, for the Lennard-Jones fluid, the results of an N,P,T simulation compare very well with those obtained in Case Study 1. In simulations of models of real molecules one would like to know whether under atmospheric conditions the model fluid has the same density as the real fluid. One would need to perform several N,V,T simulations to determine the density at which the pressure is approximately 1 atm. In an N,P,T simulation one would obtain this result in a single simulation. Furthermore, 1 atm is a relatively low pressure, and one would need long simulations to determine the pressure from an N,V,T simulation, whereas the density in general is determined accurately from an N,P,T simulation.

Case Study 8 (Phase Equilibria from Constant-Pressure Simulations)
In Case Studies 1 and 7 N,V,T or N,P,T simulations are used to determine the equation of state of a pure component. If these equation-of-state data are fitted to an analytical equation of state (for example, the van der Waals equation of state or more sophisticated forms of this equation), the vapor-liquid coexistence curve can be determined from Maxwell's equal area construction. Although this way of determining a coexistence curve is guaranteed to work for all systems, it requires many simulations and, therefore, is a rather cumbersome route. Alternative routes have been developed to determine vapor-liquid coexistence from a single simulation. In this case study we investigate one of them: zero pressure simulation.

A zero pressure simulation provides a quick (and dirty) way to obtain an

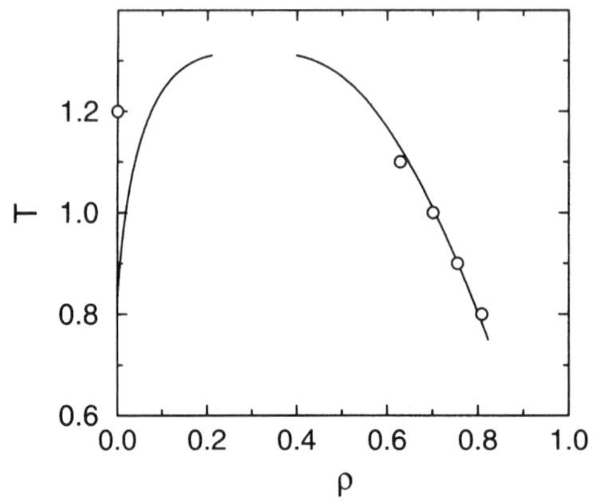

Figure 5.4: Vapor-liquid coexistence of the Lennard-Jones fluid; for each temperature the solid lines give at a given temperature the coexisting gas density (left curve) and the coexisting liquid density (right curve). The circles are the average densities obtained from N,P,T simulations at zero pressure. Important to note is that for $T = 1.2$ the zero pressure method fails to predict coexistence.

estimate of the liquid coexistence density. If we perform a simulation at zero pressure and start with a density greater than the liquid density, the average density obtained from a simulation that is not too long will be close to the coexistence density. Such a simulation should not be too long because the probability exists that the system will undergo a large fluctuation in density. If this fluctuation is towards a lower density the system size can become infinitely large, since the equilibrium density that corresponds to zero pressure is exactly zero. Figure 5.4 shows the results for the Lennard-Jones fluid. Not too close to the critical temperature, a reasonable estimate of the liquid density can be obtained via a single simulation. Important to note is that this estimate deviates systematically from the true coexistence densities and this technique should not be used to determine the coexistence curve. This technique is very useful for obtaining a first estimate of the coexistence curve.

As explained below equation (5.4.17), $\mathcal{P}(V)$, the probability density of finding a system with volume V is proportional to $\exp[-\beta(F(V)+PV)]$. This probability density can be obtained from a constant-pressure simulation by constructing a histogram of the number of times a certain volume V is observed during the simulation. Once we know $F(V)$ as a function of V, we can locate the coexistence points. In contrast to the zero pressure simulations used in Case Study 8, this histogram technique does lead to a correct estimate of the coexistence density. One of the important applications of this technique is the investigation of finite-size effects. In practice this scheme for deriving $F(V)$ from $\mathcal{P}(V)$ only works near the critical point [98–101] unless special sampling techniques are used (see section 7.4).

5.5 Isotension-Isothermal Ensemble

The NPT-MC method is perfectly adequate for homogeneous fluids. However, for inhomogeneous systems, in particular crystalline solids, it may not be sufficient that the simulation box can change size. Often we are interested in the transformation of a crystal from one structure to another or even in the change of the shape of the crystalline unit cell with temperature or with applied stress. In such cases it is essential that the shape of the simulation box has enough freedom to allow for such changes in crystal structure without creating grain boundaries or other highly stressed configurations. This problem was first tackled by Parrinello and Rahman [102, 103], who developed an extension of the constant-pressure Molecular Dynamics technique introduced by Andersen [104]. The extension of the Parrinello-Rahman method to Monte Carlo simulations is straightforward (actually, the method is quite a bit simpler in Monte Carlo than in Molecular Dynamics).

To our knowledge, the first published account of constant-stress Monte Carlo is a paper by Najafabadi and Yip [90]. At the core of the constant-stress Monte Carlo method lies the transformation from the scaled coordinates $\mathbf{s}$ to the real coordinates $\mathbf{q}$. If the simulation box is not cubic and not orthorhombic, the transformation between $\mathbf{s}$ and $\mathbf{r}$ is given by a matrix h: $r_\alpha = h_{\alpha\beta} s_\beta$. The volume of the simulation box V is equal to $|\det h|$. Without loss of generality we can choose h to be a symmetric matrix. In the constant-stress Monte Carlo procedure certain moves consist of an attempted change of one or more of the elements of h. Actually, it would be equally realistic (but not completely equivalent) to sample the elements of the metric tensor $G = h^T h$, where h^T is the transpose of h. If only hydrostatic external pressure is applied, the constant-stress Monte Carlo method is almost equivalent to constant-pressure Monte Carlo.[3] Under nonhydrostatic pressure (e.g., uniaxial stress), there is again some freedom of choice in deciding how to apply such deforming stresses. Probably the most elegant method (and the method that reflects most closely the statistical thermodynamics of deformed solids)

[3] Except that one should never use the constant-stress method for uniform fluids, because the latter offer no resistance to the deformation of the unit box and very strange (flat, elongated, etc.) box shapes may result. This may have serious consequences because simulations on systems that have shrunk considerably in any one dimension tend to exhibit appreciable finite-size effects.

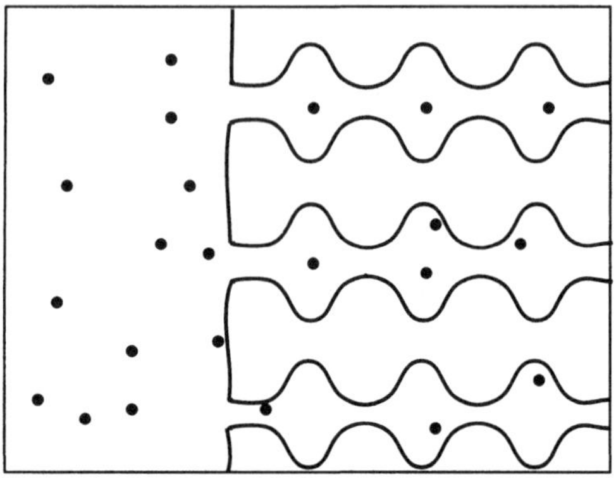

Figure 5.5: Adsorbent (for example, a zeolite) in direct contact with a gas.

is to express all external deforming stresses in terms of the so-called thermodynamic tension (see, e.g., [105]).

5.6 Grand-Canonical Ensemble

The ensembles we have discussed so far have the total number of particles imposed. For some systems, however, one would like to obtain information on the average number of particles in a system as a function of the external conditions. For example, in adsorption studies one would like to know the amount of material adsorbed as a function of the pressure and temperature of the reservoir with which the material is in contact. A naive but theoretically valid approach would be to use the Molecular Dynamics technique (microcanonical ensemble) and simulate the experimental situation; an adsorbent in contact with a gas (see Figure 5.5). Such a simulation is possible for only very simple systems. In real experiments, equilibration may take minutes or even several hours, depending on the type of gas molecules. These equilibration times would be reflected in a Molecular Dynamics simulation, the difference being that a minute of experimental time takes on the order of 10^9 seconds on a computer. Furthermore, in most cases, we are not interested in the properties of the gas phase, yet a significant amount of CPU time will be spent on the simulation of this phase, and finally, in such a simulation, there is an interface between the gas phase and the adsorbent. In the interfacial region the properties of the system are different from the bulk properties in which we are interested. Since in a simulation the system is relatively small, we have to simulate a very large system to minimize the

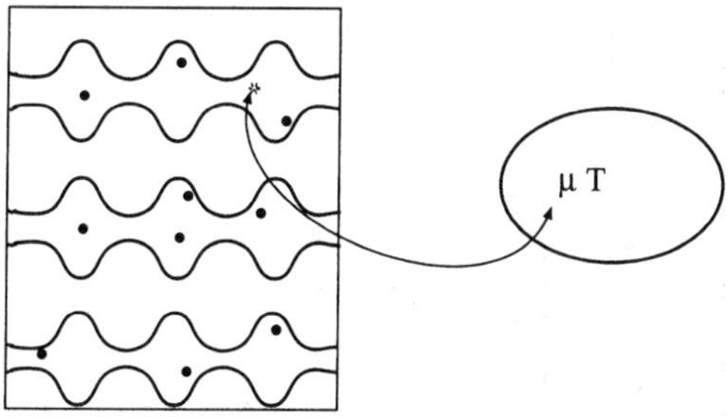

Figure 5.6: Adsorbent in contact with a reservoir that imposes constant chemical potential and temperature by exchanging particles and energy.

influence of this interfacial region.[4]

Most of these problems can be solved by a careful choice of ensembles. For adsorption studies, a natural ensemble to use is the grand-canonical ensemble (or μ,V,T ensemble). In this ensemble, the temperature, volume, and chemical potential are fixed. In the experimental setup, the adsorbed gas is in equilibrium with the gas in the reservoir. The equilibrium conditions are that the temperature and chemical potential of the gas inside and outside the adsorbent must be equal.[5] The gas that is in contact with the adsorbent can be considered as a reservoir that imposes a temperature and chemical potential on the adsorbed gas (see Figure 5.6). We therefore have to know only the temperature and chemical potential of this reservoir to determine the equilibrium concentration inside the adsorbent. This is exactly what is mimicked in the grand-canonical ensemble: the temperature and chemical potential are imposed and the number of particles is allowed to fluctuate during the simulation. This makes these simulations different from the conventional ensembles, where the number of molecules is fixed.

5.6.1 Statistical Mechanical Basis

In section 3.1.2, we introduced the Metropolis sampling scheme as a method for computing thermal averages of functions $A(\mathbf{r}^N)$ that depend explicitly on the coordinates of the molecules in the N-body system under study. Examples of such mechanical properties are the potential energy or the virial contribution to the pressure. However, the Metropolis method could not be used to determine the integral $\int d\mathbf{r}^N \exp[-\beta\mathcal{U}(\mathbf{r}^N)]$ itself. The latter quan-

[4]Such a simulation, of course, would be appropriate if the interest is in just this region.

[5]Note that the pressure is *not* defined inside the zeolite, therefore, the pressure cannot be an equilibrium quantity. However, the pressure is related to the chemical potential via an equation of state, and it is always possible to calculate the pressure of the gas that corresponds to a given chemical potential and vice versa.

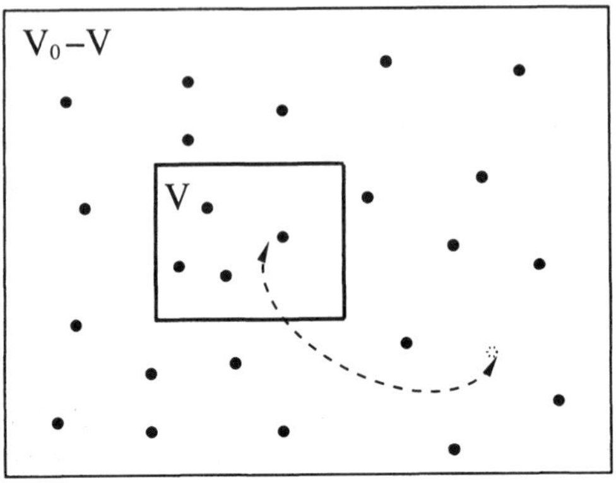

Figure 5.7: Ideal gas ($M-N$ particles, volume V_0-V) can exchange particles with a N-particle system (volume V).

tity measures the effective volume in configuration space that is accessible to the system. Hence, the original Metropolis scheme could not be used to determine those thermodynamic properties of a system that depend explicitly on the configurational integral. Examples of such thermal properties are the Helmholtz free energy F, the entropy S, and the Gibbs free energy G. However, although the Metropolis method cannot be used to measure, for instance, free energies directly, it can be used to measure the difference in free energy between two possible states of an N-body system. This fact is exploited in the grand-canonical Monte Carlo method first implemented for classical fluids by Norman and Filinov [91], and later extended and improved by a number of other groups [92, 106–113]. The basic idea of the grand-canonical Monte Carlo method is explained next.

To understand the statistical mechanical basis for the grand-canonical Monte Carlo technique, let us return to equation (5.4.5) of section 5.4. This equation gives the partition function of a combined system of N interacting particles in volume V and $M-N$ ideal gas molecules in volume $V_0 - V$:

$$Q(N, M, V, V_0, T) = \frac{V^N (V_0 - V)^{M-N}}{\Lambda^{3M} N!(M-N)!} \int d\mathbf{s}^{M-N} \int d\mathbf{s}^N \exp[-\beta \mathcal{U}(\mathbf{s}^N)].$$

Now, instead of allowing the two systems to exchange volume, let us see what happens if the systems can also exchange particles (see Figure 5.7). To be more precise, we assume that the molecules in the two subvolumes are actually identical particles. The only difference is that when they find themselves in volume V, they interact and, when they are in volume $V_0 - V$, they do not. If we transfer a molecule i from a reduced coordinate $\mathbf{s}_i$ in the volume $V_0 - V$ to the same reduced coordinate in volume V, then the potential energy function $\mathcal{U}$ changes from $\mathcal{U}(\mathbf{s}^N)$ to $\mathcal{U}(\mathbf{s}^{N+1})$. The expression for the total partition function of the system, including all possible distributions of the M particles over the two subvolumes is

$$Q(M, V, V_0, T) = \sum_{N=0}^{M} \frac{V^N (V_0 - V)^{M-N}}{\Lambda^{3M} N!(M-N)!} \int d\mathbf{s}^{M-N} \int d\mathbf{s}^N \exp[-\beta \mathcal{U}(\mathbf{s}^N)]. \tag{5.6.1}$$

Following the approach of section 5.4, we now write the probability density to find a system with $M - N$ particles at reduced coordinates $\mathbf{s}^{M-N}$ in volume $V' \equiv V_0 - V$ and N particles at reduced coordinates $\mathbf{s}^N$ in volume V:

$$\mathcal{N}(\mathbf{s}^M; N) = \frac{V^N V'^{M-N}}{Q(M, V, V', T)\Lambda^{3M} N!(M-N)!} \exp[-\beta \mathcal{U}(\mathbf{s}^N)]. \tag{5.6.2}$$

Let us now consider a trial move in which a particle is transferred from V' to the same scaled coordinate in V. First we should make sure that we construct an underlying Markov chain that is symmetric. Symmetry, in this case, implies that the *a priori* probability to move a particle from V' to V should be equal to the *a priori* probability of the reverse move. The probability of acceptance of a trial move in which we move a particle to or from volume V is determined by the ratio of the corresponding probability densities (5.6.2):

$$\alpha(N \to N+1) = \frac{V(M-N)}{V'(N+1)} \exp(-\beta[\mathcal{U}(\mathbf{s}^{N+1}) - \mathcal{U}(\mathbf{s}^N)]) \tag{5.6.3}$$

$$\alpha(N+1 \to N) = \frac{V'(N+1)}{V(M-N)} \exp(-\beta[\mathcal{U}(\mathbf{s}^{N}) - \mathcal{U}(\mathbf{s}^{N+1})]). \tag{5.6.4}$$

Now let us consider the limit that the ideal gas system is very much larger than the interacting system: $M \to \infty, V' \to \infty, (M/V') \to \rho$. Note that for an ideal gas the chemical potential μ is related to the particle density ρ by

$$\mu = k_B T \ln \Lambda^3 \rho.$$

Therefore, in the limit $(M/N) \to \infty$, the partition function (5.6.1) becomes

$$Q(\mu, V, T) \equiv \sum_{N=0}^{\infty} \frac{\exp(\beta\mu N) V^N}{\Lambda^{3N} N!} \int d\mathbf{s}^N \exp[-\beta \mathcal{U}(\mathbf{s}^N)], \tag{5.6.5}$$

and the corresponding probability density

$$\mathcal{N}_{\mu VT}(\mathbf{s}^N; N) \propto \frac{\exp(\beta\mu N) V^N}{\Lambda^{3N} N!} \exp[-\beta \mathcal{U}(\mathbf{s}^N)]. \tag{5.6.6}$$

Equations (5.6.5) and (5.6.6) are the basic equations for Monte Carlo simulations in the grand-canonical ensemble. Note that, in these equations, all explicit reference to the ideal gas system has disappeared.

5.6.2 Monte Carlo Simulations

In a grand-canonical simulation, we have to sample the distribution (5.6.6). Acceptable trial moves are

1. *Displacement of particles.* A particle is selected at random and given a new conformation (for example, in the case of atoms a random displacement). This move is accepted with a probability

$$\mathrm{acc}(s \to s') = \min\left(1, \exp\{-\beta[\mathcal{U}(\mathbf{s}'^N) - \mathcal{U}(\mathbf{s}^N)]\}\right). \tag{5.6.7}$$

2. *Insertion and removal of particles.* A particle is inserted at a random position or a randomly selected particle is removed. The creation of a particle is accepted with a probability

$$\mathrm{acc}(N \to N+1) = \min\left[1, \frac{V}{\Lambda^3(N+1)} \exp\{\beta[\mu - \mathcal{U}(N+1) + \mathcal{U}(N)]\}\right] \tag{5.6.8}$$

and the removal of a particle is accepted with a probability

$$\mathrm{acc}(N \to N-1) = \min\left[1, \frac{\Lambda^3 N}{V} \exp\{-\beta[\mu + \mathcal{U}(N-1) - \mathcal{U}(N)]\}\right]. \tag{5.6.9}$$

Appendix G demonstrates how the chemical potential of the reservoir can be related to the pressure of the reservoir. Algorithm 12 shows the basic structure of a simulation in the grand-canonical ensemble.

5.6.3 Justification of the Algorithm

It is instructive to demonstrate that the acceptance rules (5.6.7)–(5.6.9) indeed lead to a sampling of distribution (5.6.6). Consider a move in which we start with a configuration with N particles and move to a configuration with $N+1$ particles by inserting a particle in the system. Recall than we have to demonstrate that detailed balance is obeyed:

$$K(N \to N+1) = K(N+1 \to N),$$

with

$$K(N \to N+1) = \mathcal{N}(N) \times \alpha(N \to N+1) \times \mathrm{acc}(N \to N+1).$$

In Algorithm 12 at each Monte Carlo step the probability that an attempt is made to remove a particle is equal to the probability of attempting to add one:

$$\alpha_{\mathrm{gen}}(N \to N+1) = \alpha_{\mathrm{gen}}(N+1 \to N),$$

Algorithm 12 (Basic Grand-Canonical Ensemble Simulation)

```
PROGRAM mc_gc                          basic μVT ensemble
                                       simulation
do icycl=1,ncycl                       perform ncycl MC cycles
 ran=int(ranf()*(npav+nexc))+1
 if (ran.le.npart) then
   call mcmove                         displace a particle
 else
   call mcexc                          exchange a particle
 endif                                 with the reservoir
 if (mod(icycl,nsamp).eq.0)
+  call sample                         sample averages
enddo
end
```

Comments to this algorithm:

1. *This algorithm ensures that, after each MC step, detailed balance is obeyed. Per cycle we perform on average* `npav` *attempts*[6] *to displace particles and* `nexc` *attempts to exchange particles with the reservoir.*
2. *Subroutine* `mcmove` *attempts to displace a particle (Algorithm 2), subroutine* `mcexc` *attempts to exchange a particle with a reservoir (Algorithm 13), and subroutine* `sample` *samples quantities every* `nsamp` *cycle.*

where the subscript "gen" refers to the fact that α measures the probability to generate this trial move. Substitution of this equation together with equation (5.6.6) into the condition of detailed balance gives

$$\begin{aligned}\frac{\mathrm{acc}(N \to N+1)}{\mathrm{acc}(N+1 \to N)} &= \frac{\exp[\beta\mu(N+1)]V^{N+1}\exp[-\beta\mathcal{U}(\mathbf{s}^{N+1})]}{\Lambda^{3(N+1)}(N+1)!} \\ &\quad\times \frac{\Lambda^{3N}N!\exp[\beta\mathcal{U}(\mathbf{s}^N)]}{\exp(\beta\mu N)V^N} \\ &= \frac{\exp(\beta\mu)V}{\Lambda^3(N+1)}\exp\{-\beta[\mathcal{U}(\mathbf{s}^{N+1})-\mathcal{U}(\mathbf{s}^N)]\}.\end{aligned}$$

It is straightforward to show that acceptance rules (5.6.8) and (5.6.9) obey this condition.

[6] In the corresponding algorithm in the first edition of this book, we suggested that one could use the (fluctuating) actual number of particles `npart` rather than a preset number equal to the expected average number of particles `npav`. However, the resulting scheme violates detailed balance!

Algorithm 13 (Attempt to Exchange a Particle with a Reservoir)

```
  SUBROUTINE mcexc                   attempt to exchange a particle
                                     with a reservoir
  if (ranf().lt.0.5) then            decide to remove or add a particle
    if (npart.eq.0) return           test whether there is a particle
    o=int(npart*ranf())+1            select a particle to be removed
    call ener(x(o),eno)              energy particle o
    arg=npart*exp(beta*eno)          acceptance rule (5.6.9)
+        /(zz*vol)
    if (ranf().lt.arg) then
      x(o)=x(npart)                  accepted: remove particle o
      npart=npart-1
    endif
  else
    xn=ranf()*box                    new particle at a random position
    call ener(xn,enn)                energy new particle
    arg=zz*vol*exp(-beta*enn)        acceptance rule (5.6.8)
+        /(npart+1)
    if (ranf().lt.arg) then
      x(npart+1)=xn                  accepted: add new particle
      npart=npart+1
    endif
  endif
  return
  end
```

Comment to this algorithm:

1. *We have defined:* zz $= \exp(\beta\mu)/\Lambda^3$. *The subroutine* `ener` *calculates the energy of a particle at a given position.*

The most salient feature of the grand-canonical Monte Carlo technique is that in such simulations the chemical potential μ is imposed, while the number of particles N is a fluctuating quantity. During the simulation we may measure other thermodynamic quantities, such as the pressure P, the average density $\langle\rho\rangle$, or the internal energy $\langle\mathcal{U}\rangle$. As we know μ, we can derive all other thermal properties, such as the Helmholtz free energy or the entropy. This may seem surprising. After all, in section 3.1 we stated that Metropolis sampling cannot be used to sample absolute free energies and related quantities. Yet, with grand-canonical Monte Carlo we seem to be doing precisely that. The answer is that, in fact, we do not. What we measure is not an absolute but a relative free energy. In grand-canonical Monte Carlo, we are equating the chemical potential of a molecule in an ideal gas at density ρ (for the ideal gas case we know how to compute μ) and the chemical potential of the same species in an interacting system at density ρ'.

Grand-canonical Monte Carlo works best if the acceptance of trial moves by which particles are added or removed is not too low. For atomic fluids, this condition effectively limits the maximum density at which the method can be used to about twice the critical density. Special tricks are needed to extend the grand-canonical Monte Carlo method to somewhat higher densities [111]. Grand-canonical Monte Carlo is easily implemented for mixtures and inhomogeneous systems, such as fluids near interfaces. In fact, some of the most useful applications of the grand-canonical Monte Carlo method are precisely in these areas of research. Although the grand-canonical Monte Carlo technique can be applied to simple models of nonspherical molecules, special techniques are required since the method converges very poorly for all but the smallest polyatomic molecules. In section 13.6.1 some of these techniques are discussed.

5.6.4 Applications

Case Study 9 (Equation of State of the Lennard-Jones Fluid)
In Case Studies 1 and 7, we used N,V,T simulations and N,P,T simulations to determine the equation of state of the Lennard-Jones fluid. A third way to determine the equation of state is to impose the temperature and chemical

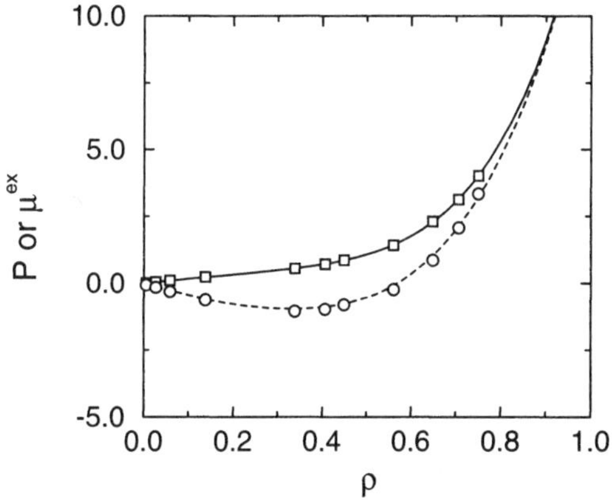

Figure 5.8: Equation of state of the Lennard-Jones fluid; isotherm at T = 2.0. The solid line is the equation of state of Johnson *et al.* [62] and the squares are the results from grand-canonical simulations (with volume V = 250.047). The dotted line is the excess chemical potential as calculated from the equation of state of Johnson *et al.* and the circles are the results of the simulations.

potential and calculate the density and pressure. An example of such a calculation is shown in Figure 5.8. This is not a very convenient method since both the pressure and density are fluctuating quantities; hence we will have an error in both quantities, while in the N,P,T ensemble either the pressure or density is imposed and therefore known without a statistical error. Of course, a grand-canonical simulation is useful if we want to have information on the chemical potential of our system.

Example 2 (Adsorption Isotherms of Zeolites)
Zeolites are crystalline inorganic polymers that form a three-dimensional network of micropores (see Figure 5.9). These pores are accessible to various guest molecules. The large internal surface, the thermal stability, and the presence of thousands of acid sites make zeolites an important class of catalytic materials for petrochemical applications. For a rational use of zeolites, it is essential to have a detailed knowledge of the behavior of the adsorbed molecules inside the pores of the zeolites. Since this type of information is very difficult to obtain experimentally, simulations are an attractive alternative. One of the first attempts to study the thermodynamic properties of a molecule adsorbed in a zeolite was made by Stroud *et al.* [114]. Reviews of the various applications of computer simulations of zeolites can be found in [115].

Besides zeolites many other porous materials exist with many interesting properties. In ref. [116] a review is given of phase separations in these materials.

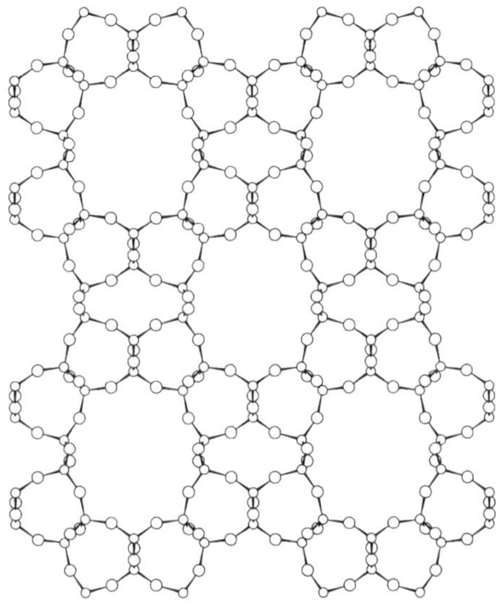

Figure 5.9: Example of a zeolite structure (Theta-1), the pore size is approx. 4.4×5.5 Å^2. The Si atoms have four bonds and the O atoms two.

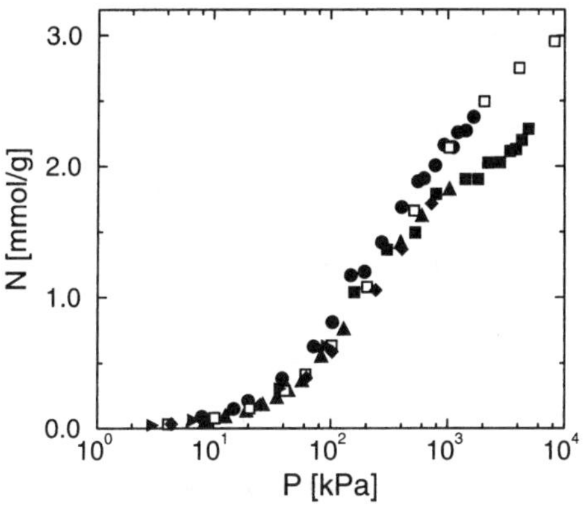

Figure 5.10: Adsorption isotherms of methane in silicalite, showing the amount of methane adsorbed as a function of the external pressure. The black symbols are experimental data (see [124] for details). The open squares are the results of grand-canonical simulations using the model of [119].

For small absorbents such as methane or the noble gases, grand-canonical Monte Carlo simulations can be applied to calculate the adsorption isotherms in the various zeolites [117–123]. An example of an adsorption isotherm of methane in the zeolite silicalite is shown in Figure 5.10. These calculations are based on the model of Goodbody *et al.* [119]. The agreement with the experimental data is very good, which shows that for these well-characterized systems simulations can give data that are comparable with experiments.

For long-chain alkanes (butane and longer) it is very difficult to perform a successful insertion; in almost all attempts one of the atoms of the molecule will overlap with one of the atoms of the zeolite. As a consequence the number of attempts has to be astronomically large to have a reasonable number of successful exchanges with the reservoir. In Chapter 13 we show how this problem can be solved.

5.7 Questions and Exercises

Question 12 (Trial Moves)

1. *If one uses a very large fraction of particle swap trial moves in the grand-canonical ensemble, one can increase the number of accepted insertions and deletions. However, is this an efficient approach? (Hint: assume that during*

these swap moves one has at least a successful deletion of a particle from the system, before the positions of the other particles are changed.)

2. *Which trial move, particle displacement, change of the volume, or particle insertion or deletion will be computationally most expensive? Why?*
3. *In a simulation of a molecule that consists of more than one interaction site, a trial move that rotates the molecule around its center of mass is usually included. Why? What is the acceptance/rejection rule for this trial move?*
4. *When a particle is added in the grand-canonical ensemble, the tail corrections to the potential energy may result in an energy change. Derive an expression for this energy change when a Lennard-Jones potential is used.*

Question 13 (Multicomponent Simulation) *We consider developing a grand-canonical Monte Carlo scheme for a mixture of two components. Assume the temperature is* T *and the chemical potential of the components* μ_1 *and* μ_2.

1. *To add or remove particles the following scheme is used:*
   - *Select at random to add or remove a particle.*
   - *Select at random a component.*
   - *Add or remove a particle of this component.*

   Derive the acceptance rules for these trial moves.
2. *An alternative scheme would be:*
   - *Select at random to add or remove a particle.*
   - *Select at random a particle, independent of its identity.*

   Is this scheme obeying detailed balance if the previous acceptance rules are used? If not can this be corrected ? Hint: you might want to see ref. [125].

Exercise 11 (Monte Carlo in the NPT **Ensemble)**
On the book's website you can find a program to simulate hard spheres (diameter 1) in the NPT ensemble using the Monte Carlo technique.

1. What problems would arise if you tried to calculate the virial for this system directly?
2. In the current code, a random walk is performed in $\ln(V)$ instead of V. Change the code in such a way that a random walk in V is performed. Check that the average densities calculated by both algorithms are equal.
3. Make a plot of the acceptance probability for volume displacements as a function of the maximum volume displacement for both algorithms.

Exercise 12 (Ising Model)

In this exercise we consider a 2D Ising model. In this model, N spins s (± 1) are arranged on a square lattice. Every spin (i) has 4 neighbors ($j = 1, 2, 3, 4$). The total energy of the system equals

$$U = -\frac{\epsilon}{2} \sum_{i=1}^{i=N} \sum_{j \in nn_i} s_i s_j,$$

in which $s_i = \pm 1$ and $\epsilon > 0$. The second summation is a summation over all nearest neighbors of spin i (nn_i). The total magnetization M equals the sum over all spins:

$$M = \sum_{i=1}^{i=N} s_i. \tag{5.7.1}$$

The 2D Ising model has a critical point close to $\beta_c \approx 0.44$.

1. Complete the given simulation code for this system (see *ising.f*).
2. Calculate the distribution of M for $N = 32 \times 32$ and $\beta = 0.5$ in the canonical ensemble. This distribution should be symmetrical:

   $$p(M) = p(-M). \tag{5.7.2}$$

   The simulation does not appear to yield such a symmetric distribution. Why not?
3. Instead of a simulation in the canonical ensemble, one can perform a biased simulation using a distribution function

   $$\pi \propto \exp[-\beta U + W(M)]. \tag{5.7.3}$$

   The average value of an observable O in the canonical ensemble is related to the "π-average", through

   $$\langle O \rangle = \frac{\langle O \exp[-W(M)] \rangle_\pi}{\langle \exp[-W(M)] \rangle_\pi},$$

   in which $\langle \cdots \rangle_\pi$ denotes an ensemble average in the biased system. Derive this relation.
4. Perform simulations with some given distributions $W(M)$ (for example, the files *w.type1.dat* and *w.type2.dat* on the book's website). Explain your results. How should one choose the function $W(M)$ to obtain the optimal efficiency?
5. What happens when $W(M)$ is a Gaussian,

   $$W(M) = A \exp\left[-\left(\frac{M}{\upsilon}\right)^2\right]$$

   with $A > 0$?
6. What happens when $W(M) = W(U) = \beta U$?

Chapter 6

Molecular Dynamics in Various Ensembles

The Molecular Dynamics technique discussed in Chapter 4 is a scheme for studying the natural time evolution of a classical system of N particles in volume V. In such simulations, the total energy E is a constant of motion. If we assume that time averages are equivalent to ensemble averages, then the (time) averages obtained in a conventional MD simulation are equivalent to ensemble averages in the microcanonical (constant-NVE) ensemble. However, as was discussed in Chapter 5, it is often more convenient to perform simulations in other ensembles (e.g., N,V,T or N,P,T). At first sight, it would seem that it is impossible to perform MD simulations in ensembles other than the microcanonical. Fortunately, it turns out that this is not the case. Two rather different solutions to this problem have been proposed. One is based on the idea that dynamical simulation of other ensembles is possible by mixing Newtonian MD with certain Monte Carlo moves. The second approach is completely dynamical in origin: it is based on a reformulation of the Lagrangian equations of motion of the system.

Both approaches occur time and again in many areas of MD simulation, and we will not attempt to list them all. In particular, the extended Lagrangian method, first introduced by Andersen in the context of constant-pressure MD simulations [104], has become one of the most important tricks to extend the applicability of MD simulations. To name but a few of the more conspicuous examples, the method is used in the Parrinello-Rahman scheme to simulate crystalline solids under conditions of constant stress [102, 103]. In this approach, both the volume and the shape of the crystal unit cell are allowed to fluctuate. As a consequence, the Parrinello-Rahman scheme is particularly useful for studying displacive phase transitions in solids.

In this chapter, we do not attempt to give a comprehensive, or even historical, presentation of non-NVE MD simulations. Rather, we have selected a single (but important) case that will be discussed in detail, namely, constant temperature simulations. This example allows us to illustrate the main features of the different approaches. The extension of this method to constant-pressure and -temperature simulations is discussed in less detail. With this background, the relevant literature on other applications of these Molecular Dynamics methods should be more accessible to the reader.

6.1 Molecular Dynamics at Constant Temperature

Before considering different schemes to perform Molecular Dynamics simulations at constant temperature, we should first specify what we mean by constant temperature. From a statistical mechanical point of view, there is no ambiguity: we can impose a temperature on a system by bringing it into thermal contact with a large heat bath (see section 2.1). Under those conditions, the probability of finding the system in a given energy state is given by the Boltzmann distribution and, for a classical system, the Maxwell-Boltzmann velocity distribution follows:

$$\mathcal{P}(p) = \left(\frac{\beta}{2\pi m}\right)^{3/2} \exp\left[-\beta p^2/(2m)\right]. \tag{6.1.1}$$

In particular, we then obtain the simple relation between the imposed temperature T and the (translational) kinetic energy per particle:

$$k_B T = m\left\langle v_\alpha^2 \right\rangle,$$

where m is the mass of the particle and v_α is the αth component of its velocity. As discussed in Chapter 4, this relation is often used to measure the temperature in a (microcanonical) MD simulation. However, the condition of constant temperature is not equivalent to the condition that the kinetic energy per particle is constant. To see this, consider the relative variance of the kinetic energy per particle in a canonical ensemble. If we constrain the kinetic energy to be always equal to its average, then the variance vanishes by construction. Now consider a system that is in thermal equilibrium with a bath. The relative variance in the kinetic energy of any given particle is simply related to the second and fourth moments of the Maxwell-Boltzmann distribution. For the second moment, $p^2 = \sum_\alpha p_\alpha^2$, we have

$$\left\langle p^2 \right\rangle = \int d\mathbf{p}\, p^2 \mathcal{P}(p) = \frac{3m}{\beta}$$

and for the fourth moment, $p^4 = \left(\sum_\alpha p_\alpha^2\right)^2$, we can write

$$\left\langle p^4 \right\rangle = \int d\mathbf{p}\, p^4 \mathcal{P}(p) = 15\left(\frac{m}{\beta}\right)^2.$$

The relative variance of the kinetic energy of that particle is

$$\frac{\sigma_{p^2}^2}{\langle p^2\rangle^2} \equiv \frac{\langle p^4\rangle - \langle p^2\rangle^2}{\langle p^2\rangle^2} = \frac{15(m/\beta)^2 - (3m/\beta)^2}{(3m/\beta)^2} = \frac{2}{3}.$$

If we would use the kinetic energy per particle as a measure of the instantaneous temperature, then we would find that, in a canonical ensemble, this temperature (denoted by T_k) fluctuates. Its relative variance is

$$\begin{aligned}
\frac{\sigma_{T_k}^2}{\langle T_k\rangle_{NVT}^2} &\equiv \frac{\langle T_k^2\rangle_{NVT} - \langle T_k\rangle_{NVT}^2}{\langle T_k\rangle_{NVT}^2} \\
&= \frac{N\langle p^4\rangle + N(N-1)\langle p^2\rangle\langle p^2\rangle - N^2\langle p^2\rangle^2}{N^2\langle p^2\rangle^2} \\
&= \frac{1}{N}\frac{\langle p^4\rangle - \langle p^2\rangle^2}{\langle p^2\rangle^2} = \frac{2}{3N}.
\end{aligned}$$

So indeed, in a canonical ensemble of a finite system, the instantaneous kinetic temperature T_k fluctuates. In fact, if we were to keep the average kinetic energy per particle rigorously constant, as is done in the so-called isokinetic MD scheme [29] or the more naive velocity-scaling schemes, then we would not simulate the true constant-temperature ensemble. In practice, the difference between isokinetic and canonical schemes is often negligible. But problems can be expected if isokinetic simulations are used to measure equilibrium averages that are sensitive to fluctuations. Moreover, one should distinguish between the isokinetic scheme of [29] and other, more or less ad hoc velocity-scaling methods. The isokinetic scheme of Evans and Morriss is well behaved in the sense that it yields the correct canonical ensemble averages for all properties that depend only on the positions of the particles [29, 126].

The ad hoc methods yield only the desired kinetic energy per particle but otherwise do not correspond to any known ensemble. Of course, any kind of temperature regulation, no matter how unphysical, can be used while preparing the system at a desired temperature (i.e., during equilibration). But, as efficient MD schemes exist that do generate a true canonical distribution, there is little need to use more suspect techniques to fix the temperature. Here we discuss two of the most widely used canonical MD schemes.

6.1.1 The Andersen Thermostat

In the constant-temperature method proposed by Andersen [104] the system is coupled to a heat bath that imposes the desired temperature. The coupling to a heat bath is represented by stochastic impulsive forces that act occasionally on randomly selected particles. These stochastic collisions with the heat bath can be considered as Monte Carlo moves that transport the system from one constant-energy shell to another. Between stochastic collisions, the system evolves at constant energy according to the normal Newtonian laws of motion. The stochastic collisions ensure that all accessible constant-energy shells are visited according to their Boltzmann weight.

Before starting such a constant-temperature simulation, we should first select the strength of the coupling to the heat bath. This coupling strength is determined by the frequency of stochastic collisions. Let us denote this frequency by ν. If successive collisions are uncorrelated, then the distribution of time intervals between two successive stochastic collisions, $P(t;\nu)$, is of the Poisson form [127,128]

$$P(t;\nu) = \nu \exp[-\nu t]. \tag{6.1.2}$$

where $P(t;\nu)dt$ is the probability that the next collision will take place in the interval $[t, t+dt]$.

A constant-temperature simulation now consists of the following steps:

1. Start with an initial set of positions and momenta $\{\mathbf{r}^N(0), \mathbf{p}^N(0)\}$ and integrate the equations of motion for a time Δt.
2. A number of particles are selected to undergo a collision with the heat bath. The probability that a particle is selected in a time step of length Δt is $\nu\Delta t$.
3. If particle i has been selected to undergo a collision, its new velocity will be drawn from a Maxwell-Boltzmann distribution corresponding to the desired temperature T. All other particles are unaffected by this collision.

The mixing of Newtonian dynamics with stochastic collisions turns the Molecular Dynamics simulation into a Markov process [47]. As shown in [104], a canonical distribution in phase space is invariant under repeated application of the Andersen algorithm. Combined with the fact that the Markov chain is also irreducible and aperiodic [104,127,128], this implies that the Andersen algorithm does, indeed, generate a canonical distribution. In Algorithms 14 and 15, we show how the Andersen method can be implemented in a Molecular Dynamics simulation.

Case Study 10 (Lennard-Jones: Andersen Thermostat)

In the present case study, we illustrate some of the strong and weak points of the Andersen thermostat. The first, and most important, thing to show is that this thermostat does produce a canonical distribution. Unfortunately, we can show this only indirectly: we can check whether the Andersen thermostat reproduces known properties of a canonical ensemble. In Figure 6.1 we compare the velocity distribution of a Lennard-Jones fluid as generated by the

Algorithm 14 (Molecular Dynamics: Andersen Thermostat)

```
program md_Andersen                    MD at constant temperature
call init(temp)                        initialization
call force(f,en)                       determine the forces
t=0
do while (t.lt.tmax)                   MD loop
   call integrate(1,f,en,temp)         first part of the eqs. of motion
   call force(f,en)                    determine the forces
   call integrate(2,f,en,temp)         second part of eqs. of motion
   t=t+dt
   call sample                         sample averages
enddo
stop
end
```

Comments to this algorithm:

1. *This part of the algorithm is very similar to the simple Molecular Dynamics program (Algorithm 3). The difference is that we use the velocity Verlet algorithm (see section 4.3) for the integration of the equations of motion:*

$$\begin{aligned} r(t+\Delta t) &= r(t) + v(t)\Delta t + \frac{f(t)}{2m}\Delta t^2 \\ v(t+\Delta t) &= v(t) + \frac{f(t+\Delta t)+f(t)}{2m}\Delta t. \end{aligned}$$

   This algorithm is implemented in two steps, in step 1, `call integrate(1,f,en,temp)`*, we know the forces and velocities at time* t*, and we update* $r(t)$ *and determine*

$$v' = v(t) + \frac{f(t)}{2m}\Delta t.$$

   Then, in `call force(f,en)` *we determine the forces at* $t+\Delta t$*; and finally we determine in step 2,* `call integrate(2,f,en,temp)`*, the velocities at time* $t+\Delta t$*,*

$$v(t+\Delta t) = v' + \frac{f(t+\Delta t)}{2m}\Delta t.$$

   The subroutine `integrate` *is described in Algorithm 15.*

2. *Subroutines* `init` *and* `force` *are described in Algorithms 4 and 5, respectively. Subroutine* `sample` *is used to calculate ensemble averages.*

Algorithm 15 (Equations of Motion: Andersen Thermostat)

```
      subroutine integrate(switch,f         integrate equations of motion:
     +                   ,en,temp)          with Andersen thermostat
      if (switch.eq.1) then                 first step velocity Verlet
         do i=1,npart
            x(i)=x(i)+dt*v(i)+              update positions current time
     +        dt*dt*f(i)/2
            v(i)=v(i)+dt*f(i)/2             first update velocity
         enddo
      else if (switch.eq.2) then            second step velocity Verlet
         tempa=0
         do i=1,npart
            v(i)=v(i)+dt*f(i)/2             second update velocity
            tempa=tempa+v(i)**2
         enddo
         tempa=tempa/(s*npart)              instantaneous temperature
         sigma=sqrt(temp)                   Andersen heat bath
         do i=1,npart
           if (ranf().lt.nu*dt) then        test for collision with bath
              v(i)=gauss(sigma)             give particle Gaussian velocity
           endif
         enddo
      endif
      return
      end
```

Comments to this algorithm:

1. *In this subroutine we use the velocity Verlet algorithm [69] (see notes to Algorithm 14).*
2. *The function* `gauss(sigma)` *returns a value taken from a Gaussian distribution with zero mean and standard deviation* `sigma` *(see Algorithm 44);* `ranf()` *is a uniform random number* $\in [0, 1]$.
3. *The collisions with the heat bath are Poisson distributed (6.1.2). The collision frequency* nu *is set at the beginning of the simulation.*
4. *In this algorithm, neither the total energy nor the total momentum is conserved.* s *depends on the mass of the particles* $s = 3/m$.

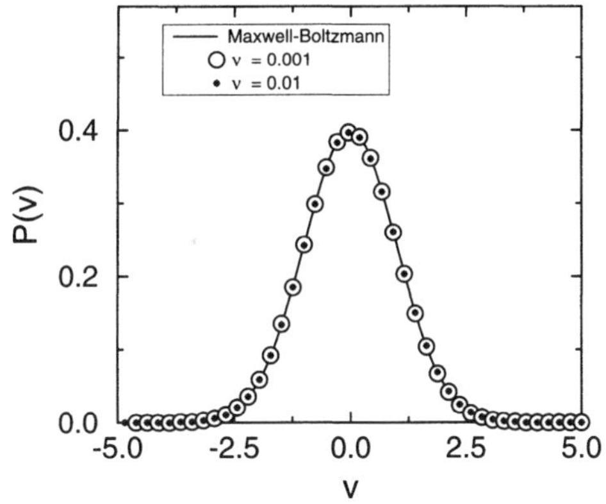

Figure 6.1: Velocity distribution in a Lennard-Jones fluid ($T = 2.0$, $\rho = 0.8442$, and $N = 108$). The solid line is the Maxwell-Boltzmann distribution (6.1.1), and the symbols are from a simulation using $\nu = 0.01$ and $\nu = 0.001$ as collision rates.

Andersen thermostat, with the exact Maxwell-Boltzmann distribution (6.1.1). The figure illustrates that the desired distribution is generated independent of the value of the collision frequency ν.

The results of constant N,V,T Molecular Dynamics simulations should be identical to those of canonical Monte Carlo simulations as presented in Figure 3.5. In making this comparison, we should be a bit careful because the Monte Carlo simulations were performed on a model with a truncated and shifted Lennard-Jones potential — the appropriate tail correction was added afterward. In our Molecular Dynamics program we simulate the Lennard-Jones model with a truncated and shifted potential. Again, the appropriate tail correction is added afterward (see section 3.2.2). For the Lennard-Jones fluid, the tail correction to the pressure is

$$P^{tail} = \frac{16}{3}\pi\rho^2\epsilon\sigma^6\left[\frac{2}{3}\left(\frac{\sigma}{r_c}\right)^9 - \left(\frac{\sigma}{r_c}\right)^3\right].$$

In Figure 6.2 the results of the Molecular Dynamics and Monte Carlo simulations are compared. In addition, we also compare them with the analytical equation-of-state data of [62]. Clearly, the canonical MD and MC simulations yield the same answer and agree with the equation-of-state data of [62]. This case study shows that the Andersen thermostat yields good results for time-independent properties, such as the equation of state. However, as the method is based on a stochastic scheme, one may wonder whether it can also be used to determine dynamic properties, such as the diffusion coeffi-

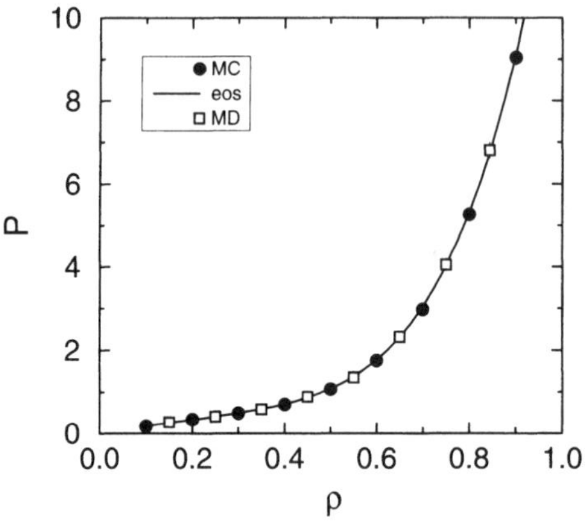

Figure 6.2: Equation of state of the Lennard-Jones fluid ($T = 2.0$ and $N = 108$); comparison of the Molecular Dynamics results using the Andersen thermostat (open symbols) with the results of Monte Carlo simulations (closed symbols) and the equation of state of Johnson *et al.* [62].

cient. In general, the answer to this question is no. The stochastic collisions disturb the dynamics in a way that is not realistic — it leads to sudden random decorrelation of particle velocities. This effect will result in an enhanced decay of the velocity autocorrelation function, and hence the diffusion constant (i.e., the time integral of the velocity autocorrelation function) is changed. Clearly this effect will be more pronounced as the collision frequency ν is increased. In fact, Tanaka *et al.* [129] have measured the diffusion coefficient of the Lennard-Jones fluid for various values of the collision frequency ν. They observed that the diffusion coefficient is independent of ν in a rather narrow frequency range. This effect is also illustrated in Figure 6.3. In practical cases, ν is usually chosen such that the decay rate of energy fluctuations in the simulation is comparable to that of energy fluctuations in a system of the same size embedded in an infinite heat bath. Typically, this can be achieved with relatively small collision rates and hence the effect of collisions on the dynamics may be small [104]. Nevertheless, one should always bear in mind that the dynamics generated by the Andersen thermostat is unphysical. It therefore is risky to use the Andersen method when studying dynamical properties. Figure 6.3 shows that the frequency of stochastic collisions has a strong effect on the time dependence of the mean-squared displacement. The mean-squared displacement becomes only independent of ν in the limit of very low stochastic collision rates. Yet, all static properties such as the pressure or potential energy are rigorously independent of the stochastic collision frequency.

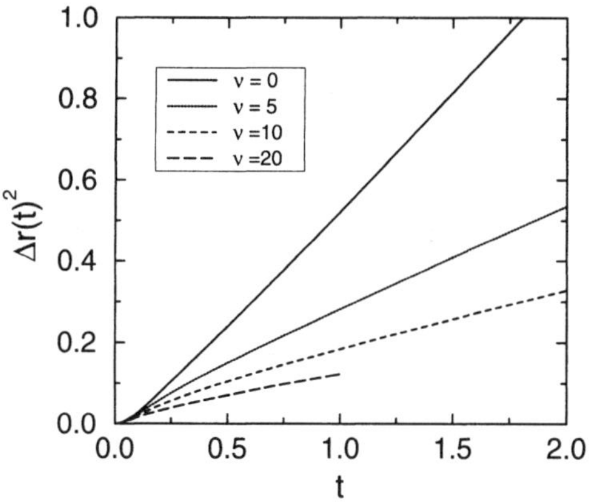

Figure 6.3: Mean-squared displacement as a function of time for various values of the collision frequency ν of the Lennard-Jones fluid ($T = 2.0$ and $N = 108$).

6.1.2 Nosé-Hoover Thermostat

In the Andersen approach to isothermal Molecular Dynamics simulation, constant temperature is achieved by stochastic collisions with a heat bath. Nosé has shown that one also can perform deterministic Molecular Dynamics at constant temperature [126, 130]. The approach of Nosé is based on the clever use of an extended Lagrangian; that is, a Lagrangian that contains additional, artificial coordinates and velocities. The extended-Lagrangian approach was introduced by Andersen [104] in the context of constant-pressure MD simulations. However, at present, extended Lagrangian methods are widely used not only for simulations in ensembles other than constant NVE, but also as a stable and efficient approach to perform simulations in which an expensive optimization has to be carried out at each time step. We discuss the Nosé thermostat as an illustration of an extended-Lagrangian method. However, for constant-temperature MD simulations, it is now more common to use the Nosé scheme in the formulation of Hoover [131, 132]. We therefore also discuss the so-called Nosé-Hoover thermostat.

In this section we assume that the reader is familiar with the Lagrangian and Hamiltonian formulation of classical mechanics. In Appendix A a review of these concepts is given.

Extended-Lagrangian Formulation

To construct isothermal Molecular Dynamics, Nosé introduced an additional coordinate s in the Lagrangian of a classical N-body system:

$$\mathcal{L}_{\text{Nose}} = \sum_{i=1}^{N} \frac{m_i}{2} s^2 \dot{\mathbf{r}}_i^2 - \mathcal{U}(\mathbf{r}^N) + \frac{Q}{2}\dot{s}^2 - \frac{L}{\beta}\ln s, \tag{6.1.3}$$

where L is a parameter that will be fixed later. Q is an effective "mass" associated to s. The momenta conjugate to $\mathbf{r}_i$ and s follow directly from equation (6.1.3):

$$\mathbf{p}_i \equiv \frac{\partial \mathcal{L}}{\partial \dot{\mathbf{r}}_i} = m_i s^2 \dot{\mathbf{r}}_i \tag{6.1.4}$$

$$p_s \equiv \frac{\partial \mathcal{L}}{\partial \dot{s}} = Q\dot{s}. \tag{6.1.5}$$

This gives for the Hamiltonian of the extended system of the N particles plus additional coordinate s:

$$\mathcal{H}_{\text{Nose}} = \sum_{i=1}^{N} \frac{\mathbf{p}_i^2}{2m_i s^2} + \mathcal{U}(\mathbf{r}^N) + \frac{p_s^2}{2Q} + L\frac{\ln s}{\beta}. \tag{6.1.6}$$

We consider a system containing N atoms. The extended system generates a microcanonical ensemble of 6N + 2 degrees of freedom. The partition function of this ensemble[1] is

$$\begin{aligned} Q_{\text{Nose}} &= \frac{1}{N!}\int dp_s ds d\mathbf{p}^N d\mathbf{r}^N \delta\left(E - \mathcal{H}_{\text{Nose}}\right) \\ &= \frac{1}{N!}\int dp_s ds d\mathbf{p}'^N d\mathbf{r}^N s^{3N} \\ &\quad \times \delta\left[\sum_{i=1}^{N}\frac{\mathbf{p}'^2_i}{2m_i} + \mathcal{U}(\mathbf{r}^N) + \frac{p_s^2}{2Q} + \frac{L}{\beta}\ln s - E\right], \end{aligned} \tag{6.1.7}$$

in which we have introduced

$$\mathbf{p}' = \mathbf{p}/s.$$

Let us define

$$\mathcal{H}(\mathbf{p}', \mathbf{r}) = \sum_{i=1}^{N}\frac{\mathbf{p}'^2_i}{2m_i} + \mathcal{U}(\mathbf{r}^N). \tag{6.1.8}$$

[1]We assume implicitly that conservation of energy is the only conservation law; in Appendix B.2 the more general case is considered.

For a δ function of a function $h(s)$, we can write

$$\delta[h(s)] = \delta(s - s_0)/|h'(s_0)|,$$

where $h(s)$ is a function that has a single root at s_0. If we substitute this expression into equation (6.1.7) and use equation (6.1.8), we find, for the partition function,

$$\begin{aligned}
Q_{\text{Nose}} &= \frac{1}{N!}\int dp_s d\mathbf{p}'^N d\mathbf{r}^N ds\ \frac{\beta s^{3N+1}}{L} \\
&\quad \times \delta\left\{s - \exp\left[-\beta\frac{\mathcal{H}(\mathbf{p}',\mathbf{r}) + p_s^2/(2Q) - E}{L}\right]\right\} \\
&= \frac{1}{N!}\frac{\beta\exp[E(3N+1)/L]}{L}\int dp_s\ \exp\left[-\beta\frac{3N+1}{L}p_s^2/(2Q)\right] \\
&\quad \times \int d\mathbf{p}'^N d\mathbf{r}^N\ \exp\left[-\beta\frac{3N+1}{L}\mathcal{H}(\mathbf{p}',\mathbf{r})\right] \\
&= C\frac{1}{N!}\int d\mathbf{p}'^N d\mathbf{r}^N\ \exp\left[-\beta\frac{3N+1}{L}\mathcal{H}(\mathbf{p}',\mathbf{r})\right].
\end{aligned} \tag{6.1.9}$$

If we perform a simulation in this extended ensemble, the average of a quantity that depends on $\mathbf{p}', \mathbf{r}$ is given by

$$\bar{A} = \lim_{\tau\to\infty}\frac{1}{\tau}\int_0^\tau dt\ A(\mathbf{p}(t)/s(t), \mathbf{r}(t)) \equiv \langle A(\mathbf{p}/s, \mathbf{r})\rangle_{\text{Nose}}. \tag{6.1.10}$$

With the choice $L = 3N + 1$, this ensemble average reduces to the canonical average:

$$\begin{aligned}
\langle A(\mathbf{p}/s, \mathbf{r})\rangle_{\text{Nose}} &\equiv \frac{\int d\mathbf{p}'^N d\mathbf{r}^N A(\mathbf{p}',\mathbf{r})\exp[-\beta\mathcal{H}(\mathbf{p}',\mathbf{r})(3N+1)/L]}{\int d\mathbf{p}'^N d\mathbf{r}^N\ \exp[-\beta\mathcal{H}(\mathbf{p}',\mathbf{r})(3N+1)/l]} \\
&= \frac{(1/N!)\int d\mathbf{p}'^N d\mathbf{r}^N A(\mathbf{p}',\mathbf{r})\exp[-\beta\mathcal{H}(\mathbf{p}',\mathbf{r})]}{Q(NVT)} \\
&= \langle A(\mathbf{p}',\mathbf{r})\rangle_{NVT}.
\end{aligned} \tag{6.1.11}$$

It is instructive to consider the role of the variable s in some detail. In the ensemble average in equation (6.1.11), the phase space is spanned by the coordinates $\mathbf{r}$ and the scaled momenta $\mathbf{p}'$. As the scaled momentum is related most directly to observable properties, we refer to $\mathbf{p}'$ as the real momentum, while $\mathbf{p}$ is interpreted as a virtual momentum. We make a similar distinction between real and virtual for the other variables. Real variables are indicated by a prime, to distinguish them from their unprimed virtual counterparts.

The real and virtual variables are related as follows:

$$r' = r \tag{6.1.12}$$
$$p' = p/s \tag{6.1.13}$$
$$s' = s \tag{6.1.14}$$
$$\Delta t' = \Delta t/s. \tag{6.1.15}$$

From equation (6.1.15) it follows that s can be interpreted as a scaling factor of the time step. This implies that the real time step fluctuates during a simulation. The sampling in equation (6.1.10) is done at integer multiples of the (virtual) time step Δt, which corresponds to real time steps that are not constant. It is also possible to sample at equal intervals in real time. In that case, we measure a slightly different average. Instead of equation (6.1.10) we define

$$\lim_{\tau'\to\infty} \frac{1}{\tau'} \int_0^{\tau'} dt'\, A\left[\mathbf{p}(t')/s(t'), \mathbf{r}(t')\right]. \tag{6.1.16}$$

Equation (6.1.15) shows that the real and virtual measuring times τ' and τ, respectively, are related through

$$\tau' = \int_0^{\tau} dt\; 1/s(t).$$

This gives, for equation (6.1.16),

$$\begin{aligned}
&\lim_{\tau'\to\infty} \frac{1}{\tau'} \int_0^{\tau'} dt'\, A\left[\mathbf{p}(t')/s(t'), \mathbf{r}(t')\right] \\
&= \lim_{\tau'\to\infty} \frac{\tau}{\tau'}\frac{1}{\tau} \int_0^{\tau} dt\, A\left[\mathbf{p}(t)/s(t), \mathbf{r}(t)\right]/s(t) \\
&= \frac{\lim_{\tau\to\infty} \frac{1}{\tau}\int^{\tau} dt\, A\left[\mathbf{p}(t)/s(t), \mathbf{r}(t)\right]/s(t)}{\lim_{\tau\to\infty} \frac{1}{\tau}\int_0^{\tau} dt\; 1/s(t)} \\
&= \langle A(\mathbf{p}/s, \mathbf{r})/s\rangle \,/\, \langle 1/s\rangle .
\end{aligned} \tag{6.1.17}$$

If we consider again the partition function (6.1.9), we can write for the ensemble average,

$$\begin{aligned}
\frac{\langle A(\mathbf{p}/s, \mathbf{r})/s\rangle}{\langle 1/s\rangle} &\equiv \frac{\left\{\frac{\int d\mathbf{p}'^N d\mathbf{r}^N\; A(\mathbf{p}',\mathbf{r})\exp[-\beta\mathcal{H}(\mathbf{p}',\mathbf{r})3N/L]}{\int d\mathbf{p}'^N d\mathbf{r}^N\; \exp[-\beta\mathcal{H}(\mathbf{p}',\mathbf{r})3(N+1)/L]}\right\}}{\left\{\frac{\int d\mathbf{p}'^N d\mathbf{r}^N\; \exp[-\beta[\mathcal{H}(\mathbf{p}',\mathbf{r})]3N/L]}{\int d\mathbf{p}'^N d\mathbf{r}^N\; \exp[-\beta[\mathcal{H}(\mathbf{p}',\mathbf{r})]3(N+1)/L]}\right\}} \\
&= \frac{\int d\mathbf{p}'^N d\mathbf{r}^N\; A(\mathbf{p}/s,\mathbf{r})\exp\left[-\beta\mathcal{H}(\mathbf{p}',\mathbf{r})3N/L\right]}{\int d\mathbf{p}'^N d\mathbf{r}^N\; \exp[-\beta[\mathcal{H}(\mathbf{p}',\mathbf{r})]3N/L]} \\
&= \langle A(\mathbf{p}/s,\mathbf{r})\rangle_{NVT}.
\end{aligned} \tag{6.1.18}$$

In the last step we have assumed that L = 3N. Therefore, if we use a sampling scheme based on equal time steps in real time, we have to use a different value for L.

From the Hamiltonian (6.1.6), we can derive the equations of motion for the virtual variables **p**, **r**, and t:

$$
\begin{aligned}
\frac{d\mathbf{r}_i}{dt} &= \frac{\partial \mathcal{H}_{\text{Nose}}}{\partial \mathbf{p}_i} = \mathbf{p}_i/(m_i s^2) \\
\frac{d\mathbf{p}_i}{dt} &= -\frac{\partial \mathcal{H}_{\text{Nose}}}{\partial \mathbf{r}_i} = -\frac{\partial \mathcal{U}(\mathbf{r}^N)}{\partial \mathbf{r}_i} \\
\frac{ds}{dt} &= \frac{\partial \mathcal{H}_{\text{Nose}}}{\partial p_s} = p_s/Q \\
\frac{dp_s}{dt} &= -\frac{\partial \mathcal{H}_{\text{Nose}}}{\partial s} = \left(\sum_i p_i^2/(m_i s^2) - \frac{L}{\beta}\right)/s.
\end{aligned}
$$

In terms of the real variables, these equations of motion can be written as

$$
\begin{aligned}
\frac{d\mathbf{r}'_i}{dt'} &= s\frac{d\mathbf{r}_i}{dt} = \mathbf{p}_i/(m_i s) = \mathbf{p}'_i/m_i && (6.1.19) \\
\frac{d\mathbf{p}'_i}{dt'} &= s\frac{d\mathbf{p}_i/s}{dt} = \frac{d\mathbf{p}_i}{dt} - \frac{1}{s}\mathbf{p}_i\frac{ds}{dt} && (6.1.20) \\
&= -\frac{\partial \mathcal{U}(\mathbf{r}'^N)}{\partial \mathbf{r}'_i} - (s'p'_s/Q)\mathbf{p}'_i \\
\frac{1}{s}\frac{ds'}{dt'} &= \frac{s}{s}\frac{ds}{dt} = s'p'_s/Q && (6.1.21) \\
\frac{d(s'p'_s/Q)}{dt'} &= \frac{s}{Q}\frac{dp_s}{dt} \\
&= \left(\sum_i p'^2_i/m_i - \frac{L}{\beta}\right)/Q. && (6.1.22)
\end{aligned}
$$

For these equations of motion, the following quantity is conserved:

$$
H'_{\text{Nose}} = \sum_{i=1}^{N} \frac{\mathbf{p}'^2_i}{2m_i} + \mathcal{U}(\mathbf{r}'^N) + \frac{s'^2 p'^2_s}{2Q} + L\frac{\ln s'}{\beta}. \qquad (6.1.23)
$$

It should be stressed, however, that this H'_{Nose} is not a Hamiltonian, since the equations of motion cannot be derived from it.

Implementation

In the previous section we showed how the introduction of an additional dynamical variable (s) in the Lagrangian can be used to perform MD simulations subject to a constraint (in this case, constant temperature). We stress once again that the importance of such extended Lagrangian techniques transcends the specific application. In addition, the problems encountered in the numerical implementation of the Nosé scheme are representative of a wider class of algorithms (namely, those where forces depend explicitly on velocities). It is for this reason that we discuss the numerical implementation of the Nosé thermostat in some detail (see also Appendix E.2).

The Nosé equations of motion can be written in terms of virtual variables or real variables. In a simulation it is not convenient to work with fluctuating time intervals. Therefore the real-variable formulation is recommended. Hoover [132] has shown that the equations derived by Nosé can be further simplified [133]. In equations (6.1.20), (6.1.21), and (6.1.22), the variables s', p'_s, and Q occur only as $s'p'_s/Q$. To simplify these equations, we can introduce the thermodynamic friction coefficient $\xi = s'p'_s/Q$. The equations of motion then become (dropping the primes and using dots to denote time derivatives)

$$\dot{\mathbf{r}}_i = \mathbf{p}_i/m_i \tag{6.1.24}$$

$$\dot{\mathbf{p}}_i = -\frac{\partial \mathcal{U}(\mathbf{r}^N)}{\partial \mathbf{r}_i} - \xi \mathbf{p}_i \tag{6.1.25}$$

$$\dot{\xi} = \left(\sum_i p_i^2/m_i - \frac{L}{\beta}\right)/Q \tag{6.1.26}$$

$$\dot{s}/s = \frac{d \ln s}{dt} = \xi. \tag{6.1.27}$$

Note that the last equation, in fact, is redundant, since equations (6.1.24)–(6.1.26) form a closed set. However, if we solve the equations of motion for s as well, we can use equation (6.1.23) as a diagnostic tool, since H' has to be conserved during the simulation. In terms of the variables used in equations (6.1.24)–(6.1.27), H reads

$$H_{\text{Nose}} = \sum_{i=1}^{N} \frac{\mathbf{p}_i^2}{2m_i} + \mathcal{U}(\mathbf{r}^N) + \frac{\xi^2 Q}{2} + L\frac{\ln s}{\beta}. \tag{6.1.28}$$

As we use the real-variable formulation in this set of equations, we have to take $L = 3N$.

An important implication of the Nosé equations is that in the Lagrangian (6.1.3) a logarithmic term (ln s) is needed to have the correct scaling of time. Any other scheme that does not have such a logarithmic term will fail to describe the canonical ensemble correctly.

An important result obtained by Hoover [132] is that the equations of motion (6.1.24)–(6.1.26) are unique, in the sense that other equations of the same form cannot lead to a canonical distribution. In Appendix E.2 we discuss an efficient way of implementing the Nosé-Hoover scheme.

The equations of motion of the Nosé-Hoover scheme cannot be derived from a Hamiltonian. This implies that one cannot use the standard methods (see Appendix A) to make the connection of the dynamics generated by solving these equations of motion with Statistical Mechanics. In Appendix B we discuss how one can analyze such non-Hamiltonian dynamics. The result of this analysis is that the conventional Nosé-Hoover algorithm only generates the correct distribution if there is a *single* constant of motion. Normally, the total energy defined by H_{Nose}, see equation (6.1.28), is always conserved. This implies that one should not have any other conserved quantity. In most conventional simulations this is the case if the momentum is not conserved, for example, if their is an external force; i.e., the sum of the forces $\sum_i \mathbf{F}_i \neq 0$. If we simulate a system without external forces, $\sum_i \mathbf{F}_i = 0$, which implies we have an additional conservation law, the Nosé-Hoover scheme is still correct provided that the center of mass remains fixed. This condition can be fulfilled easily if we ensure that during the equilibration the velocity of the center of mass is set to 0. If we simulate a system using no external field and in which the center of mass is not fixed or if we have more than one conservation law, we have to use Nosé-Hoover chains to obtain the correct canonical distribution. This method will be discussed in section 6.1.3.

Application

We illustrate some of the points discussed above in a Nosé-Hoover simulation of the Lennard-Jones fluid.

Case Study 11 (Lennard-Jones: Nosé-Hoover Thermostat)
As in Case Study 10, we start by showing that the Nosé-Hoover method reproduces the behavior of a system at constant NVT. In Figure 6.4 we compare the velocity distribution generated by the Nosé-Hoover thermostat with the correct Maxwell-Boltzmann distribution for the same temperature (6.1.1). The figure illustrates that the velocity distribution indeed is independent of the value chosen for the coupling constant Q.

It is instructive to see how the system reacts to a sudden increase in the imposed temperature. Figure 6.5 shows the evolution of the kinetic temperature of the system. After 12,000 time steps the imposed temperature is suddenly increased from $T = 1$ to $T = 1.5$. The figure illustrates the role of the coupling constant Q. A small value of Q corresponds to a low inertia of the heat bath and leads to rapid temperature fluctuations. A large value of Q leads to a slow, ringing response to the temperature jump.

Next, we consider the effect of the Nosé-Hoover coupling constant Q on the diffusion coefficient. As can be seen in Figure 6.6, the effect is much smaller than in Andersen's method. However, it would be wrong to conclude that the diffusion coefficient is independent of Q. The Nosé-Hoover method simply provides a way to keep the temperature constant more gentle than

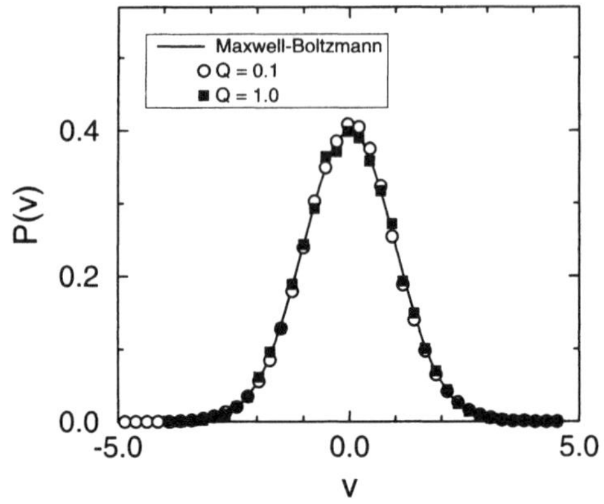

Figure 6.4: Velocity distribution in a Lennard-Jones fluid (T = 1.0, ρ = 0.75, and N = 256). The solid line is the Maxwell-Boltzmann distribution (6.1.1) the symbols were obtained in a simulation using the Nosé-Hoover thermostat.

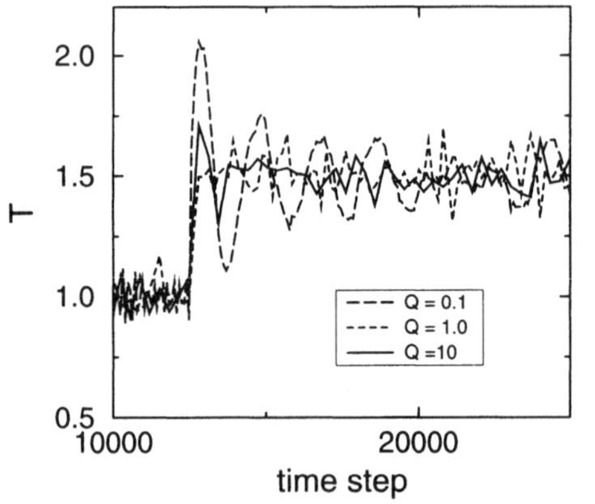

Figure 6.5: Response of the system to a sudden increase of the imposed temperature. The various lines show the actual temperature of the system (a Lennard-Jones fluid ρ = 0.75, and N = 256) as a function of the number of time steps for various values of the Nosé-Hoover coupling constant Q.

Andersen's method where particles suddenly get new, random velocities. For the calculations of transport properties, we prefer simple N,V,E simulations.

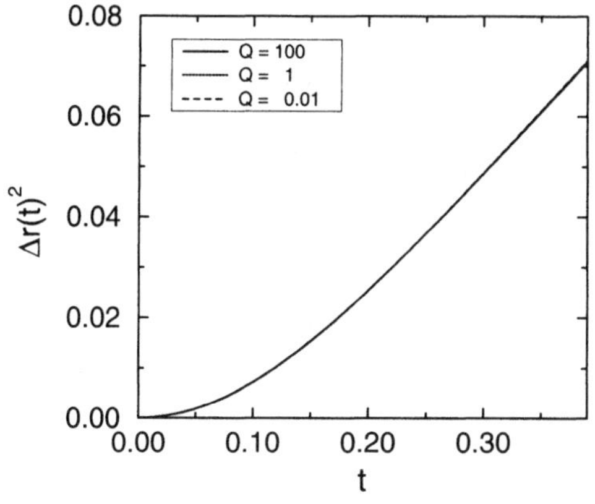

Figure 6.6: Effect of the coupling constant Q on the mean-squared displacement for a Lennard-Jones fluid ($T = 1.0$, $\rho = 0.75$, and $N = 256$).

6.1.3 Nosé-Hoover Chains

In the preceding examples, we have applied the Andersen and Nosé-Hoover thermostats to the Lennard-Jones fluid. For the Nosé-Hoover thermostat we have shown that for a system in which there are no external forces and the center of mass remains fixed, a canonical distribution will be generated. However, even though for systems with external forces the Nosé-Hoover thermostat generates the desired distribution, there can be exceptional cases in which we do not find the expected behavior. To illustrate this, we consider a particularly pathological case, namely, the one-dimensional harmonic oscillator.

Case Study 12 (Harmonic Oscillator (I))
As the equations of motion of the harmonic oscillator can be solved analytically, this model system is often used to test algorithms. However, the harmonic oscillator is also a rather atypical dynamical system. This will show up clearly when we apply our thermostating algorithms to this simple model system.

The potential energy function of the harmonic oscillator is

$$u(r) = \frac{1}{2}r^2.$$

The Newtonian equations of motion are

$$\begin{aligned} \dot{r} &= v \\ \dot{v} &= -r. \end{aligned}$$

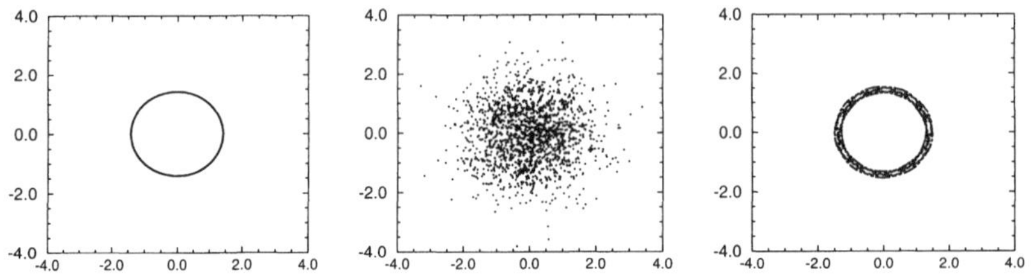

Figure 6.7: Trajectories of the harmonic oscillator: (from left to right) in the microcanonical ensemble, using the Andersen method, and using the Nosé-Hoover method. The y axis is the velocity and the x axis is the position.

If we solve the equations of motion of the harmonic oscillator for a given set of initial conditions, we can trace the trajectory of the system in phase space. Figure 6.7 shows a typical phase space trajectory of the harmonic oscillator, in a closed loop, which is characteristic of periodic motion. It is straightforward to simulate a harmonic oscillator at constant temperature using the Andersen thermostat (see section 6.1.1). A trajectory is shown in Figure 6.7. In this case the trajectories are points that are not connected by lines. This is due to the stochastic collisions with the bath. In this example, we allowed the oscillator to interact with the heat bath at each time step. As a result, the phase space density is a collection of discrete points. The resulting velocity distribution is Gaussian by construction; also for the positions we find a Gaussian distribution.

We can also perform a constant-temperature Nosé-Hoover simulation using the algorithm described in Appendix E.2. A typical trajectory of the harmonic oscillator generated with the Nosé-Hoover scheme is shown in Figure 6.7. The most striking feature of Figure 6.7 is that, unlike the Andersen scheme, the Nosé-Hoover method does not yield a canonical distribution in phase space. Even for very long simulations, the entire trajectory would lie in the same ribbon shown in Figure 6.7. Moreover, this band of trajectories depends on the initial configuration. This nonergodic behavior of the Nosé-Hoover algorithm was first discovered by Hoover [131]. Toxvaerd and Olson have shown that similar effects can also be observed in the simulation of a realistic model for butane [134]. The reason why we do not find a canonical distribution is that conservation of energy is not the only conservation law. Tuckerman *et al.* [135] have shown that an additional conservation law exists. In Appendix B.2.1 we show that in the presence of such an additional conservation law the algorithm does not generate the desired distribution.

In the previous section it is argued that the Nosé-Hoover algorithm only generates a correct canonical distribution for molecular systems in which there in only one conserved quantity or if there are no external forces and the center of mass remains fixed. The last condition can be obeyed in most practical systems by initializing the system with a zero center-of-mass velocity. However, if one is interested in simulating more general systems one cannot rely on the simple Nosé-Hoover algorithm. At this point it is important to note that the Andersen thermostat does not suffer from such problems, but its dynamics is less realistic.

To alleviate the restriction for the Nosé-Hoover thermostat, Martyna *et al.* [136] proposed a scheme in which the Nosé-Hoover thermostat is coupled to another thermostat or, if necessary, to a whole chain of thermostats. As we show in Appendix B.2.2 these chains take into account additional conservation laws. In [136] it is shown that this generalization of the original Nosé-Hoover method still generates a canonical distribution (provided that it is ergodic).

The equations of motion for a system of N particles coupled with M Nosé-Hoover chains are given (in real variables, hence $L = 3N$) by

$$\dot{\mathbf{r}}_i = \frac{\mathbf{p}_i}{m_i} \tag{6.1.29}$$

$$\dot{\mathbf{p}}_i = \mathbf{F}_i - \frac{p_{\xi_1}}{Q_1}\mathbf{p}_i \tag{6.1.30}$$

$$\dot{\xi}_k = \frac{p_{\xi_k}}{Q_k} \qquad k = 1, \ldots, M \tag{6.1.31}$$

$$\dot{p}_{\xi_1} = \left(\sum_i \frac{\mathbf{p}_i^2}{m_i} - Lk_BT\right) - \frac{p_{\xi_2}}{Q_2}p_{\xi_1} \tag{6.1.32}$$

$$\dot{p}_{\xi_k} = \left[\frac{p_{\xi_{k-1}}^2}{Q_{k-1}} - k_BT\right] - \frac{p_{\xi_{k+1}}}{Q_{k+1}}p_{\xi_k} \tag{6.1.33}$$

$$\dot{p}_{\xi_M} = \left[\frac{p_{\xi_{M-1}}^2}{Q_{M-1}} - k_BT\right]. \tag{6.1.34}$$

For these equations of motion the conserved energy is

$$H_{NHC} = \mathcal{H}(\mathbf{r}, \mathbf{p}) + \sum_{k=1}^{M} \frac{p_{\xi_k}^2}{2Q_k} + Lk_BT\xi_1 + \sum_{k=2}^{M} k_BT\xi_k. \tag{6.1.35}$$

We can use this conserved quantity to check the integration scheme. It is important to note that the additional M − 1 equations of motion form a simple one-dimensional chain and therefore are relatively simple to implement. In Appendix E.2, we describe an algorithm for a system with a Nosé-Hoover chain thermostat.

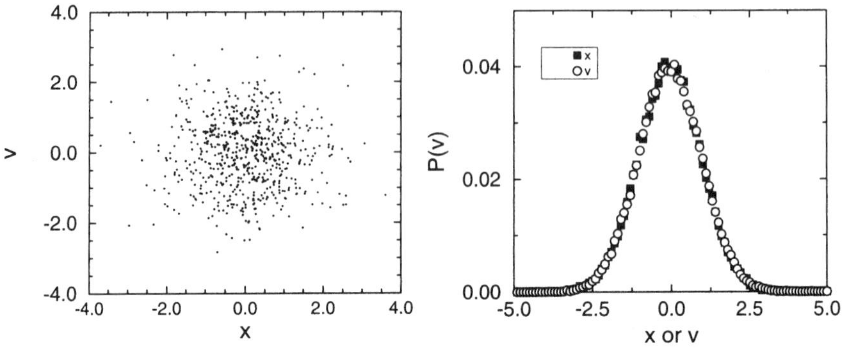

Figure 6.8: Test of the phase space trajectory of a harmonic oscillator, coupled to a Nosé-Hoover chain thermostat. The left-hand side of the figure shows part of a trajectory: the dots correspond to consecutive points separated by 10,000 time steps. The right-hand side shows the distributions of velocity and position. Due to our choice of units, both distributions should be Gaussians of equal width.

Case Study 13 (Harmonic Oscillator (II))
The harmonic oscillator is the obvious model system on which we test the Nosé-Hoover chain thermostat. If we use a chain of two coupling parameters, the equations of motion are

$$\begin{aligned}\dot{r} &= v\\ \dot{v} &= -r - \xi_1 v\\ \dot{\xi}_1 &= \frac{v^2 - T}{Q_1} - \xi_1 \xi_2\\ \dot{\xi}_2 &= \frac{Q_1 \xi_1^2 - T}{Q_2}.\end{aligned}$$

A typical trajectory generated with the Nosé-Hoover chains is shown in Figure 6.8. The distribution of the velocity and position of the oscillator are also shown in Figure 6.8. Comparison with the results obtained using the Andersen thermostat (see Case Study 12) shows that the Nosé-Hoover chains do generate a canonical distribution, even for the harmonic oscillator.

6.2 Molecular Dynamics at Constant Pressure

Most experiments are performed at constant pressure instead of constant volume. If one is interested in simulating the effect of, for example, the composition of the solvent on the properties of a system one has to adjust the volume of an N,V,T simulation to ensure that the pressure remains constant. For such a system it is therefore much more convenient to simulate at constant pressure. To simulate at constant pressure in a Molecular Dynamics simulation the volume is considered as a dynamical variable that changes during the simulation.

In Chapter 5 we have seen that one can perform Monte Carlo simulations at constant pressure by changing the volume of the simulation box. Here we consider the equivalent for a Molecular Dynamics simulation. Similar to the Monte Carlo case, this is an excellent method for homogeneous fluids. For inhomogeneous systems, however, one may need to change the shape of the simulation box as well [102, 103].

In Appendix B we have shown that the correct thermostating of a Molecular Dynamics simulation has many subtleties related to the conservation laws and whether a simulation is performed with a fixed center of mass. Similar problems arise with the isothermal-isobaric ensemble. The earlier scheme of Hoover [132] can only approximate the desired distribution [137]. Since the scheme of Martyna *et al.* does give the desired distribution, we focus on this scheme. All these schemes are based on the extended ensemble approach pioneered by Andersen [104].

The equations of motion proposed by Martyna *et al.* [138] for the positions and the momenta are

$$\dot{\mathbf{r}}_i = \frac{\mathbf{p}_i}{m_i} + \frac{p_\epsilon}{W}\mathbf{r}_i \tag{6.2.1}$$

$$\dot{\mathbf{p}}_i = \mathbf{F}_i - \left(1 + \frac{d}{dN}\right)\frac{p_\epsilon}{W}\mathbf{p}_i - \frac{p_{\xi_1}}{Q_1}\mathbf{p}_i, \tag{6.2.2}$$

where N is the number of particles. In these equations of motion we recognize a thermostat that is introduced via the variables ξ_1, p_{ξ_1}, and Q_1, similar to the N,V,T version of the Nosé-Hoover chain algorithm. A barostat is introduced via the variables ϵ, p_ϵ, and W. ϵ is defined as the logarithm of the volume V of the system

$$\epsilon = \ln(V/V(0)),$$

where $V(0)$ is the volume at $t = 0$, W is the mass parameter associated to ϵ, and p_ϵ is the momentum conjugate to ϵ.

The equations of motion (6.2.1) and (6.2.2) are complemented with an equation of motion for the volume, which reads in d dimensions

$$\dot{V} = \frac{dVp_\epsilon}{W} \tag{6.2.3}$$

$$\dot{p}_\epsilon = dV(P_{\text{int}} - P_{\text{ext}}) + \frac{1}{N}\sum_{i=1}^{N}\frac{\mathbf{p}_i^2}{m_i} - \frac{p_{\xi_1}}{Q_1}p_\epsilon. \tag{6.2.4}$$

In these equations P_{ext} is the external pressure, which is imposed (like the temperature). P_{int} is the internal pressure, which can be calculated during the simulation

$$P_{int} = \frac{1}{dV}\left[\sum_{i=1}^{N}\left(\frac{\mathbf{p}_i^2}{m_i} + \mathbf{r}_i \cdot \mathbf{F}_i\right) - dV\frac{\partial U(V)}{\partial V}\right],$$

where U is the potential. This equation differs from the conventional virial equation for a constant-volume simulation.

The equations of the chain of length M are[2]

$$\dot{\xi}_k = \frac{p_{\xi_k}}{Q_k} \quad \text{for } k = 1, \ldots, M \tag{6.2.5}$$

$$\dot{p}_{\xi_1} = \sum_{i=1}^{N}\frac{\mathbf{p}_i^2}{m_i} + \frac{p_\epsilon^2}{W} - (dN+1)k_B T - \frac{p_{\xi_2}}{Q_2}p_{\xi_1} \tag{6.2.6}$$

$$\dot{p}_{\xi_k} = \frac{p_{\xi_{k-1}}^2}{Q_{k-1}} - k_B T - \frac{p_{\xi_{k+1}}}{Q_{k+1}}p_{\xi_k} \quad \text{for } k = 2, \ldots, M-1 \tag{6.2.7}$$

$$\dot{p}_{\xi_M} = \frac{p_{\xi_{M-1}}^2}{Q_{M-1}} - k_B T. \tag{6.2.8}$$

The conserved quantity for these equations of motion is

$$\begin{aligned} H_{N,P_{ext},T} &= \mathcal{H}(\mathbf{p},\mathbf{r}) + \frac{p_\epsilon^2}{W} + \sum_{k=1}^{M}\frac{p_{\xi_k}^2}{Q_k} + (dN+1)k_B T\xi_1 \\ &\quad + k_B T\sum_{k=1}^{M}\xi_k + P_{ext}V. \end{aligned} \tag{6.2.9}$$

In Appendix B.3 we demonstrate that this method indeed generates the correct distribution. The implementation of this algorithm is described in Appendix E.2.2.

6.3 Questions and Exercises

Question 14 (Andersen Thermostat)

1. *Why is it that the static properties calculated by* NVT*-MD using the Andersen thermostat do not depend on* ν*?*

[2]Here we give the equations of motion in which the particles and the barostat are coupled to the same Nosé-Hoover chain. The more general case is to couple the particle and the barostat to two different thermostats. The advantage of having two different chains is that they can be optimized to the different time scales associated with temperature and volume fluctuations. Since in practice these times scales can be very different it is advised to use two different chains. The equations of motion can be found in Ref. [137].

2. *Why does the diffusivity decrease with increasing* ν*?*

Question 15 (Nosé-Hoover Thermostat)

1. *Explain when do we have to use* $g = 3N + 1$ *and* $g = 3N$ *in the Nosé-Hoover thermostat.*
2. *Instead of a single Nosé-Hoover thermostat, one can also use a chain of thermostats. Does this lead to a significant increase of the total CPU time for the simulation of a large system?*
3. *Another widely used thermostat is the "temperature coupling" of Berendsen et al. [139]. This method, however, does not generate the canonical ensemble exactly. In this algorithm, the temperature of the system is controlled by scaling the velocities to every time step with a factor* λ

$$\lambda = \left[1 + \frac{\Delta t}{\tau_T}\left(\frac{T_0}{T} - 1\right)\right]^{\frac{1}{2}}, \tag{6.3.1}$$

in which T_0 *is the desired temperature,* T *is the actual temperature,* Δt *is the time step of the integration algorithm and* τ_T *is a constant.*

- *Show that this scaling is equivalent to a temperature coupling of the system with a heat bath at* $T = T_0$

$$J = \alpha (T_0 - T), \tag{6.3.2}$$

in which J *is the heat flux and* α *is the heat transfer coefficient.*

- *What is the relation between* α *and* τ_T*?*

Exercise 13 (Barrier Crossing (Part 1))
Consider the movement of a single particle that moves on a 1D potential energy surface with the following functional form:

$$U(x) = \begin{cases} \epsilon B x^2 & x < 0 \\ \epsilon (1 - \cos(2\pi x)) & 0 \le x \le 1. \\ \epsilon B (x-1)^2 & x > 1 \end{cases}$$

The energy, force, and the derivative of the force are continuous functions of the position x and $\epsilon > 0$.

1. Derive an expression for B. Make a sketch of the potential energy landscape.
2. You can find a program on the book's website that integrates the equation of motion of the particle starting at $x(t = 0) = 0$ using several methods:

(a) No thermostat (NVE ensemble). What do you expect the phase space trajectories to look like?

(b) Andersen thermostat. In this method, the velocity of the particle is coupled to a stochastic heat bath, which leads to a canonical distribution.

(c) A Nosé-Hoover chain. In this method, the motion of the particle is coupled to a chain of thermostats. The equations of motion are integrated using an explicit time-reversible algorithm that might look a little bit complicated at first sight (see Appendix E.2.1), see *integrate_res.f*. One can prove that this method yields a canonical distribution provided that the system is ergodic.

(d) No molecular dynamics, but a simple Monte Carlo scheme.

The Andersen thermostat and the NVE integration algorithm are not implemented yet, so you will have to do this yourself (see *integrate_nve.f* and *integrate_and.f*). Try to use all methods for a low temperature, $T = 0.05$, for which the system behaves like a harmonic oscillator. Pay special attention to the following:

(a) Why does the distribution of the MC scheme look so much different at low temperatures?

(b) Why does the phase space distribution of the NVE scheme look like a circle (or ellipse)?

(c) Compare the phase space distributions of the Nosé-Hoover chain method with distribution generated by the Andersen thermostat. How long should the Nosé-Hoover chain be to generate a canonical distribution?

3. Investigate at which temperature the probability that the particle crosses the energy barrier becomes nonnegligible.

4. Another widely used algorithm is the "temperature coupling" of Berendsen *et al.* [139]. This method does not produce a true canonical ensemble. In this algorithm, the temperature of the system is controlled by scaling the velocities every time step with a factor λ

$$\lambda = \left[1 + \frac{\Delta t}{\tau_T}\left(\frac{T_0}{T} - 1\right)\right]^{\frac{1}{2}}, \tag{6.3.3}$$

in which T_0 is the desired temperature, T is the actual temperature, Δt is the time step of the integration algorithm, and τ_T is a constant. The temperature coupling algorithm can be used in combination with a

Leap-Frog Algorithm

$$v\left(t+\frac{\Delta t}{2}\right) = \lambda\left(t-\frac{\Delta t}{2}\right)\times\left(v\left(t-\frac{\Delta t}{2}\right)+\frac{F(t)}{m}\Delta t\right)$$
$$x(t+\Delta t) = x(t)+v\left(t+\frac{\Delta t}{2}\right)\Delta t. \quad (6.3.4)$$

Compare the distributions of the Berendsen *et al.* temperature bath with the canonical distributions.

5. Modify the program in such a way that the potential energy function

$$U=\epsilon\left[1-\cos\left(2\pi x\right)\right] \quad (6.3.5)$$

is used. Calculate the diffusion coefficient as a function of the temperature. Why is it impossible to calculate the diffusivity at low temperatures using ordinary molecular dynamics? Why is the diffusion coefficient obtained by using the Andersen thermostat a function of the collision frequency?

Part III

Free Energies and Phase Equilibria

Chapter 7

Free Energy Calculations

One of the aims of computer simulation is to predict the phase behavior of a given substance. The techniques that can be used to study phase diagrams numerically depend on the character of the phase transitions. In particular, there is quite a difference between the tools used to study first-order phase transitions and those applied to analyze critical phenomena near continuous phase transitions. In this chapter, which is devoted to free energy calculations, we discuss only first-order phase transitions. Continuous transitions are, from a computational point of view, no less challenging than first-order phase transitions (see e.g., [21,36–38,140]). However, the technical problems are not free energy related.

The most straightforward way to study phase behavior by computer simulation is to change the temperature or pressure of a given model system and then to simply wait for a phase transformation to occur. For instance, we can prepare a system in a stable crystalline phase, and then heat it until it melts. This approach can be used (and has been, in some cases [82,83,141–145]). But in general it suffers from a serious drawback: first-order phase transitions usually exhibit appreciable hysteresis. Hence, the transformation to the new stable phase, if it happens at all, will proceed irreversibly well beyond the coexistence point. The reason why hysteresis is common in first-order phase transitions is that a large free energy barrier separates the two phases at, or near, coexistence. The height of this barrier is determined by the interfacial free energy of the interface separating the two coexisting phases. The larger the area of this interface, the higher is the free energy barrier. Direct simulations of first-order phase transitions therefore either start with a system that has been prepared such that the interface is already present or by eliminating the interface altogether.

There exist several schemes to study phase coexistence without creating an interface. In fluids, the best-known method is the Gibbs ensemble method of Panagiotopoulos [94,146–148], which is discussed in Chapter 8. For what follows, it is important to point out the main limitation of the Gibbs ensemble method: it breaks down when at least one of the two phases becomes so dense that it becomes effectively impossible to exchange particles. This happens for instance, when one of the coexisting phases is a dense liquid or, *a fortiori*, a solid.

A direct method for studying phase coexistence that does not suffer from this drawback is the so-called Gibbs-Duhem integration method of Kofke [149–151] (see section 9.2). This method is designed to trace out a two-phase coexistence curve, once one point on that curve is known. As a consequence, the Kofke method can be used only in combination with another method that generates the initial point on the coexistence curve.

In the early 1980s, Parrinello and Rahman [102,103] designed a powerful Molecular Dynamics scheme specifically for studying solid-solid transitions (see section 5.5). This technique can be applied to those transitions that cause the crystal unit cell to deform without much other rearrangement of the molecules within the unit cell. Even if these conditions are met, the Parrinello-Rahman method suffers from some hysteresis. More important, when the two solids have very different unit cells then the Parrinello-Rahman method cannot be used.

From this discussion it is clear that there is a great need for robust methods to compute free energies. In what follows, we shall briefly review several such techniques. A detailed comparison of the various methods and their optimization can be found in Ref. [152].

7.1 Thermodynamic Integration

Let us briefly recall why free energies are important when we are interested in the relative stability of several phases. The second law of thermodynamics states that for a closed system with energy E, volume V, and number of particles N, the entropy S is at a maximum when the system is in equilibrium. From this formulation of the second law it is simple to derive the corresponding equilibrium conditions for systems that can exchange heat, particles, or volume with a reservoir. In particular, if a system is in contact with a heat bath, such that its temperature T, volume V, and number of particles N are fixed, then the Helmholtz free energy $F \equiv E - TS$ is at a minimum in equilibrium. Analogously, for a system of N particles at constant pressure P and temperature T, the Gibbs free energy $G \equiv F + PV$ is at a minimum.

If we wish to know which of two phases (denoted by α and β) is stable at a given temperature and density, we should simply compare the Helmholtz free energies F_α and F_β of these phases. It would seem that the obvious thing to do is simply to measure F_α and F_β by computer simulation. Unfortunately, it is not possible to measure the free energy (or entropy) directly in a simulation (see section 3.1.2). Entropy, free energy, and related quantities are not simply averages of functions of the phase space coordinates of the system. Rather they are directly related to the volume in phase space that is accessible to a system. For instance, in classical statistical mechanics, the Helmholtz free energy F is directly related to the canonical partition function $Q(N, V, T)$, (see equation (2.1.15)) :

$$F = -k_B T \ln Q(N, V, T) \equiv -k_B T \ln \left(\frac{\int d\mathbf{p}^N d\mathbf{r}^N \exp\left[-\beta \mathcal{H}(\mathbf{p}^N, \mathbf{r}^N)\right]}{\Lambda^{dN} N!} \right), \tag{7.1.1}$$

where d is the dimensionality of the system. It is clear that $Q(N, V, T)$ is not of the form of a canonical average over phase space. And this is why F or, for that matter, S or G cannot be measured directly in a simulation. We call such quantities that depend directly on the available volume in phase thermal quantities.

Nothing is strange about the fact that thermal quantities cannot be measured directly in a simulation: the same problem occurs in the real world —thermal quantities cannot be measured directly in real experiments either. When considering numerical schemes to determine the free energy, therefore, it is instructive to see how this problem is solved in the real world. Experiments always determine a derivative of the free energy, such as the derivative with respect to volume V or temperature T:

$$\left(\frac{\partial F}{\partial V}\right)_{NT} = -P \tag{7.1.2}$$

and

$$\left(\frac{\partial F/T}{\partial 1/T}\right)_{VN} = E. \tag{7.1.3}$$

As the pressure P and the energy E are mechanical quantities, they can be measured in a simulation. To compute the free energy of a system at given temperature and density, we should find a reversible path in the V-T plane that links the state under consideration to a state of known free energy. The change in F along that path can then simply be evaluated by thermodynamic integration, that is, integration of equations (7.1.2) and (7.1.3). There are only very few thermodynamic states for which the free energy of a substance is known. One state is the ideal gas phase; the other is the low-temperature harmonic crystal. A well-known example of the thermodynamic integration method is the calculation of the free energy of a liquid through integration of the equation of state. In this case the reference state is the ideal gas.

In computer simulations, the situation is quite similar. To compute the free energy of a dense liquid, one may construct a reversible path to the

Reversible path to a known state.
(in V-T plane) To determine free energy.

very dilute gas phase. It is not really necessary to go all the way to the ideal gas. But at least one should reach a state that is sufficiently dilute to ensure that the free energy can be computed accurately, from knowledge either of the first few terms in the virial expansion of the compressibility factor $PV/(Nk_BT)$ or that the chemical potential can be computed by other means (see, for example, sections 7.2.1 and 7.2.3). For the solid, the ideal gas reference state is less useful and another approach is called for. This problem is discussed in detail in Chapter 10.

In a simulation we are not limited to using a physical thermodynamic integration path, that is, a path that can also be followed in experiments. Rather, we can use all parameters in the potential energy function as thermodynamic variables. For example, if we know the free energy of the Lennard-Jones fluid, we can determine the free energy of the Stockmayer fluid[1] by calculating the reversible work required to switch on the dipolar interactions in the Lennard-Jones fluid [153]. The formalism used to calculate this free energy difference is Kirkwood's coupling parameter method [154]. Let us consider an N-particle system with a potential energy function $\mathcal{U}$. We assume that $\mathcal{U}$ depends linearly on a coupling parameter λ such that, for $\lambda = 0$, $\mathcal{U}$ corresponds to the potential energy of our reference system (denoted by I), while for $\lambda = 1$, we recover the potential energy of the system of interest (denoted by II):

$$\begin{aligned} \mathcal{U}(\lambda) &= (1-\lambda)\mathcal{U}_{\mathrm{I}} + \lambda\mathcal{U}_{\mathrm{II}} \\ &= \mathcal{U}_{\mathrm{I}} + \lambda(\mathcal{U}_{\mathrm{II}} - \mathcal{U}_{\mathrm{I}}). \end{aligned} \tag{7.1.4}$$

In our example, system I corresponds to the Lennard-Jones fluid, while system II refers to the Stockmayer fluid. In what follows, we assume that the free energy of system I is known (either analytically or numerically). The partition function for a system with a potential energy function that corresponds to a value of λ between 0 and 1 is

$$Q(N, V, T, \lambda) = \frac{1}{\Lambda^{3N} N!} \int d\mathbf{r}^N \exp[-\beta\mathcal{U}(\lambda)].$$

The derivative of the Helmholtz free energy $F(\lambda)$ with respect to λ can be written as an ensemble average:

$$\begin{aligned} \left(\frac{\partial F(\lambda)}{\partial \lambda}\right)_{N,V,T} &= -\frac{1}{\beta}\frac{\partial}{\partial\lambda}\ln Q(N,V,T,\lambda) = -\frac{1}{\beta Q(N,V,T,\lambda)}\frac{\partial Q(N,V,T,\lambda)}{\partial\lambda} \\ &= \frac{\int d\mathbf{r}^N (\partial\mathcal{U}(\lambda)/\partial\lambda)\exp[-\beta\mathcal{U}(\lambda)]}{\int d\mathbf{r}^N \exp[-\beta\mathcal{U}(\lambda)]} \\ &= \left\langle \frac{\partial\mathcal{U}(\lambda)}{\partial\lambda} \right\rangle_\lambda, \end{aligned} \tag{7.1.5}$$

[1]A Stockmayer fluid consists of Lennard-Jones particles with an embedded point dipole.

where $\langle \cdots \rangle_\lambda$ denotes an ensemble average for a system with a potential energy function $\mathcal{U}(\lambda)$ (7.1.4).

The free energy difference between systems II and I can be obtained by integrating equation (7.1.5):

$$F(\lambda = 1) - F(\lambda = 0) = \int_{\lambda=0}^{\lambda=1} d\lambda \left\langle \frac{\partial \mathcal{U}(\lambda)}{\partial \lambda} \right\rangle_\lambda . \tag{7.1.6}$$

The importance of this result is that it expresses a free energy difference in terms of an ensemble average, which, unlike a free energy, can be calculated directly in a simulation. In principle, we could perform the thermodynamic integration using any (in general, nonlinear) function $\mathcal{U}(\lambda)$, as long as this function is differentiable and satisfies the boundary condition: $\mathcal{U}(\lambda = 0) = \mathcal{U}_{\mathrm{I}}$ and $\mathcal{U}(\lambda = 1) = \mathcal{U}_{\mathrm{II}}$. However, the linear interpolation (7.1.4) is particularly convenient because in that case we know the sign of $\partial^2 F/\partial\lambda^2$. Straightforward differentiation of equation (7.1.5) shows that

$$\left(\frac{\partial^2 F}{\partial \lambda^2}\right)_{N,V,T} = -\beta \left\{ \left\langle (\mathcal{U}_{\mathrm{II}} - \mathcal{U}_{\mathrm{I}})^2 \right\rangle_\lambda - \langle \mathcal{U}_{\mathrm{II}} - \mathcal{U}_{\mathrm{I}} \rangle_\lambda^2 \right\} \leq 0.$$

In words, $(\partial F/\partial \lambda)$ can never increase with increasing λ. This (Gibbs-Bogoliubov) inequality can be used to test the validity or accuracy of the simulation results. In practice, the integration in equation (7.1.6) must be carried out numerically, such as, by Gaussian quadrature. Of course, such a numerical integration will work only if the integrand in equation (7.1.6) is a well-behaved function of λ. Occasionally, however, a linear parameterization of $\mathcal{U}(\lambda)$ may lead to a weak (and relatively harmless) singularity in equation (7.1.6) for $\lambda \to 0$. This point is discussed in more detail in section 10.3.1. Examples of applications of the thermodynamic integration technique can be found in [155–168].

Artificial thermodynamic integration is often used to compute the difference in excess free energy of similar but distinct molecules. Such calculations are of particular importance in biomolecular modeling (see e.g., [169]). For instance, one can thus compute the effect of a chemical substitution on the binding strength of a molecule to an enzyme. In such calculations, the thermodynamic integration involves a gradual replacement of part of the molecule by another building block (for instance, an H could be transformed into a CH_3 group).

It should be noted that the thermodynamic integration method based on equation (7.1.6) is intrinsically static; that is, the derivative of the free energy is obtained in a series of equilibrium simulations (either by the Monte Carlo technique or by Molecular Dynamics). An intrinsically dynamic scheme for performing free energy computations was suggested by Watanabe and Reinhardt [170]. This method also relies on the existence of a reversible path between the state point of interest and a simple reference system. However, the approach of [170] is based on the concept of adiabatic changes. If the Hamiltonian of a closed system is changed sufficiently slowly (compared to the slowest natural time scale in the system under consideration), then the entropy of the system does not change. Actually, the relation between adiabatic transformations in mechanics and in thermodynamics is subtle. A time-dependent external perturbation of the Hamiltonian of a classical many-body system will not change the total volume in phase space that is occupied by the system. However, if the external perturbation proceeds too rapidly, then the final energy of the system will depend on the initial conditions. As a consequence, a system that was initially characterized by the thermodynamic variables N, V, and E ends up in a state that does not correspond to any known ensemble. In contrast, if the change of the Hamiltonian is truly adiabatic, then the energy of the system is a unique function of time ($E(t)$). The system can then be characterized at every time t by a set of parameters $N, V, E(t)$, while the entropy of the system remains constant (because of the conservation of phase space volume).

The requirement that an adiabatic transformation converts a microcanonical system characterized by variables N, V, and E_1 into another system characterized by N, V, and E_2 can be used as a diagnostic tool in a simulation. Ideally, different configurations having an energy E_1 should on application of this adiabatic transformation all end up with the same energy E_2. If this condition is not satisfied the transformation has been carried out too rapidly and the transformation, strictly speaking, is not adiabatic. However, Watanabe and Reinhardt point out that, in practice, this condition is too strong. Rather one should verify that the average final energy E_2 becomes independent of the switching procedure.

Clearly, the method of Watanabe and Reinhardt is designed for Molecular Dynamics, rather than for Monte Carlo, simulations. However, the method is not limited to the microcanonical ensemble. It can be extended to other ensembles (e.g., NVT), in which case it becomes a technique for computing free energy changes [170]. It should be noted that several authors have used Molecular Dynamics simulations with a time-dependent Hamiltonian to compute free energy changes [171]. These simulations are close in spirit to the Monte Carlo scheme, although, as pointed out in [170], they can be derived from the adiabatic transformation scheme. All these "slow-growth" methods can be considered as special cases of the more general nonequilibrium method for measuring free energy changes that is described in section 7.4.1.

7.2 Chemical Potentials

In the previous section we discussed several direct, and usually robust, techniques for computing the thermal properties of a many-body system. However, in many cases computationally cheaper ways to arrive at the same result are available. In particular, there exist several techniques for "measuring" the chemical potential of a given species in a single Monte Carlo or Molecular Dynamics simulation. This seems to contradict the statement made in section 7.1 that thermal properties cannot be measured directly in a simulation. However, on closer inspection, it will appear that what we measure is not the chemical potential itself but the excess chemical potential, that is, the *difference* between the chemical potential of a given species and that of an ideal gas under the same conditions.

7.2.1 The Particle Insertion Method

A particularly simple and elegant method for measuring the chemical potential μ of a species in a pure fluid or in a mixture is the particle insertion method (often referred to as the Widom method [172]). The statistical mechanics that is the basis for this method is quite simple. Consider the definition of the chemical potential μ_α of a species α. From thermodynamics, we know that μ is defined as

$$\begin{aligned}\mu_\alpha &= \left(\frac{\partial G}{\partial N_\alpha}\right)_{PTN_{b\neq\alpha}}\\ &= \left(\frac{\partial F}{\partial N_\alpha}\right)_{VTN_{b\neq\alpha}}\\ &= -T\left(\frac{\partial S}{\partial N_\alpha}\right)_{VEN_{b\neq\alpha}},\end{aligned} \tag{7.2.1}$$

where G, F, and S are the Gibbs free energy, the Helmholtz free energy, and the entropy, respectively. Here and in the next few paragraphs we focus on a one-component system; hence we drop the subscript α. Moreover, to keep the notation compact, we shall initially assume that we deal with a system of N atoms in a cubic volume with diameter L and volume $V = L^d$, at constant temperature T. The classical partition function of such a system is given by

$$Q(N,V,T) = \frac{V^N}{\Lambda^{dN}N!}\int_0^1 \cdots \int_0^1 d\mathbf{s}^N \exp[-\beta\mathcal{U}(\mathbf{s}^N;L)], \tag{7.2.2}$$

in which the scaled coordinates $\mathbf{s}^N = \mathbf{r}^N/L$ are introduced. In equation (7.2.2), we have written $\mathcal{U}(\mathbf{s}^N;L)$ to indicate that $\mathcal{U}$ depends on the real rather than the scaled distances between the particles. The expression for the Helmholtz free energy of the system is

$$\begin{aligned}F(N,V,T) &= -k_BT\ln Q\\ &= -k_BT\ln\left(\frac{V^N}{\Lambda^{dN}N!}\right) - k_BT\ln\left\{\int d\mathbf{s}^N \exp[-\beta\mathcal{U}(\mathbf{s}^N;L)]\right\}\\ &= F_{id}(N,V,T) + F_{ex}(N,V,T).\end{aligned} \tag{7.2.3}$$

In the last line of this equation we have identified the two contributions to the Helmholtz free energy on the previous line as the ideal gas expression plus an excess part. It is obvious from equation (7.2.1) that, for sufficiently large N, the chemical potential is given by

$$\mu = -k_B T \ln(Q_{N+1}/Q_N).$$

If we use the explicit form, equation (7.2.3) for Q_N, we find that

$$\begin{aligned} \mu &= -k_B T \ln(Q_{N+1}/Q_N) \\ &= -k_B T \ln\left(\frac{V/\Lambda^d}{N+1}\right) - k_B T \ln\left\{\frac{\int ds^{N+1} \exp[-\beta \mathcal{U}(\mathbf{s}^{N+1})]}{\int ds^N \exp[-\beta \mathcal{U}(\mathbf{s}^N)]}\right\} \\ &\equiv \mu_{id}(\rho) + \mu_{ex}. \end{aligned} \tag{7.2.4}$$

In the last line of equation (7.2.4), we have separated the chemical potential in an ideal gas contribution μ_{id} and the excess part μ_{ex}. As $\mu_{id}(\rho)$ can be evaluated analytically, we focus on μ_{ex}. We now separate the potential energy of the $N+1$-particle system into the potential energy function of the N-particle system, $\mathcal{U}(\mathbf{s}^N)$, and the interaction energy of the $(N+1)$th particle with the rest: $\Delta\mathcal{U} \equiv \mathcal{U}(\mathbf{s}^{N+1}) - \mathcal{U}(\mathbf{s}^N)$. Using this separation, we can write μ_{ex} as

$$\mu_{ex} = -k_B T \ln \int d\mathbf{s}_{N+1} \left\langle \exp(-\beta \Delta\mathcal{U}) \right\rangle_N, \tag{7.2.5}$$

where $\langle \cdots \rangle_N$ denotes canonical ensemble averaging over the configuration space of the N-particle system. The important point to note is that equation (7.2.5) expresses μ_{ex} as an ensemble average that can be sampled by the conventional Metropolis scheme [6]. One aspect of the average in equation (7.2.5) makes it different from the quantities usually sampled in a computer simulation: we have to compute the average of an integral over the position of particle $N+1$. This last integral can be sampled by brute force (unweighted) Monte Carlo sampling. In practice the procedure is as follows: we carry out a conventional constant-NVT Monte Carlo simulation on the system of N particles. At frequent intervals during this simulation, we randomly generate a coordinate $\mathbf{s}_{N+1}$, uniformly over the unit cube. For this value of $\mathbf{s}_{N+1}$, we then compute $\exp(-\beta\Delta\mathcal{U})$. By averaging the latter quantity over all generated trial positions, we obtain the average that appears in equation (7.2.5). So, in effect, we are computing the average of the Boltzmann factor associated with the random insertion of an additional particle in an N-particle system, but we never accept any such trial insertions, because then we would no longer be sampling the average needed in equation (7.2.5). The Widom method provides us with a very powerful scheme for computing the chemical potential of (not too dense) atomic and simple molecular liquids. In Algorithm 16, we demonstrate how this method can be implemented in a simulation.

Algorithm 16 (Widom Test Particle Insertion)

```
      subroutine Widom              excess chemical potential
                                    via the addition of test particles
      xtest=box*ranf()              generate a random position
      call ener(xtest,entest)       determine energy
      wtest=wtest                   update Boltzmann factor in (7.2.5)
     +        +exp(-beta*entest)
      return
      end
```

Comments to this algorithm:

1. *This algorithm shows the basic structure of the Widom test particle method for the N,V,T ensemble. This subroutine is usually called in the sampling step of a Monte Carlo simulation, for example, in subroutine* `sample` *in Algorithm 1. Usually, many such test particle insertions are needed to obtain reliable statistics.*
2. *The excess chemical potential follows from* $\beta\mu_{ex} = -\ln(\texttt{wtest}/M)$, *where* M *is the total number of test particle insertions. The accuracy of* $\beta\mu_{ex}$ *can be estimated using* $\sigma_{\beta\mu_{ex}} = \sigma_{\texttt{wtest}}/\texttt{wtest}$.
3. *Subroutine* `ener` *calculates the energy of the test particle. Note that the test particle insertion is virtual and is never accepted.*
4. *For pairwise additive interactions, we can approximately correct for the effect of the truncation of the intermolecular interactions on the value of the chemical potential by evaluating a tail correction. This correction turns out to be a factor of 2 larger than that for the potential energy per particle (see Case Study 14).*

These equations were derived for a spatially homogeneous system. Widom [173] also considered the case of a spatially inhomogeneous system. In that case, μ_{ex} depends explicitly on the position **r**. However, in equilibrium the chemical potential itself is constant throughout the system. In other words

$$\mu = k_B T \ln\left(\frac{\rho(\mathbf{r})}{\langle \exp\left[-\beta\Delta\mathcal{U}(\mathbf{r})\right]\rangle_N}\right)$$

is constant.

Case Study 14 (Chemical Potential: Widom Method)

In this case study, we use the Widom test particle insertion method to determine the excess chemical potential of the Lennard-Jones fluid. The algorithm we use is a combination of the basic algorithm for performing Monte Carlo simulations in the NVT ensemble (Algorithms 1 and 2) and determining the excess chemical potential (Algorithm 16).

We stress that the tail correction for the chemical potential is similar, but not identical, to that for the potential energy. In the Widom test particle method we determine the energy difference:

$$\Delta\mathcal{U} = \mathcal{U}(\mathbf{s}^{N+1}) - \mathcal{U}(\mathbf{s}^{N}).$$

The tail correction is

$$\begin{aligned}
\beta\mu^{\text{tail}} &= \mathcal{U}(\mathbf{s}^{N+1})^{\text{tail}} - \mathcal{U}(\mathbf{s}^{N})^{\text{tail}} \\
&= (N+1)u^{\text{tail}}((N+1)/V) - Nu^{\text{tail}}(N/V) \\
&= \left((N+1)\frac{N+1}{V} - N\frac{N}{V}\right)\frac{1}{2}4\pi\int_{r_c}^{\infty} dr\, r^2 u(r) \\
&\approx \frac{2N}{V}\frac{1}{2}4\pi\int_{r_c}^{\infty} dr\, r^2 u(r) \\
&= 2u^{\text{tail}}(\rho).
\end{aligned} \tag{7.2.6}$$

In Case Study 9, we performed a grand-canonical Monte Carlo simulation to determine the equation of state of the Lennard-Jones fluid. In the grand-canonical ensemble the chemical potential and the temperature are imposed (the density is determined during the simulation). Of course, we can also calculate the chemical potential during the simulation, using the Widom method. Figure 7.1 shows a comparison of the imposed and measured chemical potentials.

7.2.2 Other Ensembles

The extension of the Widom method to other ensembles, in particular NPT and NVE, is relatively straightforward. However, it would be incorrect simply to apply equation (7.2.5) to these other ensembles. As this point is not always fully appreciated in the literature, we shall briefly discuss the application of the Widom method to the NPT ensemble (see [174, 175]) and to the NVE ensemble (see [156, 176, 177]). To derive the expression for the chemical potential in the NPT ensemble, we start from the expression for the Gibbs free energy:

$$G(N,P,T) = -k_B T \ln\left\{\int dV \frac{V^N \exp(-\beta PV)}{\Lambda^{dN} N!}\int d\mathbf{s}^N \exp\left[-\beta\mathcal{U}(\mathbf{s}^N; V)\right]\right\}.$$

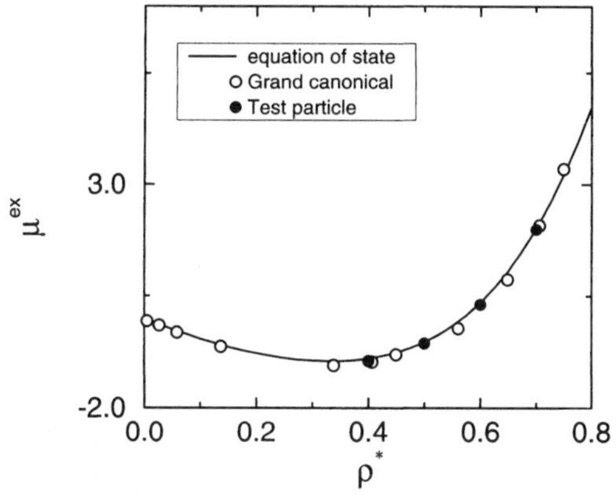

Figure 7.1: Excess chemical potential of the Lennard-Jones fluid (T = 2.0) as calculated from the equation of state, grand-canonical Monte Carlo, and the test particle insertion method.

We must evaluate $\mu = (\partial G/\partial N)_{PT}$. Entirely analogous to the NVT case, we find that $\mu = G(N+1,P,T) - G(N,P,T)$ equals

$$\begin{aligned} \mu &= -k_B T \ln \left\langle \frac{V}{\Lambda^d (N+1)} \int ds_{N+1} \exp(-\beta \Delta \mathcal{U}) \right\rangle \\ &= -k_B T \ln(k_B T / P \Lambda^d) - k_B T \ln \left\langle \frac{PV}{(N+1) k_B T} \int ds_{N+1} \exp(-\beta \Delta \mathcal{U}) \right\rangle \\ &= \mu_{id}(P) + \mu_{ex}(P). \end{aligned} \quad (7.2.7)$$

Two points should be noted. First of all, we now define the ideal gas reference state at the same pressure, rather than at the same average density as the system under study. And second, the fluctuating quantity that we are averaging is no longer $\exp(-\beta\Delta\mathcal{U})$, but $V\exp(-\beta\Delta\mathcal{U})$. In practice, one should only expect the fluctuating volume term in equation (7.2.7) to be important if large volume fluctuations are possible, for instance, in the vicinity of critical points. But chemical potentials are often calculated precisely to locate phase transitions near such points.

In the constant-NVE ensemble, that is, the one probed by conventional Molecular Dynamics simulations, we start from the relation

$$\mu/T = -(\partial S/\partial N)_{VE}.$$

In the microcanonical ensemble, the entropy S is related to $\Omega(N,V,E)$, the total number of accessible states, by $S = k_B \ln \Omega(N,V,E)$. The classical expression for $\Omega(N, V, E)$ is

$$\Omega(N, V, E) = \frac{1}{h^{3N}N!} \int d\mathbf{p}^N d\mathbf{r}^N \delta(\mathcal{H}(\mathbf{p}^N, \mathbf{r}^N) - E). \tag{7.2.8}$$

Again the derivation proceeds much as before, but that we must now compute $\Omega(N+1, V, E)/\Omega(N, V, E)$. This is slightly more cumbersome (see [156]), and we quote only the final result:

$$\beta\mu_{ex} = -\ln\left\{\langle T\rangle^{-3/2}\left\langle T^{3/2}\exp(-\Delta\mathcal{U}/k_B T)\right\rangle\right\}, \tag{7.2.9}$$

where T is the (fluctuating) temperature (as determined from the instantaneous kinetic energy of the particles). Such fluctuations tend to be large where the heat capacity of the system is large (see [80]).

Extensive numerical studies of the excess chemical potential of finite periodic systems have shown that this quantity is rather strongly dependent on the size of the system [92, 178]. Computer simulations are typically carried out for periodic systems in which the fundamental cell contains on order of 10^2 to 10^3 particles, and the correction needed to give the infinite system result therefore can be large. Of course, it is possible to estimate the finite-size correction empirically by carrying out simulations for different values of N (the number of particles), but this is very time consuming. It clearly would be much more convenient if the finite-size correction could be estimated directly, since it would then be possible to estimate the chemical potential in the thermodynamic limit on the basis of results obtained from simulations of relatively small systems. In fact, Siepmann *et al.* [179] have derived an expression for the leading ($\mathcal{O}(1/N)$) system-size dependence of the excess chemical potential:

$$\Delta\mu_{ex}(N) = \frac{1}{2N}\left(\frac{\partial P}{\partial \rho}\right)\left[1 - k_B T\left(\frac{\partial \rho}{\partial P}\right) - \rho k_B T\frac{(\partial^2 P/\partial \rho^2)}{(\partial P/\partial \rho)^2}\right]. \tag{7.2.10}$$

As shown in [179], this expression agrees with the exact result for hard rods in one dimension and is in excellent agreement with numerical results for hard disks in two dimensions.

The particle insertion method can be modified to measure the difference in chemical potential between two species α and β in a mixture. In this case a trial move consists of an attempt to transform a particle of species α into species β (without, of course, ever accepting such trial moves). For more details, the reader is referred to [175, 180]. Finally it should be stressed that particle insertion and swapping techniques are not limited to the measurement of chemical potentials. In fact, a wide class of partial molar quantities (such as the partial molar enthalpy h_α or the partial molar volume v_α) can be measured in this way. For details, see [175, 181].

7.2.3 Overlapping Distribution Method

The reader may wonder why, in the previous section, we have been discussing only a trial move that attempts to add a particle to the system and not the reverse move. After all, the chemical potential can also be written as

$$\begin{aligned}\mu &= +k_BT\ln(Q_N/Q_{N+1}) \\ &= \mu_{id} + k_BT\ln\langle\exp(+\beta\Delta\mathcal{U})\rangle_{N+1}, \end{aligned} \qquad (7.2.11)$$

where $\Delta\mathcal{U}$ denotes the interaction energy of particle $N+1$ with the remaining N particles. It would seem that equation (7.2.11) can be sampled by straightforward Metropolis Monte Carlo. In general, however, this is not true. The reason is that the function $\exp(\beta\Delta\mathcal{U})$, in principle, is not bounded. It can become arbitrarily large, as $\Delta\mathcal{U}$ grows. (Incidentally, this is not true for $\exp(-\beta\Delta\mathcal{U})$, because one of the conditions that a system must satisfy to be describable by classical statistical mechanics is that its potential energy function has a lower bound.) The problem with equation (7.2.11) is that very large values of the integrand coincide with very small values $\mathcal{O}(\exp(-\beta\Delta\mathcal{U}))$ of the Boltzmann factor (which determines how often a configuration is sampled during a Monte Carlo run). As a consequence, an appreciable contribution to the average in equation (7.2.11) comes from a part of configuration space that is hardly ever, or indeed never, sampled during a run. Hard spheres offer a good illustration. As the potential energy function of nonoverlapping hard spheres is always zero, a simple Monte Carlo sampling of equation (7.2.11) for a dense fluid of hard spheres would always yield the nonsensical estimate $\mu_{ex} = 0$ (whereas, in fact, at freezing, $\mu_{ex}/k_BT \sim 15$). The correct way to obtain chemical potentials from simulations involving both particle insertions and particle removals has been indicated by Shing and Gubbins [182,183]. However, we find it convenient to discuss this problem in the context of a more general technique for measuring free energy differences, first introduced by Bennett [184], called the *overlapping distribution method*.

Consider two N-particle systems, labeled 0 and 1 with partition functions Q_0 and Q_1. For convenience we assume that both systems have the same volume V, but this is not essential. From equation (5.4.4) it follows that the free energy difference $\Delta F = F_1 - F_0$ can be written as

$$\begin{aligned}\Delta F &= -k_BT\ln(Q_1/Q_0) \\ &= -k_BT\ln\left(\frac{\int d\mathbf{s}^N \exp[-\beta\mathcal{U}_1(\mathbf{s}^N)]}{\int d\mathbf{s}^N \exp[-\beta\mathcal{U}_0(\mathbf{s}^N)]}\right). \end{aligned} \qquad (7.2.12)$$

Suppose that we are carrying out a (Metropolis) sampling of the configuration space of system 1. For every configuration visited during this sampling of system 1 we can compute the potential energy of system 0 ($\mathcal{U}_0(\mathbf{s}^N)$) for the same configuration and, hence, the potential energy difference $\Delta\mathcal{U} = \mathcal{U}_1(\mathbf{s}^N) - \mathcal{U}_0(\mathbf{s}^N)$. We use this information to construct a histogram that measures the probability density for the potential energy difference $\Delta\mathcal{U}$. Let us denote this probability density by $p_1(\Delta\mathcal{U})$. In the canonical ensemble, $p_1(\Delta\mathcal{U})$ is given by

$$p_1(\Delta\mathcal{U}) = \frac{\int d\mathbf{s}^N \exp(-\beta\mathcal{U}_1)\,\delta(\mathcal{U}_1 - \mathcal{U}_0 - \Delta\mathcal{U})}{q_1}, \tag{7.2.13}$$

where we have denoted the scaled, configurational part of the partition function by a q (e.g., $q_1 = \int d\mathbf{s}^N \exp[-\beta\mathcal{U}_1(\mathbf{s}^N)]$). The δ-function in equation (7.2.13) allows us to substitute $\mathcal{U}_0 + \Delta\mathcal{U}$ for $\mathcal{U}_1$ in the Boltzmann factor; hence,

$$\begin{aligned} p_1(\Delta\mathcal{U}) &= \frac{\int d\mathbf{s}^N \exp[-\beta(\mathcal{U}_0 + \Delta\mathcal{U})]\,\delta(\mathcal{U}_1 - \mathcal{U}_0 - \Delta\mathcal{U})}{q_1} \\ &= \frac{q_0}{q_1}\exp(-\beta\Delta\mathcal{U})\frac{\int d\mathbf{s}^N \exp(-\beta\mathcal{U}_0)\,\delta(\mathcal{U}_1 - \mathcal{U}_0 - \Delta\mathcal{U})}{q_0} \\ &= \frac{q_0}{q_1}\exp(-\beta\Delta\mathcal{U})p_0(\Delta\mathcal{U}), \end{aligned} \tag{7.2.14}$$

where $p_0(\Delta\mathcal{U})$ is the probability density of finding a potential energy difference $\Delta\mathcal{U}$ between systems 1 and 0, while Boltzmann sampling the available configurations of system 0. As the free energy difference between systems 1 and 0 is simply $\Delta F = -k_B T \ln(q_1/q_0)$, we find from equation (7.2.14) that

$$\ln p_1(\Delta\mathcal{U}) = \beta(\Delta F - \Delta\mathcal{U}) + \ln p_0(\Delta\mathcal{U}). \tag{7.2.15}$$

To obtain ΔF from equation (7.2.15) in practical cases, it is convenient to define two functions f_0 and f_1 by

$$f_0(\Delta\mathcal{U}) = \ln p_0(\Delta\mathcal{U}) - \frac{\beta\Delta\mathcal{U}}{2}$$

and

$$f_1(\Delta\mathcal{U}) = \ln p_1(\Delta\mathcal{U}) + \frac{\beta\Delta\mathcal{U}}{2}$$

such that

$$f_1(\Delta\mathcal{U}) = f_0(\Delta\mathcal{U}) + \beta\Delta F.$$

Suppose that we have measured f_0 and f_1 in two separate simulations (one sampling system 0, the other system 1). We can then obtain ΔF by fitting the functions f_0 and f_1 to two polynomials in $\Delta\mathcal{U}$ that are identical but for the constant term. The constant offset between the two polynomials yields our estimate for ΔF. Note that, to perform such a fit, it is not necessary that there even exists a range of $\Delta\mathcal{U}$ where both f_0 and f_1 can be measured.

However, in the absence of such a range of overlap, the statistical accuracy of the method is usually poor.

Now consider the particle insertion-removal problem. Let us assume that system 1 is a system with N interacting particles, while system 0 contains N − 1 interacting particles and 1 ideal gas particle. The difference in free energy between these two systems is obviously equal to μ_{ex}. Applying equation (7.2.15) to this particular case, we find that

$$\beta\mu_{ex} = f_1(\Delta\mathcal{U}) - f_0(\Delta\mathcal{U}). \tag{7.2.16}$$

Equation (7.2.16) is equivalent to the result obtained by Shing and Gubbins. Using the overlapping distribution method it is possible to combine the results of simulations with trial insertions and trial removals to arrive at a more accurate estimate for the chemical potential. In section 7.3.1, we discuss the extension of the Bennett method to multiple histograms and indicate the relation with recent developments in this field [37, 185].

Case Study 15 (Chemical Potential: Overlapping Distribution)
In Case Study 14, we used the Widom test particle method to determine the chemical potential of the Lennard-Jones fluid. This method breaks down at high densities, where most of the test particles have such a high energy that the Boltzmann factor in equation (7.2.5) is negligible. Once in a while, a hole in the fluid is found that gives a nonzero contribution to this Boltzmann factor. However, since these nonzero contributions are only rarely observed, the statistical accuracy with which we sample the excess chemical potential will be low. The overlapping distribution method provides a good diagnostic tool for detecting such sampling problems.

For the overlapping distribution method, we have to perform two simulations: one simulation using a system of N + 1 particles (system 1) and a system with N particles and one ideal gas particle (system 0). For each of these systems we determine the distribution of energy differences, equations (7.2.13) and (7.2.14). For system 1, this energy difference $\Delta\mathcal{U}$ is defined as the change of the total energy of the system when one of the particles would be transformed into an ideal gas particle. We have to make a histogram of the energies of the particles in this system. This can be done conveniently using Algorithm 2, in which the energy of a randomly selected particle is calculated before an attempt is made to displace this particle; the distribution of these energies yields $p_1(\Delta\mathcal{U})$.

For system 0, we have to determine the energy difference $\Delta\mathcal{U}$, which is the difference in total energy when the ideal gas particle would be turned into an interacting particle. This energy difference equals the energy of a test particle in the Widom method (section 7.2.1). When we determine $p_0(\Delta\mathcal{U})$, at the same time we can obtain an estimate of the excess chemical potential from the Widom particle insertion method.

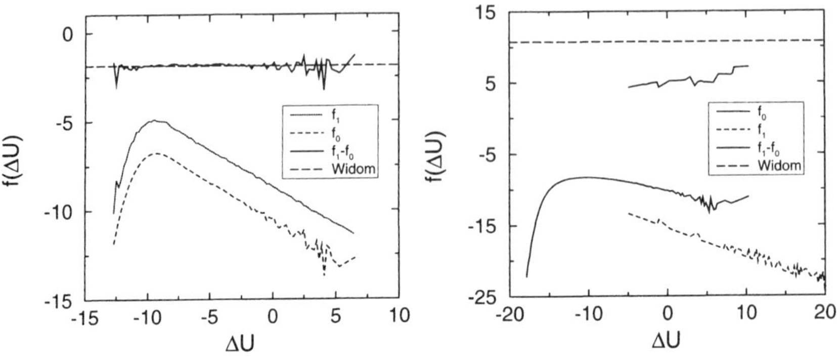

Figure 7.2: Comparison of the overlapping distribution function method and the Widom particle insertion scheme for measuring the chemical potential of the Lennard-Jones fluid at $T = 1.2$. The solid curve is the particle-insertion result, the dashed curve was obtained using the overlapping distribution method ($\beta\mu_{ex} = f_1 - f_0$). The units for $\beta\mu_{ex}$ are the same as for $f(\Delta\mathcal{U})$. The figure on the left corresponds to a moderately dense liquid ($\rho = 0.7$). In this case, the distributions overlap and the two methods yield identical results. The right-hand figure corresponds to a dense liquid ($\rho = 1.00$). In this case, the insertion probability is very low. The distributions f_0 and f_1 hardly overlap, and the two different estimates of $\beta\mu_{ex}$ do not coincide.

In Figure 7.2 we have plotted the functions $f_0(\Delta\mathcal{U})$, $f_1(\Delta\mathcal{U})$, and $\mu_{ex}(\Delta\mathcal{U})$ (as defined by equation (7.2.16)) as a function of $\Delta\mathcal{U}$ for the Lennard-Jones fluid at $\rho = 0.7$ and $\rho = 1.00$. For the sake of comparison, we have also plotted the results obtained using the Widom particle insertion method. The figure shows that at $\rho = 0.7$ there is a sufficiently large range of energy differences for which the two functions overlap ($-10 < \Delta\mathcal{U} < -5$). The result of the overlapping distribution function therefore is in good agreement with the results from the Widom method. At $\rho = 1.00$, the range of overlap is limited to the wings of the histograms p_0 and p_1, where the statistical accuracy is low. As a consequence, our estimate for $\mu_{ex}(\Delta\mathcal{U})$ is not constant (as it should) but appears to depend on $\Delta\mathcal{U}$. Moreover, the results from the overlapping distribution method are not consistent with the result of the Widom particle insertion method.

Note that two separate simulations are needed to determine the excess chemical potential from the overlapping distribution method. One might think that the particle addition and particle removal histograms could be measured in a single simulation of an N-particle system. This would indeed be correct if there were no difference between the histograms for particle removal from N and N + 1 particle systems. However, the overlapping distribution method normally is used for systems of only a few hundred particles and at relatively high densities. Under those conditions, the system-size dependence of μ_{ex} can be appreciable. Nevertheless, it is useful to keep track of both $p_0(\Delta\mathcal{U})$ and $p_1(\Delta\mathcal{U})$ in a single simulation, simply by checking whether overlap between the two distributions is sufficient.

7.3 Other Free Energy Methods

The present section describes free energy methods of particular importance for barrier crossing problems (see also Chapter 16). By this we mean either the computation of the free energy barrier between two states (phases, conformations) or the computation of the free energy of a system that cannot be equilibrated by conventional means because its free energy landscape consists of many valleys separated by high barriers. The numerical techniques described in this section therefore are not designed exclusively for free energy calculations but also for equilibration of glassy systems. This is a rather common problem and deserves special attention. We start, however, with a simple extension of the overlapping distribution method of section 7.2.3. In that section, we focused on the application of the technique for computing the free energy difference between two systems. Now we will discuss the method, in its generalized form, in a different context, namely, as a technique for studying high free energy barriers. Such barriers cannot be studied using conventional Monte Carlo or Molecular Dynamics simulations because the probability that a spontaneous fluctuation will bring the system to the top of the barrier is vanishingly small.

7.3.1 Multiple Histograms

As was shown in section 7.2.3, the overlapping distribution function method can work even if the two distributions do not really overlap. However, they should not be too far apart. In case there is a large gap between f_0 and f_1, it is often useful to perform additional simulations for intermediate values of the Hamiltonian. We thus obtain a sequence of distribution functions f_0, f_1, f_2, $\cdots$ f_n, such that f_0 overlaps with f_1, f_1 with f_2, etc. The free energy difference between system 0 and system n can then be obtained by adding $\Delta F_{0,1}$ + $\Delta F_{1,2}$ + $\cdots$ + $\Delta F_{n-1,n}$. The problem with a naive implementation of this approach is that the statistical errors in the individual free energy differences $\Delta F_{i,i+1}$ add up quadratically in the final result. Fortunately, techniques exist that prevent such error propagation. One approach, proposed by Ferrenberg and Swendsen [185], is based on the idea that it is possible to construct a self-consistent estimate for the histograms without assuming any specific functional form.

For what follows, it is convenient to define the distributions that we are considering in a slightly more general form. We assume that the original system that we are studying is characterized by a potential energy function $\mathcal{U}_0$. We now define a sequence of n related models characterized by a potential energy function $\mathcal{U}_i \equiv \mathcal{U}_0 + W_i$ $(i = 1, \cdots, n)$. For example, W_i could be $\lambda_i(\mathcal{U}_1 - \mathcal{U}_0)$, with $0 < \lambda_i \leq 1$. Alternatively, W_i could be a function of some order parameter or reaction coordinate $Q(\mathbf{r}^N)$. For the sake of generality, we take the latter point of view, because $W_i = \lambda_i(\mathcal{U}_1 - \mathcal{U}_0)$ can be interpreted as a function of the order parameter $Q(\mathbf{r}^N) \equiv \mathcal{U}_1 - \mathcal{U}_0$. We now consider histograms $p_i(Q)$ defined as follows:

$$p_i(Q) = \frac{\int d\mathbf{r}^N \exp[-\beta(\mathcal{U}_0 + W_i)]\delta[Q - Q(\mathbf{r}^N)]}{Z_i},$$

where Z_i is given by

$$Z_i \equiv \int d\mathbf{r}^N \exp[-\beta(\mathcal{U}_0 + W_i)]. \qquad (7.3.1)$$

The original system is characterized by an order parameter distribution

$$p_0(Q) = \frac{\int d\mathbf{r}^N \exp(-\beta\mathcal{U}_0)\delta[Q - Q(\mathbf{r}^N)]}{Z_0}.$$

In some cases, we are interested in the ratio Z_0/Z_n, because that determines the free energy difference between systems 0 and n. But, often, the distribution function $p_0(Q)$ itself is of central interest, because it allows us to compute the Landau free energy of the original system as a function of the order parameter Q:

$$F_{\text{Landau}}(Q) = -k_B T \ln p_0(Q).$$

Self-Consistent Histogram Method

In a simulation, the histograms $p_i(Q)$ are estimated by measuring $H_i(Q)$, the number of times that a system with potential energy function $\mathcal{U}_0 + W_i$ has a value on the order of the parameter between Q and $Q + \Delta Q$. If the total number of points collected in histogram i is denoted by M_i, then

$$p_i(Q)\Delta Q = \langle H_i(Q) \rangle / M_i,$$

where the $\langle \cdots \rangle$ denotes ensemble averaging. Of course, in a simulation of finite length, the ratio $H_i(Q)/M_i$ will fluctuate around $p_i(Q)\Delta Q$. If we assume that the number of points in bin i is determined by a Poisson distribution, then it follows that the variance in $p_i^{\text{est}}(Q)\Delta Q$, that is, our estimate of

$p_i(Q)\Delta Q$, is given by

$$\begin{aligned}\frac{\langle H_i^2(Q)\rangle - \langle H_i(Q)\rangle^2}{M_i^2} &= \frac{\langle H_i(Q)\rangle}{M_i^2} \\ &= \frac{p_i(Q)\Delta Q}{M_i}.\end{aligned} \tag{7.3.2}$$

In what follows, we shall assume, without loss of generality, that our units are chosen such that $\Delta Q = 1$. This choice is not essential, of course, but it simplifies the notation and does not affect the final results. Once we have measured a set of histograms, we can try to combine this information to arrive at an estimate of $p_0(Q)$ or, equivalently, of $F_{\text{Landau}}(Q)$. First of all, we should note that, in principle, although not in practice, p_0 can be reconstructed from every individual histogram $p_i(Q)$:

$$p_0(Q) = \exp(+\beta W_i)\frac{Z_i}{Z_0}p_i(Q). \tag{7.3.3}$$

In practice, this approach will hardly ever work because the range of Q values where $p_0(Q)$ and $p_i(Q)$ differ significantly from zero need not overlap. Hence the most important contribution to $p_0(Q)$ would come from a range of Q values where $\exp(+\beta W_i)$ is very large but $p_i(Q)$ is vanishingly small. We therefore shall attempt to construct our estimate $p_0^{\text{est}}(Q)$ by a linear combination of the estimates based on the different histograms:

$$p_0^{\text{est}}(Q) = \sum_{i=1}^{n} w_i(Q)\exp(+\beta W_i)\frac{Z_i}{Z_0}p_i^{\text{est}}(Q), \tag{7.3.4}$$

where $w_i(Q)$ is an as-yet undetermined weight function, subject to the condition

$$\sum_{i=1}^{n} w_i(Q) = 1. \tag{7.3.5}$$

Note that, at this stage, the values of the ratios Z_i/Z_0 are also still unknown. Let us now choose the weights $w_i(Q)$ such that the estimated variance in $p_0^{\text{est}}(Q)$ is minimized. Using the fact that fluctuations in different simulations are uncorrelated, this variance is given by

$$\begin{aligned}
&\langle p_0^{est}(Q)^2\rangle - \langle p_0^{est}(Q)\rangle^2 \\
&= \sum_{i=1}^{n} w_i^2(Q)\exp(+2\beta W_i)\left(\frac{Z_i}{Z_0}\right)^2 \langle p_i^{est}(Q)^2\rangle - \langle p_i^{est}(Q)\rangle^2 \\
&= \sum_{i=1}^{n} w_i^2(Q)\exp(+2\beta W_i)\left(\frac{Z_i}{Z_0}\right)^2 p_i(Q)/M_i \\
&= p_0(Q)\sum_{i=1}^{n} w_i^2(Q)\exp(+\beta W_i)\frac{Z_i}{Z_0}/M_i,
\end{aligned} \tag{7.3.6}$$

where, in the last line, we have used equations (7.3.2) and (7.3.3). Next we must determine the values for $w_i(Q)$ that minimize the variance of $p_0^{est}(Q)$. Differentiating equation (7.3.2), under the constraint given by equation (7.3.5), yields

$$w_i(Q) = \alpha \exp(-\beta W_i) M_i \frac{Z_0}{Z_i}, \tag{7.3.7}$$

where α is an undetermined multiplier, fixed by the condition that $w_i(Q)$ is normalized:

$$\alpha = \frac{1}{\sum_{i=1}^{n}\exp(-\beta W_i) M_i Z_0/Z_i}. \tag{7.3.8}$$

Inserting equation (7.3.7) in equation (7.3.4) yields

$$p_0^{est}(Q) = \frac{\sum_{i=1}^{n} H_i(Q)}{\sum_{i=1}^{n}\exp(-\beta W_i) M_i Z_0/Z_i}. \tag{7.3.9}$$

Finally, we must determine Z_i. We do this by inserting equations (7.3.9) and (7.3.3) in the definition of Z_i (equation (7.3.1)):

$$\begin{aligned}
Z_i &= \int d\mathbf{r}^N \exp[-\beta(\mathcal{U}_0 + W_i)] \\
&= \int dQ Z_0 p_0(Q)\exp(-\beta W_i) \\
&= \int dQ \exp(-\beta W_i)\frac{\sum_{j=1}^{n} H_j(Q)}{\sum_{k=1}^{n}\exp(-\beta W_k) M_k/Z_k}.
\end{aligned} \tag{7.3.10}$$

This is an implicit equation for Z_i that must be solved self-consistently. In fact, we cannot determine the absolute value of all Z_i but only their ratio. Therefore, we can arbitrarily fix one of the Z_i (e.g., Z_1) at a constant value. The (nonlinear) set of equations (7.3.10) is then solved for the remaining Z_i until self-consistency is reached. The free energy difference between system n and system 1 then simply can be computed as

$$\Delta F = -k_B T \ln(Z_n/Z_1).$$

To give a specific example, assume that we have simulated only two systems 1 and 2, with potential energy functions $\mathcal{U}_1$ and $\mathcal{U}_2 = \mathcal{U}_1 + \Delta\mathcal{U}$. Moreover, let us assume, for the sake of simplicity, that both simulations resulted in the same number of histogram entries ($M = M_1 = M_2$). Then equation (7.3.10) becomes

$$\begin{aligned} Z_2 &= \int d\Delta\mathcal{U} \exp(-\beta\Delta\mathcal{U}) \frac{H_1(\Delta\mathcal{U}) + H_2(\Delta\mathcal{U})}{M[1/Z_1 + \exp(-\beta\Delta\mathcal{U})]/Z_2} \\ \frac{Z_2}{Z_1} &= \int d\Delta\mathcal{U} \exp(-\beta\Delta\mathcal{U}) \frac{p_1(\Delta\mathcal{U}) + p_2(\Delta\mathcal{U})}{[1 + \exp(-\beta\Delta\mathcal{U})](Z_1/Z_2)}. \end{aligned} \tag{7.3.11}$$

Using $\beta\Delta F = -\ln(Z_2/Z_1)$, we can rewrite equation (7.3.11) as

$$\int d\Delta\mathcal{U} \frac{p_1(\Delta\mathcal{U})}{1 + \exp[-\beta(\Delta\mathcal{U} - \Delta F)]} = \int d\Delta\mathcal{U} \frac{p_2(\Delta\mathcal{U})}{1 + \exp[-\beta(\Delta F - \Delta\mathcal{U})]},$$

which is equivalent to the equation that must be solved self-consistently for ΔF in Bennett's acceptance ratio method [184], to be discussed in section 7.3.2. This is somewhat surprising because the acceptance ratio method was devised to minimize the estimated error in the free energy difference between two systems. In contrast, equation (7.3.10) minimizes the error in our estimate of $p_0(Q)$. If the number of histograms is larger than 2, this expression is not completely equivalent to the one obtained by minimizing the error in the free energy difference between system 1 and n (say). But it is straightforward to derive the equations for the set $\{Z_i\}$ that minimize the error in the estimate of a particular free energy difference.

Example 3 (Ideal Gas Molecule in an External Field)
To illustrate the self-consistent histogram method, let us consider a trivial example, namely, the simulation of an ideal gas molecule in an external field:

$$u(z) = \begin{cases} z & z > 0 \\ \infty & z \leq 0 \end{cases}.$$

For this system, the probability of finding an ideal gas molecule at a position z is given by the barometric distribution:

$$p_0(z) = C \exp[-\beta u(z)].$$

The Landau free energy as a function of the coordinate z is, in this case, simply equal to the potential energy:

$$F(z) = -k_B T \ln[p_0(z)] = u(z) = z,$$

where we have chosen our reference point at $z = 0$. A direct simulation of the barometric height distribution yields poor statistics if $\beta u(z) \gg 1$. This is

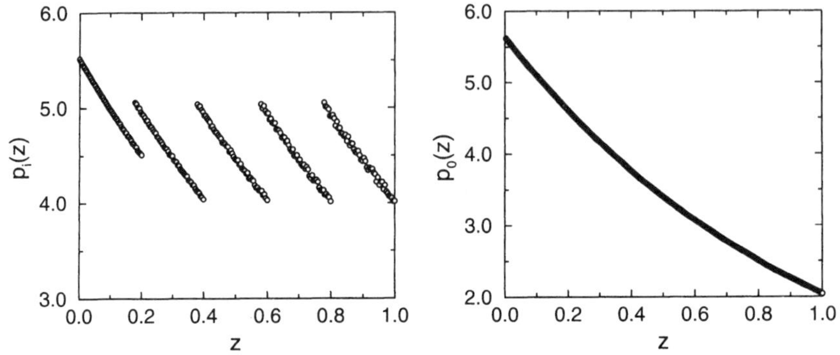

Figure 7.3: The probability of finding an ideal gas particle at position z. The figure on the left shows the results for the various windows, and the right figure shows the reconstructed distribution function as obtained from the self-consistent histogram method.

why we use the self-consistent histogram method. For window i we use the following window potential:

$$W_i(z) = \begin{cases} \infty & z < z_i^{\min} \\ 0 & z_i^{\min} < z < z_i^{\max} \\ \infty & z > z_i^{\max} \end{cases} .$$

We allow only neighboring windows to overlap:

$$\begin{array}{lll} z_{i-2}^{\max} & < z_i^{\min} & < z_{i-1}^{\max} \\ z_{i+2}^{\min} & > z_i^{\max} & > z_{i+1}^{\min}. \end{array}$$

For each window we perform M samples to estimate the probability $p_i(z)$ to find an ideal gas particle at a position z. The results of such a simulation are shown in Figure 7.3. The self-consistent histogram method is used to reconstruct the desired distribution $p_0(z)$. According to equations (7.3.9) and (7.3.1) this distribution is given by

$$p_0^{\text{est}}(z) = \frac{\sum_{i=1}^{n} p_i(z)}{\sum_{i=1}^{n} \exp(-\beta W_i) Z_0 / Z_i},$$

with

$$
\begin{aligned}
Z_i &= \int dz \, \exp(-\beta W_i) \frac{\sum_{j=1}^{n} p_j(z)}{\sum_{k=1}^{n} \exp(-\beta W_k)/Z_k} \\
&= \int_{z_i^{\min}}^{z_i^{\max}} dz \, \frac{\sum_{j=1}^{n} p_j(z)}{\sum_{k=1}^{n} \exp(-\beta W_k)/Z_k} \\
&= A_i \frac{Z_i Z_{i-1}}{Z_{i-1} + Z_i} + (B_i + 1) Z_i + C_i \frac{Z_i Z_{i+1}}{Z_{i+1} + Z_i}.
\end{aligned}
$$

We therefore obtain the following recursive relation for Z_{i+1}

$$
Z_{i+1} = -\frac{Z_i Z_{i-1} A_i + (Z_i Z_{i-1} + Z_i Z_i) B_i}{Z_{i-1} A_i + (Z_{i-1} + Z_i) B_i + (Z_{i-1} + Z_i) C_i}, \tag{a}
$$

in which A_i, B_i, and C_i are defined as

$$
\begin{aligned}
A_i &= \int_{z_i^{\min}}^{z_{i-1}^{\max}} dz \, [p_{i-1}(z) + p_i(z)] \\
B_i &= -1 + \int_{z_{i-1}^{\max}}^{z_{i+1}^{\min}} dz \, p_i(z) \\
C_i &= \int_{z_{i+1}^{\min}}^{z_i^{\max}} dz \, [p_i(z) + p_{i+1}(z)].
\end{aligned}
$$

If we take $Z_1 = 1$ and assume that the first window overlaps only with window 2 ($A_1 = 0$) then from equation (a) it follows that Z_2 can be written as

$$
Z_2 = -\frac{B_1}{B_1 + C_1}.
$$

This result can be substituted into equation (a) to give Z_3, which can in its turn be used to calculate Z_4, etc. The result of this calculation is shown in Figure 7.3.

For this example, it is easy to verify that, if we use the analytical expressions for A_i, B_i, and C_i, we do indeed recover $p_0(z)$ exactly. In practice, the histograms (and hence A_i, B_i, and C_i) will be subject to statistical fluctuations. Moreover, more than two windows may be overlapping simultaneously. In this more general case equation (7.3.1) needs to be solved iteratively. Still, an initial estimate for the various Z_i may be obtained with the recursive scheme just described.

7.3.2 Acceptance Ratio Method

The acceptance ratio method is a scheme, derived by Bennett [184], for estimating the free energy difference between two systems (0 and 1) from two simulations: one of system 0 and one of system 1. To derive this scheme, consider the following identity:

$$\begin{aligned}\frac{Q_0}{Q_1} &= \frac{Q_0 \int d\mathbf{r}^N w(\mathbf{r}^N) \exp[-\beta(\mathcal{U}_0 + \mathcal{U}_1)]}{Q_1 \int d\mathbf{r}^N w(\mathbf{r}^N) \exp[-\beta(\mathcal{U}_0 + \mathcal{U}_1)]} \\ &= \frac{\langle w \exp(-\beta\mathcal{U}_0)\rangle_1}{\langle w \exp(-\beta\mathcal{U}_1)\rangle_0}. \end{aligned} \tag{7.3.12}$$

Equation (7.3.12) is valid for an arbitrary w. The question is this: what choice of w yields the highest statistical accuracy for $\beta\Delta F = \ln(Q_0/Q_1)$? Let us first write ΔF in terms of w:

$$\beta\Delta F = \ln \langle w \exp(-\beta\mathcal{U}_0)\rangle_1 - \ln \langle w \exp(-\beta\mathcal{U}_1)\rangle_0 . \tag{7.3.13}$$

Next we compute the estimated statistical error in $\beta\Delta F$ from the variance in the two terms on the right-hand side of equation (7.3.13), divided by the number of (statistically independent) samples (n_0 and n_1, respectively):

$$\begin{aligned}\sigma^2_{\beta\Delta F} &= \frac{\langle [w\exp(-\beta\mathcal{U}_1)]^2\rangle_0 - \langle w\exp(-\beta\mathcal{U}_1)\rangle_0^2}{n_0 \langle w\exp(-\beta\mathcal{U}_1)\rangle_0^2} \\ &\quad + \frac{\langle [w\exp(-\beta\mathcal{U}_0)]^2\rangle_1 - \langle w\exp(-\beta\mathcal{U}_0)\rangle_1^2}{n_1 \langle w\exp(-\beta\mathcal{U}_0)\rangle_1^2} \\ &= \int d\mathbf{r}^N \{[(Q_0/n_0)\exp(-\beta\mathcal{U}_1) + (Q_1/n_1)\exp(-\beta\mathcal{U}_0)] \\ &\quad \times w^2 \exp[-\beta(\mathcal{U}_0 + \mathcal{U}_1)]\} \\ &\quad \times \frac{1}{\{\int d\mathbf{r}^N w \exp[-\beta(\mathcal{U}_0 + \mathcal{U}_1)]\}^2} - \frac{1}{n_0} - \frac{1}{n_1}. \end{aligned} \tag{7.3.14}$$

Note that the right-hand side of equation (7.3.14) does not change if we multiply w by a constant factor. Therefore, without loss of generality, we can choose the following normalization for w:

$$\int d\mathbf{r}^N w \exp[-\beta(\mathcal{U}_0 + \mathcal{U}_1)] = \text{constant}. \tag{7.3.15}$$

Next, we minimize the statistical error in $\beta\Delta F$ with respect to w, with the constraint (7.3.15). This is done most conveniently using Lagrange multipliers:

$$\begin{aligned} 0 &= [(Q_0/n_0)\exp(-\beta\mathcal{U}_1) + (Q_1/n_1)\exp(-\beta\mathcal{U}_0)]\exp[-\beta(\mathcal{U}_0 + \mathcal{U}_1)]w\delta w \\ &\quad - \lambda \exp[-\beta(\mathcal{U}_0 + \mathcal{U}_1)]\delta w \end{aligned} \tag{7.3.16}$$

or

$$w = \frac{\text{constant}}{(Q_0/n_0)\exp(-\beta\mathcal{U}_1) + (Q_1/n_1)\exp(-\beta\mathcal{U}_0)}. \tag{7.3.17}$$

If we now insert this expression for w in equation (7.3.12), we obtain

$$\frac{Q_0}{Q_1} = \frac{\langle\{1+\exp[\beta(\mathcal{U}_0 - \mathcal{U}_1 + C)]\}^{-1}\rangle}{\langle\{1+\exp[\beta(\mathcal{U}_1 - \mathcal{U}_0 - C)]\}^{-1}\rangle} \exp(\beta C), \tag{7.3.18}$$

where we have defined $\exp(\beta C) \equiv (Q_0 n_1)/(Q_1 n_0)$. We can express equation (7.3.18) in terms of the Fermi-Dirac function $f(x) \equiv 1/(1+\exp(\beta x))$:

$$\frac{Q_0}{Q_1} = \frac{\langle f(\mathcal{U}_0 - \mathcal{U}_1 + C)\rangle_1}{\langle f(\mathcal{U}_1 - \mathcal{U}_0 - C)\rangle_0} \exp(\beta C). \tag{7.3.19}$$

Note that equation (7.3.19) is valid for any choice of C. However, the choice $C = \ln[(Q_0 n_1)/(Q_1 n_0)]$ is optimal.

At first sight this choice of C seems problematic because it presupposes knowledge of the very quantity we wish to compute, (Q_0/Q_1). In practice, C is determined by a self-consistency requirement, described next.

Suppose that we have obtained numerical estimates for $\langle f(\mathcal{U}_0 - \mathcal{U}_1 + C)\rangle_1$ and $\langle f(\mathcal{U}_1 - \mathcal{U}_0 - C)\rangle_0$ for a range of values of C:

$$\begin{aligned} \langle f(\mathcal{U}_0 - \mathcal{U}_1 + C)\rangle_1 &= \frac{1}{n_1}\sum_m f_m(\mathcal{U}_0 - \mathcal{U}_1 + C) \\ \langle f(\mathcal{U}_1 - \mathcal{U}_0 - C)\rangle_0 &= \frac{1}{n_0}\sum_{m'} f_{m'}(\mathcal{U}_1 - \mathcal{U}_0 - C), \end{aligned} \tag{7.3.20}$$

where $\sum_m$ ($\sum_{m'}$) stands for the sum over all configurations sampled in a Monte Carlo simulation of system 1 (0). Inserting equations (7.3.20) and (7.3.19) in equation (7.3.13), we obtain

$$\beta\Delta F = \ln\frac{\sum_1 f(\mathcal{U}_0 - \mathcal{U}_1 + C)}{\sum_0 f(\mathcal{U}_1 - \mathcal{U}_0 - C)} - \ln(n_1/n_0) + \beta C, \tag{7.3.21}$$

while the optimal choice for C can be rewritten as

$$\beta\Delta F = -\ln(n_1/n_0) + \beta C. \tag{7.3.22}$$

Clearly, equations (7.3.21) and (7.3.22) are consistent only if

$$\sum_m f(\mathcal{U}_0 - \mathcal{U}_1 + C) = \sum_{m'} f(\mathcal{U}_1 - \mathcal{U}_0 - C). \tag{7.3.23}$$

In practical situations, C will be treated as an adjustable parameter that is varied until equation (7.3.23) is satisfied. For that value of C, $\beta\Delta F$ then follows immediately from equation (7.3.22).

7.4 Umbrella Sampling

In section 7.2.3, the distribution functions $p_0(\Delta\mathcal{U})$ and $p_1(\Delta\mathcal{U})$ were introduced. The functions measure the probability of finding system 0 (1) in an equilibrium configuration $\mathbf{s}^N$ where the potential energy of system 1 and 0 differ by an amount $\Delta\mathcal{U}$. At first sight it would seem that knowledge of either p_0 or p_1 is sufficient for estimating the free energy difference between systems 0 and 1. After all, equation (7.2.15) states that

$$p_1(\Delta\mathcal{U}) = p_0(\Delta\mathcal{U})\exp[\beta(\Delta F - \Delta\mathcal{U})].$$

If we integrate over $\Delta\mathcal{U}$ on both sides of this equation, we obtain

$$\begin{aligned}\int_{-\infty}^{\infty} d\Delta\mathcal{U}\, p_1(\Delta\mathcal{U}) &= \exp(\beta\Delta F)\int_{-\infty}^{\infty} d\Delta\mathcal{U}\, p_0(\Delta\mathcal{U})\exp(-\beta\Delta\mathcal{U}) \\ 1 &= \exp(\beta\Delta F)\left\langle \exp(-\beta\Delta\mathcal{U})\right\rangle_0, \end{aligned} \tag{7.4.1}$$

or

$$\exp(-\beta\Delta F) = \left\langle \exp(-\beta\Delta\mathcal{U})\right\rangle_0 .$$

Although equation (7.4.1) is very useful to estimate free energy differences between two systems that are not too dissimilar, its applicability is limited. The problem is that, in many cases of practical interest, the largest contributions to the average $\langle \exp(-\beta\Delta\mathcal{U})\rangle_0$ come from the region of configuration space, where $p_0(\Delta\mathcal{U})$ is very small while $\exp(-\beta\Delta\mathcal{U})$ is very large. As a result, the statistical error in ΔF is large.

One method to achieve a more accurate estimate of ΔF is the "umbrella sampling" scheme suggested by Torrie and Valleau [186]. The basic idea behind umbrella sampling is that, to obtain an accurate estimate of the free energy difference between two systems (0 and 1), one should sample both the part of configuration space accessible to system 1 and the part accessible to 0. To achieve this in a single simulation, one should modify the Markov chain that governs the sampling of configuration space. This is achieved by replacing the Boltzmann factor of the system by a nonnegative weight function $\pi(\mathbf{r}^N)$. As a result, the probability of visiting a point $\mathbf{r}^N$ in configuration space is now proportional to $\pi(\mathbf{r}^N)$. The expression for $\langle \exp(-\beta\Delta\mathcal{U})\rangle_0$ becomes (see equation (7.2.12))

$$\left\langle \exp(-\beta\Delta\mathcal{U})\right\rangle_0 = \frac{\int d\mathbf{r}^N \pi(\mathbf{r}^N)\exp[-\beta\mathcal{U}_1(\mathbf{r}^N)]/\pi(\mathbf{r}^N)}{\int d\mathbf{r}^N \pi(\mathbf{r}^N)\exp[-\beta\mathcal{U}_0(\mathbf{r}^N)]/\pi(\mathbf{r}^N)}, \tag{7.4.2}$$

or introducing the notation $\langle\cdots\rangle_\pi$ to denote an average over a probability distribution proportional to $\pi(\mathbf{r}^N)$,

$$\left\langle \exp(-\beta\Delta\mathcal{U})\right\rangle_0 = \frac{\langle \exp(-\beta\mathcal{U}_1)/\pi\rangle_\pi}{\langle \exp(-\beta\mathcal{U}_0)/\pi\rangle_\pi}. \tag{7.4.3}$$

For both the numerator and the denominator in this equation to be nonzero, $\pi(\mathbf{r}^N)$ should have an appreciable overlap with both regions of configuration space that are sampled by system 0 and system 1. This bridging property of w is responsible for the name *umbrella sampling.*

At first sight, it might seem advantageous to refine the computation of w in such a way that all relevant configurations can be sampled in one run. Surprisingly, this is not the case. It is usually better to perform several umbrella sampling runs in (partially overlapping) windows. To see this, let us define an order parameter Φ that is a measure for the location of a given configuration between systems 0 and 1. For instance, Φ might be the potential energy or (in constant-pressure simulations) the density. Let us denote the average value of Φ in system 1 by Φ_{max} and the value in 0 by Φ_{min}. Let us assume that we sample an interval $\Phi_{max} - \Phi_{min} \equiv \Delta\Phi$ in n umbrella sampling simulations. The optimum choice of n is clearly the one that samples the complete Φ interval in the minimum computing time. To estimate this time, let us assume that the system performs a random walk in Φ space within the window $\Delta\Phi/n$. Associated with the random walk in Φ space is a diffusion constant D_Φ. The characteristic time needed to sample one interval $\Delta\Phi/n$ is

$$\tau_n = \frac{(\Delta\Phi/n)^2}{D_\Phi}.$$

Clearly, the total time to sample all n windows is

$$\tau_{tot} = n\tau_n = \frac{(\Delta\Phi)^2}{nD_\Phi}.$$

The important point to note is that the computing time decreases with increasing n. It would be incorrect, however, to assume that n should be chosen as large as possible. The actual equilibration time of a run in one of the Φ windows also depends on the rate at which all coordinates orthogonal to Φ are sampled. Let us denote this time by $\tau_\perp$. Clearly, once $\tau_\perp$ becomes appreciably larger than τ_n, the total computation will scale as $n \times \tau_\perp$. This suggests that the optimum choice of n is the one for which $\tau_n \approx \tau_\perp$. For a more detailed discussion, see [187].

The following simple example is meant to demonstrate the power of the umbrella sampling technique. Consider a model for n-butane, where all bond lengths and bond angles are fixed, except the torsional angle ϕ. Let us assume that we know the intramolecular energy function $\mathcal{U}_{intra}(\phi)$ associated with changes of the conformation of the molecule. In the dilute gas, the probability of finding a value of the torsion angle ϕ is proportional to $\exp[-\beta\mathcal{U}_{intra}(\phi)]$. For n-butane, this distribution has a maximum at $\phi = 0°$ (the trans conformation) and two lower maxima at $\phi \approx \pm 120°$, corresponding to the gauche conformation. Let us suppose that we wish to know what happens to the probability of finding a molecule at the transition state between the two gauche conformations, when the molecule is dissolved in an atomic liquid. The total potential energy function for the molecule plus solvent is

$$\mathcal{U}_{\text{tot}} = \mathcal{U}_{\text{inter}}(\mathbf{r}^N, \phi_1) + \mathcal{U}_{\text{intra}}(\phi_1).$$

The probability density $P(\phi)$ of finding a particular value of the angle ϕ is given by[2]

$$P(\phi) = \frac{\int \exp(-\beta\mathcal{U}_{\text{tot}})\delta(\phi - \phi_1)d\mathbf{r}^N d\phi_1}{\int \exp(-\beta\mathcal{U}_{\text{tot}})d\mathbf{r}^N d\phi_1}.$$

Let us now choose the weighting function w equal to $\exp(+\beta\mathcal{U}_{\text{intra}})$. With this choice, we can rewrite $P(\phi)$ as

$$\begin{aligned} P(\phi) &= \frac{\int \exp(-\beta\mathcal{U}_{\text{tot}})w(\delta(\phi - \phi_1)/w)d\mathbf{r}^N d\phi_1}{\int \exp(-\beta\mathcal{U}_{\text{tot}})ww^{-1}d\mathbf{r}^N d\phi_1} \\ &= \frac{\int \exp(-\beta\mathcal{U}_{\text{inter}})[\delta(\phi - \phi_1)\exp(-\beta\mathcal{U}_{\text{intra}})]d\mathbf{r}^N d\phi_1}{\int \exp(-\beta\mathcal{U}_{\text{inter}})\exp(-\beta\mathcal{U}_{\text{intra}})d\mathbf{r}^N d\phi_1} \\ &= \frac{\langle \delta(\phi - \phi_1)\exp(-\beta\mathcal{U}_{\text{intra}})\rangle_{\text{inter}}}{\langle \exp(-\beta\mathcal{U}_{\text{intra}})\rangle_{\text{inter}}}. \end{aligned} \tag{7.4.4}$$

But, as $\mathcal{U}_{\text{intra}}$ depends only on ϕ, we can rewrite equation (7.4.4) as

$$P(\phi) = \frac{\exp[-\beta\mathcal{U}_{\text{intra}}(\phi)]}{\langle \exp[-\beta\mathcal{U}_{\text{intra}}(\phi)]\rangle_{\text{inter}}} P_{\text{inter}}(\phi),$$

where $P_{\text{inter}}(\phi)$ is the probability of finding a conformation with internal angle ϕ in the absence of the intramolecular torsion barrier. $P_{\text{inter}}(\phi)$ can be computed accurately, even for values of ϕ that are very unlikely in the real system, due to the presence of the internal potential energy barrier $\mathcal{U}_{\text{intra}}$.

Although umbrella sampling, in principle, is a powerful technique, one drawback is that the function π is not known *a priori*. Rather, it must be constructed using the available information about the Boltzmann weights of systems 0 and 1. Constructing a good sampling distribution used to require skill and patience. However, during the past few years, several developments should have taken some of the black magic out of umbrella sampling. In fact, a fairly large number of independent papers arrive at similar results. Next, we briefly review some of these developments.

First of all, Valleau [188–190] has proposed a systematic scheme for computing the weight function π that is needed if umbrella sampling is used to compute the properties of a system over a wide range of densities and temperatures. This problem was already considered in 1967 by McDonald and

[2]For convenience we ignore that a Jacobian is associated with the transformation from Cartesian to generalized coordinates (see Chapter 15).

Singer [15, 191, 192], who developed a precursor of the umbrella sampling scheme.

In the case of temperature scaling alone, the umbrella sampling scheme is fairly straightforward. Suppose that we wish to gather statistics on the configurational properties of a system in a temperature range between upper and lower limits T_U and T_L. We can then use, as our sampling function, a linear combination of Boltzmann weights at a discrete set of points in the temperature interval of interest:

$$\pi(\mathbf{r}^N) = \sum_{i=i_L}^{i_U} w_i \exp\left[-\beta_i \mathcal{U}(\mathbf{r}^N)\right], \tag{7.4.5}$$

where $\beta_i = 1/k_B T_i$ and w_i is a nonnegative weight that remains to be specified. To get the sampling as uniform as possible, a logical choice for w_i is

$$w_i \sim \exp[+\beta_i F_{ex}(T_i, V)],$$

because this choice ensures that every term equation (7.4.5) yields the same contribution to the configuration space integral of π. Of course, we do not know $F_{ex}(T_i, V)$ *a priori* and hence we have to make a reasonable guess (on basis of the equation of state). But this is not particularly difficult.

Things become more subtle when we try to apply the same approach to include sampling over a range of densities. At first sight, it would seem that we can use almost the same expression as in equation (7.4.5) (but for the need to use scaled coordinates, $\mathbf{s} = \mathbf{r}/V^{1/d}$). The weight function π would then become

$$\pi(\mathbf{s}^N) = \sum_{i=i_L, j=j_L}^{i_U, j_U} w_{i,j} \exp\left[-\beta_i \mathcal{U}([V_j^{1/d}\mathbf{s}]^N)\right], \tag{7.4.6}$$

where V_j is a volume between the limits V_U and V_L. Again, we could choose

$$w_{i,j} \sim \exp[+\beta_i F_{ex}(T_i, V_j)],$$

but this approach will not always work. The reason is that the sampling function is too tolerant, in the sense that it allows particles to approach one another up to a distance that corresponds to a hard-core diameter in the system with the lowest density. Let us call this distance $s_{min} = \sigma/(V_U^{1/d})$. If we sample configuration space with the function given in equation (7.4.6), then for all but the smallest systems, there will be at least one pair of particles with a separation near s_{min}. For systems with a volume $V < V_U$, this separation will correspond to a very high-energy configuration because the real pair distance would be $r_{min} = \sigma(V/V_U)^{1/d}$. As a consequence, the distribution π is not really a bridging distribution. Valleau [188] has suggested taking care of this problem by making the weight function $\pi(\mathbf{r}^N)$ depend explicitly on the shortest pair distance in the system. Although this solution works for the example discussed by Valleau, it is not a general purpose solution, and other tricks may be needed in other situations.

Umbrella sampling is a very general technique and, in principle, will give the correct answer independently of the umbrella potential that is used. However, the efficiency of this method does depend very much on a clever choice of this umbrella potential. Here, we describe a class of algorithms that achieve essentially the same as umbrella sampling but are more robust (i.e., require less thought).

7.4.1 Nonequilibrium Free Energy Methods

Above, we discussed a range of techniques for computing free energy differences. All these techniques assume either that the system under study is in thermodynamic equilibrium or, at least, that the system is changing slowly in time. Surprisingly, it is also possible to relate the free energy difference between two systems to the *nonequilibrium* work needed to transform one system into the other in an arbitrarily short "switching" time t_s. In what follows, we briefly describe the nonequilibrium free energy expression due to Jarzynski [193, 194] and some of the generalizations proposed by Crooks [195–197]. As before, we consider two N-particle systems: one with a Hamiltonian $H_0(\Gamma)$ and the other with a Hamiltonian $H_1(\Gamma)$, where $\Gamma \equiv \{\mathbf{p}^N, \mathbf{r}^N\}$ represents the phase space coordinates of the system. We assume that we can switch the Hamiltonian of the N-particle system from H_0 to H_1 — that is, we introduce a Hamiltonian H_λ that is a function of a time-dependent switching parameter $\lambda(t)$, such that for $\lambda = 0$, $H_{\lambda=0} = H_0$, while for $\lambda = 1$, $H_{\lambda=1} = H_1$. Clearly, we can then write

$$H_1[\Gamma(t_s)] = H_0[\Gamma(0)] + \int_0^{t_s} dt\, \dot{\lambda} \frac{\partial H_\lambda[\Gamma(t)]}{\partial \lambda}. \tag{7.4.7}$$

Clearly,

$$\int_0^{t_s} dt\, \dot{\lambda} \frac{\partial H_\lambda[\Gamma(t)]}{\partial \lambda}$$

is equal to the work W performed on the system due to the switching of the Hamiltonian. If the switching takes place very slowly, the system remains in equilibrium during the transformation, and W reduces to the *reversible* work needed to transform system 0 into system 1. Hence $W(t_s \to \infty) = F_1 - F_0 \equiv \Delta F$. However, for a finite switching time, the average amount of work $\overline{W}$ that must be expended to transform the system from state 0 to state 1 is *larger* than the free energy difference ΔF

$$\overline{W}(t_s) \geq \Delta F.$$

The work $W(t_s)$ depends on the path through phase space and, for a Hamiltonian system, this path itself depends on the initial phase space coordinate $\Gamma(0)$ (later we shall consider more general situations where many paths connect $\Gamma(0)$ with $\Gamma(t_s)$). Let us next consider the average of $\exp[-\beta W(t_s)]$. The work $W(t_s)$ is a function of the initial phase space position $\Gamma(0)$. We assume that at time $t = 0$, the system is in thermal equilibrium. Then the probability of finding the system 0 with phase space position $\Gamma(0)$ is given by the canonical distribution

$$P_0[\Gamma(0)] = \frac{\exp\{-\beta H_0[\Gamma(0)]\}}{Q_0},$$

where Q_0 is the canonical partition function of the system 0. The average of $\exp[-\beta W(t_s)]$ is then given by

$$\begin{aligned}
&\overline{\exp[-\beta W(t_s)]}\\
&= \int d\Gamma(0)\, P_0[\Gamma(0)] \exp\{-\beta W[t_s, \Gamma(0)]\}\\
&= \int d\Gamma(0)\, \frac{\exp\{-\beta H_0[\Gamma(0)]\}}{Q_0} \exp\{-\beta W[t_s, \Gamma(0)]\}\\
&= \int d\Gamma(0)\, \frac{\exp\{-\beta H_0[\Gamma(0)]\}}{Q_0} \exp\{-\beta [H_1(\Gamma(t_s)) - H_0(\Gamma(0))]\}\\
&= \int d\Gamma(0)\, \frac{\exp\{-\beta H_1[\Gamma(t_s)]\}}{Q_0},
\end{aligned} \tag{7.4.8}$$

where we have used the fact that $W(t_s) = H_1[\Gamma(t_s)] - H_0[\Gamma(0)]$. Finally, we use the fact that the Hamiltonian equations of motion are area preserving. This implies that $d\Gamma(t_s) = d\Gamma(0)$. We then obtain Jarzynski's central result

$$\begin{aligned}
\overline{\exp[-\beta W(t_s)]} &= \int d\Gamma(t_s)\, \frac{\exp\{-\beta H_1[\Gamma(t_s)]\}}{Q_0}\\
&= \frac{Q_1}{Q_0} = \exp(-\beta \Delta F).
\end{aligned} \tag{7.4.9}$$

This is a surprising result, because it tells us that we can obtain information about equilibrium free energy differences from a *nonequilibrium* simulation. In fact, we already know two limiting cases of this result. First of all, in the limit of infinitely slow switching, we recover the relation between ΔF and the *reversible* work W_s, written in the form

$$\exp(-\beta \delta F) = \exp(-\beta W_s).$$

The other limit is instantaneous switching. In that case, W is simply equal to $H_1[\Gamma(0)] - H_0[\Gamma(0)]$ and we get

$$\exp(-\beta \Delta F) = \langle \exp(-\beta \Delta H) \rangle .$$

Crooks [195] has given a more general derivation of equation (7.4.9) that is not limited to Hamiltonian systems. In fact, Crooks has shown that equation (7.4.9) is valid, provided that the dynamics of the system is Markovian and microscopically reversible. This implies that equation (7.4.9) is also valid if the "time-evolution" of the system is determined by a Metropolis Monte Carlo scheme.

Although equation (7.4.9) is both surprising and elegant, it would be wrong to conclude that it necessarily leads to great reductions in the cost of computing free energy differences. In science, as in life, there is no such thing as a free lunch. The problem is that, for strongly nonequilibrium transformations, the statistical accuracy of equation (7.4.9) may be poor. This we already know for the limit of instantaneous switching. The dominant contribution to the average that we wish to compute may come from initial configurations that are rarely sampled. This was the reason why, for the measurement of chemical potentials, the "particle-removal" method was not a viable alternative to the "particle-insertion" scheme. For instance, we could consider a change in the Hamiltonian of the system that does not change the free energy of the system. An example would be a Monte Carlo move that displaces one molecule in a liquid over a distance $+X$. If X is not small compared to the typical molecular dimensions, then the displacement of the particle will most likely require positive work to be performed. The same holds for the reverse situation where we move the particle back over a distance $-X$ from its new position to its starting point. However, the free energies of the initial and final situations are the same. Hence, ΔF should be zero. This implies that the very small fraction of all configurations for which the work is *negative* makes a large contribution to the average of $\exp(-\beta W)$. In fact, as in the particle-insertion/particle-removal case, the resolution of the problem lies in a combination of the forward and reverse schemes. We illustrate this by considering the Hamiltonian system. However, the result is general. We now consider two nonequilibrium processes: one transforms the Hamiltonian from H_0 to H_1 in a time interval t_s, and the other process does the reverse. For both processes, we can make a histogram of the work that is expended during the transformation. For the forward process, we can write

$$p_0(W) = \int d\Gamma(0) \frac{\exp\{-\beta H_0[\Gamma(0)]\}}{Q_0} \delta[W - W(t_s)]. \tag{7.4.10}$$

If we multiply both sides of this equation by $\exp(-\beta W)$ and use the fact that

$W(t_s) = H_1[\Gamma(t_s)] - H_0[\Gamma(0)]$, we get

$$\begin{aligned}
\exp(-\beta W)p_0(W) &= \int d\Gamma(0)\frac{\exp\{-\beta H_0[\Gamma(0)]\}}{Q_0}\exp(-\beta W)\delta[W - W(t_s)] \\
&= \int d\Gamma(0)\frac{\exp\{-\beta H_0[\Gamma(0)]\}}{Q_0} \\
&\quad \times \exp\{-\beta H_1[\Gamma(t_s)] - H_0[\Gamma(0)]\}\delta[W - W(t_s)] \\
&= \int d\Gamma(0)\frac{\exp\{-\beta H_1[\Gamma(t_s)]\}}{Q_0}\delta[W - W(t_s)] \\
&= \frac{Q_1}{Q_0}\int d\Gamma(t_s)\frac{\exp\{-\beta H_1[\Gamma(t_s)]\}}{Q_1}\delta[W - W(t_s)] \\
&= \exp(-\beta\Delta F)p_1(-W).
\end{aligned} \qquad (7.4.11)$$

In the last line, we have used the fact that the work that is performed on going from 1 to 0 is equal to $H_0[\Gamma(0)] - H_1[\Gamma(t_s)] = -W(t_s)$. Hence, just as in the overlapping distribution method (7.2.3), we can obtain ΔF reliably if the histograms of the forward and reverse work show some overlap. This illustrates that the approach of Jarzynski and Crooks provides us not only with a correct statistical mechanical procedure to determine free energy changes in nonequilibrium processes, but also with a powerful diagnostic tool for testing the reliability of the numerical results.

7.5 Questions and Exercises

Question 16 (Free Energy)

1. *Why does equation (7.2.11) not work for hard spheres?*
2. *Derive an expression for the error in estimate of the chemical potential obtained by Widom's test particle method for a system of hard spheres. Assume that the probability of generating a trail position with at least one overlap is equal to* p.
3. *An alternative method for calculating the free energy difference between state* A *and state* B *is to use an expression involving the difference of the two Hamiltonians:*

$$F_A - F_B = \frac{-\ln\left[\langle \exp\left[-\beta\left(H_A - H_B\right)\right]\rangle_{N,V,T,B}\right]}{\beta}. \qquad (7.5.1)$$

Derive this equation. What are the limitations of this method? Show that the Widom's test particle method is just a special case of this equation.

Question 17 (Ghost Volume) *The virial equation is not very convenient for computing the pressure of a hard sphere fluid. Why? It is more convenient to perform a constant-pressure simulation and compute the density. An alternative way to compute the pressure of a hard sphere fluid is to use a trial volume change. In this method, a virtual displacement of the volume is performed and the probability that such a (virtual) move is accepted has to be computed. Derive that this is indeed a correct way of calculating the pressure. (Hint: consider the analogy of Widom's test particle method.)*

Consider now a system of hard-core chain molecules. Can we still use the same scheme to compute the pressure?

Chapter 8

The Gibbs Ensemble

In many respects, computer simulations resemble experiments. Yet, in the study of first-order phase transitions, there seems to be a difference. In experiments, a first-order phase transition is easy to locate: at the right density and temperature, we will observe that an initially homogeneous system will separate into two distinct phases, divided by an interface. Measurement of the properties of the coexisting phases is then quite straightforward. In contrast, in a simulation we often locate a first-order phase transition by computing the thermodynamic properties of the individual phases, then finding the point where the temperature, pressure, and chemical potential(s) of the two bulk phases are equal.

The reason we are usually forced to follow this more indirect route in a simulation is related to the small size of the system studied. If two phases coexist in such systems, a relatively large fraction of all particles resides in or near the interface dividing the phases. To estimate this effect, consider the idealized case that we have a cubic domain of one phase, surrounded by the other. We assume that the outermost particles in the cube belong to the interface and that the rest is bulk-like. The fraction of particles in the interface depends on the system size. As can be seen in Table 8.1, systems with fewer than 1000 particles are interface dominated. And, even for quite large systems, the fraction of particles in the interface is nonnegligible. It is essential therefore to use relatively large systems to calculate reliable coexistence properties. Unfortunately, for large systems long equilibration times are needed, not only because the systems contain many particles, but also because equilibration times in two-phase systems tend to be longer than those in homogeneous systems.

N	125	1,000	64,000	1,000,000
P_{int}	78%	49%	14%	6%

Table 8.1: Percentage of particles (P_{int}) in the interface of a cubic domain containing N particles. Only the outermost particles are assumed to belong to the interface.

Direct simulations of first-order phase coexistence therefore often are computationally rather expensive.[1] However, in the mid-1980s, Panagiotopoulos [94] devised a new computational scheme for studying first-order phase transitions. This scheme has many of the advantages of a direct simulation of coexistence yet few of its disadvantages. Where applicable, this scheme (usually referred to as the *Gibbs ensemble method*) results in a very significant reduction of the computer time required for phase equilibrium calculations. With this method, phase equilibria can be studied in a single simulation. At present, the Gibbs ensemble method has become the technique par excellence to study vapor-liquid and liquid-liquid equilibria. However, like simulations in the grand-canonical ensemble, the method does rely on a reasonable number of successful particle insertions to achieve compositional equilibrium. As a consequence, the Gibbs ensemble method is not very efficient for studying equilibria involving very dense phases. However, there is a technique that greatly facilitates the numerical study of phase equilibria of dense phases. This is the so-called *semigrand ensemble method* of Kofke and Glandt [198], which is discussed in Chapter 9.

The success of the Gibbs ensemble method relies on the possibility of exchanging particles between the two coexisting phases. If one of the coexisting phases is a crystal, one would need to find a vacancy in order to insert a particle. However, the equilibrium concentration of such defects is usually so low that the conventional Gibbs ensemble method becomes impractical. Tilwani and Wu [199] suggested an alternative approach in which an atom is added to the unit box of the solid and this new unit box is used to fill up (tile) space. In this way particles can be added or removed from the system, while the crystal structure is maintained. Tilwani and Wu showed that for the two-dimensional square-well fluid, their method agrees well with the results from free energy calculations [200].

[1]It should be pointed out, though, that, as computers become more powerful, the direct method becomes increasingly attractive because of its simplicity.

8.1 The Gibbs Ensemble Technique

The condition for coexistence of two or more phases I, II, $\cdots$ is that the pressure of all coexisting phases must be equal ($P_I = P_{II} = \cdots = P$), as must be the temperature ($T_I = T_{II} = \cdots = T$) and the chemical potentials of all species ($\mu_I^\alpha = \mu_{II}^\alpha = \cdots = \mu^\alpha$). Hence, one might be inclined to think that the best ensemble for studying would be the "constant-μPT ensemble". The quotation marks around the name of this "ensemble" are intentional, because, strictly speaking, no such ensemble exists. The reason is simple: if we specify only intensive parameters, such as P, T, and μ, the extensive variables (such as V) are unbounded. Another way to say the same thing is that the set P, T, μ is linearly dependent. To get a decent ensemble, we must fix at least one extensive variable. In the case of constant-pressure Monte Carlo simulations this variable is the number of particles N, while in grand-canonical Monte Carlo the volume V of the system is fixed.

After this introduction, it may come as a surprise that the Gibbs ensemble method of Panagiotopoulos [94, 147] comes very close to achieving the impossible: simulating phase equilibria under conditions where the pressure, temperature, and chemical potential(s) of the coexisting phases are equal. The reason this method can work is that, although the *difference* between chemical potentials in different phases is fixed (namely, at $\Delta\mu = 0$), the absolute values are still undetermined. Here, we show how the Gibbs ensemble method can be derived, following the approach developed in the previous chapters.

At this stage, we focus on the version of the Gibbs ensemble where the total number of particles and the total volume of the two boxes remain constant; that is, the total system is at N, V, T conditions. The description of the N, P, T version can be found in [147]. This constant-P method can be applied only to systems containing two or more components because in a one-component system, the two-phase region is a *line* in the P-T plane. Hence, the probability that any specific choice of P and T will actually be *at* the phase transition is vanishingly small. In contrast, for two-component systems, the two-phase region corresponds to a finite *area* in the P-T plane.

Note that in either formulation of the Gibbs method, the total number of particles is fixed. The method can be extended to study inhomogeneous systems [146] and is particularly suited to study phase equilibria in multi-component mixtures [147]. A review of applications of the Gibbs ensemble technique is given in [201]. The great advantage of the Gibbs method over the conventional techniques for studying phase coexistence is that, in the Gibbs method, the system spontaneously "finds" the densities and compositions of the coexisting phases. Hence, there is no need to compute the relevant chemical potentials as a function of pressure at a number of different compositions and then *construct* the coexistence line.

8.2 The Partition Function

In his original article [94], Panagiotopoulos introduced the Gibbs ensemble as a combination of the N, V, T ensemble, N, P, T ensemble, and μ, V, T ensemble. In the previous section we stated that the Gibbs ensemble is *not* a "constant-μ, P, T ensemble", but we did not say what ensemble it actually corresponds to. This point is considered in detail in Appendix H, where we demonstrate that, in the thermodynamic limit, the (constant-V) Gibbs ensemble is rigorously equivalent to the canonical ensemble.

We start our discussion with the expression for the partition function for a system of N particles distributed over two volumes V_1 and $V_2 = V - V_1$, where the particles interact with each other in volume 1 but behave like an ideal gas in volume 2 (see equation (5.6.1)):

$$Q(N, V_1, V_2, T) = \sum_{n_1=0}^{N} \frac{V_1^{n_1}(V - V_1)^{N-n_1}}{\Lambda^{3N} n_1!(N - n_1)!} \int d\mathbf{s}_2^{N-n_1} \int d\mathbf{s}_1^{n_1} \exp[-\beta \mathcal{U}(\mathbf{s}_1^{n_1})].$$

To derive the partition function of the grand-canonical ensemble (section 5.6), we assumed that the particles in volume V_2 behaved as ideal gas molecules. Now we consider the case that the particles in both volumes are subject to the same intermolecular interactions and that the volumes V_1 and V_2 can change in such a way that the total volume $V = V_1 + V_2$ remains constant (see Figure 8.1). In this case, we have to integrate over the volume V_1, which gives, for the partition function [201–203],

$$\begin{aligned} Q_G(N, V, T) \equiv & \sum_{n_1=0}^{N} \frac{1}{V \Lambda^{3N} n_1!(N - n_1)!} \int_0^V dV_1 \, V_1^{n_1}(V - V_1)^{N-n_1} \\ & \times \int d\mathbf{s}_1^{n_1} \exp[-\beta \mathcal{U}(\mathbf{s}_1^{n_1})] \int d\mathbf{s}_2^{N-n_1} \exp[-\beta \mathcal{U}(\mathbf{s}_2^{N-n_1})]. \end{aligned} \tag{8.2.1}$$

From the preceding expressions, it follows that the probability of finding a configuration with n_1 particles in box 1 with a volume V_1 and positions $\mathbf{s}_1^{n_1}$ and $\mathbf{s}_2^{N-n_1}$ is given by

$$\mathcal{N}(n_1, V_1, \mathbf{s}_1^{n_1}, \mathbf{s}_2^{N-n_1}) \propto \frac{V_1^{n_1}(V - V_1)^{N-n_1}}{n_1!(N - n_1)!} \exp\left\{-\beta[\mathcal{U}(\mathbf{s}_1^{n_1}) + \mathcal{U}(\mathbf{s}_2^{N-n_1})]\right\}. \tag{8.2.2}$$

We shall use equation (8.2.2) to derive the acceptance rules for trial moves in Gibbs ensemble simulations.

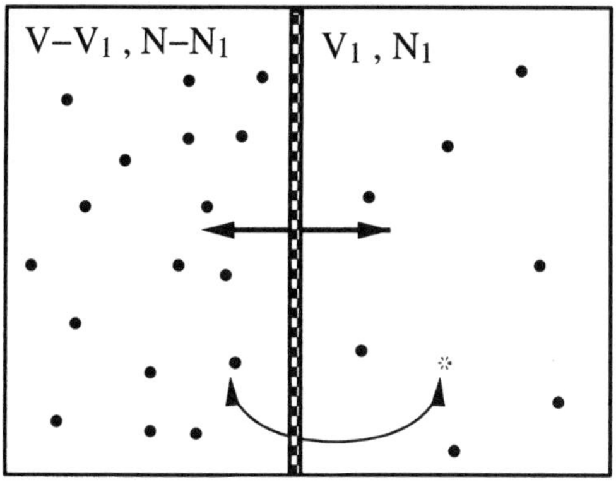

Figure 8.1: Schematic sketch of the "Gibbs ensemble" in which two systems can exchange both volume and particles in such a way that total volume V and the total number of particles N are fixed.

8.3 Monte Carlo Simulations

Equation (8.2.2) suggests the following Monte Carlo scheme for sampling all possible configurations of two systems that can exchange particles and volume. In this scheme, we consider the following trial moves (see Figure 8.2):

1. Displacement of a randomly selected particle.
2. Change of the volume in such a way that the total volume remains constant.
3. Transfer of a randomly selected particle from one box to the other.

The acceptance rules for these steps in the Gibbs ensemble can be derived from the condition of detailed balance

$$K(o \to n) = K(n \to o),$$

where $K(o \to n)$ is the flow of configuration o to n, which is equal to the product of the probability of being in configuration o, the probability of generating configuration n, and the probability of accepting this move:

$$K(o \to n) = \mathcal{N}(o) \times \alpha(o \to n) \times \text{acc}(o \to n).$$

8.3.1 Particle Displacement

We assume that state n is obtained from state o via the displacement of a randomly selected particle in box 1. The ratio of the statistical weights of

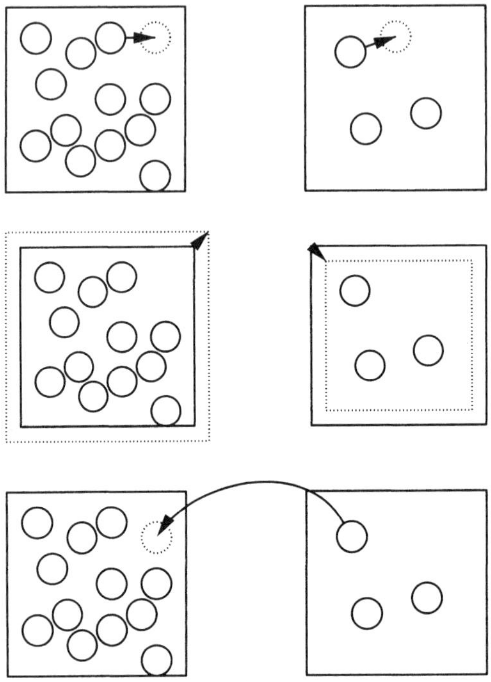

Figure 8.2: Monte Carlo steps in the Gibbs ensemble method: particle displacement, volume change, and exchange of particles.

these two configurations is given by

$$\frac{\mathcal{N}(n)}{\mathcal{N}(o)} = \frac{\exp[-\beta\mathcal{U}(\mathbf{s}_n^{n_1})]}{\exp[-\beta\mathcal{U}(\mathbf{s}_o^{n_1})]}.$$

Substitution of this ratio into the condition of detailed balance (5.1.1) gives, as an acceptance rule,

$$\text{acc}(o \to n) = \min\left(1, \exp\{-\beta[\mathcal{U}(\mathbf{s}_n^{n_1}) - \mathcal{U}(\mathbf{s}_o^{n_1})]\}\right). \tag{8.3.1}$$

This acceptance rule is identical to that used in a conventional N,V,T ensemble simulation (see section 5.2).

8.3.2 Volume Change

For a change of the volume of box 1 by an amount ΔV, $V_1^n = V_1^o + \Delta V$, the ratio of the statistical weights of the configurations after and before the move is given by

$$\frac{\mathcal{N}(n)}{\mathcal{N}(o)} = \frac{(V_1^n)^{n_1}(V - V_1^n)^{N-n_1}}{(V_1^o)^{n_1}(V - V_1^o)^{N-n_1}} \frac{\exp[-\beta \mathcal{U}(\mathbf{s}_n^N)]}{\exp[-\beta \mathcal{U}(\mathbf{s}_o^N)]}.$$

Imposing the condition of detailed balance gives, as an acceptance rule for this volume change,

$$\mathrm{acc}(o \to n) = \min\left\{1, \frac{(V_1^n)^{n_1}(V - V_1^n)^{N-n_1}}{(V_1^o)^{n_1}(V - V_1^o)^{N-n_1}} \exp\{-\beta[\mathcal{U}(\mathbf{s}_n^N) - \mathcal{U}(\mathbf{s}_o^N)]\}\right\}. \tag{8.3.2}$$

This way of changing the volume was originally proposed by Panagiotopoulos *et al.* [94, 147]. A more natural choice for generating a new configuration in the volume change step is to make a random walk in $\ln[V_1/(V - V_1)]$ instead of in V_1 (see also Chapter 5 for the N, P, T ensemble). This has the advantage that the domain of this random walk coincides with all possible values of V_1. Furthermore, the average step size turns out to be less sensitive to the density. To adopt this method to the Gibbs ensemble, the acceptance rule for the volume has to be modified.

If we perform a random walk in $\ln[V_1/(V - V_1)]$, it is natural to rewrite equation (8.2.1) as

$$\begin{aligned} Q_{N,V,T} &= \frac{1}{\Lambda^{3N} N!} \sum_{n_1=0}^{N} \binom{N}{n_1} \\ &\quad \times \int_{-\infty}^{\infty} d\ln\left(\frac{V_1}{V - V_1}\right) \frac{V_1(V - V_1)}{V} V_1^{n_1}(V - V_1)^{N-n_1} \\ &\quad \times \int d\mathbf{s}_1^{n_1} \exp[-\beta \mathcal{U}(\mathbf{s}_1^{n_1})] \int d\mathbf{s}_2^{N-n_1} \exp\left[-\beta \mathcal{U}(\mathbf{s}_2^{N-n_1})\right]. \end{aligned}$$

The statistical weight of a configuration n with volume V_1 is proportional to

$$\mathcal{N}(n) \propto \frac{(V_1^n)^{n_1+1}(V - V_1^n)^{N-n_1+1}}{V n_1!(N - n_1)!} \exp[-\beta \mathcal{U}({\mathbf{s}_n}^N)].$$

Imposing detailed balance for this move leads to the acceptance rule

$$\begin{aligned} \mathrm{acc}(o \to n) &= \min\left\{1, \left(\frac{V_1^n}{V_1^o}\right)^{n_1+1} \left(\frac{V - V_1^n}{V - V_1^o}\right)^{N-n_1+1}\right. \\ &\quad \left. \times \exp\left\{-\beta\left[\mathcal{U}(\mathbf{s}_n^N) - \mathcal{U}(\mathbf{s}_o^N)\right]\right\}\right\}. \end{aligned} \tag{8.3.3}$$

Note that this modification does not affect the acceptance rules for the particle displacement or particle exchange.

8.3.3 Particle Exchange

Let us assume that we generate configuration n from configuration o (n_1 particles in box 1) by removing a particle from box 1 and inserting this particle in box 2. The ratio of statistical weights of the configurations is given by

$$\frac{\mathcal{N}(n)}{\mathcal{N}(o)} = \frac{n_1!(N-n_1)!V_1^{n_1-1}(V-V_1)^{N-(n_1-1)}}{(n_1-1)!(N-(n_1-1))!V_1^{n_1}(V-V_1)^{N-n_1}} \exp\{-\beta[\mathcal{U}(\mathbf{s}_n^N)-\mathcal{U}(\mathbf{s}_o^N)]\}.$$

Imposing detailed balance for this move leads to the following acceptance rule:

$$\text{acc}(o \to n) = \min\left\{1, \frac{n_1(V-V_1)}{(N-n_1+1)V_1} \exp\{-\beta[\mathcal{U}(\mathbf{s}_n^N)-\mathcal{U}(\mathbf{s}_o^N)]\}\right\}. \quad (8.3.4)$$

8.3.4 Implementation

A convenient method for generating trial configurations is to perform a simulation in cycles. One cycle consists of (on average) N_{part} attempts to displace a (randomly selected) particle in one of the (randomly chosen) boxes, N_{vol} attempts to change the volume of the subsystems, and N_{swap} attempts to exchange particles between the boxes. It is important to ensure that at each step of the simulation the condition of microscopic reversibility is fulfilled. Possible Gibbs ensemble algorithms are shown in Algorithms 17–19.

In the original implementation of a Gibbs ensemble simulation the calculations were performed slightly differently [94]; instead of making a random choice of the type of trial move (particle displacement, volume change, or particle exchange) at every Monte Carlo step, the different trial moves were carried out in a fixed order. First, N attempts were made to move every particle in succession (the N, V, T part), then one attempt was made to change the volume (the N, P, T part), and finally N_{try} attempts were made to exchange particles (the μ, V, T part). However, if in a simulation it is possible to choose from a repertoire of trial moves, random selection of the type of trial move is recommended, because this way microscopic reversibility is guaranteed. An additional disadvantage of performing trial moves in a fixed order is that it may make a difference at what point in the program the measurement of the physical properties is performed (e.g., after the N, V, T part, the N, P, T part, or the μ, V, T part). If trial moves are selected at random, all trial moves are on average equivalent and one can simply perform measurements after a predetermined number of MC cycles.

Algorithm 17 (Basic Gibbs Ensemble Simulation)

`PROGRAM mc_Gibbs`	Gibbs ensemble simulation
`do icycl=1,ncycl`	perform `ncycl` MC cycles
`  ran=ranf()*(npart+nvol+nswap)`	
`  if (ran.le.npart) then`	
`    call mcmove`	attempt to displace a particle
`  else if (ran.le.(npart+nvol))`	
`    call mcvol`	attempt to change the volume
`  else`	
`    call mcswap`	attempt to swap a particle
`  endif`	
`  call sample`	sample averages
`enddo`	
`end`	

Comments on this algorithm:

1. *This algorithm ensures that, in each Monte Carlo step, detailed balance is obeyed. On average, we perform per cycle* `npart` *attempts to displace particles,* `nvol` *attempts to change the volume, and* `nswap` *attempts to swap particles between the two boxes.*
2. *Subroutine* `mcmove` *attempts to displace a randomly selected particle; this algorithm is very similar to Algorithm 2 (but remember that particles are in* two *different boxes). The subroutine* `mcvol` *attempts to change the volume of the two boxes (see Algorithm 18), the subroutine* `mcswap` *attempts to swap a particle between the two boxes (see Algorithm 19), and subroutine* `sample` *samples the ensemble averages.*

The implementation of trial moves for particle displacement and volume change in Gibbs ensemble simulations is very similar to that of the corresponding trial moves in a normal N, V, T or N, P, T simulation. However, the attempts to exchange particles require some care. To ensure that detailed balance is obeyed, it is important to first select at random from which box a particle will be removed and subsequently select a particle at random in this box. An alternative would be to first select one particle at random (from all N particles) and then try to move this particle to the other simulation box. However, in that case, acceptance rule (8.3.4) has to be replaced by a slightly different one [125].

The number of attempts to exchange a particle will depend on the condi-

Algorithm 18 (Attempt to Change the Volume in the Gibbs Ensemble)

```
      SUBROUTINE mcvol                        attempt to change
                                              the volume
      call toterg(box1,en1o)                  energy old conf. box 1
      call toterg(box2,en2o)                  and 2 (box1: box length)
      vo1=box1**3                             old volume box 1 and 2
      vo2=v-vo1
      lnvn=log(vo1/vo2)+                      random walk in ln V1/V2
     +    (ranf()-0.5)*vmax
      v1n=v*exp(lnvn)/(1+exp(lnvn))           new volume box 1 and 2
      v2n=v-v1n
      box1n=v1n**(1/3)                        new box length box 1
      box2n=v2n**(1/3)                        new box length box 2
      do i=1,npart
        if (ibox(i).eq.1) then                determine which box
          fact=box1n/box1o
        else
          fact=box2n/box2o
        endif
        x(i)=x(i)*fact                        rescale positions
      enddo
      call toterg(box1n,en1n)                 total energy box 1
      call toterg(box2n,en2n)                 total energy box 2
      arg1=-beta*((en1n-en1o)+
     + (npbox(1)+1)*log(v1n/v1o)/beta)        appropriate weight function
      arg2=-beta*((en2n-en2o)+                acceptance rule (8.3.3)
     + (npbox(2)+1)*log(v2n/v2o)/beta)
      if (ranf().gt.exp(arg1+arg2)) then
        do i=1,npart                          REJECTED
          if (ibox(i).eq.)  then              determine which box
            fact=box1o/box1n
          else
            fact=box2o/box2n
          endif
          x(i)=x(i)*fact                      restore old configuration
        enddo
      endif
      return
      end
```

Comments to this algorithm:

1. *The term* `ibox(i) = 1` *indicates that particle* `i` *is in box 1;* `npart = npbox(1) + npbox(2)` *where* `npbox(i)` *gives the number of particles in box* `i`.
2. *In this algorithm we perform a random walk in* $\ln V$ *and we use acceptance rule (8.3.3).*
3. *The subroutine* `toterg` *calculates the total energy of one of the two boxes. In most cases the energy of the old configuration is known, and therefore it is not necessary to determine this energy at the beginning of the volume step.*

tions of the system. For example, it can be expected that close to the critical temperature, the percentage of accepted exchanges will be higher than close to the triple point. As a possible check whether the number of attempts is sufficient, calculate the chemical potential. Since the calculated energy of a particle that is to be inserted corresponds to just the test particle energy, the chemical potential can be calculated without additional costs. Appendix H shows that, in the Gibbs ensemble, the chemical potential can be obtained from

$$\mu_1 = -k_B T \ln \frac{1}{\Lambda^3} \left\langle \frac{V_1}{n_1 + 1} \exp\left[-\beta \Delta \mathcal{U}_1^+\right] \right\rangle_{\text{Gibbs, box 1}}, \tag{8.3.5}$$

where $\Delta\mathcal{U}_1^+$ is the energy of a (ghost) particle in box 1 and $\langle \cdots \rangle_{\text{Gibbs, box } i}$ denotes an ensemble average in the Gibbs ensemble restricted to box i. It is important to note that this ensemble average is valid only if the boxes do not change identity during a simulation.

Inspection of the partition function (8.2.1) shows that one must allow for $n_1 = 0$ (box 1 empty) and $n_1 = N$ (box 2 empty) to calculate ensemble averages correctly. It is important therefore to ensure that the program can handle the case that one of the boxes is empty. As is clear from equation (8.3.4), the acceptance rule is constructed such that it indeed rejects trial moves that would attempt to remove particles from a box already empty. However, if one also calculates the chemical potential during the exchange step one should be careful. To calculate the chemical potential correctly (see Appendix H) one should continue to add *test particles* when one of the boxes is full.

Case Study 16 (Phase Equilibria of the Lennard-Jones Fluid)

To illustrate the use of the Gibbs ensemble technique, we determine the vapor-liquid curve of the Lennard-Jones fluid. In Case Studies 1, 7, and 9 we already determined parts of the equation of state of this fluid and in

Algorithm 19 (Attempt to Swap a Particle between the Two Boxes)

```
  SUBROUTINE mcswap                   attempts to swap a particle
                                      between the two boxes
  if (ranf().lt.0.5) then             which box to add or remove
    in=1
    out=2
  else
    in=2
    out=1
  endif
  xn=ranf()*box(in)                   new particle at a random position
  call ener(xn,enn,in)                energy new particle in box in
  w(in)=w(in)+vol(in)*                update chemical potential (8.3.5)
+ exp(-beta*enn)/(npbox(in)+1)
  if (npbox(out).eq.0) return         if box empty return
  ido=0                               find a particle to be removed
  do while (ido.ne.out)
     o=int(npart*ranf())+1
     ido=ibox(o)
  enddo
  call ener(x(o),eno,out)             energy particle o in box out
  arg=exp(-beta*(enn-eno +
+ log(vol(out)*(npbox(in)+1)/         acceptance rule (8.3.4)
+ (vol(in)*npbox(out)))/beta))
  if (ranf().lt.arg) then
    x(o)=xn                           add new particle to box in
    ibox(o)=in
    nbox(out)=npbox(out)-1
    nbox(in)=npbox(in)+1
  endif
  return
  end
```

Comments to this algorithm:

1. *The acceptance rule (8.3.4) is used in this algorithm.*
2. *The subroutine* `ener` *calculates the energy of a particle at the given position and box.*
3. *At the end of the simulation, the chemical potential can be calculated from* `w(box)` *using* $\mu_{\text{box}} = -\ln\langle w_{\text{box}}\rangle/\beta$.

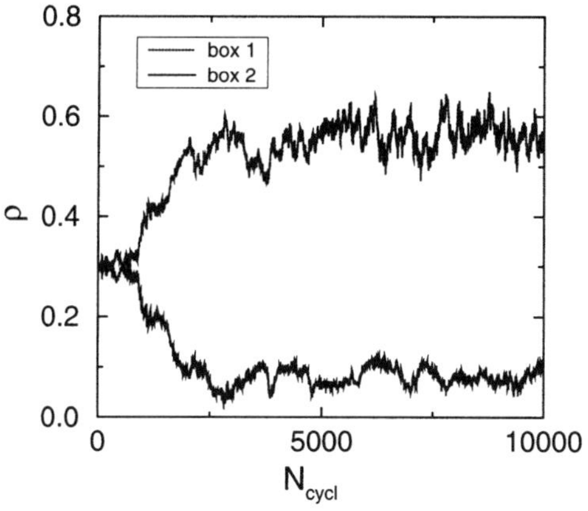

Figure 8.3: Density of the two boxes of the Gibbs ensemble as a function of the number of Monte Carlo cycles for a system of Lennard-Jones particles; the number of particles was $N = 256$ and temperature $T = 1.2$.

Case Study 8 we have made an estimate of the liquid coexistence density from a zero pressure simulation.

In Figure 8.3, the density of the fluid in the two boxes is plotted as a function of the number of Monte Carlo cycles (as defined in Algorithm 17). The simulation was started with equal density in both boxes. During the first 1000 Monte Carlo cycles, the system has not yet "decided" which box would evolve to a liquid density and which box to a gas density. After 5000 Monte Carlo cycles, the system already has reached equilibrium and the coexistence properties can be determined.

In Figure 8.4, the phase diagram of the Lennard-Jones as obtained from Gibbs ensemble simulations is compared with the phase diagram obtained from the equation of state of Nicolas *et al.* [61]. The Gibbs ensemble data are in very good agreement with the equation of state of Nicolas *et al.* Close to the critical point the results deviate because Nicolas *et al.* fitted the equation of state in such a way that the critical point coincides well with the estimate of Verlet [13]: $T_c = 1.35$ and $\rho_c = 0.35$. The Gibbs ensemble simulations [53] give as the estimate for the critical point $T_c = 1.316 \pm 0.006$ $\rho_c = 0.304 \pm 0.006$. Lofti *et al.* [204] used a combination of N,P,T-simulations and test particle insertion to determine the coexistence curve. The estimate of the critical point of Lofti *et al.* ($T_c = 1.310$ and $\rho_c = 0.314$) is in good agreement with the estimate obtained from the Gibbs ensemble. The Gibbs ensemble simulations and the simulation of Lofti *et al.* indicate that the estimate of the critical temperature of Verlet is too high. Johnson *et al.* [62] used this new estimate of the critical point together with additional equation-of-state data

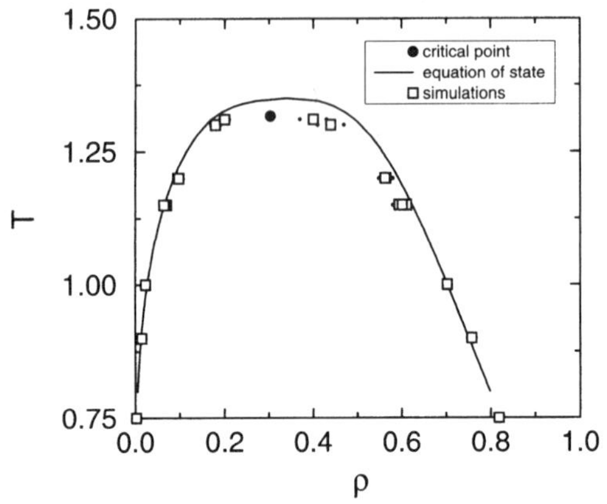

Figure 8.4: Phase diagram of the Lennard-Jones fluid as calculated with the Gibbs ensemble technique (squares) and equation of state of Nicolas *et al.* (solid lines). The solid circle gives the estimate of the critical point.

to improve the equation of state of Nicolas *et al.*

8.3.5 Analyzing the Results

Assuming that we have a working algorithm to perform a simulation in the Gibbs ensemble, we must now address the question whether the numbers generated in a simulation are reliable. First of all, the equilibrium conditions should be fulfilled:

- The pressure in both subsystems must be equal.
- The chemical potential must be equal in both phases.

Unfortunately, both the chemical potential and the pressure of the liquid phase are subject to relatively large fluctuations. Hence, the observation that the equilibrium conditions have been fulfilled *within the statistical error* is not always sufficient. It is convenient therefore to use additional methods to analyze the data and judge whether a simulation has been successful.

Graphical Analysis of Simulation Results

In Appendix H, we describe a graphical technique for analyzing the results of a Gibbs ensemble simulation. In this scheme, the fraction of all particles (n_i/N) in box i is plotted versus the fraction of the total volume (v_i/V) taken up by this box. In the x-y plane, where $x = n_i/N$ and $y = V_i/V$, every dot represents a point sampled in the simulation.

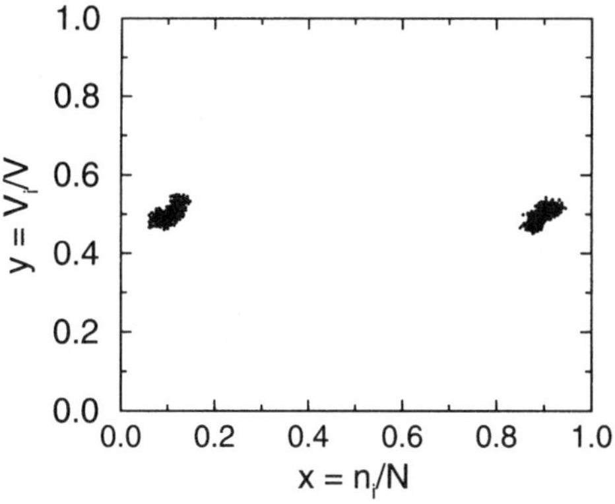

Figure 8.5: Probability plot in the $x - y$ plane of a successful simulation of a Lennard-Jones fluid well below the critical temperature ($T = 1.15$ and $N = 500$).

In the thermodynamic limit, only two points in the x-y plane are sampled; namely, those that correspond to the coexisting liquid and gas density (see Appendix H). For a finite system, we expect to observe fluctuations around these points. Figure 8.5 shows an x-y plot for a simulation of two-phase coexistence well below the critical temperature. The fact that the simulation results cluster around the two points that correspond to the coexisting liquid and vapor indicates that the system was well equilibrated. If a simulation in the Gibbs ensemble is performed far below the critical temperature, it is in general no problem to analyze the results. After the equilibration, it becomes clear which of the boxes contains the vapor phase and which the liquid phase. The densities of the coexisting phases can simply be obtained by sampling the densities at regular intervals. When estimating the accuracy of the simulation one should be careful since the "measured" densities are not sampled independently: in estimating the standard deviations of the results one should take this into account (this aspect is discussed in more detail in Appendix A of [205]).

Close to the critical point, however, it is possible that the boxes continuously change "identity" during a simulation. In Figure 8.6 the evolution of the density in such a simulation close to the critical point is shown. In such a system, the average density in any one of the two boxes will tend to the overall density (N/V). In those circumstances, it is more convenient to construct a histogram of the probability density $P(\rho)$ to observe a density ρ in either box. Even when the boxes change identity during the simulation, the maxima of $P(\rho)$ are still well defined. And, as shown in Appendix H,

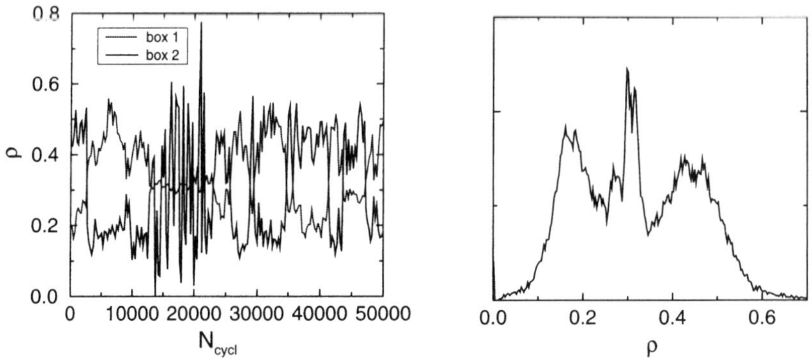

Figure 8.6: Density in the two boxes in a Gibbs ensemble simulation close to the critical temperature. The left figure shows the evolution of the density of the two boxes during a simulation. The right figure gives the corresponding probability density. The simulations were performed on a Lennard-Jones fluid with $N = 256$ at $T = 1.30$.

in the thermodynamic limit, the two maxima of $P(\rho)$ correspond to coexisting vapor and liquid densities, except precisely at the critical point. (For a discussion of the critical behavior of $P(\rho)$, see the article by Allen and Tildesley [41].) Because $P(\rho)$ is obtained by sampling the density in both boxes, the results are not influenced when the boxes change identity. In Figure 8.6 an example of such a density distribution is shown. In this particular example, the simulation was carried out rather close to the critical point. Under those conditions, the interpretation of the density histogram is complicated because interfaces may form in both boxes. As a consequence, three peaks are observed; the two outside peaks correspond to the coexisting liquid and gas phase. A simple model that accounts for the existence of the middle peak is discussed in [148].

Determining the Critical Point

Close to the critical point, the free energy associated with the formation of the liquid-vapor interface becomes very small. As a consequence, the penalty on the creation of an interface in either box becomes small, while the formation of such interfaces is entropically favorable. For this reason, just below the critical point, vapor-liquid coexistence can no longer be observed in a Gibbs ensemble simulation [148]. Therefore, the highest temperature at which the coexistence can be observed is not a proper estimate of the critical temperature of the system. To estimate the critical temperature, the results can be fitted to the law of rectilinear diameters [206]:

$$\frac{\rho_l + \rho_g}{2} = \rho_c + A(T - T_c), \tag{8.3.6}$$

where $\rho_l(\rho_g)$ is the density of the liquid (gas) phase, ρ_c the critical density, and T_c the critical temperature. Furthermore, the temperature dependence of the density difference of the coexisting phases is fitted to a scaling law [207]

$$\rho_l - \rho_g = B(T - T_c)^{\beta}, \tag{8.3.7}$$

where β is the critical exponent[2] (for three-dimensional systems $\beta \approx 0.32$ and for two-dimensional systems $\beta = 0.125$ [207]). A and B depend on the system and are obtained from the fit.

These equations should be used with care. Strictly speaking, they cannot be used for a simulation of a finite system. The reason is that, at the critical point, the correlation length that measures the spatial extent of spontaneous density fluctuations diverges. In a finite system, however, these fluctuations are constrained by the size of the simulation box. If we suppress long-range fluctuations, we in fact are modeling a classical system, which has mean field critical exponents. We therefore can expect to observe a crossover temperature; below this temperature we sample all relevant fluctuations and we expect to observe nonclassical behavior. Above this temperature we expect classical behavior. The crossover temperature will depend on the kind of ensemble used in the simulation.

For the Lennard-Jones fluid in three and two dimensions the finite-size effects for the Gibbs ensemble have been analyzed by Panagiotopoulos [208] (see Example 4). The results of this study indicate that for off-lattice systems this crossover temperature is very close to the critical temperature. This suggests that, for applications in which we want to obtain an estimate of the critical temperature, it is safe to use equations (8.3.6) and (8.3.7). In cases where finite-size effects are nevertheless thought to be significant, it is always possible to perform some simulations using different system sizes (although it seems natural to perform such additional simulations on larger systems, an estimate of the importance of finite-size effects can usually be obtained with much less effort from simulations on smaller systems). Of course, if one is interested specifically in finite-size effects or in the accurate determination of critical exponents then one has to be more careful and a proper finite-size scaling analysis should be performed (see, for example, the work of Rovere *et al.* [99, 101, 209] and Wilding and Bruce [100]). For such calculations, the Gibbs ensemble technique is not particularly well suited.

---

[2] Strictly speaking, the use of a scaling law with nonclassical critical exponents is not consistent with the use of law of rectilinear diameters. However, within the accuracy of the simulations, deviations from the law of rectilinear diameters will be difficult to observe.

Example 4 (Finite-Size Effects in the Gibbs Ensemble)
Most Gibbs ensemble simulations are performed on relatively small systems ($64 \leq N \leq 500$). One therefore would expect to see significant finite-size effects, in particular, close to the critical point. Indeed, in simulations of a system of 100 Ising spins on a lattice,[3] phase coexistence is observed at temperatures as much as 25% above the critical temperature of the infinite system. In contrast to what is found in lattice gases, the first Gibbs ensemble studies of the phase diagram of the Lennard-Jones fluid (in two and three dimensions) [52, 94, 147, 148] did not show significant finite-size effects. This striking difference with the lattice models motivated Mon and Binder [210] to investigate the finite-size effects in the Gibbs ensemble for the two-dimensional Ising model in detail. For the two-dimensional Ising model the critical exponents and critical temperature are known exactly. Mon and Binder determined for various system sizes L the order parameter $M_L(T)$ (see equation (8.3.7)):

$$M_L(T) = \frac{\rho_l(T) - \rho_c}{\rho_c} = (1 - T/T_c)^{\beta},$$

where $\rho_l(T)$ is the density of the liquid phase, ρ_c and T_c are the critical density and temperature, respectively, and β is the critical exponent.

The results of the simulations of Mon and Binder are shown in Figure 8.7, in which the order parameter $M_L(T)$ is plotted as $M_L^{1/\beta}(T)$ versus T/T_c. Such a plot of the order parameter allows us to determine the effective critical exponent of the system. If the system behaves classically, the critical exponent has the mean field value $\beta = 1/2$ and we would expect a linear behavior of $M_L^2(T)$. On the other hand, if the system shows nonclassical behavior, with exponent $\beta = 1/8$, we would expect a straight line for $M_L^8(T)$. Figure 8.7 shows that, away from the critical point, the temperature dependence of the order parameter is best described with an exponent $\beta = 1/8$. Closer to the critical point, the mean field exponent $\beta = 1/2$ fits the data better. This behavior is as expected. Away from the critical point the system can accommodate all relevant fluctuations and exhibits nonclassical behavior. But close to the critical point the system is too small to accommodate all fluctuations and, as a consequence, mean field behavior is observed. In addition, Figure 8.7 shows that we still can observe vapor-liquid coexistence at temperatures 20% above the critical temperature of the infinite system, which implies significant finite-size effects. The study of Mon and Binder shows that, in a lattice model of a fluid, finite-size effects on the liquid-vapor coexistence curve are very pronounced. It is important to note that, in this lattice version of the Gibbs ensemble, we do not change the volume and therefore fewer fluctuations are possible than in the off-lattice version.

---

[3] The Ising model is equivalent to a lattice-gas model of a fluid. The latter model is the simplest that exhibits a liquid-vapor transition.

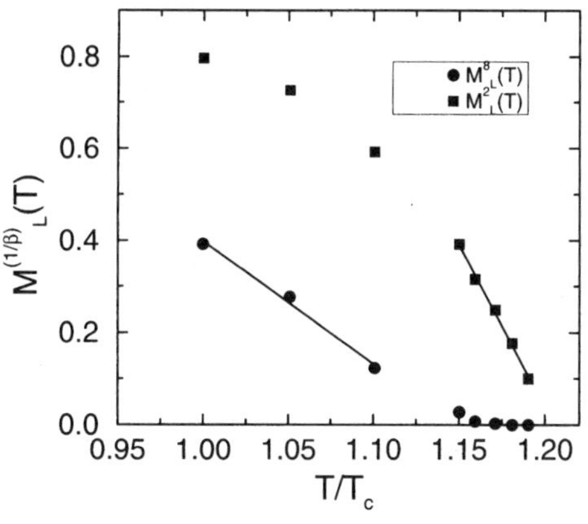

Figure 8.7: Finite-size effects in a Gibbs ensemble simulation of the two-dimensional Ising model. Order parameter $M_L^{1/\beta}(T)$ for $L = 10$ (i.e., $L \times L = 100$ spins) versus T/T_c, where T_c is the exact critical temperature for the infinite system. The lines are fitted through the points. The simulation data are taken from [210].

The striking differences between the findings of Mon and Binder and the results of the early simulations of the Lennard-Jones fluid motivated Panagiotopoulos to reinvestigate in some detail the finite-size effects of Gibbs ensemble simulations of the two- and three-dimensional Lennard-Jones fluid [208]. The results of the simulations of Panagiotopoulos are shown in Figure 8.8. For the Lennard-Jones fluid, the order parameter is defined as

$$M_L(T) = \rho_l - \rho_g.$$

The results for the two-dimensional Lennard-Jones fluid are qualitatively similar to the results of Mon and Binder. At low temperatures, Ising-like behavior is observed and close to the critical point mean-field-like behavior. An important difference is the magnitude of the finite-size effects. Figure 8.8 shows that, for the two system sizes, the results are very similar; the finite size effects are at most 5%. In addition, Figure 8.8 also indicates why the initial Gibbs ensemble studies on the Lennard-Jones fluids did not show significant finite-size effects. All these studies used equations (8.3.6) and (8.3.7) to determine the critical point. If we use these equations we implicitly assume nonclassical behavior up to the critical point. In Figure 8.8, this corresponds to extrapolating the lines, fitted to the data point, for $\beta = 1/8$. Extrapolation of these lines to $M_L^8(T) = 0$ gives a critical point that is not only independent of the system size but also very close to the true critical point of the infinite system.

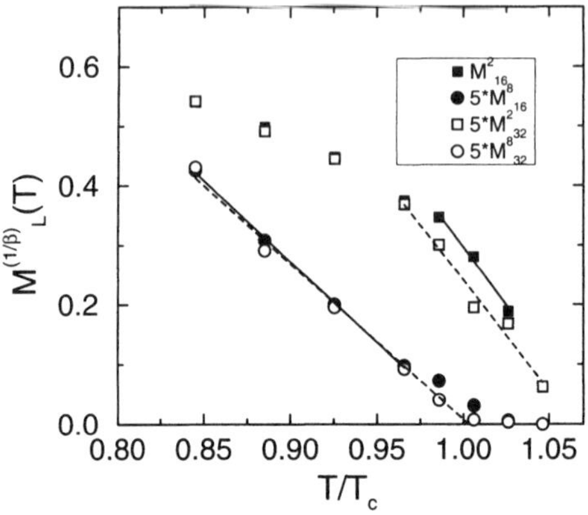

Figure 8.8: Finite-size effects in the liquid-vapor coexistence curve of the two-dimensional Lennard-Jones fluid (truncated potential $r_c = 5.0\sigma$) studied by Gibbs ensemble simulation. The order parameter M corresponds to the density difference between the coexisting liquid and vapor phases. The figure shows $M_L^{1/\beta}(T)$ versus T/T_c for various system sizes L. T_c is the estimated critical temperature for the infinite system ($T_c = 0.497 \pm 0.003$). The simulation data are taken from [208].

For the three-dimensional Lennard-Jones fluid Panagiotopoulos did not observe a crossover from Ising-like to mean field behavior in the temperature regime that could be studied conveniently in the Gibbs ensemble ($T < 0.98T_c$). Also for liquid-liquid equilibria for the square well fluid, Recht and Panagiotopoulos [211] and de Miguel *et al.* [212] did not observe such a crossover. Moreover, for the three-dimensional Lennard-Jones fluid, the finite-size effects were negligible away from T_c and very small close to T_c.

8.4 Applications

The Gibbs ensemble technique has been used to study the phase behavior of a variety of systems. The results of these simulations are reviewed in [201, 213, 214]. Here we discuss a few applications of the Gibbs ensemble for which the algorithm differs significantly from the one described in section 8.3.4.

Example 5 (Dense Liquids)

At high densities, the number of exchange steps can become very large and the simulation requires a significant amount of CPU time. This problem occurs also in conventional grand-canonical Monte Carlo simulations. Various methods, which are used to extend simulations in the grand-canonical ensemble to higher densities, can also be used in the Gibbs ensemble. An example of such a technique is the so-called excluded volume map sampling. This technique, based on the ideas of Deitrick *et al.* [215] and Mezei [111], has been applied to the Gibbs ensemble by Stapleton and Panagiotopoulos [216]. Before calculating the energy of the particle that has to be inserted, a map is made of the receiving subsystem, by dividing this subsystem into small boxes that can contain at most one particle. Each box carries a label that indicates whether it is empty or contains a particle. This map can then be used as a lookup table to check whether there is "space" for the particle to be inserted. If such a space is not available, the trial configuration can be rejected immediately. When using the excluded-volume map, some additional bookkeeping is needed to guarantee detailed balance (see [111] for further details).

Example 6 (Polar and Ionic Fluids)
Because of the long range of the dipolar and Coulombic interactions, the dipolar and Coulombic potential cannot simply be truncated. Special techniques, such as the Ewald summation or reaction field (see Chapter 12.1), have been developed to take into account the long-range nature of the potential in a simulation. A simple truncation of long-range intermolecular interactions at half the box diameter can lead to an incorrect estimate of the phase coexistence curve. In addition to the truncation of dipolar or Coulomb interactions never being admissible, there is a problem if the potential is truncated at half the diameter of the periodic box. As the size of the simulation box fluctuates during the simulation, the effective potential is also changing. As a consequence, a particle in a large simulation box feels a very different interaction potential than a particle in a small box. The result is that a Gibbs ensemble simulation with simulation boxes of different sizes may find apparent phase coexistence between two systems that are described by *different* potentials. In fact, this problem is not limited to Coulomb or dipolar interactions. Even for the relatively short-range Lennard-Jones potential, the phase diagram is very sensitive to the details of the truncation of the potential (see section 3.2.2). An example of an application of the Gibbs ensemble technique to a fluid with Coulomb interactions is the study by Panagiotopoulos [217] of a simple model for an ionic solution, namely, the restricted primitive model.[4] The estimate of the location of the critical point that follows from this simulation differed appreciably from an earlier estimate [188, 218] obtained by truncating the Coulomb potential at half the box diameter.

When the Ewald summation method is used to account for Coulomb or dipolar interactions, then the system size dependence of the results of a Gibbs ensemble simulation is usually quite small. Such weak system size

[4] The restricted primitive model is a hard-core potential with a point charge.

dependence, for instance, was found in a Gibbs ensemble simulation of the liquid-vapor transition in a Stockmayer fluid[5] [219, 220].

In the case of a closely related system, the dipolar hard-sphere fluid, the Gibbs ensemble simulations have even resulted in new insights into an old problem, namely, the location of the liquid-vapor critical point. At first sight, nothing seems to be special about the liquid-vapor transition in the dipolar hard-sphere fluid. As the orientationally averaged interaction between two dipoles results in a van der Waals-like $1/r^6$ attraction, de Gennes and Pincus conjectured that the vapor-liquid coexistence would be similar to that of a conventional van der Waals fluid [221]. Kalikmanov [222] used this conjecture to estimate the critical point. More sophisticated liquid-state theories [223] gave qualitatively similar (though quantitatively different) results. And, indeed, early constant N,V,T Monte Carlo simulations of Ng *et al.* [224] found evidence for such liquid-vapor coexistence, supporting the theoretical predictions of the existence of liquid-vapor coexistence in a dipolar hard-sphere fluid.

However, more recent simulations found no evidence for a liquid-vapor transition [225, 226]. To be more precise, these simulations found no evidence for liquid-vapor coexistence in the temperature range predicted by the different theories. Even at the lowest temperatures that could be studied, liquid-vapor coexistence was not observed. Rather, it was found that at low temperatures the dipoles align nose to tail and form chains [226, 227]. These chains make it very difficult to equilibrate the system. Hence, it becomes difficult to distinguish between two possibilities: either the system is in a thermodynamically stable, homogeneous phase or the system would like to phase separate into a liquid phase and a vapor phase, but this phase separation is kinetically inhibited. In either case, the simulations show that the theoretical description of the liquid-vapor transition in a dipolar hard-sphere fluid needs to be revised. Camp *et al.* [228] have performed extensive (NPT and NVT) Monte Carlo simulations to determine the equation-of-state of the dipolar hard-sphere fluid. These simulations suggest that, in the dipolar hard-sphere system, there is a phase transition between a dilute and a more concentrated isotropic fluid phase. Both phases appear to consist of a network of chains of dipolar molecules. A possible explanation for the occurrence of this phase transition, in terms of a defect-induced critical phase separation, was suggested by Tlusty and Safran [229]. A normal liquid-gas transition is driven by the isotropic aggregation of particles. This conventional mechanism is absent in dipolar hard spheres because it is preempted by chain formation. In the language of ref. [229], the two coexisting phases of the dipolar hard-sphere fluid can be thought of as a dilute gas of chain ends and a high-density liquid of chain branching points. The formation of branching points in the dipolar chains costs entropy but lowers the energy.

[5]The Stockmayer potential is a Lennard-Jones potential plus a point dipole.

Which of the two factors dominates depends on the density of the fluid. For details, we refer the reader to ref. [229].

Example 7 (Mixtures)
An important application of the Gibbs ensemble technique is the simulation of the phase behavior of mixtures (see, e.g., [147,230,231]). One of the main problems in studying liquid-liquid phase coexistence is that both phases are usually quite dense. It is difficult therefore to exchange particles between the two phases. This problem is more serious for the larger of the two species. Fortunately, it is not necessary to carry out such exchanges for all species, to impose equality of the chemical potentials in the coexisting phases. It is sufficient that the chemical potential of only one of the components, label i, is equal in both phases. For the other components, j, we impose that $\mu_j - \mu_i$ should be equal in the two phases. Of course, this implies that, when μ_i is the same in both phases, then so are all μ_j. However, the condition that $\mu_j - \mu_i$ is fixed is much easier to impose in a simulation. In practice, this is achieved by performing Monte Carlo trial moves in which change is attempted on the identity of a particle (e.g., from i to j). The imposed chemical potential difference determines the acceptance probability of such trial moves. This approach was first applied to Gibbs ensemble simulation of mixtures by Panagiotopoulos [232]. In these simulations, only the smaller particles are swapped between the two simulation boxes, while for the larger particles only identity change moves are attempted.

The situation becomes even simpler when studying symmetric mixtures. In such systems, the densities of the coexisting phases are equal, while the molar compositions in boxes I and II are symmetry related ($x_I = 1 - x_{II}$). As a result, in Gibbs ensemble simulations of such symmetric systems, it is not necessary to perform volume changes [233, 234] or particle exchanges between the boxes [212].

8.5 Questions and Exercises

Question 18 (Gibbs Ensemble) *When one of the boxes in the Gibbs ensemble is infinitely large and the molecules in this box do not have intermolecular interactions, the acceptance/rejection rule for particle swap becomes identical to the acceptance/rejection rule for particle swap in the grand-canonical ensemble. Derive this result.*

Question 19 (Scaling of the Potential) *When an attempt is made to change the volume in the Gibbs ensemble, the energy of the new configuration can be calculated efficiently if scaling properties of the potential can be used. Consider a system of*

Lennard-Jones particles. The total energy $\mathcal{U}$ *of the system is equal to*

$$\mathcal{U} = \sum_{i<j} 4\epsilon \left[\left(\frac{\sigma}{r_{ij}} \right)^{12} - \left(\frac{\sigma}{r_{ij}} \right)^{6} \right]. \tag{8.5.1}$$

Suppose that the box size of the system is changed from L *to* L' *and* $s = L'/L$.

1. *Why is this scheme so efficient?*
2. *What is the expression for the total virial of this system?*
3. *Why does this method only work when the cutoff radius is scaled as well?*
4. *Derive expressions for the new energy* $\mathcal{U}'$ *and new virial* V' *as a function of* s, *the old energy* ($\mathcal{U}$) *and virial* (V).
5. *How does the tail corrections scale?*

Exercise 14 (Vapor-Liquid Equilibrium)
In this exercise, we use Widom's test particle method (see section 7.2.1) to locate a vapor-liquid equilibrium. The results are compared with a Gibbs ensemble simulation.

1. Modify the Monte Carlo program of Lennard-Jones particles in the NVT ensemble (only in the file *mc_nvt.f*) in such a way that the chemical potential can be calculated using Widom's test particle method:

$$\mu = \mu_0 - \frac{\ln \left(\rho^{-1} \left\langle \exp\left[-\beta \Delta \mathcal{U}^+\right] \right\rangle \right)}{\beta}, \tag{8.5.2}$$

   in which ρ is the number of particles per volume, $\mathcal{U}^+$ is the energy of a test particle, and

$$\mu_0 = \frac{-\ln\left(\Lambda^3\right)}{\beta}. \tag{8.5.3}$$

   - Make a plot of the chemical potential and pressure as a function of the density for $T = 0.8$.
   - Why is it more difficult to calculate the chemical potential at high densities than at low densities?
   - How can you locate the vapor-liquid coexistence densities?

2. Perform a Gibbs ensemble simulation of the system at $T = 0.8$. In the Gibbs ensemble, the chemical potential of box i is equal to [203]

$$\mu_i = \mu_0 - \frac{\ln \left\langle \frac{V_i}{n_i+1} \exp\left[-\beta \Delta \mathcal{U}_i^+\right] \right\rangle}{\beta}, \tag{8.5.4}$$

   in which n_i is the number of particles in box i and V_i is the volume of box i. Do the vapor-liquid density and chemical potential agree with your previous results?

Chapter 9

Other Methods to Study Coexistence

The great advantage of the Gibbs ensemble method is that we can study coexistence between two phases without creating an interface. The present chapter describes two alternative techniques to study phase coexistence: semigrand ensemble simulations and Gibbs-Duhem integration. These methods also avoid the creation of an interfaces separating the coexisting phases.

9.1 Semigrand Ensemble

There is an alternative way to study the phase behavior of mixtures. This method is also based on the observation that, once the chemical potential of one component in a mixture is fixed, the chemical potential of all other components can be imposed by allowing trial moves that attempt to change the identity of particles. Such simulations are called *semigrand-canonical ensemble simulations* [198].

To explain the basic idea behind semigrand ensemble simulations, it is useful to recall a result derived in section 7.2.1. There we showed that the excess chemical potential of a molecule in a fluid is related to the average Boltzmann factor associated with the random addition of such a molecule to a fluid with N particles present (equation (7.2.5)):

$$\mu_{ex} = -k_B T \ln \int ds_{N+1} \ \langle \exp(-\beta \Delta \mathcal{U}) \rangle_N \,.$$

Suppose that we wish to simulate the phase behavior of a binary mixture. In that case, we must compute the Gibbs free energy per mole of the mixture, as a function of the composition:

$$G(x_A) = x_A \mu_A + x_B \mu_B, \tag{9.1.1}$$

where x_A ($= 1 - x_B$) denotes the mole fraction of species A and μ_A (μ_B) denotes the chemical potentials of the component in the mixture. Let us assume that we have already computed the Gibbs free energy of one of the pure phases (for instance, by one of the thermodynamic integration methods described in section 7.1). At first sight it might seem that, to compute G as a function of x_A, we have to repeat such a thermodynamic integration for a large number of x_A values. Fortunately, this is not usually the case. Rather than recomputing $G(x)$ for a number of compositions, we can study the *change* in $G(x)$ with x. In other words, we need to have a "microscopic" expression for

$$\begin{aligned} \left(\frac{\partial G(x)}{\partial x}\right)_{P,T,N} &= \mu_A - \mu_B \\ &= (\mu_A - \mu_B)_{id} + (\mu_A - \mu_B)_{ex}. \end{aligned} \tag{9.1.2}$$

In the first line of equation (9.1.2), we have used the Gibbs-Duhem relation. We assume that the ideal gas contributions to the chemical potential of both A and B are known. The quantity that we must compute is $\Delta\mu_{ex} \equiv (\mu_A - \mu_B)_{ex}$. Naively, we might try to measure this quantity by using the particle insertion method to obtain μ_{ex} of species A and B separately and then subtracting the result. Although such an approach would be correct *in principle*, it is time consuming and not very accurate. Fortunately, $\Delta\mu_{ex}$ can be obtained much more directly by measuring the Boltzmann factor associated with a virtual trial move, where a randomly selected particle of type B is transformed into a particle of type A [174, 175, 235] (see Figure 9.1). We leave it as an exercise to the reader to derive that the resulting expression for $\Delta\mu_{ex}$ is

$$\Delta\mu_{ex} = -k_B T \ln \left\langle \frac{N_B}{N_A + 1} \exp(-\beta \Delta\mathcal{U}^{+-}) \right\rangle, \tag{9.1.3}$$

where $\Delta\mathcal{U}^{+-}$ denotes the change in potential energy of the system if one particle of type B is changed into type A; $-k_B T \ln(N_B/[N_A + 1])$ is simply the ideal mixing contribution to the chemical potential. The point to note about equation (9.1.3) is that, for a perfect mixture (i.e., A and B have the same intermolecular interactions), $\ln \langle \exp(-\beta\Delta\mathcal{U}^{+-}) \rangle$ is identically equal to 0. In other words, we may obtain very good statistics on $\Delta\mu_{ex}$ even when the direct measurement of the excess chemical potential of the individual species would yield poor statistics.

The aim of this introduction to the semigrand ensemble is twofold. First of all, equation (9.1.3) shows that the Boltzmann factor associated with the

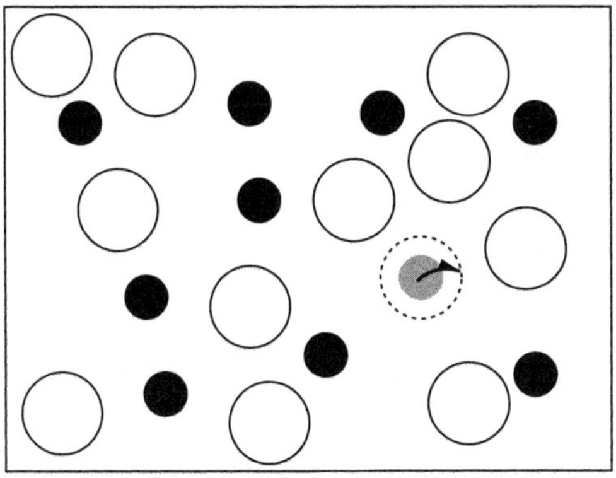

Figure 9.1: Schematic sketch of the semigrand ensemble, in which Monte Carlo moves are attempted in which the identity of the molecules can be interchanged. The figure shows an attempt to transfer the small molecule into a big one.

change of identity of a particle in a mixture is related to the difference in excess chemical potential of the two species that take part in the interchange. Second, we have made it plausible that we can get good statistics on $\Delta\mu_{ex}$ even when the particle insertion method for measuring the excess chemical potential of the individual species would fail, for instance, in a crystalline solid [236]. We recall that the grand-canonical Monte Carlo (GCMC) method has about the same range of applicability as the particle insertion method. It is logical therefore to infer that it should be possible to construct a simulation scheme based on particle *interchanges* that should work under conditions where the GCMC scheme fails. The semigrand-canonical Monte Carlo (SGCMC) method is such a scheme.

How does the SGCMC method work? Let us first consider the expression for Ξ, the grand-canonical partition function for an n-component mixture:

$$\begin{aligned}\Xi(\mu_1,\cdots,\mu_n,T,V) &= \sum_{N_1,N_2,\cdots,N_n}^{\infty}\prod_{i=1}^{n}\frac{q_i^{N_i}\exp(\beta\mu_i N_i)}{N_i!}V^N \\ &\quad\times\int d\mathbf{s}^N \exp[-\beta\mathcal{U}(\mathbf{s}^N)], \end{aligned} \tag{9.1.4}$$

where $N \equiv \sum_i N_i$, $\mathcal{U}(\mathbf{s}^N)$ denotes the potential energy function of the n-component mixture and q_i is the "kinetic" contribution to the partition function due to species i. Next, we consider a related partition function Ξ', identical to Ξ, except for the constraint that $N = \sum_i N_i$ is fixed. If N is fixed, we can eliminate one of the N_i, for instance N_1, from the sum in equation (9.1.4), and we obtain

$$\begin{aligned}\Xi' &= \sum_{N_2,\cdots,N_n}{}' q_1^N \exp(\beta\mu_1 N) \prod_{i=1}^{n} \left(\frac{q_i}{q_1}\right)^{N_i} \frac{\exp[\beta(\mu_i - \mu_1)N_i]}{N_i!} V^N \\ &\times \int ds^N \exp[-\beta\mathcal{U}(\mathbf{s}^N)]. \end{aligned} \tag{9.1.5}$$

We now multiply this equation on both sides by $\exp(-\beta\mu_1 N)$ and we define a new partition function $\mathcal{Y} \equiv \Xi' \exp(-\beta\mu_1 N)$:

$$\mathcal{Y} = \sum_{N_2,\cdots,N_n}{}' q_1^N \prod_{i=1}^{n} \left(\frac{q_i}{q_1}\right)^{N_i} \frac{\exp[\beta(\mu_i - \mu_1)N_i]}{N_i!} V^N \int ds^N \exp[-\beta\mathcal{U}(\mathbf{s}^N)]. \tag{9.1.6}$$

The next step is subtle. We shall reinterpret the sum over all N_i in equations (9.1.4) and (9.1.6). In these equations, we had assumed that, to every composition $N_1, N_2, \cdots, N_n$, there corresponds *one* term in the sum. Let us now take a somewhat different point of view: we assume that these different species are all manifestations of the same "particle". This sounds strange, so we shall use an analogy to explain what we mean. Let us consider that we have a group of 100 people, made up of three groups: eaters, drinkers, and sleepers. In fact, we want to consider all possible combinations of these groups, with the constraint that the total number is fixed. One such combination would be 30 eaters, 30 drinkers, and 40 sleepers. We make a discovery: the same person can be an eater, a drinker, or a sleeper but not simultaneously. Now our sum over all combinations becomes different: we have 100 "persons" who can all take on any one out of the three possible identities. In that case, we have many more ways in which we can make a group of 30 eaters, 30 drinkers, and 40 sleepers, namely $100!/(30!30!40!)$. If we wish to have the same total number of terms in our sum as before, we have to divide by this factor.

Let us now translate this example back to the sum over particles in equation (9.1.6). We replace the sum over numbers of particles of species i by a sum over all possible identities of all particles. But then we must correct for double counting by dividing by $N!/(N_1! \times \cdots \times N_n!)$. If we do that, equation (9.1.6) becomes

$$\mathcal{Y} = \sum_{\text{identities}} \frac{q_1^N}{N!} \prod_{i=1}^{n} \left(\frac{q_i}{q_1}\right)^{N_i} \exp[\beta(\mu_i - \mu_1)N_i] V^N \int ds^N \exp[-\beta\mathcal{U}(\mathbf{s}^N)]. \tag{9.1.7}$$

Finally, it turns out that it is more convenient to consider the corresponding ensemble at constant pressure. In that case, the partition function changes to

$$\begin{aligned}\mathcal{Y}' &= \beta P \int dV \exp(-\beta PV) \frac{(Vq_1)^N}{N!} \\ &\times \left(\sum_{\text{identities}} \prod_{i=1}^{n} \left(\frac{q_i}{q_1}\right)^{N_i} \exp[\beta(\mu_i - \mu_1)N_i] \int d\mathbf{s}^N \exp[-\beta \mathcal{U}(\mathbf{s}^N)] \right).\end{aligned} \tag{9.1.8}$$

For cosmetic reasons, we rewrite equation (9.1.8) in terms of the *fugacities* f_i, rather than the chemical potentials μ_i. Recall that the fugacity of a species i is defined by the expression

$$\mu_i(P, T, \{x_i\}) \equiv \mu_i^0(T) + k_B T \ln(f_i), \tag{9.1.9}$$

where $\mu_i^0(T)$ is the chemical potential of the ideal gas reference state ($P = 1$) of species i. Using the expression for the chemical potential of an ideal gas at pressure P and recalling that, for an ideal gas $f = P$, it is easy to show that

$$\mu_i^0(T) = -k_B T \ln(k_B T q_i). \tag{9.1.10}$$

Inserting equation (9.1.10) in equation (9.1.8), we obtain

$$\mathcal{Y}' = \beta P \int dV \exp(-\beta PV) \frac{(Vq_1)^N}{N!} \sum_{\text{identities}} \prod_{i=1}^{n} \left(\frac{f_i}{f_1}\right)^{N_i} \int d\mathbf{s}^N \exp[-\beta \mathcal{U}(\mathbf{s}^N)]. \tag{9.1.11}$$

What have we achieved by this sequence of transformations from one ensemble to the next? To answer this, it is instructive to look at the characteristic thermodynamic function associated with the various partition functions we have introduced. We started with a grand-canonical ensemble. The link with thermodynamics is given by

$$\beta PV = \ln \Xi(V, T, \{\mu_i\}). \tag{9.1.12}$$

The transformation to $\mathcal{Y}$, equation (9.1.6), corresponds to a change to the thermodynamic variable

$$\beta(PV - \mu_1 N) = \ln \mathcal{Y}(V, T, N, \{\mu_i | i \neq 1\}). \tag{9.1.13}$$

Finally, we change to constant pressure, which means that the characteristic thermodynamic function becomes

$$\beta \mu_1 N = -\ln \mathcal{Y}' [P, T, N, \{\ln(f_i/f_1) | i \neq 1\}]. \tag{9.1.14}$$

Rather than use $\ln(f_i/f_1)$ as the independent variable, it is more convenient to follow [198] and use the *fugacity fraction* ξ_i, defined as

$$\xi_i \equiv \frac{f_i}{\sum_{j=1}^{n} f_j} \tag{9.1.15}$$

The advantage is that, while $\ln(f_i/f_1)$ varies between $-\infty$ and $+\infty$ as we go from pure 1 to pure i, ξ_i varies between 0 and 1. Clearly,

$$\begin{aligned}\beta\mu_1 N &= -\ln\left\{\beta P \int dV\ \exp(-\beta PV)\frac{(Vq_1)^N}{N!}\right.\\ &\quad \times \left.\sum_{\text{identities}}\prod_{i=1}^{n}\left(\frac{\xi_i}{\xi_1}\right)^{N_i}\int d\mathbf{s}^N \exp\left[-\beta\mathcal{U}(\mathbf{s}^N)\right]\right\}. \qquad (9.1.16)\end{aligned}$$

How does μ_1, the chemical potential of the reference species, change with the fugacity fractions of the other species? To see this we consider the derivative of equation (9.1.16) with respect to ξ_i:

$$\left(\frac{\partial\beta\mu_1 N}{\partial\xi_i}\right)_{N,P,T,\{\xi_j|j\neq i\}} = -\left\langle\frac{N_i}{\xi_i}+\frac{N-N_1}{\xi_1}\right\rangle, \qquad (9.1.17)$$

where we have used the fact that $d\xi_i = -\sum_{j\neq i} d\xi_j$. Equation (9.1.17) tells us how we can measure the change in μ_1 changes as we change the chemical potential difference between species 1 and the other species. Let us consider the application to phase coexistence in a binary mixture. In that case we vary only ξ_2. First we measure the excess chemical potential in phase I consisting of pure species 1 and in phase II consisting of pure species 2 (for instance, by thermodynamic integration). Next we compute the change in μ_1 in phase I as we increase ξ_2 from 0 and the corresponding change in μ_1 in phase II as we lower ξ_2 from 1. The point where $\mu_1^{(I)}(\xi_2) = \mu_1^{(II)}(\xi_2)$ is the coexistence point, because at that point $f_1^{(I)} = f_1^{(II)}$ and $f_2^{(I)} = f_2^{(II)}$. Note that we have not specified the nature of phase I and II. They could be liquid, solid, or liquid crystalline.

The only practical problem that remains is the Monte Carlo sampling of $-\langle N_i/\xi_i + (N-N_1)/\xi_1\rangle$, equation (9.1.17). Note that the N_i are the dependent variables. The ξ_i are imposed during a given simulation. In addition to the usual particle moves and volume changes, we must now also consider a move where a particle changes identity. To this end, we select one of the N particles at random and with equal probability assign it one of the n possible identities. The probability of accepting such a trial move is

$$\text{acc}(\xi_i \to \xi_i') = \min\left\{1, \frac{\xi_{i'}}{\xi_i}\exp\left[-\beta\Delta\mathcal{U}(\mathbf{s}^N)\right]\right\}, \qquad (9.1.18)$$

where $\Delta\mathcal{U}(\mathbf{s}^N)$ denotes the change in potential energy of the system if we change the identity of a randomly selected particle from i into i′.

We conclude this discussion of the semigrand ensemble with three comments. First of all, SGCMC is very well suited to study phase equilibria in multicomponent systems that are also in chemical equilibrium. The reason is that every chemical equilibrium simply imposes a relation between the fugacities of the reacting species. Hence, the only effect of a chemical equilibrium is that the number of *independent* fugacities is reduced by one (see Example 8). In its simplest form, the semigrand ensemble method can be used only to study chemical equilibria that involve reactions in which the total number of molecules is conserved. For reactions in which the total number of molecules is not conserved one can use the approach described by Johnson *et al.* [237] and Smith and Triska [238]. This approach, however, does require, in addition to identity changes, Monte Carlo moves that involve the insertion and deletion of particles.

Example 8 (Vapor-Liquid Equilibria of Br_2-Cl_2-BrCl)
The vapor-liquid curve of the ternary system Br_2-Cl_2-BrCl is an example of a phase equilibrium problem in which the components are also in chemical equilibrium. The chemical reaction of interest is

$$Br_2 + Cl_2 \leftrightarrow 2BrCl,$$

with equilibrium constant

$$K(T) = \frac{f^2_{BrCl}}{f_{Br_2} f_{Cl_2}}. \tag{a}$$

This equilibrium constant is approximately 10 (at $T = 273$ K). Since in this chemical reaction the total number of molecules is conserved, we can use the standard semigrand ensemble technique to locate the liquid-vapor coexistence curve.

Let us first consider what the approach would be if we were to perform ordinary N,P,T simulations to determine the vapor-liquid coexistence curve. In that case, we would determine the fugacities (chemical potentials) of the three components in both phases and then find the points for which the fugacity of each is the same in both phases, subject to the constraint imposed by equation (a). Kofke and Glandt [198] have shown that the semigrand ensemble can simplify this procedure significantly.

Let us take Br_2 as the reference component. In the constant-pressure version of the semigrand ensemble, we have as independent variables: the pressure, the temperature, the total number of particles, and the chemical potential differences of two of the three components. However, these two chemical potential differences are not independent, since the fugacities must satisfy equation (a). Substitution of equation (a) in the normalization of the fugacity fraction (9.1.15),

$$\xi_{BrCl} + \xi_{Br_2} + \xi_{Cl_2} \equiv 1, \tag{b}$$

yields a quadratic equation that allows us to express both ξ_{BrCl} and ξ_{Cl_2} as a function of ξ_{Br_2}. The next step is the calculation of the fugacity of Br_2 along the path defined by equations (a) and (b) for both the liquid and the vapor phases. The change in chemical potential of species 1 (in this case, Br_2) along an arbitrary path described by the functions $\xi_i = \xi_i(\nu)$ is given by

$$\beta\mu_1^{(b)} - \beta\mu_1^{(a)} = \int_{\nu^{(a)}}^{\nu^{(b)}} d\nu \sum_{i=1}^{n} \left(\frac{\partial \beta\mu_1 N}{\partial \xi_i}\right)_{N,P,T,\{\xi_j|j\neq i\}} \frac{d\xi_i}{d\nu}. \qquad (c)$$

Using equation (9.1.17), we can write this expression as

$$\beta\mu_1^{(b)} = \beta\mu_1^{(a)} + \int_{\nu^{(a)}}^{\nu^{(b)}} d\nu \left\langle \left[\frac{1}{\xi_1}\frac{d\xi_1}{d\nu} - \sum_{i=2}^{n} \frac{x_i}{\xi_i}\frac{d\xi_i}{d\nu}\right]\right\rangle. \qquad (d)$$

For our system the integration variable ν is $\nu = \xi_{Br_2}$. Equation (d) can be used to determine the change in the chemical potential of the reference compound fugacity along the path defined by equations (a) and (b).

In practice, the simulation proceeds as follows. For the liquid phase the following steps are performed:

1. We start the integration of equation (d) from a state point where the chemical potential of the reference compound can be computed relatively easily. The most natural starting point would be to determine the fugacity of pure liquid Br_2 using one of the methods described in Chapter 7.
2. Next, we must integrate equation (d) from $\nu^{(a)} = \xi_{Br_2} = 1$ to $\nu^{(b)} = \xi_{Br_2} = 0$ and evaluate the integrand in equation (d). This integrand is an ensemble average that is conveniently determined in a semigrand ensemble simulation. Once ξ_{Br_2} is specified, ξ_{Cl_2} and ξ_{BrCl} follow. During the simulation, a trial move may involve either the attempted displacement of a particle or an attempt to change its chemical identity. Attempted identity changes are accepted with a probability given by equation (9.1.18).

In principle, the same scheme can be used to compute the chemical potential of Br_2 in the vapor phase. However, if the vapor phase is dilute, it is often more convenient to compute the lowest few virial coefficients of the mixture. The chemical potential of Br_2 can then be computed analytically from knowledge of these virial coefficients.

Once the dependence of the chemical potential of Br_2 on the fugacity fraction ξ_{Br_2} is known for both phases, we can determine the point where μ_{Br_2} is equal in the vapor and the liquid. By construction, the chemical potentials of the other species are then also equal in both phases.

When compared to the Gibbs ensemble technique, the disadvantage of the semigrand ensemble method is that it is necessary to perform (expensive) free energy calculations for (at most) two reference points. The advantage is that, once this information is known, the semigrand scheme can be applied to dense phases, such as solids.

The SGCMC scheme can also be used to simulate phase equilibria in continuously polydisperse systems, including polydisperse solids. And finally, it can be quite advantageous to combine the SGCMC method with the Gibbs ensemble method for mixtures. In that case the fugacity ratios in both simulation boxes are kept the same. In other words, we allow particles in either box to change identity while remaining in the same box. But in addition we allow trial moves where we attempt to move a particle of the reference species 1 from one box to the other. Now the selection of the particle to be swapped goes as follows. First select box I or box II with equal probability. Next, select any molecule of type 1 in the selected box and try to insert it in the other box. The acceptance probability of such a move is given by equation (8.3.2).[1] The natural choice for the reference species 1 is clearly the one that can be swapped most efficiently, that is, the smallest molecule in the system. In Example 10 we describe an application of the semigrand ensemble for polydisperse systems.

9.2 Tracing Coexistence Curves

Once a single point on the coexistence curve between two phases is known, the rest of that curve can be computed without further free energy calculations. A numerical technique for achieving this has been proposed by Kofke [149,150]. In its simplest form, Kofke's method is equivalent to the numerical integration of the Clausius-Clapeyron equation (although Kofke refers to his approach as *Gibbs-Duhem integration*). Let us briefly recall the derivation of the Clausius-Clapeyron equation. When two phases α and β coexist at a given temperature T and pressure P, their chemical potentials must be equal. If we change both the pressure and the temperature by infinitesimal amounts dP and dT, respectively, then the difference in chemical potential of the two phases becomes

$$d\mu_\alpha - d\mu_\beta = -(s_\alpha - s_\beta)dT + (v_\alpha - v_\beta)dP. \tag{9.2.1}$$

Along the coexistence curve $\mu_\alpha = \mu_\beta$, and hence

$$\frac{dP}{dT} = \frac{s_\alpha - s_\beta}{v_\alpha - v_\beta} = \frac{\Delta h}{T\Delta v}, \tag{9.2.2}$$

where we have used the fact that, at coexistence, $T\Delta s = \Delta h$, where h_α (h_β) denotes the molar enthalpy of phase α (β). As Δh, T, and Δv all can be computed directly in a simulation, dP/dT can be computed from equation (9.2.2).

[1]We point out that the implementation that we suggest here is slightly different from the one advocated in [198] and closer to the approach of Stapleton *et al.* [239].

Kofke used a predictor-corrector algorithm to solve equation (9.2.2). If one of the two coexisting phases is the (dilute) vapor phase, it is convenient to cast equation (9.2.2) in a slightly different form:

$$\frac{d \ln P}{d 1/T} = -\frac{\Delta h}{P \Delta v / T}. \tag{9.2.3}$$

Kofke and co-workers have applied this method to locate the vapor-liquid [149,150] and solid-liquid coexistence curve of the Lennard-Jones fluid [240]. Other applications of the Kofke method can be found in [241–244]. It should be stressed that Gibbs-Duhem integration is in no way limited to the computation of coexistence curves in the P, T plane. A particularly important class of problems that can be treated in an analogous fashion is that where one studies the location of a phase transition as a function of the intermolecular interaction potential. For instance, Agrawal and Kofke have investigated the effect of a change of the steepness of the intermolecular potential in atomic systems on the melting point (see Example 9). In the same spirit, Dijkstra and Frenkel [245, 246] studied the effect of a change in flexibility of rodlike polymers on the location of the isotropic nematic transition, Bolhuis and Kofke [247] the freezing of polydisperse hard spheres, and Bolhuis and Frenkel [248] the isotropic-solid coexistence curve of spherocylinders.

Although Gibbs-Duhem integration is potentially a very efficient technique for tracing a coexistence curve, it is not very robust, as it lacks built-in diagnostics. By this we mean that the numerical errors in the integration of equation (9.2.2) may result in large deviations of the computed coexistence points from the true coexistence curve. Similarly, any error in the location of the initial coexistence points will lead to an incorrect estimate of the coexistence curve. For this reason, it is important to check the numerical stability of the scheme. This can be achieved by performing additional free energy calculations to fix two or more points where the two phases are in equilibrium. Meijer and El Azhar [249] have developed such a scheme in which the estimates of the coexistence densities are systematically improved by combining the Gibbs-Duhem scheme with a free energy difference calculation. In addition, the stability of the integration procedure can be checked by integrating backward and forward in the same interval. There is some evidence [244] that other integration schemes may be preferable to the predictor-corrector method used by Kofke.

In fact, use of the predictor-corrector scheme to estimate phase-coexistence curves can lead to unphysical oscillations. These oscillations occur because of inevitable inaccuracies in our estimate of the initial coexistence point. This problem can be reduced by exploiting the fact that we know that the coexistence curve is smooth. For instance, in ref. [244], it was assumed that the coexistence curve can be fitted to a polynomial in T, of the form

$$\frac{d \ln P}{d1/T} = \sum_{i=0}^{3} \alpha_i T^i.$$

The Gibbs-Duhem integration is initiated as follows. First we perform the original Gibbs-Duhem integration to obtain an estimate for the coexistence curve. At every state point, the right-hand side of equation (9.2.3) is computed. We then fit the coefficients of our polynomial to these numerical data. This provides us with a new estimate of the coexistence pressures. The old and the new pressures are mixed together to improve the stability. This procedure is iterated to convergence.

In some cases, for example, for systems containing long-chain molecules, percolating systems, or lattice models it is very difficult to perform volume changes. Escobedo and de Pablo [250] have shown that, under those conditions, it may be preferable to combine Gibbs-Duhem integration with the grand-canonical ensemble. In this scheme, μ and T are the independent variables, rather than P and T. The variation in the pressure difference of phases α and β is given by

$$dP_\alpha - dP_\beta = (\rho_\alpha - \rho_\beta) d\mu + \left(\frac{s_\alpha}{v_\alpha} - \frac{s_\beta}{v_\beta} \right) dT. \tag{9.2.4}$$

Along coexistence, we have $P_\alpha = P_\beta$, which gives

$$\frac{d\beta\mu}{d\beta} = \frac{\rho_\alpha h_\alpha - \rho_\beta h_\beta}{\rho_\alpha - \rho_\beta} = \frac{\rho_\alpha u_\alpha - \rho_\beta u_\beta}{\rho_\alpha - \rho_\beta}. \tag{9.2.5}$$

Implementing this equation in a Gibbs-Duhem integration scheme implies that the volume changes of the constant pressure simulations are replaced by particle exchanges and removals.

Escobedo [251] developed extensions of the Gibbs-Duhem integration technique for multicomponent fluid mixtures.

Example 9 (Freezing of Soft Spheres)

The earliest simulations of freezing were performed by Alder and Wainwright [17] and Wood and Jacobson [18]. The exact location of this freezing transition was first determined by Hoover and Ree [252]. Subsequently, several authors studied the dependence of the freezing transition on the "softness" of the intermolecular potential. This is done most conveniently by considering a class of model systems of variable softness that contains the hard-sphere model as a limiting case. In this context so-called soft-sphere models have been studied extensively. The soft-sphere model is characterized by a pair potential of the form

$$u(r) = \epsilon \left(\frac{\sigma}{r} \right)^n.$$

Limiting cases are the hard-sphere model ($n \to \infty$) and the one-component plasma (n=1). Before the advent of the Gibbs-Duhem integration scheme, individual simulation studies had been performed to locate the freezing point for soft spheres with $n = 1$ [253], $n = 4, 6, 9$ [254, 255], $n = 12$ [143, 256–258], and $n = \infty$ [252,259]. Actually, the crystal structure at melting changes from fcc (face-centered cubic) (or possibly hcp, hexagonal close-packed) for large n, to bcc (body-centered cubic) for small n. Hoover and Ree [252] have argued that the change from fcc to bcc takes place around n=6. Agrawal and Kofke [151, 240] showed that the Gibbs-Duhem integration technique can be used to locate the melting points of *all* soft-sphere models in one single simulation. The quantity that is changed in this Gibbs-Duhem integration is softness parameter s, defined by $s \equiv 1/n$. We can interpret s as a thermodynamic variable, on the same footing as the pressure P and the temperature T. An infinitesimal variation in the thermodynamic variables T, P, and s results in a variation of the Gibbs free energy G:

$$dG = -SdT + VdP + \frac{N\lambda}{\beta} ds,$$

where we have defined λ as the thermodynamic "force" conjugate to λ (the factor N/β has been introduced to keep our notation consistent with that of [151,240]). We now consider phase coexistence at constant temperature. If we vary both P and s, the difference in chemical potential of the two phases will change:

$$\beta\mu_\alpha - \beta\mu_\beta = \beta(\nu_\alpha - \nu_\beta)dP + (\lambda_\alpha - \lambda_\beta)ds,$$

where ν_α (ν_β) is the molar volume of phase α (β). Along the coexistence curve, μ_α = $\beta\mu_\beta$ and hence

$$\left(\frac{\partial \ln P}{\partial s}\right)_{\text{coex}} = -\frac{\Delta\lambda}{\beta P \Delta\nu}.$$

To use this equation in a simulation, we need the statistical mechanical expression for λ. The partition function of a system at constant pressure and temperature is given by equation (5.4.7)

$$\begin{aligned} Q(N,P,T) &= \frac{\beta P}{\Lambda^{3N} N!} \int dV \exp(-\beta PV) \int d\mathbf{r}^N \exp[-\beta \mathcal{U}(\mathbf{r}^N; s)] \\ &= \frac{\beta P}{\Lambda^{3N} N!} \int dV \exp(-\beta PV) \int d\mathbf{r}^N \prod_{i>j} \exp\left[-\beta\epsilon\left(\sigma/\mathbf{r}_{ij}\right)^{(1/s)}\right]. \end{aligned}$$

The thermodynamic definition of λ is written

$$\lambda \equiv \left(\frac{\beta \partial G}{N \partial s}\right)_{T,P}.$$

Using $G = -k_B T \ln Q(N, P, T)$, we obtain

$$\begin{aligned}\lambda &= -\frac{\beta}{NQ(N,P,T)}\left[\frac{\partial Q(N,P,T)}{\partial s}\right]_{T,P} \\ &= -\frac{\beta\epsilon}{s^2}\left\langle\left(\frac{\sigma}{r}\right)^{1/s}\ln(\sigma/r)\right\rangle \\ &= -\frac{\beta\epsilon}{s^2}\langle u(r)\ln(\sigma/r)\rangle .\end{aligned}$$

The preceding expression is used to measure λ in the coexisting solid and liquid phases. This makes it possible to compute the melting curve in the (P, s) plane. Following this approach, Agrawal and Kofke were able to obtain the melting pressure of the soft-sphere model for all n between 1 and ∞. They were also able to locate the fluid-fcc-bcc triple point at $s \approx 0.16$.

Example 10 (Freezing of Polydisperse Hard Spheres)
One of the early successes of molecular simulation was the discovery that a hard-sphere fluid has a freezing transition [17,18]. Certain colloidal solutions can be considered to be excellent experimental realizations of hard-sphere fluids. However, real colloidal solutions are never perfectly monodisperse. The polydispersity of colloidal suspensions has a strong effect on the location of the freezing transition. It is therefore of considerable interest to study the effect of polydispersity on the location of the freezing transition.

It would seem that the grand-canonical ensemble is the natural one to study polydisperse systems. In this ensemble we can impose the chemical potential distribution that generates a continuous size distribution. For the numerical study of freezing, this approach is not useful, as the probability of a successful insertion/deletion of a particle in the solid phase is very low. To avoid this problem, Bolhuis and Kofke [260,261] used the semigrand ensemble to study the freezing curve of polydisperse hard spheres. To trace out the solid-fluid coexistence curve, they combined the semigrand ensemble (see section 9.1) with the Gibbs-Duhem integration technique (see section 9.2).

In experiments, the polydispersity of a suspension is characterized by the probability density $p(\sigma)$, where σ is the diameter of a hard sphere. In contrast, in a grand-canonical simulation, one would impose $\mu(\sigma)$, i.e., the chemical potential as a function of sigma. This implies that $p(\sigma)$ is not imposed *a priori* but follows from the simulation. In a semigrand-ensemble simulation the chemical potentials are fixed with respect to the chemical potential μ_1 of an arbitrarily chosen reference component. A typical semigrand-ensemble Monte Carlo move is to select a particle at random and change the diameter of this particle. In this way, the insertion and deletion of entire particles are avoided. If we simulate in a similar way a second system at the same pressure and temperature (we use the NPT version for both systems), and we somehow manage to ensure that the reference chemical potentials are matched, then we know that the two systems are in equilibrium. This makes this method ideal for phase equilibrium calculations.

To perform a Gibbs-Duhem integration scheme, we need to derive the equivalent of the Clausius-Clapeyron equation for a polydisperse system. For an ordinary mixture under isobaric conditions, the Gibbs free energy has its minimum value at equilibrium. The Gibbs free energy is a natural function of the temperature, pressure, and number of particles of the various components: $G = G(T, P, N_1, N_2, \ldots, N_m)$. In the semigrand ensemble, however, we keep the total number of particles fixed ($N = \sum_{i=1}^{m} N_i$), and we impose the chemical potential differences with respect to a reference chemical potential, say component 1. To find the fundamental thermodynamic function of state that corresponds to this ensemble, we have to perform a Legendre transformation to change the N_i dependence into a $\mu_i - \mu_1$ dependence for all components except the reference component 1:

$$\begin{aligned} Y &= G - \sum_{i=1}^{m} N_i(\mu_i - \mu_1) + N_1\mu_1 \\ &= \sum_{i=1}^{m} N_i\mu_i - \sum_{i=1}^{m} N_i(\mu_i - \mu_1) = \sum_{i=1} N_i\mu_1 \\ &= N\mu_1. \end{aligned}$$

For a polydisperse system we have a chemical potential distribution, which implies that we can replace the summation by an integration over σ:

$$Y = G - N\int d\sigma\, p(\sigma)\,[\mu(\sigma) - \mu_1(\sigma)] = N\mu(\sigma).$$

The differential form of this equation gives the fundamental thermodynamic equation [198]

$$d(\beta Y) = Hd\beta + \beta VdP - N\int d\sigma\, p(\sigma)\beta\delta\,[\mu(\sigma) - \mu_1(\sigma)] + \beta N\mu(\sigma_1)dN,$$

where δ represents a functional differential.

Bolhuis and Kofke used a quadratic form for the chemical potential to express the polydispersity:

$$\beta\,[\mu(\sigma) - \mu_1(\sigma)] = -\frac{(\sigma - \sigma_1)^2}{2\nu}. \qquad (9.2.6)$$

In the limit $\nu \to 0$ we recover a system of purely monodisperse hard-spheres diameter σ_1. With this equation for the chemical potential, the fundamental thermodynamic equation can be written as

$$d(\beta Y) = Hd\beta + \beta VdP\beta N\mu(\sigma_1)dN - \frac{Nm_1}{\nu}d\sigma_1 - \frac{Nm_2}{2\nu^2}d\nu \qquad (9.2.7)$$

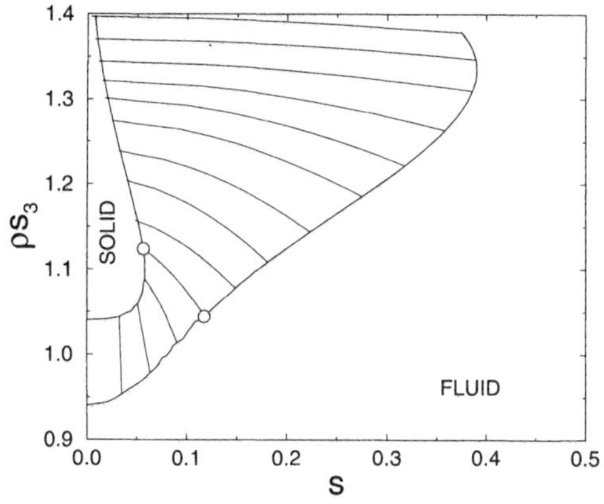

Figure 9.2: Coexisting fluid and solid phases in the volume fraction polydispersity plane. For high polydispersity in equation (9.2.7) higher order terms in σ have been added, see ref. [261] for details.

with

$$m_n \equiv \int d\sigma \, (\sigma - \sigma_1)^n p(\sigma).$$

The conventional form of the Clausius-Clapeyron equation expresses how, at coexistence, the pressure changes if one changes the temperature. In the present case, we are interested in the dependence of the coexistence curve on the polydispersity parameter ν. It is straightforward to derive the corresponding Clausius-Clapeyron equation. Since $\beta Y = \beta\mu_1 N$, we have

$$d(\beta Y) = N d[\beta\mu(\sigma_1)] + \beta\mu(\sigma_1) dN.$$

If we combine this equation with equation (9.2.7), we obtain

$$d[\beta\mu(\sigma_1)] = h d\beta + \beta v dP - \frac{m_1}{\nu} d\sigma_1 - \frac{m_2}{2\nu^2} d\nu.$$

If two phase are in equilibrium the chemical potentials must be the same in the two phases. This is the case if $\mu(\sigma_1)$ is the same in the two phases and the potential difference functions are the same; i.e., we should use equation (9.2.6) with the same ν and σ_1 for the two phases. Hence, for the two phases we have $d\sigma_1 = 0$ and $d\beta = 0$, giving the desired Clausius-Clapeyron equation:

$$\frac{dP}{d\nu} = \frac{\Delta m_2}{2\nu^2 \beta \Delta v}.$$

Assuming that we know one point on the coexistence curve, we can use this equation to trace the coexistence curve as a function of the ν.

In Figure 9.2 the solid-fluid coexistence curve is plotted as a function of the polydispersity s, which is determined during the simulations using

$$s \equiv \frac{\langle \sigma^2 \rangle}{\langle \sigma \rangle^2} - 1.$$

It turns out that the original (fcc) hard-sphere crystal structure cannot support a polydispersity larger than 5.7% of the average sphere diameter.

Chapter 10

Free Energies of Solids

On cooling or compression, almost all liquids undergo a first-order phase transition to the solid state. It is of considerable practical importance to be able to predict the location of the freezing point. In this chapter, we describe various methods for locating solid-liquid coexistence by simulation. The reader may wonder why a special chapter is devoted to numerical schemes for locating the solid-liquid coexistence curve. After all, techniques for locating a first-order coexistence curve were already discussed in Chapter 8 in the context of the liquid-vapor transition. The reason is, of course, that most of the techniques that work for moderately dense liquids and gases do not work for solids. Consider, for example, the Gibbs ensemble method. For this method to work, it is essential that it be possible to exchange particles between the two coexisting phases. The introduction of a particle into the solid phase requires the presence of a vacancy in the lattice. Such defects do occur in real solids, but their concentration is so low (for example, in the case of a hard-sphere crystal near melting, there is on average one defect in a system of 8000 particles) that one would need a very large crystal to observe a reasonable number of holes in a simulation. Hence, the Gibbs ensemble technique, although still valid in principle, would not be very practical for the study of solid-liquid or solid-solid coexistence.[1]

For more details and theoretical background concerning the numerical study of solid-fluid coexistence, we refer the readers to the review of Monson and Kofke [262].

---

[1] In some special case the method of Tilwani and Wu [199] (see Chapter 8) might make a direct Gibbs ensemble simulation feasible.

10.1 Thermodynamic Integration

Thermodynamic integration is the method most commonly used in the study of the solid-liquid transition. For the liquid phase, this calculation is straightforward and was already discussed in section 7.1: the Helmholtz free energy F of the liquid is determined by integrating this equation of state, starting at low densities where the fluid behaves effectively as an ideal gas:

$$\frac{F(\rho)}{Nk_BT} = \frac{F^{id}(\rho)}{Nk_BT} + \frac{1}{k_BT}\int_0^\rho d\rho' \left(\frac{P(\rho') - \rho' k_BT}{\rho'^2}\right), \tag{10.1.1}$$

where the equation of state as a function of the density (ρ) is denoted by $P(\rho)$, and $F^{id}(\rho)$ is the free energy of an ideal gas at density ρ. An important condition is that the integration path in equation (10.1.1) should be reversible. If the integration path crosses a strong first-order phase transition, hysteresis may occur, and equation (10.1.1) can no longer be used. For a liquid phase, this problem can be avoided by performing the integration in two steps. Start the simulation at a temperature well above the critical temperature and determine the equation of state for compression along an isotherm to the desired density. In the second step, the system is cooled at constant density to the desired temperature. The free energy change in this step is given by

$$\frac{F(T = T_{II})}{k_BT_{II}} - \frac{F(T = T_I)}{k_BT_I} = \int_{T_I}^{T_{II}} d(1/T)U(T, N, V). \tag{10.1.2}$$

The solid-liquid coexistence curve itself does not end in a critical point, and hence there exists no "natural" reversible path from the solid to the ideal gas that does not cross a first-order phase transition. It is usually possible, however, to construct a reversible path to other states of known free energy. The construction of such paths is the main topic of the present chapter.

Various routes arrive at a state of known free energy. In the mid-1960s, Hoover and Ree introduced the so-called *single-occupancy cell method* [252, 263]. In the single-occupancy cell method, the solid is modeled as a lattice gas; each particle is assigned to a single lattice point and is allowed to move only in its "cell" around this lattice point. The lattice sites coincide with the average positions of the atoms of the unconstrained solid. If the density is sufficiently high—such that the walls of the cells have a negligible influence on the properties of the system—the free energy of this lattice model is identical to that of the original solid. The single-occupancy cell model can be expanded uniformly without melting (or, more precisely, without loosing its translational order). In this way, we obtain a (presumably reversible) integration path to a dilute lattice gas, the free energy of which can be calculated analytically. The earliest application of the single-occupancy cell method was the calculation by Hoover and Ree of the free energy of the hard-disk [263] and hard-sphere solid [252].

An alternative for the single-occupancy cell method was also developed by Hoover and co-workers [254, 256]. In this approach the solid is cooled to a temperature sufficiently low for it to behave as a harmonic crystal. The Helmholtz free energy of a harmonic crystal can be calculated analytically, using lattice dynamics. The free energy of the solid at higher temperatures then follows from integration of equation (10.1.2).[2]

In practice, both the single-occupancy cell method and the method using the harmonic solid have some limitations that make a more general scheme desirable. For example, there is some evidence that the isothermal expansion of the single-occupancy cell model may not be completely free of hysteresis [258]: at the density where the solid would become mechanically unstable in the absence of the artificial cell walls, the equation of state of the single-occupancy cell model appears to develop a cusp or possibly even a weak first-order phase transition. This makes the accurate numerical integration of equation (10.1.1) difficult.

The harmonic-solid method can work only if the solid phase under consideration can be cooled reversibly all the way down to the low temperatures where the solid becomes effectively harmonic. However, many molecular solids undergo one or more first-order phase transitions on cooling. Even more problematic are model systems for which the particles interact via a discontinuous (e.g., hard-core) potential. The crystalline phase of such model systems can never be made to behave like a harmonic solid. For complex molecular systems the problem is of a different nature. Even if these materials can be cooled to become a harmonic crystal, often it is a highly nontrivial matter to compute the Helmholtz free energy in this limit.

In the present chapter, we discuss a method that does not suffer from these limitations and can be applied to arbitrary solids [264]. Although the method is generally applicable, it is advantageous to make small modifications depending on whether we study an atomic solid with a discontinuous potential [259], or with a continuous potential [265], or a molecular solid [155].

10.2 Free Energies of Solids

The method discussed in this section is a thermodynamic integration technique for computing the Helmholtz free energy of an atomic solid. The basic idea is to transform the solid under consideration reversibly into an Einstein crystal. To this end, the atoms are coupled harmonically to their lattice sites. If the coupling is sufficiently strong, the solid behaves as an Einstein crystal,

[2] If we use equation (10.1.2) directly, the integration will diverge in for the limit $T \to 0$. This divergence can be avoided if we determine the difference in free energy of the solid of interest and the corresponding harmonic crystal.

the free energy of which can be calculated exactly. The method was first used for continuous potentials by Broughton and Gilmer [266], while Frenkel and Ladd [259] used a slightly different approach to compute the free energy of the hard-sphere solid. Subsequent applications to atomic and molecular substances can be found in [155,265].

10.2.1 Atomic Solids with Continuous Potentials

Let us first consider a system that interacts with a continuous potential, $\mathcal{U}(\mathbf{r}^N)$. We shall use thermodynamic integration (7.1.6) to relate the free energy of this system to that of a solid of known free energy. For our reference solid, we choose an Einstein crystal, i.e., a solid of noninteracting particles that are all coupled to their respective lattice sites by harmonic springs. During the thermodynamic integration we switch on these spring constants and switch off the intermolecular interactions. To this end we consider a potential energy function

$$\tilde{\mathcal{U}}(\mathbf{r}^N) = \mathcal{U}(\mathbf{r}_0^N) + (1-\lambda)\left[\mathcal{U}(\mathbf{r}^N) - \mathcal{U}(\mathbf{r}_0^N)\right] + \lambda \sum_{i=1}^{N} \alpha_i (\mathbf{r}_i - \mathbf{r}_{0,i})^2, \quad (10.2.1)$$

where $\mathbf{r}_{0,i}$ is the lattice position of atom i and $\mathcal{U}(\mathbf{r}_0^N)$ is the static contribution to the potential energy (i.e., the potential energy of a crystal with all atoms at their lattice positions), λ is the switching parameter, and α_i is the Einstein-crystal spring constant coupling atom i to its lattice site. Note that for $\lambda = 0$ we recover the original interactions; for $\lambda = 1$, we have switched off the intramolecular interactions completely (except for the constant static term) and the system behaves like an ideal (noninteracting) Einstein crystal. The free energy difference is calculated using equation (7.1.6):

$$\begin{aligned} F &= F_{Ein} + \int_{\lambda=1}^{\lambda=0} d\lambda \left\langle \frac{\partial \mathcal{U}(\lambda)}{\partial \lambda} \right\rangle_\lambda & (10.2.2) \\ &= F_{Ein} + \int_{\lambda=1}^{\lambda=0} d\lambda \left\langle \sum_{i=1}^{N} \alpha_i (\mathbf{r}_i - \mathbf{r}_{0,i})^2 - \left[\mathcal{U}(\mathbf{r}^N) - \mathcal{U}(\mathbf{r}_0^N)\right] \right\rangle_\lambda . \end{aligned}$$

The configurational free energy of the noninteracting Einstein crystal is given by

$$F_{Ein} = \mathcal{U}(\mathbf{r}_0^N) - \frac{d}{2\beta} \sum_{i=1}^{N} \ln(\pi/\alpha_i \beta). \quad (10.2.3)$$

As we shall see later, it is computationally more convenient to consider a crystal with fixed center of mass. This will result in a slight modification of equation (10.2.3) (see section 10.3.2). The "spring constants" α_i can be adjusted to optimize the accuracy of the numerical integration of equation (10.2.2). It is reasonable to assume that the integration is optimal if the fluctuations of the quantity $\sum_{i=1}^{N} \alpha_i (\mathbf{r}_i - \mathbf{r}_{0,i})^2 - \mathcal{U}(\mathbf{r}^N)$ are minimal, which implies that the interactions in the pure Einstein crystal should differ as little as possible from those in the original system. This suggests that α_i should be chosen such that the mean-squared displacements for $\lambda = 1$ and $\lambda = 0$ are equal:

$$\left\langle \sum_{i=1}^{N} (\mathbf{r}_i - \mathbf{r}_{0,i})^2 \right\rangle_{\lambda=0} = \left\langle \sum_{i=1}^{N} (\mathbf{r}_i - \mathbf{r}_{0,i})^2 \right\rangle_{\lambda=1} .$$

Using the expression for the mean-squared displacement in an Einstein crystal (10.3.29) we find the following condition for α:

$$\frac{3}{2\beta\alpha_i} = \left\langle (\mathbf{r}_i - \mathbf{r}_{0,i})^2 \right\rangle_{\lambda=0} . \tag{10.2.4}$$

For systems with diverging short-range repulsions (such as, for instance, the Lennard-Jones potential), the integrand in equation (10.2.2) will exhibit a weak divergence. This is due to the finite probability that, in an Einstein crystal, two particles may overlap. In practice, this causes few problems since the divergence is integrable. Furthermore, the amplitude of the diverging contribution can be strongly suppressed by increasing the value of the α's or by truncating the potential for small values of r.

10.3 Free Energies of Molecular Solids

A molecular solid has internal degrees of freedom in addition to the translation degrees of freedom. These other degrees of freedom can give rise to a wide variety of crystal structures. For example, a simple molecule such as nitrogen has at least seven different solid phases [267].

The orientational degrees of freedom of a molecular solid usually complicate the numerical calculation of the free energy. Although we can still use an Einstein crystal as reference system, it is often nontrivial to find a path to this reference system that is free of phase transitions and not plagued by divergences of the integrand of equation (10.2.2). In such cases it may be advantageous to use an alternative method that is more robust than the conventional coupling-parameter method described previously.

Let us consider an orientationally disordered molecular solid. We transform this solid into a state of known free energy in two stages [155,264]. First we couple the molecules in the solid with harmonic springs to their lattice sites. But in contrast to the method described earlier, we leave the original intramolecular interactions unaffected. Subsequently, we expand this "interacting Einstein crystal" to zero density (see Figure 10.1). Due to the coupling

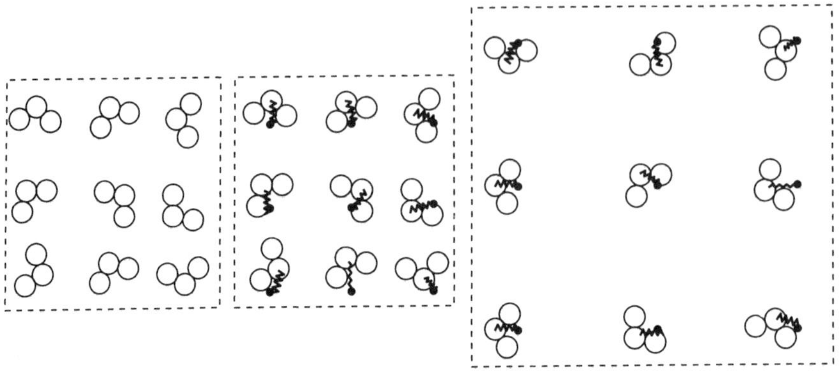

Figure 10.1: Schematic drawing of the lattice-coupling-expansion method for calculating the free energy of a molecular solid: the first step is the coupling to an Einstein crystal (denoted by the black dots) and the second step the expansion to zero density.

to the lattice, the crystal cannot melt on expansion but keeps its original structure. In the low-density limit, all intermolecular interactions vanish and the system behaves as an ideal Einstein crystal. This scheme of calculating an absolute free energy is referred to as the *lattice-coupling-expansion method* [155, 264].

During the first stage of the thermodynamic integration, the potential energy function $\tilde{\mathcal{U}}_I$ contains both the original intermolecular potential and the harmonic coupling to the lattice:

$$\tilde{\mathcal{U}}_I(\mathbf{r}^N, \boldsymbol{\Omega}^N; \lambda) = \mathcal{U}(\mathbf{r}^N, \boldsymbol{\Omega}^N) + \lambda \sum_{i=1}^{N} \alpha(\mathbf{r}_i - \mathbf{r}_{0,i})^2, \tag{10.3.1}$$

where $\boldsymbol{\Omega}_i$ denotes the orientation of particle i, $\mathbf{r}_i$ its center-of-mass position, and $\mathbf{r}_{0,i}$ the lattice site of particle i. For convenience, we have assumed that all lattice sites are equivalent. We therefore use the same value of the coupling constant α for all sites. In most molecular solids, several nonequivalent molecules may be in a unit cell. In that case different coupling constants may be chosen for all distinct lattice sites.

The change in free energy associated with switching on the harmonic springs is given by equation (7.1.6):

$$\Delta F_I = F(\lambda = 1) - F_{\text{mol sol}} = \int_0^1 d\lambda \left\langle \sum_{i=1}^{N} \alpha(\mathbf{r}_i - \mathbf{r}_{0,i})^2 \right\rangle_\lambda . \tag{10.3.2}$$

It is reasonable to expect that the integrand in equation (10.3.2) is a smooth function of λ, as the mean-squared displacement decreases monotonically with increasing λ.

During the second stage of the thermodynamic integration, all molecules remain harmonically coupled to their (Einstein) lattice sites, but this reference lattice is expanded uniformly to zero density. In what follows, we assume for convenience that the intermolecular potential is pairwise additive:

$$\mathcal{U}(\mathbf{r}^N, \mathbf{\Omega}^N) = \sum_{i<j}^{N} \mathcal{U}_{\text{pair}}(r_{ij}, \mathbf{\Omega}_i, \mathbf{\Omega}_j),$$

where $r_{ij} = |\mathbf{r}_i - \mathbf{r}_j|$ is the distance between the centers of mass of the molecules i and j. When we expand the system uniformly with a factor γ, the coordinates of lattice sites are given by $\gamma\mathbf{r}_{0,i}$. When the lattice is expanded, the intermolecular interactions between the molecules change. This is best seen by expressing the center-of-mass coordinate of particle i as

$$\mathbf{r}_i = \gamma\mathbf{r}_{0,i} + \Delta\mathbf{r}_i,$$

where $\Delta\mathbf{r}_i$ is the position with respect to a lattice site of the expanded Einstein crystal. For the distance between two particles, we can write

$$r_{ij} = |\gamma(\mathbf{r}_{0,i} - \mathbf{r}_{0,j}) + \Delta\mathbf{r}_i - \Delta\mathbf{r}_j|. \tag{10.3.3}$$

In terms of these coordinates, the potential energy of the expanded crystal is

$$\tilde{\mathcal{U}}_{\text{II}}(\mathbf{r}^N, \mathbf{\Omega}^N; \gamma) = \sum_{i<j}^{N} \mathcal{U}_{\text{pair}}(r_{ij}, \mathbf{\Omega}_i, \mathbf{\Omega}_j) + \sum_{i=1}^{N} \alpha(\Delta\mathbf{r}_i)^2. \tag{10.3.4}$$

Note that only the *intermolecular* interactions depend on γ (through r_{ij}). The derivative of the free energy with respect to γ is

$$\left\langle \frac{\partial\tilde{\mathcal{U}}}{\partial\gamma} \right\rangle = \left\langle \sum_{i<j}^{N} \frac{\partial\mathcal{U}_{\text{pair}}(r_{ij}, \mathbf{\Omega}_i, \mathbf{\Omega}_j)}{\partial\gamma} \right\rangle = \left\langle \sum_{i<j}^{N} \frac{\partial\mathcal{U}_{\text{pair}}(r_{ij}, \mathbf{\Omega}_i, \mathbf{\Omega}_j)}{\partial r_{ij}} \frac{\partial r_{ij}}{\partial\gamma} \right\rangle.$$

Only at this stage do we make use of the assumption that the intermolecular potential is pairwise additive. We stress, however, that the assumption is not essential—it just yields a simpler form for the final expression.

From equation (10.3.3) it follows that

$$\frac{\partial r_{ij}}{\partial\gamma} = \frac{\mathbf{r}_{ij} \cdot \mathbf{r}_{ij}^0}{r_{ij}}.$$

The change in free energy due to the uniform expansion is

$$\begin{aligned} \Delta F_{II} &= F^{mol}_{Ein} - F(\lambda = 1, \gamma = 1) \\ &= \int_1^\infty d\gamma \left\langle \frac{\partial \tilde{\mathcal{U}}}{\partial \gamma} \right\rangle \\ &= \int_1^\infty d\gamma \left\langle \sum_{i<j}^{N} \frac{\partial \mathcal{U}_{pair}(r_{ij}, \boldsymbol{\Omega}_i, \boldsymbol{\Omega}_j)}{\partial r_{ij}} \frac{\mathbf{r}_{ij} \cdot \mathbf{r}^0_{ij}}{r_{ij}} \right\rangle . \end{aligned} \tag{10.3.5}$$

Note that if, in the preceding equation, we replace $\mathbf{r}^0_{ij}$ by r_{ij}, the expression in angular brackets reduces to the virial. We therefore refer to the integrand in equation (10.3.5) as a modified virial.

Finally, we must evaluate the free energy of the reference state: an Einstein crystal consisting of noninteracting molecules. As the intramolecular contribution to the free energy is a constant that depends only on temperature, we shall ignore it. The expression for the total free energy of the molecular solid then becomes

$$F_{mol\ sol} = F^{mol}_{Ein} - \Delta F_I - \Delta F_{II}. \tag{10.3.6}$$

For molecular solids with (partial) orientational order, a similar scheme can be used. We transform the solid by imposing a coupling of the centers of mass of the molecules together with a coupling of the *ordered* degrees of freedom. With the combined coupling the solid is expanded. The details of this scheme depend on the nature of the orientational order.

10.3.1 Atomic Solids with Discontinuous Potentials

Let us finally consider a system of atoms that interact via a hard-core potential $\mathcal{U}_0$. As before, we try to construct a reversible path between this system and a noninteracting Einstein crystal. The problem is that, in this case, it is not possible to use a *linear* coupling scheme that simultaneously switches on the Einstein spring constants and switches off the hard-core interactions. One solution would be to use the lattice-expansion technique described in section 10.3. An alternative is to consider a system where we can switch on the spring constants, while leaving the hard-core interactions between the particles unaffected:

$$\mathcal{U}(\lambda) = \mathcal{U}_0 + \lambda\mathcal{U} = \mathcal{U}_0 + \lambda \sum_{i=1}^{N} (\mathbf{r}_i - \mathbf{r}_{0,i})^2, \tag{10.3.7}$$

where N denotes the total number of particles and $\mathbf{r}_{0,i}$ the position of the lattice site to which particle i is assigned. The free energy difference between the system with coupling λ and the hard-sphere fluid is then

$$F_{HS} = F(\lambda_{max}) - \int_0^{\lambda_{max}} d\lambda \left\langle \mathcal{U}(\mathbf{r}^N, \lambda) \right\rangle_\lambda . \qquad (10.3.8)$$

At sufficiently high values of λ_{max}, the hard particles do not "feel" each other and the free energy reduces to that of a noninteracting Einstein crystal. Clearly, the value of the spring constant λ should be sufficiently large to ensure that the harmonically bound crystal is indeed behaving as an Einstein crystal. At the same time, λ should not be too large, because this would make the numerical integration of equation (10.3.8) less accurate. In general, the choice of the optimal value for λ depends on the details of the model. In Case Study 17, we show how to choose λ for a particular model system and we discuss other practical issues that are specific for hard-core interactions.

10.3.2 General Implementation Issues

If all particles are coupled to the Einstein lattice, the crystal as a whole does not move. However, in the limit $\lambda \to 0$, there is no penalty for moving the particles away from their "Einstein" lattice position. As a consequence, the crystal as a whole may start to drift and the mean-squared particle displacement $\langle r^2 \rangle$ becomes on the order L^2. If this happens, the integrand in equation (10.3.8) becomes sharply peaked around $\lambda = 0$. This would seem to imply that the numerical integration of equation (10.3.8) requires many simulations for low values of λ. This problem can be avoided if we perform the simulation under the constraint that the center of mass of the solid remains fixed. In this case, $\langle r^2 \rangle$ tends to $\langle r^2 \rangle_0$, the mean-squared displacement of a particle from its lattice site in the normal (i.e., interacting) crystal.

To perform a Monte Carlo simulation under the constraint of a fixed center of mass we have to ensure that, if a particle is given a random displacement, all particles are subsequently shifted in the opposite direction such that the center of mass remains fixed. In practice, it is not very convenient to keep the center of mass in place by moving all particles every time a single-particle trial move is carried out. Rather, we update the center-of-mass position every time a single-particle trial move is accepted. We need to correct for the shift of the center of mass only when computing the potential energy of the harmonic springs connecting the particles to their lattice sites. In contrast, the calculation of the intermolecular potential can be carried out without knowledge of the position of the center of mass, as a shift of the center of mass does not change the distance *between* particles.

It is convenient to distinguish between the "absolute" coordinates ($\mathbf{r}$) of a particle (i.e., those that have been corrected for center of mass motion) and the uncorrected coordinates ($\mathbf{r}^{(U)}$). When computing the potential energy of the harmonic springs, we need to know $\sum_{i=1}^N (\mathbf{r}_i - \mathbf{r}_{0,i})^2$. To compute the distance of a particle i to its lattice site, $\mathbf{r}_i - \mathbf{r}_{0,i}$, we must keep track of the shift of the center of mass:

$$\Delta\mathbf{r}_i \equiv \mathbf{r}_i - \mathbf{r}_{0,i} = \mathbf{r}_i^{(u)} - \mathbf{r}_{0,i}^{(u)} - \Delta\mathbf{R}_{CM},$$

where $\Delta\mathbf{R}_{CM}$ denotes the accumulated shift of the center of mass of the system. Every time a particle is moved from $\mathbf{r}^{(u)} \to \mathbf{r}^{(u)} + \Delta\mathbf{r}$, $\Delta\mathbf{R}_{CM}$ changes to $\Delta\mathbf{R}_{CM} + \Delta\mathbf{r}/N$.

The computation of the change in energy of the harmonic interaction between all particles and their lattice site is quite trivial. Suppose that we attempt to move particle i that is at a distance $\Delta\mathbf{r}_i$ from its lattice site $r_{0,i}$, by an amount $\boldsymbol{\Delta}_i$. This causes a shift $\boldsymbol{\Delta}_i/N$ in the center of mass. The change in the harmonic potential energy is

$$\begin{aligned}\Delta\mathcal{U}_{\text{Harm}}(\lambda) &= \lambda\sum_{j\neq i}^{N}\left[(\Delta\mathbf{r}_j - \boldsymbol{\Delta}_i/N)^2 - \Delta\mathbf{r}_j^2\right] \\ &\quad +\lambda\left[(\Delta\mathbf{r}_i + (1-1/N)\boldsymbol{\Delta}_i)^2 - \Delta\mathbf{r}_i^2\right] \\ &= \lambda\left(2\Delta\mathbf{r}_i\cdot\boldsymbol{\Delta}_i + \frac{N-1}{N}\Delta_i^2\right), \end{aligned} \tag{10.3.9}$$

where, in the last line, we used the fact that $\sum_{i=1}^{N}\Delta\mathbf{r}_i = 0$.

One more caveat should be considered: normally, when a particle moves out of the periodic box, the particle is put back at the other side of the box. However, when simulating a system with a fixed center of mass, moving a particle back into the original simulation box creates a discontinuous change in the position of the center of mass and hence a sudden change of the energy of the Einstein lattice. Therefore, in a simulation with a fixed center of mass, particles that move out of the original simulation box should *not* be put back in. In any event, the excursion that a harmonically bound particle can make is small and therefore there is no real need to put the particles back in the simulation box. Algorithms 20 and 21 sketch how the Einstein-crystal method is implemented in a Monte Carlo simulation.

Constraints and Finite-Size Effects

The constraint that the center of mass of the system is fixed eliminates a number of degrees of freedom from the system, and this has an effect on the free energy. Strictly speaking, the change in free energy due to any hard constraint is infinite. However, as we shall always consider *differences* in free energy, the infinities drop out. The remaining change in the free energy becomes negligible in the thermodynamic limit. However, as simulations are necessarily performed on finite systems, it is important to have an estimate

Algorithm 20 (Monte Carlo Simulation with Fixed Center of Mass)

```
      subroutine mcmove                    attempts to move a particle
                                           keeping the center of mass fixed
      call setlat                          set up the reference lattice
      o=int(ranf()*npart)+1                select particle at random
      dis=(ranf()-0.5)*delx                give particle random displ.
      xn=x(o)+dis
      dx=x(o)-x0(o)-dxcm                   calculate Δr_i
      del=lambda*(2*dx*dis+                energy difference with lattice
     +    dis*dis*(npart-1)/npart)         equation (10.3.9)
      arg1=-beta*del
      if (ranf().lt.exp(arg1)) then
       call ener(x(o),eno)                 energy old configuration
       call ener(xn,enn)                   energy new configuration
       arg2=-beta*(enn-eno)
       if(ranf().lt.exp(arg2)) then
         dxcm=dxcm+(xn-x(o))/npart         new shift center of mass
         x(o)=xn                           accepted: replace x(o) by xn
       endif
      endif
      return
      end
```

Comments on this algorithm:

1. *Subroutine* `setlat` *sets up the reference lattice and calculates the centers of mass (Algorithm 21). This subroutine normally is called only once during the simulation, and* `ener` *calculates the energy with the other particles.*
2. *If a move is accepted, the shift in the center of mass of the system is updated.*
3. *The term* λ (`lambda`) *is the coupling constant as defined in equation (10.3.7) and* `dxcm`= $\Delta \mathbf{R}_{CM}$ *is the accumulated shift of the center of mass.*
4. *For hard-core systems, it is important to compute* first *the Boltzmann factor associated with the potential-energy change of the harmonic springs and apply the Metropolis rule to see if the move should be rejected. Only if this test is passed should we attempt to perform the more expensive test for overlaps.*

of the finite size. Below, we describe in some detail how the free energy of an unconstrained crystal is computed using simulations of a system with a fixed center of mass. To keep the discussion general, we will consider a d-dimensional crystal system of N_{mol} molecules composed of a total of N

Algorithm 21 (Generate an Einstein Crystal)

```
      subroutine setlat(nx,ny,nz)          generates the reference lattice
                                           3D-fcc structure: nx*ny atoms in
                                           close-packed plane, nz planes
      a1=(4*vol/(nx*ny*nz))**(1/3)         lattice vectors
      a0=sqrt(a1*a1/2)
      i=0
      xcm0=0
      xcm=0
      do iz=0,nz-1
       do iy=0,ny-1
        do ix=0,nx-1
         i=i+1
         x0(i)=a0*ix+(a0/2.)               lattice point of particle i
     +        *mod(iz,2)
         y0(i)=a0*iy+(a0/2.)
     +        *mod(iz,2)
         z0(i)=(a1/2)*iz
         xcm0=xcm0+x0(i)                   center of mass of lattice
         xcm=xcm+x(i)                      center of mass of solid
        enddo
       enddo
      enddo
      xcm0=xcm0/npart
      xcm=xcm/npart
      dxcm=xcm-xcm0                        Shift centers of mass
      return
      end
```

Comments on this algorithm:

1. *This algorithm generates an fcc (face-centered cubic) lattice and calculates the centers of mass of the Einstein lattice and the solid.*
2. *Note that normally the center of mass includes the y and z coordinates as well.*

atoms. The partition function for the unconstrained solid is given by

$$Q = c_N \int d\mathbf{r}^{dN} d\mathbf{p}^{dN} \exp[-\beta \mathcal{H}(\mathbf{r}_i, \mathbf{p}_i)], \tag{10.3.10}$$

where $c_N=(h^{dN_{mol}}N_1!N_2!...N_m!)^{-1}$, where N_1 denotes the number of indistinguishable particles of type 1, N_2 the number of particles of type 2, etc., and $N_1 + N_2 + ... + N_m = N_{mol}$. In all calculations of phase equilibria between systems that obey classical statistical mechanics, Planck's constant h drops out of the result. Hence, in what follows, we omit all factors h. As discussed in ref. [268], one can write the partition function Q^{con} of a constrained system as

$$\begin{aligned} Q^{con} &= c_N \int d\mathbf{r}^{dN} d\mathbf{p}^{dN} \exp\left[-\beta\mathcal{H}(\mathbf{r}_i, \mathbf{p}_i)\right] \\ &\times \delta\left[\boldsymbol{\sigma}(\mathbf{r})\right] \delta(\mathbf{G}^{-1} \cdot \dot{\boldsymbol{\sigma}}), \end{aligned} \quad (10.3.11)$$

where $\boldsymbol{\sigma}(\mathbf{r})$ and $\dot{\boldsymbol{\sigma}}$ are the constraints and time derivatives of the constraints, respectively, and

$$G_{kl} = \sum_{i=1}^{N} \frac{1}{m_i} \nabla_{\mathbf{r}_i} \sigma_k \cdot \nabla_{\mathbf{r}_i} \sigma_l. \quad (10.3.12)$$

In order to constrain the center of mass (CM), we take $\boldsymbol{\sigma}(\mathbf{r}) = \sum_{i=1}^{N} \mu_i \mathbf{r}_i$, and, thus, $\dot{\boldsymbol{\sigma}} = \sum_{i=1}^{N} (\mu_i/m_i)\mathbf{p}_i$, where $\mu_i \equiv m_i/\sum_i m_i$. To simplify matters, we have assumed that there are no additional internal molecular constraints, such as fixed bond lengths or bond angles.

We first consider the case of an Einstein crystal, which has a potential energy function given by

$$U_{Ein} = \frac{1}{2} \sum_{i=1}^{N} \alpha_i \left(\mathbf{r}_i - \mathbf{r}_{0,i}\right)^2,$$

where $\mathbf{r}_{0,i}$ are the equilibrium lattice positions. Note that the particles in a crystal are associated with specific lattice points and therefore behave as if they are distinguishable—thus, $c_N = 1$ (as we omit the factor $1/h^{d(N-1)}$). It is easy to show that

$$Q_{Ein}^{CM} = Z_{Ein}^{CM} P_{Ein}^{CM}, \quad (10.3.13)$$

with

$$Z_{Ein}^{CM} = \int d\mathbf{r}^{dN} \prod_{i=1}^{N} \exp\left[-(\beta\alpha_i/2)r_i^2\right] \delta\left(\sum_{i=1}^{N} \mu_i \mathbf{r}_i\right) \quad (10.3.14)$$

and

$$
\begin{aligned}
P_{\text{Ein}}^{\text{CM}} &= \int d\mathbf{p}^{dN} \prod_{i=1}^{N} \exp\left[-(\beta/2m_i)p_i^2\right] \delta\left(\sum_{i=1}^{N} \mathbf{p}_i\right) \\
&= \left(\frac{\beta}{2\pi M}\right)^{d/2} \prod_{i=1}^{N} \left(\frac{2\pi m_i}{\beta}\right)^{d/2} \\
&= \left(\frac{\beta}{2\pi M}\right)^{d/2} P_{\text{Ein}}, \qquad (10.3.15)
\end{aligned}
$$

where $M = \sum_i m_i$ and Z_{Ein} and P_{Ein} are the configurational and kinetic contributions to Q_{Ein}, the partition function of the unconstrained Einstein crystal. It then follows that

$$
Q_{\text{Ein}}^{\text{CM}} = \left(\frac{\sum_i m_i}{\sum_i m_i^2/\alpha_i}\right)^{\frac{d}{2}} (\beta^2/4\pi^2)^{\frac{d}{2}} Q_{\text{Ein}}. \qquad (10.3.16)
$$

In fact, this expression can be further simplified if we make the specific choice $\alpha_i = \alpha m_i$. In that case,

$$
Q_{\text{Ein}}^{\text{CM}} = (\beta^2\alpha/4\pi^2)^{d/2} Q_{\text{Ein}}. \qquad (10.3.17)
$$

There is a good reason for making this choice for α_i: in this case the net force on the center of mass of the crystal, due to the harmonic springs is always zero, provided that it is zero when all particles are on their lattice sites. This makes it easier to perform MD simulations on Einstein crystals with fixed center of mass. The free energy difference between the constrained and the unconstrained Einstein crystals is then

$$
F_{\text{Ein}}^{\text{CM}} = F_{\text{Ein}} - k_B T \ln \left(\frac{\beta^2\alpha}{4\pi^2}\right)^{d/2}. \qquad (10.3.18)
$$

For an arbitrary crystalline system in the absence of external forces, the partition function subject to the CM constraint is given by

$$
Q^{\text{CM}} = Z^{\text{CM}} (\beta/2\pi M)^{d/2} \prod_{i=1}^{N} (2\pi m_i/\beta)^{d/2}, \qquad (10.3.19)
$$

with

$$
Z^{\text{CM}} = \int d\mathbf{r}^{dN} \exp[-\beta U(\mathbf{r}_i)] \delta\left(\sum_{i=1}^{N} \mu_i \mathbf{r}_i\right), \qquad (10.3.20)
$$

while the partition function of the unconstrained crystal is given by

$$
Q = Z \prod_{i=1}^{N} (2\pi m_i/\beta)^{d/2}, \qquad (10.3.21)
$$

with

$$Z = \int d\mathbf{r}^{dN} \exp[-\beta U(\mathbf{r}_i)]. \tag{10.3.22}$$

Note that, as far as the kinetic part of the partition function is concerned, the effect of the fixed center-of-mass constraint is the same for an Einstein crystal as for an arbitrary "realistic" crystal. Using equations (10.3.19) and (10.3.21), the Helmholtz free energy difference between the constrained and unconstrained crystal is given by

$$F^{CM} = F - k_B T \ln(Z^{CM}/Z) - k_B T \ln\left(\beta/2\pi M\right)^{d/2} \tag{10.3.23}$$

We note that

$$\begin{aligned} \frac{Z^{CM}}{Z} &= \frac{\int d\mathbf{r}^{dN} \exp[-\beta U(\mathbf{r}_i)]\delta\left(\sum_i \mu_i \mathbf{r}_i\right)}{\int d\mathbf{r}^{dN} \exp[-\beta U(\mathbf{r}_i)]} \\ &= \left\langle \delta\left(\sum_i \mu_i \mathbf{r}_i\right)\right\rangle \\ &= \mathcal{P}(\mathbf{r}_{CM} = \mathbf{0}), \end{aligned} \tag{10.3.24}$$

where $\mathbf{r}_{CM} \equiv \sum_i \mu_i \mathbf{r}_i$, and $\mathcal{P}(\mathbf{r}_{CM})$ is the probability distribution function of the center of mass, $\mathbf{r}_{CM}$. To calculate $\mathcal{P}(\mathbf{r}_{CM})$ we exploit the fact that the probability distribution of the center of mass of the lattice is evenly distributed over a volume equal to that of the Wigner-Seitz cell[3] of the lattice. The reason the integration over the center-of-mass coordinates is limited to a single Wigner-Seitz cell is that if the center of mass were to another Wigner-Seitz cell, we would have created a copy of the crystal that simply corresponds to another permutation of the particles. Such configurations are not to be counted as independent. It then follows that $\mathcal{P}(\mathbf{r}_{CM}) = 1/V_{WS} = N_{WS}/V$, where V_{WS} is the volume of a Wigner-Seitz cell, and N_{WS} is the number of such cells in the system. Thus, $Z^{CM}/Z = \mathcal{P}(\mathbf{r}_{CM} = \mathbf{0}) = N_{WS}/V$. In the case of one molecule per cell, this implies $Z^{CM}/Z = N_{mol}/V$, where N_{mol} is the number of molecules in the system.

In numerical free energy calculations, the actual simulation involves computing the free energy difference between the Einstein crystal and the normal crystal, both with constrained centers of mass. We denote this free energy difference by

$$\Delta F^{CM} \equiv F^{CM} - F^{CM}_{Ein}.$$

The free energy per particle of the unconstrained crystal (in units of $k_B T$) is then

$$\frac{\beta F}{N} = \frac{\beta \Delta F^{CM}}{N} + \frac{\beta F_{Ein}}{N} + \frac{\ln(N_{mol}/V)}{N} - \frac{d}{2N}\ln(\beta\alpha M/2\pi). \tag{10.3.25}$$

[3]A Wigner-Seitz cell is constructed by drawing lines to connect a given lattice point to all nearby lattice points. At the midpoints of these lines, surfaces normal to these lines are constructed. The smallest enclosed volume defines the Wigner-Seitz cell.

If we consider the special case of a system of identical atomic particles ($m_i = m$ and $N = N_{mol}$), we obtain the following:

$$\frac{\beta F}{N} = \frac{\beta \Delta F^{CM}}{N} + \frac{\beta F_{Ein}}{N} + \frac{\ln \rho}{N} - \frac{d}{2N} \ln N - \frac{d}{2N} \ln\left(\frac{\beta \alpha m}{2\pi}\right). \qquad (10.3.26)$$

In practice, we usually calculate the excess free energy, $F^{ex} \equiv F - F^{id}$, where F^{id} is the ideal gas free energy. Let us therefore compute the finite-size corrections to the latter quantity: Given that

$$\beta F^{id}/N = -\ln[V^N (2\pi m/\beta)^{dN/2}/N!]/N,$$

we find that

$$\begin{aligned} \frac{\beta F^{ex}}{N} &= \frac{\beta \Delta F^{CM}}{N} + \frac{\beta F_{Ein}}{N} + \frac{\ln \rho}{N} - \frac{d}{2N} \ln\left(\frac{\beta \alpha m}{2\pi}\right) \\ &\quad - \frac{d+1}{2}\frac{\ln N}{N} - \ln \rho + 1 - \frac{\ln 2\pi}{2N}, \end{aligned} \qquad (10.3.27)$$

where we have used the Stirling approximation:

$$\ln N! \approx N \ln N - N + (\ln 2\pi N)/2.$$

Hoover has analyzed the system-size dependence of the entropy of a classical harmonic crystal with periodic boundaries [269]. In this study, it was established that the leading finite-size correction to the free energy per particle of a harmonic crystal is equal to $k_B T \ln N/N$. Assuming that this result can be generalized to arbitrary crystals, we should expect that $\beta F^{ex}/N + (d-1)\ln N/(2N)$ will scale as N^{-1}, plus correction terms of order $\mathcal{O}(1/N^2)$. Figure 10.4 shows the N-dependence of $\beta F^{ex}/N + (d-1)\ln N/(2N)$ for three-dimensional hard spheres. The figure clearly suggests that the remaining system-size dependence scales as $1/N$. This is a useful result, because it provides us with a procedure to extrapolate free energy calculations for a finite system to the limit $N \to \infty$. For more details, see ref. [270].

Case Study 17 (Solid-Liquid Equilibrium of Hard Spheres)

In this case study, we locate the solid-liquid coexistence densities of the hard-sphere model. We determine these densities by equating the chemical potential and the pressure of the two phases.

For the liquid phase, we use the equation of state of Speedy [271], which is based on a Padé approximation to simulation data on both the equation of state and the virial coefficients of hard spheres:

$$z_{liquid} = \frac{P\beta}{\rho} = 1 + \frac{x + 0.076014x^2 + 0.019480x^3}{1 - 0.548986x + 0.075647x^2}.$$

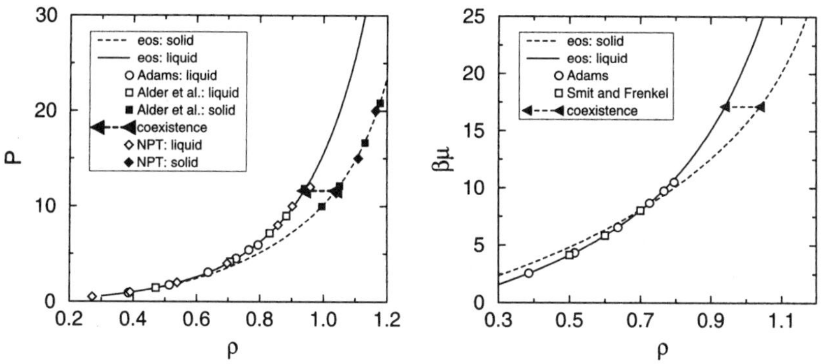

Figure 10.2: Pressure P (left) and chemical potential μ (right) as a function of the density ρ. The solid curves, showing the pressure and chemical potential of the liquid phase, are obtained from the equation of state of Speedy [271]. The dashed curve gives the pressure of the solid phase as calculated from the equation of state of ref. [272]. The open and filled symbols are the results of computer simulations for the liquid [92, 273, 274] and solid phases [273], respectively. The coexistence densities are indicated with horizontal lines.

For the solid phase of the hard-sphere model, Speedy proposed the following equation of state [272]:

$$z_{\text{solid}} = \frac{3}{1-\rho^*} - 0.5921 \frac{\rho^* - 0.7072}{\rho^* - 0.601}, \tag{10.3.28}$$

where $\rho^* = \sigma^3\rho/\sqrt{2}$. In Figure 10.2, we compare the predictions of this equation of state for the liquid and solid phases with the results from computer simulations of Alder and Wainwright [273] and Adams [92]. As can be seen, the empirical equations of state reproduce the simulation data quite well. To calculate the chemical potential of the liquid phase, we integrate the equation of state (see (10.1.1)) starting from the dilute gas limit. This yields the Helmholtz free energy as a function of the density. The chemical potential then follows from

$$\beta\mu(\rho) = \frac{\beta G}{N} = \frac{\beta F}{N} + \frac{P}{\rho k_B T}.$$

The free energy per particle of the ideal gas is given by

$$\beta f^{id}(\rho) = \frac{F^{id}(\rho)}{N k_B T} = \ln \rho\Lambda^3 - 1,$$

where Λ is the de Broglie thermal wavelength. In what follows we shall write

$$\beta f^{id}(\rho) = \ln \rho - 1.$$

That is, we shall work with the usual reduced densities and ignore the additive constant $3\ln(\Lambda/\sigma)$, as it plays no role in the location of phase equilibria for classical systems.

Figure 10.2 compares the chemical potential that follows from the Hall equation of state with some of the available simulation data (namely, grand-canonical ensemble simulations of [92] and direct calculations of the chemical potential, using the Widom test-particle method [274] (see Chapter 7)).

These results show that we have an accurate equation of state for the liquid phase and the solid phase. Since we know the absolute free energy of the ideal gas phase, we can calculate the free energy and hence the chemical potential of the liquid phase. For the solid phase we can use the equation of state to calculate only free energy *differences*; to calculate the absolute free energy we have to determine the free energy at a particular density. To perform this calculation we use the lattice coupling method.

We must now select the upper limit of the coupling parameter λ (λ_{max}) and the values of λ for which we perform the simulation. For sufficiently large values of λ we can calculate $\sum_{i=1}^{N}(\mathbf{r}_i - \mathbf{r}_{0,i})^2$ analytically, using

$$\langle r^2 \rangle_\lambda = \frac{1}{N}\frac{\partial F(\lambda)}{\partial \lambda}.$$

For the noninteracting Einstein crystal, the mean-squared displacement is given by

$$\langle r^2 \rangle_\lambda = \frac{3}{2\beta\lambda}. \tag{10.3.29}$$

For a noninteracting Einstein crystal with fixed center of mass, the free energy is given by equation (10.3.18), which gives

$$\langle r^2 \rangle_{\text{Eins},\lambda} = \frac{1}{\beta}\frac{3}{2}\frac{N-1}{N}\frac{1}{\lambda}. \tag{10.3.30}$$

In [259] an analytical expression is derived for the case of an interacting Einstein crystal, which reads

$$\begin{aligned}\langle r^2 \rangle_\lambda &= \langle r^2 \rangle_{\text{Eins},\lambda} - \frac{\beta n}{2}\frac{1}{2a(2\pi\beta\lambda)^{(1/2)}\left(1 - \left\langle P^{nn}_{\text{overlap}}\right\rangle_\lambda\right)} \\ &\quad \times \left\{\left[\sigma a - \sigma^2 - 1/(\beta\lambda)\right]\exp\left[-\beta\lambda(a-\sigma)^2/2\right]\right. \\ &\quad \left. + \left[\sigma a + \sigma^2 - 1/(\beta\lambda)\right]\exp\left[-\beta\lambda(a+\sigma)^2/2\right]\right\}, \end{aligned} \tag{10.3.31}$$

where a is the separation of two nearest neighbors i and j, $a = \mathbf{r}_{0,i} - \mathbf{r}_{0,j}$, σ is the hard-core diameter, and n is the number of nearest neighbors (for

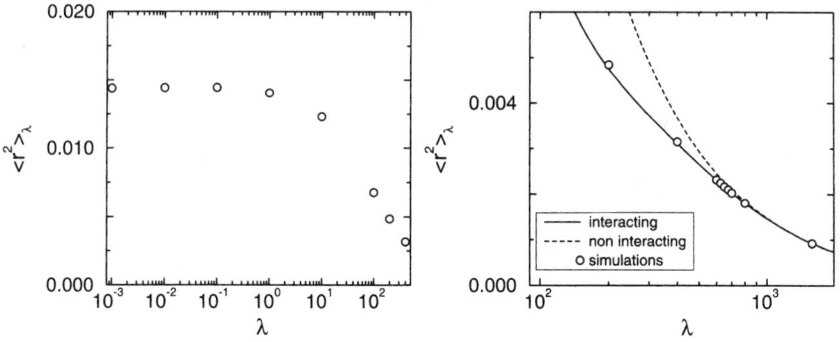

Figure 10.3: The mean-squared displacement $\langle r^2 \rangle_\lambda$ as a function of the coupling parameter λ for a hard-sphere (fcc) solid of 54 particles (6 layers of 3 × 3 close-packed atoms at a density $\rho = 1.04$). The figure on the left shows the simulation results for low values of λ, the figure on the right for high values. The solid line takes into account nearest-neighbor interactions (10.3.31); the dashed line assumes a noninteracting Einstein crystal (10.3.30). The open symbols are the simulation results.

example, $n = 12$ for fcc (face-centered cubic) and hcp (hexagonal close-packed) solids or $n = 8$ for bcc (body-centered cubic)); $\left\langle P^{nn}_{overlap} \right\rangle_\lambda$ is the probability that two nearest neighbors overlap. Such probability is given by

$$\left\langle P^{nn}_{overlap} \right\rangle_\lambda = \frac{\operatorname{erf}\left[(\beta\lambda/2)^{1/2}(\sigma + a)\right] + \operatorname{erf}\left[(\beta\lambda/2)^{1/2}(\sigma - a)\right]}{2} - \frac{\exp[-\beta\lambda(\sigma - a)^2/2] - \exp[-\beta\lambda(\sigma + a)^2/2]}{(2\pi\beta\lambda)^{1/2} a} \tag{10.3.32}$$

This equation can also be used to correct the free energy of a noninteracting Einstein crystal (10.3.18):

$$\frac{\beta F_{Einst}(\lambda)}{N} = \frac{\beta F_{Einst}}{N} + \frac{n}{2} \ln\left(1 - \left\langle P^{nn}_{overlap} \right\rangle_\lambda\right). \tag{10.3.33}$$

We choose λ_{max} such that, for values of λ larger than this maximum value, $\langle r^2 \rangle_\lambda$ obeys the analytical expression. Typically, this means that the probability of overlap of two harmonically bound particles should be considerably less than 1%. The results of these simulations are presented in Figure 10.3. This figure shows that if we rely only on the analytical results of the noninteracting Einstein crystal we have to take a value for $\lambda_{max} \approx 1000$–$2000$. If we use equation (10.3.31) for $\langle r^2 \rangle_\lambda$ $\lambda_{max} = 500$–1000 is sufficient.

We now should integrate

$$\frac{\Delta F}{N} = \int_0^{\lambda_{max}} d\lambda \left\langle r^2 \right\rangle_\lambda .$$

In practice, this integration is carried out by numerical quadrature. We therefore must specify the values of λ for which we are going to compute $\left\langle r^2 \right\rangle_\lambda$. To improve the accuracy of the numerical quadrature, it is convenient to transform to another integration variable:

$$\frac{\Delta F}{N} = \int_0^{\lambda_{max}} \frac{d\lambda}{g(\lambda)} g(\lambda) \left\langle r^2 \right\rangle_\lambda = \int_{G^{-1}(0)}^{G^{-1}(\lambda_{max})} d\left[G^{-1}(\lambda)\right] \, g(\lambda) \left\langle r^2 \right\rangle_\lambda ,$$

where $g(\lambda)$ is an as-yet arbitrary function of λ and $G^{-1}(\lambda)$ is the primitive of the function $1/g(\lambda)$. If we can find a function $g(\lambda)$ such that the integrand, $g(\lambda) \left\langle r^2 \right\rangle_\lambda$, is a slowly varying function, we need fewer function evaluations to arrive at an accurate estimate. To do this we need to have an idea about the behavior of $\left\langle r^2 \right\rangle_\lambda$.

For $\lambda \to 0$, $\left\langle r^2 \right\rangle_\lambda \to \left\langle r^2 \right\rangle_0$, which is the mean-squared displacement of an atom around its lattice site in the normal hard-sphere crystal. At high values of λ, where the system behaves like an Einstein crystal, we have $\left\langle r^2 \right\rangle_\lambda \to 3k_B T/(2\lambda)$. This leads to the following guess for the functional form of $g(\lambda)$:

$$g(\lambda) \approx k_B T / \left\langle r^2 \right\rangle_\lambda \approx c + \lambda,$$

where $c = k_B T / \left\langle r^2 \right\rangle_0$. Here, $\left\langle r^2 \right\rangle_0$ can be estimated from Figure 10.3. The value of c clearly depends on density (and temperature). For $\rho = 1.04$, extrapolation to $\lambda \to 0$ gives $\left\langle r^2 \right\rangle_0 \approx 0.014$, which gives $c = 70$. If we use this function $g(\lambda)$ the free energy difference is calculated from

$$\frac{\Delta F}{N} = \int_{\ln c}^{\ln(\lambda_{max}+c)} d\left[\ln(\lambda + c)\right] \, (\lambda + c) \left\langle r^2 \right\rangle_\lambda .$$

For the numerical integration we use a n-point Gauss-Legendre quadrature [275]. As the integrand is a smooth function, a 10-point quadrature is usually adequate. As discussed in section 10.3.2, the resulting free energy still depends (slightly) on the system size. An example of the system-size dependence of the excess free energy of a hard-sphere crystal is shown in Figure 10.4 [270]. From this figure, we can estimate the excess free energy of the infinite system to be $\beta f^{ex} = 5.91889(4)$. This is in good agreement with the estimate of Frenkel and Ladd, $\beta f^{ex} = 5.9222$ [259].

Once we have one value of the absolute free energy of the solid phase at a given density, we can compute the chemical potential of the solid phase at any other density, using the equation of state of Speedy (see Figure 10.2).

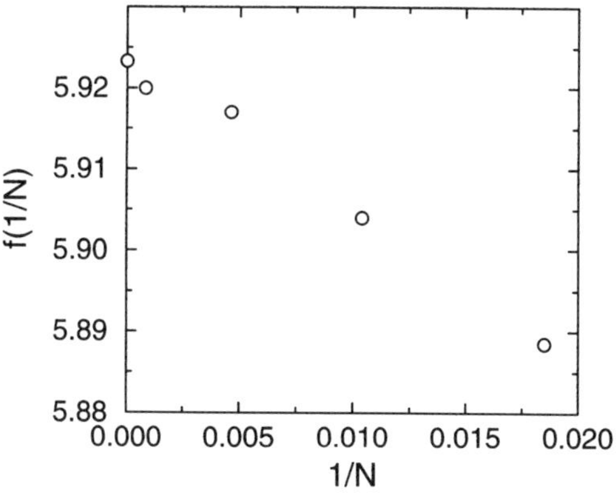

Figure 10.4: $\beta F^{ex}/N + \ln(N)/N$ versus $1/N$ for an fcc crystal of hard spheres at a density $\rho\sigma^3 = 1.0409$. The solid line is a linear fit to the data. The coefficient of the $1/N$ term is $-6.0(2)$ and the intercept (i.e., the infinite system limit of $\beta F^{ex}/N$) is equal to $5.91889(4)$.

The coexistence densities follow from the condition that the chemical potentials and pressures in the coexisting phases should be equal.

Using the value of $5.91889(4)$ from [270] for the solid at $\rho = 1.04086$, we arrive at a freezing density $\rho_l = 0.9391$ and a melting density $\rho_S = 1.0376$. At coexistence, the pressure is $P_{coex} = 11.567$ and the chemical potential is $\mu_{coex} = 17.071$. In fact, as we shall argue below, the presence of vacancies in the equilibrium crystal lowers the coexistence pressure slightly: $P_{coex} = 11.564$. These results are in surprisingly good agreement with the original data of Hoover and Ree [252], who obtained as estimate for the solid-liquid coexistence densities $\rho_s = 1.041 \pm 0.004$ and $\rho_l = 0.943 \pm 0.004$ at a pressure 11.70 ± 0.18.

The free energy difference between the fcc and hcp for large hard-sphere crystals at melting is very close to 0, but the fcc structure appears to be the more stable phase [248, 276–278].

Example 11 (Fcc or Hcp?)
Hard-sphere crystals can occur in different crystal phases. The best known among these are the face-centered cubic (fcc) and hexagonal close-packed (hcp) structures. It is not easy to determine which phase is thermodynamically most stable. The reason is that the free energy differences between the various structures are on the order of 10^{-3} k_BT per particle, or less. As a consequence, the earliest numerical studies aimed at computing this free energy difference [250] were not conclusive. More recent studies [248, 278]

showed conclusively that the fcc structure is most stable. While one of the latter simulations used the Einstein-crystal method of ref. [259], the others were based on a different approach. Here, we briefly discuss the so-called *lattice-switch Monte Carlo* method of Bruce *et al.* [278].

A close-packed crystal consists of hexagonally close-packed two-dimensional planes that are stacked up in the vertical direction. Assume that we construct the crystal by stacking planes. For every new plane there are two distinct possibilities of stacking it on the previous plane in such a way that all the atoms fit in the triangular holes between the atoms of the previous plane. Let us denote these two positions of the new plane by B and C, and the position of the original plane by A. With this notation, the fcc stacking obeys the following sequence $\cdots$ABCABCABC$\cdots$, while the hcp structure is characterized by $\cdots$ABABABA$\cdots$. In addition, many hybrid close-packed structures are possible, as long as we never stack two identical planes on top of one another (i.e., BAAB is forbidden).

At any given instant, the atoms in a layer are not exactly on a lattice point. We can therefore write

$$\mathbf{r}_i = \mathbf{R}_i(\alpha) + \mathbf{u}_i,$$

where $\mathbf{R}_i(\alpha)$ is the ideal reference lattice position of particle i in structure α, where α labels the crystal structure (e.g., fcc or hcp). We can now perform a Monte Carlo simulation where, in addition to the usual particle displacement moves, we also attempt moves that do not affect the displacement vectors, $\mathbf{u}_i$, but that switch the reference lattice, $\mathbf{R}_i(\alpha)$, from fcc to hcp. In principle, the free energy difference between these two structures would follow directly from the relative probabilities of finding the two structures in such a Monte Carlo simulation:

$$F_{hcp} - F_{fcc} = k_B T \ln\left(\frac{P(fcc)}{P(hcp)}\right).$$

However, in practice, such a lattice switch has a very low acceptance probability. The usual solution for such a problem is to decompose the large trial move into many small steps, each of which has a reasonable acceptance probability. The lattice-switch method of Bruce *et al.* employs the multicanonical method of Berg and Neuhaus [279]. This method is a version of the umbrella-sampling scheme described in section 7.4. The first step in this procedure is to define a convenient "order parameter" that connects the two states. To this end, Bruce *et al.* defined an overlap order parameter $\mathcal{M}$:

$$\mathcal{M}(\mathbf{u}^N) = M(\mathbf{u}^N, fcc) - M(\mathbf{u}^N, hcp),$$

where $M(\mathbf{u}^N, \alpha)$ is the number of pairs of hard spheres that overlap for configuration $\mathbf{u}^N$ if the α lattice is used as a reference. For example, $M(\mathbf{u}^N, hcp)$ is zero for a set of displacement vectors, $\mathbf{u}^N$, that do not yield a single overlap if we choose an hcp reference lattice. Of particular interest are those configurations for which $\mathcal{M}(\mathbf{u}^N) = 0$, since for these configurations lattice switches are always accepted. Let us define the biased distribution

$$P(\mathbf{u}^N, \alpha|\{\eta\}) \propto P(\mathbf{u}^N, \alpha) \exp\{\eta [\mathcal{M}(\mathbf{u}^N)]\},$$

where $P(\mathbf{u}^N, \alpha)$ is the unweighted distribution and $\eta[\mathcal{M}(\mathbf{u}^N)]$ are the weights that have to be set. These weights should be chosen such that all relevant values of $\mathcal{M}$ are sampled. From a given simulation one can make an estimate of these weights and these are then subsequently used and updated in the next (longer) simulation until the desired accuracy has been achieved.

Bruce *et al.* [278] used this method to compute the free energy difference between the hcp and the fcc structures with a statistical error of 10^{-5} k_BT. These calculations of Bruce *et al.* gave further support for the observation that the fcc structure is more stable then the hcp structure. Mau and Huse [280] showed that all hybrids of fcc and hcp stacking have a free energy higher than that of a pure fcc structure.

10.4 Vacancies and Interstitials

Thus far, we have described crystals as if they were free of imperfections. However, any real crystal will contain point defects, such as vacancies and interstitials. In addition, one may find extended defects such as dislocations and grain boundaries. In equilibrium, point defects are the most common. Clearly, to have a realistic description of a crystal, it is important to have an expression for the equilibrium concentration of vacancies and interstitials, and their contribution to the free energy. This is not completely trivial, as the concept of a point defect is inextricably linked to that of a lattice site. And lattice sites lose their meaning in a disordered state. So, we should first address the question: when is it permissible to count states with a different number of lattice sites as distinct? The answer is, of course, that this is only true if these different states can be assigned to distinct volumes in phase space. This is possible if we impose that every particle in a crystal is confined to its Wigner-Seitz cell. In three-dimensional crystals, this constraint on the positions of all particles has little effect on the free energy (in contrast, in a *liquid* it is not at all permissible). Below, we derive an expression for the vacancy concentration in a crystal, following the approach first given by Bennett and Alder [281].

10.4.1 Free Energies

The equilibrium concentration of vacancies in a crystal is usually very low. We shall therefore make the approximation that vacancies do not interact.

This assumption is not as reasonable as it seems, as the interaction of vacancies through their stress fields is quite long range. The assumption that vacancies are ideal implies that $F^{(n)}$, the Helmholtz free energy of a crystal with n vacancies *at specified positions*, can be written as

$$F^{(n)} = F^{(0)} - nf_1 = Mf_0 - nf_1, \tag{10.4.1}$$

where M is the number of lattice sites of the crystal, f_0 is the free energy per particle in the defect-free crystal, and $-f_1$ is the change in free energy of a crystal due to the creation of a single vacancy at a specific lattice point. Let us now consider the effect of vacancies on the Gibbs free energy of a system of N particles at constant pressure and temperature. First, we define g^{vac} as the variation in the Gibbs free energy of a crystal of M particles due to the introduction of a single vacancy *at a specific lattice position*

$$\begin{aligned} g^{\text{vac}} &\equiv G_{M+1,1}(N,P,T) - G_{M,0}(N,P,T) \\ &= F_{M+1,1}(V_{M+1,1}) - F_{M,0}(V_{M,0}) + P(V_{M+1,1} - V_{M,0}). \end{aligned} \tag{10.4.2}$$

In the above equation, the first subscript refers to the number of lattice sites in the system, and the second subscript to the number of vacancies. Clearly, the number of particles N is equal to the difference between the first and second subscripts. The next step is to write

$$\begin{aligned} F_{M+1,1}(V_{M+1,1}) - F_{M,0}(V_{M,0}) &= F_{M+1,1}(V_{M+1,1}) - F_{M+1,1}(V_{M+1,0}) \\ &\quad + F_{M+1,1}(V_{M+1,0}) - F_{M+1,0}(V_{M+1,0}) \\ &\quad + F_{M+1,0}(V_{M+1,0}) - F_{M,0}(V_{M,0}). \end{aligned} \tag{10.4.3}$$

The first line on the right-hand side of this equation is equal to $-P\Delta v$, where $\Delta v \equiv v^{\text{vac}} - v^{\text{part}}$ is the difference in the volume of the crystal as one particle is replaced by a vacancy, at constant pressure and constant number of lattice sites. The second line on the right-hand side is simply equal to $-f_1$, defined in equation (10.4.1):

$$-f_1 \equiv F_{M+1,1}(V_{M+1,0}) - F_{M+1,0}(V_{M+1,0}).$$

To rewrite the third line on the right-hand side of equation (10.4.3), we note that the Helmholtz free energy is extensive. We express this by introducing f_0, the Helmholtz free energy per particle of a defect-free crystal, and writing $F_{M,0}(V_{M,0}) = Mf_0$. Obviously, $F_{M+1,0}(V_{M+1,0}) - F_{M,0}(V_{M,0}) = f_0$. Combining these three terms, we find that

$$F_{M+1,1}(V_{M+1,1}) - F_{M,0}(V_{M,0}) = -P\Delta v - f_1 + f_0. \tag{10.4.4}$$

The volume is also an extensive quantity; hence

$$V_{M,0} = \frac{M}{M+1} V_{M+1,0}.$$

It then follows that

$$\begin{aligned} P(V_{M+1,1} - V_{M,0}) &= P(V_{M+1,1} - V_{M+1,0} + V_{M+1,0} - V_{M,0}) \\ &= P(\Delta v + V/N). \end{aligned}$$

Hence, the Gibbs free energy difference associated with the formation of a vacancy at a specific lattice site, equation (10.4.2), is then

$$\begin{aligned} g^{vac} &= P(V_{M+1,1} - V_{M,0}) - f_1 + (\Delta v + V/N)P + f_0 \\ &= P(V/N) - f_1 + f_0 \\ &= (P/\rho + f_0) - f_1 \\ &= \mu_0 - f_1, \end{aligned}$$

where we have defined $\mu_0 \equiv (P/\rho + f_0)$. Now we have to include the entropic contribution due to the distribution of n vacancies over M lattice sites. The total Gibbs free energy then becomes

$$\begin{aligned} G &= G_0(N) + n g^{vac} + M k_B T \left[\frac{n}{M} \ln\left(\frac{n}{M}\right) + \left(1 - \frac{n}{M}\right) \ln\left(1 - \frac{n}{M}\right) \right] \\ &\approx G_0(N) + n g^{vac} + n k_B T \ln \frac{n}{M} - n k_B T. \end{aligned}$$

If we minimize the Gibbs free energy with respect to n, we find that

$$\langle n \rangle \approx M \exp(-\beta g^{vac}),$$

where we have ignored a small correction due to the variation of $\ln M$ with n. If we insert this value in the expression for the total Gibbs free energy, we find that

$$G = G_0(N) + \langle n \rangle g^{vac} - \langle n \rangle g^{vac} - \langle n \rangle k_B T = G_0 - \langle n \rangle k_B T.$$

The total number of particles is $M - \langle n \rangle$. Hence the Gibbs free energy *per particle* is

$$\begin{aligned} \mu &= \frac{G_0 - \langle n \rangle k_B T}{N} = \mu_0 - \frac{\langle n \rangle k_B T}{N} \\ &\approx \mu_0 - x_v k_B T. \end{aligned} \tag{10.4.5}$$

Hence the change in chemical potential of the solid due to the presence of vacancies is

$$\Delta \mu = -x_v k_B T \tag{10.4.6}$$

from which it follows that the change in *pressure* of the solid at fixed chemical potential is equal to

$$\Delta P = x_v \rho_s k_B T. \tag{10.4.7}$$

10.4.2 Numerical Calculations

Vacancies

Numerically, it is straightforward to compute the equilibrium vacancy concentration. The central quantity that needs to be computed is $-f_1$, the change in free energy of a crystal due to the creation of a single vacancy at a specific lattice point. In fact, it is more convenient to consider $+f_1$, the change in free energy due to the removal of a vacancy at a specific lattice point. This quantity can be computed in several ways. For instance, we could use a particle-insertion method. We start with a crystal containing one single vacancy and attempt a trial insertion in the Wigner-Seitz cell surrounding that vacancy. Then f_1 is given by

$$f_1 = -k_B T \ln \left(\frac{V_{WS} \langle \exp(-\beta \Delta U \rangle}{\Lambda^d} \right), \tag{10.4.8}$$

where V_{WS} is the volume of the Wigner-Seitz cell, and ΔU is the change in potential energy associated with the insertion of a trial particle. For hard particles

$$f_1 = -k_B T \ln \left(\frac{V_{WS} P_{acc}(V_{WS})}{\Lambda^d} \right),$$

where $P_{acc}(V_{WS})$ is the probability that the trial insertion in the Wigner-Seitz cell will be accepted. As most of the Wigner-Seitz cell is not accessible, it is more efficient to attempt insertion in a subvolume (typically on the order of the cell volume in a lattice-gas model of the solid). However, then we also should consider the reverse move—the removal of a particle from a subvolume ν of the Wigner-Seitz cell, in a crystal without vacancies. The only thing we need to compute in this case is $P_{rem}(\nu)$, the probability that a particle happens to be inside this volume. The expression for f_1 is then

$$f_1 = -k_B T \ln \left(\frac{\nu P_{acc}(\nu)}{P_{rem}(\nu) \Lambda^d} \right).$$

Of course, in the final expression for the vacancy concentration, the factor Λ^d drops out (as it should), because it is cancelled by the same term in the ideal part of the chemical potential. A direct calculation of the vacancy concentration [281,282] suggests that this concentration in a hard-sphere solid near coexistence is approximately 2.6×10^{-4}. Let us assume that the defect-free crystal is in equilibrium with the liquid at a pressure P and chemical potential μ. Then it is easy to verify that the shift in the coexistence pressure due to the presence of vacancies is

$$\delta P_{coex} = \frac{-x(0) k_B T}{\nu_l - \nu_s},$$

where v_l (v_s) is the molar volume of the liquid (solid). The corresponding shift in the chemical potential at coexistence is

$$\delta\mu_{coex} = \frac{\delta P_{coex}}{\rho_l}.$$

Inserting the numerical estimate $x(0) \approx 2.6\times10^{-4}$, the decrease in the coexistence pressure due to vacancies is $\delta P_{coex} \approx -2.57 \times 10^{-3}$. The corresponding shift in the chemical potential at coexistence is $\delta\mu_{coex} = -2.74 \times 10^{-3}$. Note that these shifts are noticeable when compared to the accuracy of absolute free energy calculations of the crystalline solid.

Interstitials

Thus far, we have ignored interstitials. However, it is not *a priori* obvious that these can be ignored. The only new ingredient in the calculation of the interstitial concentration is the determination of f_I. This is best done by thermodynamic integration. To this end, we first simulate a crystal with one interstitial. We then determine the excursions of the interstitial from its average position. Next, we define a volume v_0 such that the interstitial is (with overwhelming probability) inside this volume. The probability that a point particle inserted at random in a Wigner-Seitz cell will be inside this volume is

$$P_{acc} = \frac{v_0}{V_{WS}}. \tag{10.4.9}$$

Next, we "grow" the particle to the size of the remaining spheres. This will require a reversible work w. The latter quantity can easily be calculated, because the simulation yields the pressure exerted on the surface of this sphere. The total free energy change associated with the addition of an interstitial in a given octahedral hole is then

$$f_I = -k_B T \ln\left(P_{acc}\frac{V_{WS}}{\Lambda^3}\right) + w \tag{10.4.10}$$

and

$$x_I = \exp\left\{-\beta\left[w - k_B T \ln\left(P_{acc}\frac{V_{WS}}{\Lambda^3}\right) - \mu\right]\right\}. \tag{10.4.11}$$

As before, the Λ^3 term drops out of the final result (as it should). For more details, see refs. [283–287].

Chapter 11

Free Energy of Chain Molecules

In Chapter 7, we introduced the test particle insertion scheme as an efficient method for determining the chemical potential. However, this method fails when the probability of accepting a trial insertion becomes very small. One consequence is that the simple particle insertion method is less well suited for molecular than for atomic systems. This is so because the probability of accepting the random trial insertion of a large molecule in a fluid is usually extremely small.

Fortunately, it is possible to overcome this problem, at least partially, by performing nonrandom sampling. Here, we discuss several of the techniques that have been proposed to compute the chemical potential of chain molecules. Three different approaches have been proposed to improve the efficiency of the original Widom scheme. The most direct of these techniques, in essence, is thermodynamic integration schemes. Next, we discuss a method based on (generalizations of) the Rosenbluth algorithm for generating polymer conformations. And, finally, we mention a recursive algorithm.

11.1 Chemical Potential as Reversible Work

The excess chemical potential of a (chain) molecule is simply the reversible work needed to add such a molecule to a liquid in which N other (possibly identical) molecules are already present. If we choose to break up the insertion of the molecule into a number of steps, then clearly the reversible work needed to insert the whole molecule is equal to the sum of the contributions of the substeps. At this stage, we are still free to choose the elementary steps, just as we are free to choose whatever reversible path we wish when performing thermodynamic integration. One obvious possibility is to start with an ideal (noninteracting) chain molecule and then slowly switch on the interaction of this molecule with the surrounding particles (and the nonbonded intramolecular interactions). This could be done in the way described in section 7.1. In fact, this approach was followed by Müller and Paul [288], who performed a simulation in which the polymer interaction is switched on gradually. Although this simulation could have been performed with straightforward thermodynamic integration, a multiple-histogram method (see section 7.3.1) was used instead, but this does not change the overall nature of the calculation. As stated before, the advantage of thermodynamic integration (and related techniques) is that it is robust. The disadvantage is that it is no longer possible to measure the excess chemical potential in a *single* simulation.

A closely related method for measuring the chemical potential of a chain molecule was proposed somewhat earlier by Kumar *et al.* [289,290]. In this scheme, the chain molecule is built up monomer by monomer. The method of Kumar *et al.* resembles the gradual insertion scheme for measuring excess chemical potentials that had been proposed by Mon and Griffiths [291]. The reversible work involved in the intermediate steps is measured using the Widom method; that is, the difference in excess free energy of a chain of length ℓ and $\ell + 1$ is measured by computing $\Delta\mathcal{U}(\ell \to \ell + 1)$, the change in potential energy associated with the addition of the $(\ell + 1)$th monomer. The change in free energy is then given by

$$\begin{aligned} \Delta F_{ex}(\ell \to \ell + 1) &\equiv \mu_{ex}^{incr}(\ell \to \ell + 1) \\ &= -k_B T \ln \langle \exp[-\beta \Delta\mathcal{U}(\ell \to \ell + 1)] \rangle . \end{aligned} \tag{11.1.1}$$

This equation defines the incremental excess chemical potential $\mu_{ex}^{incr}(\ell \to \ell + 1)$. The excess chemical potential of the complete chain molecule is simply the sum of the individual incremental excess chemical potentials. As the latter contributions are measured using the Widom method, the scheme of Kumar *et al.* is referred to as *the modified Widom method.* This method is subject to the same limitations as the original Widom method (i.e., the insertion probability of the individual monomers should be appreciable). In this respect, it is slightly less general than thermodynamic integration. As with the multiple-histogram method used by Müller and Paul [288], the computation of the excess chemical potential may require many individual simulations [290,292].

11.2 Rosenbluth Sampling

Several proposals for measuring the chemical potential of a chain molecule in a *single* simulation have been made. Harris and Rice [293] and Siepmann [294] showed how to compute the chemical potential of chain molecules with discrete conformations using an algorithm to generate polymer conformations due to Rosenbluth and Rosenbluth [295]. A generalization to continuously deformable molecules was proposed by Frenkel *et al.* [296,297] and by de Pablo *et al.* [298]. As the extension of the sampling scheme from molecules with discrete conformations to continuously deformable molecules is nontrivial, we shall discuss the two cases separately. The approach followed here, in many respects, is similar to the conformational-bias Monte Carlo scheme described in section 13.2.1. However, we have attempted to make the presentation self-contained.

11.2.1 Macromolecules with Discrete Conformations

It is instructive to recall how we compute μ_{ex} of a chain molecule with the Widom technique. To this end, we introduce the following notation: the position of the first segment of the chain molecule is denoted by $\mathbf{q}$ and the conformation of the molecule is described by Γ. The configurational part of the partition function of a system of chain molecules can be written as[1]

$$Q_{\text{chain}}(N,V,T) = \frac{1}{N!}\int d\mathbf{q}^N \sum_{\Gamma_1,\cdots,\Gamma_N} \exp[-\beta\mathcal{U}(\mathbf{q}^N,\Gamma^N)]. \tag{11.2.1}$$

The excess chemical potential of a chain molecule is obtained by considering the ratio

$$Q(N+1,V,T)/[Q(N,V,T)Q_{\text{noninteracting}}(1,V,T)],$$

where the numerator is the (configurational part of) the partition function of a system of N + 1 interacting chain molecules while the denominator is the partition function for a system consisting of N interacting chains and one chain that does not interact with the others. The latter chain plays the role of the ideal gas molecule (see section 7.2.1). Note, however, that although this molecule does not interact with any of the other molecules, it *does* interact with itself, through both bonded and nonbonded interactions. Unfortunately, this is not a particularly useful reference state, as we do not know *a priori* the configurational part of the partition function of such an isolated self-avoiding chain. We therefore use another reference state, that of

[1]We assume that there are no hard constraints on the intramolecular degrees of freedom.

the isolated nonself-avoiding chain (i.e., a molecule in which all nonbonded interactions have been switched off). It should be stressed that the choice of another reference system makes no difference whatsoever for the computation of any observable property (see Appendix G).

Let us consider the case of a molecule that consists of ℓ segments. Starting from segment 1, we can add segment 2 in k_2 equivalent directions, and so on. Clearly, the total number of nonself-avoiding conformations is $\Omega_{id} = \prod_{i=2}^{\ell} k_i$. For convenience, we have assumed that, for a given i, all k_i directions are equally likely (i.e., we ignore gauche-trans potential energy differences and we even allow the ideal chain to fold back on itself). These limitations are not essential but they simplify the notation. Finally, we assume for convenience that all k_i are the same. Hence, for the simple model that we consider, $\Omega_{id} = k^{\ell-1}$. Using this ideal chain as our reference system, the expression for the excess chemical potential becomes

$$\begin{aligned}\beta\mu_{ex} &= -\ln\left(\frac{Q_{chain}(N+1,V,T)}{Q(N,V,T)Q_{ideal}(1,V,T)}\right)\\ &= -\ln\left\langle \exp[-\beta\Delta\mathcal{U}(\mathbf{q}^N,\Gamma^N;\mathbf{q}_{N+1},\Gamma_{N+1})]\right\rangle, \qquad (11.2.2)\end{aligned}$$

where $\Delta\mathcal{U}$ denotes the interaction of the test chain with the N chains already present in the system *and with itself*, while $\langle\cdots\rangle$ indicates averaging over all starting positions and all ideal chain conformations of a randomly inserted chain.

The problem with the Widom approach to equation (11.2.2) is that almost all randomly inserted ideal chain conformations will overlap either with particles already present in the system or internally. The most important contributions to μ_{ex} will come from the extremely rare cases, where the trial chain happens to be in just the right conformation to fit into the available space in the fluid. Clearly, it would be desirable if we could restrict our sampling to those conformations that satisfy this condition. If we do that, we introduce a bias in our computation of the insertion probability and we must somehow correct for that bias.

The Rosenbluth approach used in [293,294] consists of two steps: in the first step a chain conformation is generated with a bias that ensures that "acceptable" conformations are created with a high probability. The next step corrects for this bias by multiplying with a weight factor. A scheme that generates acceptable chain conformations with a high probability was developed by Rosenbluth and Rosenbluth in the early 1950s [295]. In the Rosenbluth scheme, a conformation of a chain molecule is constructed segment by segment. For every segment, we have a choice of k possible directions. In the Rosenbluth scheme, this choice is not random but favors the direction with the largest Boltzmann factor. To be more specific, the following scheme is used to generate a configuration of one polymer with ℓ monomers:

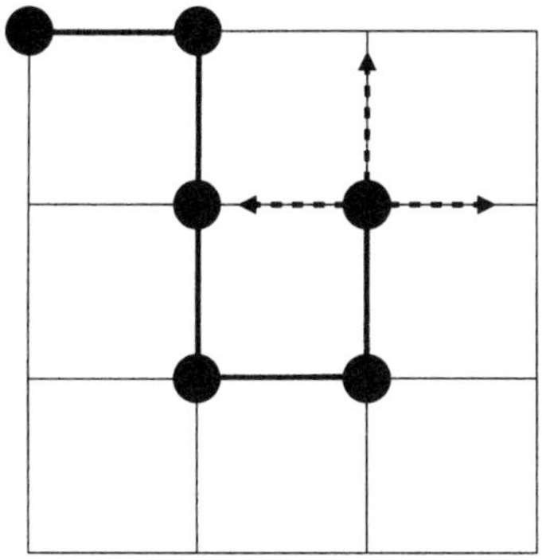

Figure 11.1: Rosenbluth scheme to insert a polymer segment by segment. The arrows indicate the trial positions for the next segment.

1. The first monomer is inserted at a random position and its energy is denoted by $u^{(1)}(n)$. We define the Rosenbluth weight of this monomer as $w_1 = k\exp[-\beta u^{(1)}(n)]$.[2]
2. For all subsequent segments $i = 2, 3, \cdots, \ell$, we consider all k trial positions adjacent to segment $i-1$ (see Figure 11.1). The energy of the jth trial position is denoted by $u^{(i)}(j)$. From the k possibilities, we select one, say, n, with a probability

$$p^{(i)}(n) = \frac{\exp[-\beta u^{(i)}(n)]}{w_i}, \tag{11.2.3}$$

where w_i is defined as

$$w_i = \sum_{j=1}^{k} \exp[-\beta u_i(j)]. \tag{11.2.4}$$

The energy $u^{(i)}(j)$ excludes the interactions with the subsequent segments $i+1$ to ℓ. Hence, the total energy of the chain is given by $\mathcal{U}(n) = \sum_{i=1}^{\ell} u^{(i)}(n)$.

3. Step 2 is repeated until the entire chain is grown, and we can compute the normalized Rosenbluth factor of configuration n:

$$\mathcal{W}(n) = \prod_{i=1}^{\ell} \frac{w_i}{k}. \tag{11.2.5}$$

[2]The factor k is included in the definition of w_1 only to keep the notation consistent with that of section 13.2.1.

We use this scheme to generate a large number of configurations, and ensemble averaged properties of these chains are calculated as follows:

$$\langle A \rangle_{\mathcal{R}} = \frac{\sum_{n=1}^{M} \mathcal{W}(n) A(n)}{\sum_{n=1}^{M} \mathcal{W}(n)}, \tag{11.2.6}$$

where $\langle \cdots \rangle_{\mathcal{R}}$ indicates that the configurations have been generated by the Rosenbluth scheme. This label is important, because the Rosenbluth algorithm does not generate chains with the correct Boltzmann weight. We refer to the distribution generated with the Rosenbluth procedure as the Rosenbluth distribution. In the Rosenbluth distribution, the probability of generating a particular conformation n is given by

$$P(n) = \prod_{i=1}^{\ell} \frac{\exp[-\beta u^{(i)}(n)]}{w_i} = k^{\ell} \frac{\exp[-\beta \mathcal{U}(n)]}{\mathcal{W}(n)}. \tag{11.2.7}$$

An important property of this probability is that it is normalized; that is,

$$\sum_{n} P(n) = 1,$$

where the sum runs over all possible conformations of the polymer. We can recover canonical averages from the Rosenbluth distribution by attributing different weights to different chain conformations. And this is precisely what is done in equation (11.2.6):

$$\langle A \rangle_{\mathcal{R}} = \frac{\sum_{n} \mathcal{W}(n) A(n) P(n)}{\sum_{n} \mathcal{W}(n) P(n)}. \tag{11.2.8}$$

Substitution of equations (11.2.5) and (11.2.7) gives

$$\begin{aligned} \langle A \rangle_{\mathcal{R}} &= \frac{\sum_{n} \mathcal{W}(n) k^{\ell} A(n) \exp[-\beta \mathcal{U}(n)]/\mathcal{W}(n)}{\sum_{n} \mathcal{W}(n) k^{\ell} \exp[-\beta \mathcal{U}(n)]/\mathcal{W}(n)} \\ &= \frac{\sum_{n} A(n) \exp[-\beta \mathcal{U}(n)]}{\sum_{n} \exp[-\beta \mathcal{U}(n)]} \\ &= \langle A \rangle, \end{aligned} \tag{11.2.9}$$

which shows that equation (11.2.6) indeed yields the correct ensemble average.

Here, we introduced the Rosenbluth factor as a correction for the bias in the sampling scheme. The Rosenbluth factor itself is also of interest, since it can be related to the excess chemical potential. To see this, let us assume that we use the Rosenbluth scheme to generate a large number of chain conformations while keeping the coordinates of all other particles in the system fixed. For this set of conformations, we compute the average of the Rosenbluth weight factor $\mathcal{W}$, $\overline{\mathcal{W}}$. Subsequently, we also perform an ensemble average over all coordinates and conformations of the N particles in the system, and we obtain

$$\langle \mathcal{W} \rangle = \left\langle \sum_{\Gamma} P_{\Gamma}(\mathbf{q}^N, \Gamma^N) \mathcal{W}_{\Gamma}\left(\mathbf{q}^N, \Gamma^N\right) \right\rangle, \tag{11.2.10}$$

where the angular brackets denote the ensemble average over all configurations of the system $\{\mathbf{q}^N, \Gamma^N\}$ of the solvent. Note that the test polymer does not form part of the N-particle system. Therefore the probability of finding the remaining particles in a configuration $\mathbf{q}^N$ does not depend on the conformation Γ of the polymer.

To simplify the expression for the average in equation (11.2.10), we first consider the average of the Rosenbluth factor for a given configuration $\{\mathbf{q}^N, \Gamma^N\}$ of the solvent:

$$\overline{\mathcal{W}}(\{\mathbf{q}^N, \Gamma^N\}) = \sum_{\Gamma} P_{\Gamma}(\mathbf{q}^N) \mathcal{W}_{\Gamma}(\{\mathbf{q}^N, \Gamma^N\}). \tag{11.2.11}$$

Substitution of equations (11.2.3) and (11.2.5) yields

$$\begin{aligned} \overline{\mathcal{W}} &= \sum_{\Gamma} \left[k \exp\left[-\beta u^{(1)}(\Gamma_1)\right] \prod_{i=2}^{\ell} \frac{\exp\left[-\beta u^{(i)}(\Gamma_i)\right]}{w_i} \right] \left[\prod_{i=1}^{\ell} \frac{w_i}{k} \right] \\ &= \sum_{\Gamma} k \exp\left[-\beta u^{(1)}(\Gamma_1)\right] \prod_{i=2}^{\ell} \frac{1}{k} \exp\left[-\beta u^{(i)}(\Gamma_i)\right] \\ &= \sum_{\Gamma} \frac{1}{k^{\ell-1}} \exp\left[-\beta \mathcal{U}_{\Gamma}\right], \end{aligned} \tag{11.2.12}$$

where we have dropped all explicit reference to the solvent coordinates $\{\mathbf{q}^N, \Gamma^N\}$. Note that equation (11.2.12) can be interpreted as an average over all *ideal* chain conformations of the Boltzmann factor $\exp\left[-\beta \mathcal{U}_{\Gamma}\right]$. If we now substitute equation (11.2.12) in equation (11.2.11), we obtain

$$\langle \mathcal{W} \rangle = \frac{\sum_{\Gamma} \left\langle \exp[-\beta \Delta \mathcal{U}(\mathbf{q}^N, \Gamma^N; \mathbf{q}_{N+1}, \Gamma_{N+1})] \right\rangle}{\sum_{\Gamma}}. \tag{11.2.13}$$

If we compare equation (11.2.13) with equation (11.2.2), we see that the ensemble average of the Rosenbluth factor is directly related to the excess chemical potential of the chain molecule.

$$\beta \mu_{ex} = -\ln \langle \mathcal{W} \rangle . \tag{11.2.14}$$

This completes our demonstration that a measurement of the average Rosenbluth factor of a trial chain can indeed be used to estimate the excess chemical potential of a polymer in a dense fluid. We should stress that the preceding method for measuring the chemical potential is in no way limited to chain molecules in a lattice. What *is* essential is that the number of possible directions for each segment (k) relative to the previous one be finite.

11.2.2 Extension to Continuously Deformable Molecules

The numerical computation of the (excess) chemical potential of a flexible chain (with or without elastic forces that counteract bending) is rather different from the corresponding calculation for a chain molecule that has a large but fixed number of undeformable conformations.

Here, we consider the case of a flexible molecule *with* intramolecular potential energy. Fully flexible chains, of course, are included as a special case. Consider a semi-flexible chain of ℓ linear segments. Like in the conformational-bias Monte Carlo scheme (see section 13.2.3), the potential energy is divided into two contributions: the "internal" potential energy $\mathcal{U}_{\text{bond}}$, which includes the bonded intramolecular interactions, and the "external" potential energy $\mathcal{U}_{\text{ext}}$, which accounts for the remainder of the interactions. A chain in the absence of the external interactions is defined as an *ideal* chain.

The conformational partition function of the ideal chain is equal to

$$Q_{\text{id}} = c \int \cdots \int d\Gamma_1 \cdots d\Gamma_\ell \prod_{i=1}^{\ell} \exp[-\beta u_{\text{bond}}(\theta_i)], \qquad (11.2.15)$$

where c is a numerical constant. We assume that Q_{id} is known. Our aim is to compute the effect of the external interactions on the conformational partition function. Hence, we wish to evaluate Q/Q_{id}, where Q denotes the partition function of the interacting chain. The excess chemical potential of the interacting chain is given by

$$\mu_{\text{ex}} = -k_B T \ln(Q/Q_{\text{id}}).$$

Before considering the "smart" approach to computing μ_{ex}, let us briefly review two not-so-smart methods.

The most naive way to compute the excess chemical potential of the interacting chain is to generate a very large number of completely random conformations of the freely jointed chain. For every conformation we compute both $\exp(-\beta \mathcal{U}_{\text{bond}})$ and $\exp[-\beta(\mathcal{U}_{\text{bond}} + \mathcal{U}_{\text{ext}})]$. The average of the former quantity is proportional to Q_{id}, while the average of the latter Boltzmann factor is proportional to Q. The ratio of these two averages therefore should yield Q/Q_{id}.

The problem with this approach is that the overwhelming majority of randomly generated conformations correspond to semi-flexible chains with a very high internal energy (and therefore very small Boltzmann weights). Hence, the statistical accuracy of this sampling scheme will be very poor.

The second scheme is designed to alleviate this problem. Rather than generating conformations of a freely jointed chain, we now sample the internal angles in the chain in such a way that the probability of finding a given angle θ_i is given by the Boltzmann weight

$$P(\theta_i) = \frac{\exp[-\beta u(\theta_i)]}{\int d\Gamma_i \exp[-\beta u(\theta_i)]}.$$

Such sampling can be performed quite easily using a rejection method (see, e.g., [19]). In what follows, we use the symbol Γ_i to denote the unit vector that specifies the orientation of the ith segment of the chain molecule. For every conformation thus generated, we compute the Boltzmann factor $\exp(-\beta\mathcal{U}_{\text{ext}})$. The average of this Boltzmann weight is then equal to

$$\begin{aligned} \langle \exp(-\beta\mathcal{U}_{\text{ext}}) \rangle &= \frac{\int \prod d\Gamma \exp[-\beta(\mathcal{U}_{\text{bond}} + \mathcal{U}_{\text{ext}})]}{\int \prod d\Gamma \exp(-\beta\mathcal{U}_{\text{bond}})} \\ &= Q/Q_{\text{id}}. \end{aligned} \tag{11.2.16}$$

This approach is obviously superior to the first scheme. However, in many practical situations it will still yield poor statistics, because most ideal chain conformations will not correspond to energetically favorable situations for the interacting chain. Hence the Boltzmann weights, again, will be small for most conformations and the statistical accuracy will not be very good.

The problem with both these schemes is that neither allows us to focus on those conformations that should contribute most to Q, namely, those for which the sum of the internal and external potential energies is not much larger than a few k_BT per degree of freedom. It would clearly be desirable to *bias* the sampling toward such favorable conformations.

It turns out that we can use a procedure similar to that used in section 11.2.1 to compute the excess chemical potential of a chain molecule with many fixed conformations. To compute μ_{ex}, we apply the following recipe for constructing a conformation of a chain of ℓ segments. The construction of chain conformations proceeds segment by segment. Let us consider the addition of one such segment. To be specific, let us assume that we have already grown i segments and we are trying to add segment $i+1$. This is done as follows:

1. Generate a fixed number (say, k) trial segments with orientations distributed according to the Boltzmann weight associated with the internal potential energy $u(\theta)$. We denote the different trial segment by indices $1, 2, \cdots, k$.

2. For all k trial segments, we compute the external Boltzmann factor $\exp[-\beta u_{ext}^{(i)}(j)]$.
3. Select one of the trial segments, say, n, with a probability

$$p^{(i)}(n) = \frac{\exp[-\beta u_{ext}^{(i)}(n)]}{w_i^{ext}}, \tag{11.2.17}$$

where we have defined

$$w_i^{ext} \equiv \sum_{j=1}^{k} \exp[-\beta u_{ext}^{(i)}(j)].$$

4. Add this segment as segment $i+1$ to the chain and repeat this procedure until the entire chain is completed. The normalized Rosenbluth factor of the entire chain is given by

$$\mathcal{W}^{ext}(n) = \prod_{i=1}^{\ell} \frac{w_i^{ext}}{k},$$

where, for the first segment, $w_1^{ext} = k \exp\left[-\beta u_{ext}^{(1)}(1)\right]$.

The desired ratio Q/Q_{id} is then equal to the average value (over many trial chains) of the product of the partial Rosenbluth weights:

$$Q/Q_{id} = \langle \mathcal{W}^{ext} \rangle . \tag{11.2.18}$$

To show that equation (11.2.18) is correct, let us consider the probability with which we generate a given chain conformation. This probability is the product of a number of factors. Let us first consider these factors for one segment and then later extend the result to the complete chain. The probability of generating a given set of k trial segments with orientations Γ_1 through Γ_k is

$$P_{id}(\Gamma_1)P_{id}(\Gamma_2)\cdots P_{id}(\Gamma_k)d\Gamma_1\cdots d\Gamma_k. \tag{11.2.19}$$

The probability of selecting any one of these trial segments follows from equation (11.2.17):

$$p^{(i)}(j) = \frac{\exp[-\beta u_{ext}^{(i)}(\Gamma_j)]}{w_i^{ext}(\Gamma_1,\cdots,\Gamma_k)}, \tag{11.2.20}$$

for $j = 2, 3, \cdots, \ell$. We wish to compute the average of w_i^{ext} over all possible sets of trial segments and all possible choices of the segment. To this end, we must sum over all j and integrate over all orientations $\prod_{j=1}^{k} d\Gamma_j$ (i.e., we average over the normalized probability distribution for the orientation of segment $i+1$):

$$\begin{aligned}\left\langle \frac{w_i^{ext}}{k} \right\rangle &= \int \prod_{j=1}^{k} d\Gamma_j P_{id}(\Gamma_j) \sum_{j'=1}^{k} \frac{\exp[-\beta u_{ext}(j')]}{w_i^{ext}(\Gamma_1, \cdots, \Gamma_k)} \frac{w_i^{ext}(\Gamma_1, \cdots, \Gamma_k)}{k} \\ &= \int \prod_{j=1}^{k} d\Gamma_j P_{id}(\Gamma_j) \sum_{j'=1}^{k} \frac{\exp[-\beta u_{ext}(j')]}{k}. \qquad (11.2.21)\end{aligned}$$

But the labeling of the trial segments is arbitrary. Hence, all k terms in the sum in this equation yield the same contribution, and this equation simplifies to

$$\left\langle \frac{w_i^{ext}}{k} \right\rangle = \int d\Gamma P_{id}(\Gamma) \exp[-\beta u_{ext}(\Gamma)] \qquad (11.2.22)$$

$$= \frac{\int d\Gamma \exp\{-\beta [u_{bond}(\Gamma) + u_{ext}(\Gamma)]\}}{\int d\Gamma \exp[-\beta u_{bond}(\Gamma)]} \qquad (11.2.23)$$

$$= \frac{Q^{(i)}}{Q_{id}^{(i)}}, \qquad (11.2.24)$$

which is indeed the desired result but for the fact that the expression in equation (11.2.24) refers to segment i (as indicated by the superscript in $Q^{(i)}$). The extension to a chain of ℓ segments is straightforward, although the intermediate expressions become a little unwieldy.

The final result is a relation between the normalized Rosenbluth factor and the excess chemical potential:

$$\beta \mu^{ex} = -\ln \frac{\langle \mathcal{W}^{ext} \rangle}{\langle \mathcal{W}_{ID}^{ext} \rangle}, \qquad (11.2.25)$$

where $\mathcal{W}_{ID}^{ext}$ is the normalized Rosenbluth factor of an isolated chain *with* nonbonded intramolecular interactions. This Rosenbluth factor has to be determined from a separate simulation.

In principle, the results of the Rosenbluth sampling scheme are exact in the sense that, in the limit of an infinitely long simulation, the results are identical to those of a Boltzmann sampling. In practice, however, there are important limitations. In contrast to the conformational-bias Monte Carlo scheme (see Chapter 13), the Rosenbluth scheme generates an unrepresentative sample of all polymer conformations as the probability of generating a given conformation is *not* proportional to its Boltzmann weight. Accurate values can be calculated only if these distributions have a sufficient overlap. If the overlap is small, then the tail of the Rosenbluth distribution makes the largest contribution to the ensemble average (11.2.6); configurations that have a very low probability of being generated in the Rosenbluth scheme may have Rosenbluth factors so large that they tend to dominate the ensemble average. Precisely because such conformations are generated very infrequently, the statistical accuracy may be poor. If the relevant conformations are never generated during a simulation, the results will even deviate systematically from the true ensemble average. This drawback of the Rosenbluth sampling scheme is well known, in fact (see, the article of Batoulis and Kremer [299,300] and Example 13).

Example 12 (Henry Coefficients in Porous Media)
For many practical applications of porous media, it is important to know the adsorption isotherm that specifies the dependence of the number of adsorbed molecules on the external pressure (or chemical potential) at a given temperature. In Examples 2 and 19, it is demonstrated how to compute the complete adsorption isotherm via Monte Carlo simulations in the grand-canonical ensemble. If the external pressures of interest are sufficiently low, a good estimate of the adsorption isotherm can be obtained from the Henry coefficient K_H. Under these conditions, the number of adsorbed molecules per unit volume (ρ_a) is proportional to the Henry coefficient and external pressure P:

$$\rho_a = K_H P.$$

The Henry coefficient is directly related to the excess chemical potential of the adsorbed molecules. To see this, consider the ensemble average of the average density in a porous medium. In the grand-canonical ensemble, this ensemble average is given by (see section 5.6)

$$\begin{aligned}
\left\langle \frac{N}{V} \right\rangle &= \frac{1}{Q} \sum_{N=0}^{\infty} \frac{q(T)^N V^N \exp(\beta\mu N)}{N!} \int d\mathbf{s}^N \, \exp[-\beta\mathcal{U}(\mathbf{s}^N)] N/V \\
&= \frac{q(T)\exp(\beta\mu)}{Q} \sum_{N=0}^{\infty} (q(T)V)^{N-1} \exp[\beta\mu(N-1)]/(N-1)! \\
&\quad \times \int d\mathbf{s}^{N-1} \, \exp\left[-\beta\mathcal{U}(\mathbf{s}^{N-1})\right] \int d\mathbf{s}_N \, \exp\left[-\beta\mathcal{U}(\mathbf{s}_N)\right] \\
&= q(T)\exp(\beta\mu)\left\langle \exp(-\beta\Delta\mathcal{U}^+) \right\rangle,
\end{aligned}$$

where $\Delta\mathcal{U}^+$ is defined as the energy of a test particle and $q(T)$ has been defined as in equation (13.6.5). In the limit $P \to 0$, the reservoir can be considered an ideal gas

$$\beta\mu = \ln\left(\frac{\beta P}{q(T)}\right).$$

Substitution of this equation and using equation (11.2.2) gives

$$\exp(\beta\mu^{ex}) = \left\langle \exp(-\beta\Delta\mathcal{U}^+) \right\rangle = \frac{\langle N/V \rangle}{\beta P}.$$

This gives, for the Henry coefficient,

$$K_H = \beta \exp(-\beta\mu^{ex}).$$

Maginn *et al.* [301] and Smit and Siepmann [302,303] used the approach described in this section to compute the Henry coefficients of linear alkanes adsorbed in the zeolite silicalite. The potential describing the alkane interactions is divided into an external potential and an internal potential. The internal potential includes bond bending and torsion:

$$u^{int} = u_{bend} + u_{tors}.$$

The alkane model uses a fixed bond length. The external interactions include the remainder of the intramolecular interactions and the interactions with the zeolite:

$$u^{ext} = u_{intra} + u_{zeo}.$$

Since the Henry coefficient is calculated at infinite dilution, there are no intramolecular alkane-alkane interactions. Smit and Siepmann use the internal interactions to generate the trial conformations (see section 13.3) and determine the normalized Rosenbluth factor using the external interactions only; this Rosenbluth factor is related to the excess chemical potential according to

$$\beta\mu^{ex} = -\ln \frac{\langle \mathcal{W}^{ext} \rangle}{\langle \mathcal{W}^{ext}_{IG} \rangle},$$

where $\langle \mathcal{W}^{ext}_{IG} \rangle$ is the Rosenbluth factor of a molecule in the ideal gas phase (no interactions with the zeolite) [304]. For an arbitrary alkane, the calculation of the Henry coefficient requires two simulations: one in the zeolite and one in the ideal gas phase. However, for butane and the shorter alkanes, all isolated (ideal gas) molecules are ideal chains, as there are no nonbonded interactions. For such chains the Rosenbluth factors in the ideal gas phase are by definition equal to 1.

In Figure 11.2 the Henry coefficients of the n-alkanes in silicalite as calculated by Smit and Siepmann are compared with those of Maginn *et al.* If we take into account that the models considered by Maginn *et al.* and Smit and Siepmann are slightly different, the results of these two independent studies are in very good agreement.

Example 13 (Rosenbluth Sampling for Polymers)

Batoulis and Kremer [300] have made a detailed analysis of the Rosenbluth algorithm for self-avoiding walks on a lattice. The Rosenbluth scheme is used to generate one walk on a lattice. Batoulis and Kremer found that, with a random insertion scheme, the probability of generating a walk of 100 steps without overlap is on the order of 0.022% (fcc-lattice). If we use the Rosenbluth scheme, on the other hand, this probability becomes almost 100%.

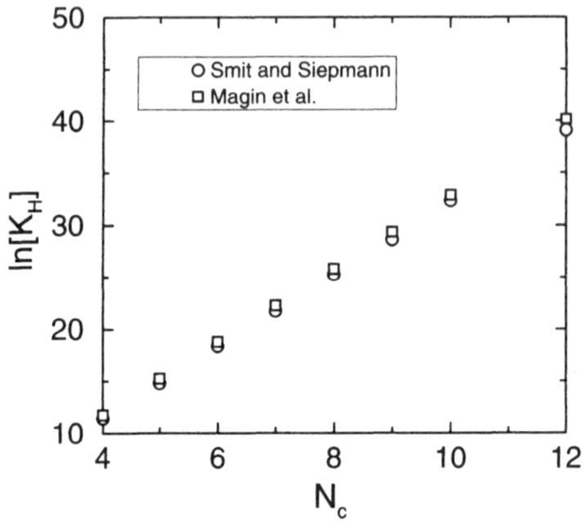

Figure 11.2: Henry coefficients K_H of n-alkanes in the zeolite silicalite as a function of the number of carbon atoms N_c as calculated by Maginn *et al.* [301] and Smit and Siepmann [303].

In Figure 11.3, the distribution of the radius of gyration of the polymer as calculated with the corrected ensemble average (11.2.6) is compared with the uncorrected average (i.e., using the Rosenbluth scheme to generate the schemes and using $\langle A \rangle = (1/M) \sum_{n=1}^{M} A(n)$ instead of equation (11.2.6) to calculate the ensemble averages). The figure shows that the Rosenbluth scheme generated chains that are more compact. Batoulis and Kremer showed that, for longer chain lengths, this difference increases exponentially. One therefore should be extremely careful when using such a non-Boltzmann sampling scheme.

11.2.3 Overlapping Distribution Rosenbluth Method

Although the Rosenbluth particle-insertion scheme described in section 11.2 is correct in principle, it may run into practical problems when the excess chemical potential becomes large. Fortunately, it is possible to combine the Rosenbluth scheme with the overlapping distribution method to obtain a technique with built-in diagnostics. This scheme is explained in Appendix I. As with the original overlapping distribution method (see section 7.2.3), the scheme described in Appendix I constructs two histograms (but now as a function of the logarithm of the Rosenbluth weight rather than the potential energy difference). If the sampled distributions do not overlap, then one should expect the estimate of the excess chemical potential of chain molecules to become unreliable and the Rosenbluth method should not be used. In fact, recent simulations by Mooij and Frenkel [305], using this over-

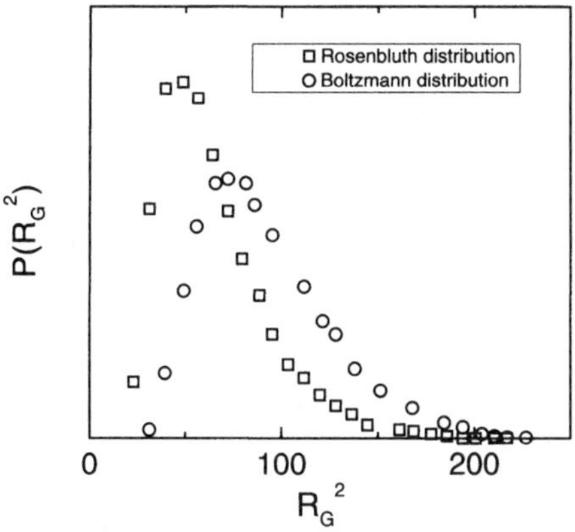

Figure 11.3: Probability distribution of the radius of gyration R_G. The circles show the Boltzmann distribution and the squares the Rosenbluth distribution. The results are of an fcc-lattice for a walk of 120 steps (data taken from ref. [300]).

lapping distribution method, show that indeed there is a tendency for the two distributions to move apart when long chains are inserted into a moderately dense fluid. Yet, these simulations also show that, at least in this case, the statistical errors in μ_{ex} become important before the systematic errors due to inadequate sampling show up.

11.2.4 Recursive Sampling

In view of the preceding discussion, it would seem attractive to have unbiased sampling schemes to measure the chemical potential. Of course, thermodynamic integration methods are unbiased and the modified Widom scheme, although biased at the level of the insertion of a single monomer (like the original Widom scheme), is less biased than the Rosenbluth method. Yet, these methods cannot be used to measure μ_{ex} in a single simulation (see section 11.1).

It turns out that nevertheless it is possible to perform unbiased sampling of μ_{ex} in a single simulation. Here, we briefly sketch the basic idea behind this method. In our description, we follow the approach proposed by Grassberger and Hegger [306–308]. Their technique is quite similar to a Monte Carlo scheme developed a few years earlier by Garel and Orland [309].

Like the Rosenbluth and modified Widom schemes, the recursive sampling approach is based on a segment-by-segment growth of the polymer. But that is about where the similarity ends. In recursive sampling the aim is to generate a *population* of trial conformations. The excess chemical potential of a chain molecule is directly related to the average number of molecules that have survived the growth process.

The first step of the procedure is to attempt a trial insertion of a monomer in the system. Suppose that the Boltzmann factor associated with this trial insertion is $b_0 \equiv \exp[-\beta u_0(\mathbf{r}^N)]$. We now allow the monomer to make multiple copies of itself, such that the average number of copies, $\langle n_0 \rangle$, is equal to

$$\langle n_0 \rangle = \pi_0 b_0,$$

where π_0 is a constant multiplicative factor that remains to be specified. A convenient rule for determining how many copies should be made is the following. Denote the fractional part of $\pi_0 b_0$ by f_0 and the integer part by i_0. Our rule is then to generate i_0 ($i_0 + 1$) copies of the inserted particle with a probability $1 - f_0$ (f_0). Clearly if $i_0 = 0$, there is a probability $1 - f_0$ that the monomer will "die." Assume that we have generated at least one copy of the monomer. Every copy from now on proceeds independently to generate offspring. For instance, to generate a dimer population, we add a segment to every surviving monomer. We denote the Boltzmann weight associated with these trial additions by $b_1(i)$, where the index i indicates that every surviving monomer will give rise to a different dimer. As before, we have to decide how many copies of the dimers should survive. This is done in exactly the same way as for the monomer; that is, the average number of dimers that descends from monomer i is given by

$$\langle n_1(i) \rangle = \pi_1 b_1(i),$$

where π_1, just like π_0 before, is a constant to be specified later. The number of dimers generated may either be larger or smaller than the original number of monomers. We now proceed with the same recipe for the next generation (trimers) and so on. In fact, as with the semi-flexible molecules discussed in section 11.2.1, it is convenient to include the intramolecular bond-bending, bond-stretching, and torsional energies in the probability distribution that determines with what orientation new segments should be added.

The average number of surviving molecules at the end of the ℓth step is

$$\langle N_\ell \rangle = \left(\prod_{i=0}^{\ell} \pi_i \right) \left\langle \exp[-\beta \mathcal{U}_\ell(\mathbf{r}^N)] \right\rangle,$$

where $\mathcal{U}_\ell(\mathbf{r}^N)$ is the total interaction of the chain molecule with the N solvent molecules (and the nonbonded intramolecular interactions). The angular brackets denote a canonical average over the coordinates and over the intramolecular Boltzmann factors of the ideal (nonself-avoiding) chain. In other words,

$$\langle N_\ell \rangle = \left(\prod_{i=0}^{\ell} \pi_i \right) \exp[-\beta \mu_{ex}(\ell)].$$

Hence, the excess chemical potential is given by

$$\mu_{ex}(\ell) = -k_B T \ln \left(\frac{\langle N_\ell \rangle}{\prod_{i=0}^{\ell} \pi_i} \right). \tag{11.2.26}$$

The constants π_i should be chosen such that there is neither a population explosion nor mass extinction. If we have a good guess for $\mu_{ex}(\ell)$ then we can use this to estimate π_i. In general, however, π_i must be determined by trial and error.

This recursive algorithm has several nice features. First of all, it is computationally quite efficient (in some cases, more than an order of magnitude faster than the Rosenbluth scheme, for the same statistical accuracy). In fact, in actual calculations, the algorithm searches in *depth* first, rather than in *breadth*. That is to say, we try to grow a polymer until it has been completed (or has died). We then continue from the last branch of the tree from where we are allowed to grow another trial conformation. In this way, we work our way back to the root of the tree. The advantage of this scheme is that the memory requirements are minimal. Moreover, the structure of the program is very simple indeed (in languages that allow recursive calls). Last but not least, the recursive scheme generates an *unbiased* (i.e., Boltzmann) population of chain conformations [310].

11.2.5 Pruned-Enriched Rosenbluth Method

An important extension of the Rosenbluth scheme has been proposed by Grassberger [311]. It is called the *pruned-enriched* Rosenbluth method (PERM). One of the reasons why the conventional Rosenbluth method fails for long chains or at high densities is that the distribution of Rosenbluth weights becomes very broad. As a consequence, it can happen that a few conformations with a high Rosenbluth weight completely dominate the average. If this is the case, we should expect to see large statistical fluctuations in the average. It would, of course, be desirable to focus the simulations on those classes of conformations that contribute most to the average, and spend little time on conformations that have a very low Rosenbluth weight. The PERM algorithm is a generalization of the recursive-sampling scheme discussed above. It also generates a *population* of chains with different conformations. And it shares the advantage that, due to the recursive nature of the algorithm, we need not keep more than one conformation (plus a set of pointers) in memory. The "birth" and "death" rules of this algorithm are such that it generates many copies of conformations with a high Rosenbluth weight, while low-weight structures have a high probability of "dying." The Rosenbluth weight of the remaining conformations is adjusted in such a way that our birth-death rules do not affect the desired average. As the PERM algorithm is recursive, it uses little memory. To summarize the algorithm in a few words: conformations with a high Rosenbluth weight are multiplied by a factor k and their weight is reduced by the same factor. Conformations with a low weight are "pruned"—half the low-weight conformations are discarded, while the weight of the remainder is doubled. Once all chains that have started from a common "ancestor" have been grown to completion (or have been discarded), we simply add the (rescaled) Rosenbluth weights of all surviving chains.

Below, we briefly sketch how the algorithm is implemented. Let us introduce an upper and a lower threshold of the Rosenbluth weight of a chain with length l, $\mathcal{W}_i^{max}$ and $\mathcal{W}_i^{min}$, respectively. If the partial Rosenbluth weight of a particular chain conformation of lenght i, $\mathcal{W}_i$, exceeds the threshold, $\mathcal{W}_i > \mathcal{W}_i^{max}$, then the single conformation is replaced by k copies. The partial Rosenbluth weight of every copy is set equal to $\mathcal{W}_i/k$. If, on the other hand, the partial Rosenbluth weight of a particular conformation, $\mathcal{W}_i$, is below the lower threshold, $\mathcal{W}_i < \mathcal{W}_i^{min}$, then we "prune." With a probability of 50% we delete the conformation. But if the conformation survives, we double its Rosenbluth weight. There is considerable freedom in the choice of $\mathcal{W}_i^{max}$, $\mathcal{W}_i^{min}$, and k. In fact, all of them can be chosen "on the fly" (as long as this choice does not depend on properties of conformations that have been grown from the same ancestor). A detailed discussion of the algorithm can be found in refs. [311,312].

The limitation of the recursive growth algorithm is that it is intrinsically a *static* Monte Carlo technique; every new configuration is generated from scratch. This is in contrast to *dynamic* (Markov-chain) MC schemes in which the basic trial move is an attempt to modify an existing configuration. Dynamic MC schemes are better suited for the simulations of many-particle systems than their static counterparts. The reason is simple: it is easy to *modify* a many-particle configuration to make other "acceptable" configurations (for instance, by displacing one particle over a small distance). In contrast, it is very difficult to generate such configurations from scratch. On the other hand, once a new configuration is successfully generated in a static scheme, it is completely independent from all earlier configurations. In contrast, successive configurations in dynamic MC are strongly correlated.

CBMC is, in a sense, a hybrid scheme: it is a dynamic (Markov-chain) MC method. But the chain-regrowing step is more similar to a static MC scheme. However, in this step it is less "smart" than the recursive algorithms discussed above, because it is rather "myopic." The scheme looks only one step ahead. It may happen that we spend a lot of time growing a chain almost to completion, only to discover that there is simply no space left for the last few monomers. This problem can be alleviated by using a scanning method of the type introduced by Meirovitch [313]. This is basically a static, Rosenbluth-like method for generating polymer configurations. But, in contrast to the Rosenbluth scheme, the scanning method looks several steps ahead. If this approach is transferred naively to a conformational-bias Monte Carlo program, it would yield an enhanced generation of acceptable trial conformations, but the computational cost would rise steeply (exponentially) with the depth of the scan. This second drawback can be avoided by incorporating a recursive scanning method that cheaply eliminates doomed trial configurations, within a dynamic Monte Carlo scheme. In section 13.7 we discuss a dynamic MC algorithm (recoil growth), that is based on this approach.

Part IV

Advanced Techniques

Chapter 12

Long-Range Interactions

As computing power continues to increase, we can simulate ever larger systems. For instance, a typical model of a biological system may contain as many as 10^5 particles. In such systems it becomes crucial to avoid computing all pair interactions, as otherwise the computational effort would be proportional to the square of the number of particles. This issue is particularly relevant for long-range interactions (e.g., Coulombic and dipolar potentials) as, for such models, truncation of the potential is never allowed. It then becomes essential to find an efficient technique for computing the long-range part of the intermolecular interactions. In section 3.2.2, we showed that, if we truncate the potential at a distance r_c, the contribution of the tail of the potential $u(r)$ can be estimated (in three dimensions) using

$$\mathcal{U}^{\text{tail}} = \frac{N\rho}{2} \int_{r_c}^{\infty} dr\, u(r) 4\pi r^2,$$

where ρ is the average number density. This equation shows that the tail correction to the potential energy diverges, unless the potential energy function $u(r)$ decays faster than r^{-3}. This is why one cannot use truncation plus tail correction for Coulombic and dipolar interactions.

In the literature, one can find numerous examples where the computational cost of evaluating long-range interactions is reduced in a rather drastic way. It is simply assumed that the long-range part of the potential is not important. The problem of the long-range interactions is then "solved" by truncation. This gets rid of the expensive part of the calculation, but it gives rise to serious inaccuracies. A discussion of the artifacts that are introduced by the various truncation schemes is presented in some detail by Steinbach and Brooks [314]. In the present chapter we discuss some of the less draconian (and more reliable) techniques for handling long-range interactions. Such schemes are more expensive than simple truncation, but the advantage is that they do respect the long-range character of the forces.

We discuss three techniques: (1) Ewald summation, (2) fast multipole methods, and (3) particle-mesh-based techniques. Of these, the Ewald summation is (still) the most widely used. As the computational effort for the Ewald summation scales as $\mathcal{O}(N^{3/2})$, this approach becomes prohibitively expensive for large systems. To overcome these restrictions, several alternative algorithms, such as the particle-particle/particle-mesh (PPPM) method of Eastwood and Hockney [315], which scales as $\mathcal{O}(N \log N)$, and the fast multipole method of Greengard and Rokhlin [316], which scales as $\mathcal{O}(N)$, have been proposed. However, these $\mathcal{O}(N)$ algorithms only become more efficient than the Ewald summation for systems containing on the order of 10^5 particles—where it should be noted that the precise location of the break-even point depends on the desired accuracy. For intermediate-size systems ($N \approx 10^3$–10^4) the so-called particle-mesh Ewald summation [317] is an attractive alternative. The complexity of the latter method also scales as $\mathcal{O}(N \log N)$.

12.1 Ewald Sums

In this section we present a simple (and nonrigorous) discussion of the Ewald method [318] for computing long-range contributions to the potential energy in a system with periodic boundary conditions. Readers who are interested in more detail (or more rigor) are referred to a series of articles by De Leeuw *et al.* [319–321] and an introductory paper by Hansen [322]. A discussion of the Ewald sum in the context of solid state physics can be found in [323]. We also refer the reader to the literature for a discussion of an ingenious alternative to the Ewald method that can be used for relatively small fluid systems [324].

12.1.1 Point Charges

Let us first consider a system consisting of positively and negatively charged particles. These particles are assumed to be located in a cube with diameter L (and volume $V = L^3$). We assume periodic boundary conditions. The total number of particles in the fundamental simulation box (the unit cell) is N. We assume that all particles repel each another at sufficiently short distances. In addition we assume that the system as a whole is electrically neutral; that is, $\sum_i q_i = 0$. We wish to compute the Coulomb contribution to the potential energy of this N-particle system,

$$\mathcal{U}_{\text{Coul}} = \frac{1}{2} \sum_{i=1}^{N} q_i \phi(r_i), \tag{12.1.1}$$

where $\phi(r_i)$ is the electrostatic potential at the position of ion i:

$$\phi(r_i) = \sum_{j,\mathbf{n}}{}' \frac{q_j}{|\mathbf{r}_{ij} + \mathbf{n}L|}, \tag{12.1.2}$$

where the prime on the summation indicates that the sum is over all periodic images $\mathbf{n}$ and over all particles j, except $j = i$ if $\mathbf{n} = \mathbf{0}$. Note that we assume that particle i interacts with *all* its periodic images, but not, of course, with itself. Here, and in what follows, we use Gaussian units, because it makes the notation more compact. Equation (12.1.2) cannot be used to compute the electrostatic energy in a simulation, because it contains a poorly converging sum (in fact, the sum is only conditionally convergent). To improve the convergence of the expression for the electrostatic potential energy, we rewrite the expression for the charge density. In equation (12.1.2) we have represented the charge density as a sum of δ-functions. The contribution to the electrostatic potential due to these point charges decays as $1/r$. Now consider what happens if we assume that every particle i with charge q_i is surrounded by a diffuse charge distribution of the opposite sign, such that the total charge of this cloud exactly cancels q_i. In that case the electrostatic potential due to particle i is due exclusively to the fraction of q_i that is not screened. At large distances, this fraction rapidly goes to 0. How rapidly depends on the functional form of the screening charge distribution. In what follows, we shall assume a Gaussian distribution for the screening charge cloud.

The contribution to the electrostatic potential at a point r_i due to a set of screened charges can be easily computed by direct summation, because the electrostatic potential due to a screened charge is a rapidly decaying function of r. However, it was not our aim to evaluate the potential due to a set of *screened* charges but due to *point* charges. Hence, we must correct for the fact that we have added a screening charge cloud to every particle. This is shown schematically in Figure 12.1. This compensating charge density varies smoothly in space. We wish to compute the electrostatic energy at the site of ion i. Of course, we should exclude the electrostatic interaction of the ion with itself. We have three contributions to the electrostatic potential: first of all, the one due to the point charge q_i, secondly, the one due to the (Gaussian) *screening* charge cloud with charge $-q_i$, and finally the one due to the *compensating* charge cloud with charge q_i. In order to exclude Coulomb self-interactions, we should not include any of these three contributions to the electrostatic potential at the position of ion i. However, it turns out that it is convenient to retain the contribution due to the *compensating* charge distribution and correct for the resulting spurious interaction afterwards. The reason we retain the compensating charge cloud for ion i is that, if we do so, the compensating charge distribution is not only a smoothly varying function, but it is also periodic. Such a function can be represented by a (rapidly

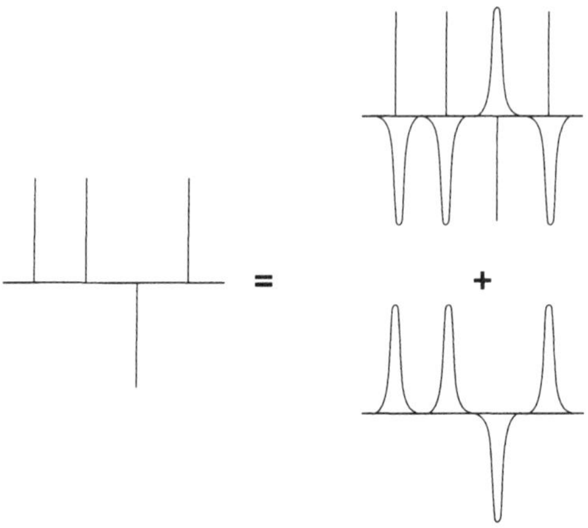

Figure 12.1: A set of point charges may be considered a set of screened charges minus the smoothly varying screening background.

converging) Fourier series, and this will turn out to be essential for the numerical implementation. Of course, in the end we should correct for the inclusion of a spurious "self" interaction between ion i and the compensating charge cloud.

After this brief sketch of the method for evaluating the electrostatic contribution to the potential energy, let us now consider the individual terms. We assume that the compensating charge distribution surrounding an ion i is a Gaussian with width $\sqrt{2/\alpha}$:

$$\rho_{\text{Gauss}}(r) = -q_i(\alpha/\pi)^{\frac{3}{2}} \exp(-\alpha r^2).$$

The choice of α will be determined later by considerations of computational efficiency. We shall first evaluate the contribution to the Coulomb energy due to the continuous background charge, then the spurious "self" term, and finally the real-space contribution due to the screened charges.

Fourier Transformation

This chapter relies heavily on the use of the Fourier transformation. It is therefore instructive to recall some of the basic equations in the context of electrostatics.

The central problem is to compute the energy of a given charge distribution $\rho(\mathbf{r})$. Formally, this corresponds to solving Poisson's equation for the electrostatic potential. Using the Gaussian notation,

$$-\nabla^2\phi(\mathbf{r}) = 4\pi\rho(\mathbf{r}), \tag{12.1.3}$$

where $\phi(\mathbf{r})$ is the electrostatic potential at point $\mathbf{r}$. For a single charge z at the origin the solution of this equation is the Coulomb potential

$$\phi(\mathbf{r}) = \frac{z}{4\pi|\mathbf{r}|}. \tag{12.1.4}$$

For a collection of N point charges we can define a charge density

$$\rho_P(\mathbf{r}) = \sum_{i=1}^{N} q_i \delta(\mathbf{r} - \mathbf{r}_i), \tag{12.1.5}$$

where $\mathbf{r}_i$ and q_i are the position and charge of particle i, respectively. The potential in a point $\mathbf{r}$ follows from a summation of the contributions of the particles

$$\phi(\mathbf{r}) = \sum_{i=1}^{N} \frac{q_i}{4\pi|\mathbf{r} - \mathbf{r}_i|}.$$

A different representation of these equations can be given in Fourier space. Let us consider a periodic system with a cubic box of length L and volume V. Any function $f(\mathbf{r})$ that depends on the coordinates of our system can be represented by a Fourier series:

$$f(\mathbf{r}) = \frac{1}{V} \sum_{l=-\infty}^{\infty} \tilde{f}(k) e^{i\mathbf{k}\cdot\mathbf{r}}, \tag{12.1.6}$$

where $\mathbf{k} = (2\pi/L)\mathbf{l}$ with $\mathbf{l} = (l_x, l_y, l_z)$ are the lattice vectors in Fourier space. The Fourier coefficients $\tilde{f}(k)$ are calculated using

$$\tilde{f}(\mathbf{k}) = \int_V d\mathbf{r}\, f(\mathbf{r}) e^{-i\mathbf{k}\cdot\mathbf{r}}. \tag{12.1.7}$$

In Fourier space Poisson's equation (12.1.3) has a much simpler form. We can write for the Poisson equation:

$$\begin{aligned} -\nabla^2\phi(\mathbf{r}) &= -\nabla^2 \left(\frac{1}{V} \sum_{\mathbf{k}} \tilde{\phi}(\mathbf{k}) e^{i\mathbf{r}\cdot\mathbf{k}} \right) \\ &= \frac{1}{V} \sum_{\mathbf{k}} k^2 \tilde{\phi}(\mathbf{k}) e^{i\mathbf{r}\cdot\mathbf{k}}. \end{aligned} \tag{12.1.8}$$

For the Fourier transform of the charge density we have

$$\rho(\mathbf{r}) = \frac{1}{V} \sum_{\mathbf{k}} \tilde{\rho}(\mathbf{k}) e^{i\mathbf{r}\cdot\mathbf{k}}. \tag{12.1.9}$$

Substitution of equations (12.1.9) and (12.1.8) into equation (12.1.3) yields the Poisson equation in Fourier space

$$k^2\tilde{\phi}(k) = 4\pi\tilde{\rho}(k). \tag{12.1.10}$$

To find the solution of Poisson's equation for a point charge of strength z at the origin, we have to perform the Fourier transform of a delta function:

$$\begin{aligned}\tilde{\rho}(\mathbf{k}) &= \int_V d\mathbf{r}\, z\delta(\mathbf{r})e^{-i\mathbf{k}\cdot\mathbf{r}} \\ &= z.\end{aligned}$$

This yields as solution for the Poisson equation

$$\tilde{\phi}(k) = \frac{4\pi z}{k^2}.$$

The solution for a unit charge is often called the Green's function:

$$\tilde{g}(k) = \frac{4\pi}{k^2}. \tag{12.1.11}$$

For a collection of point charges with charge density given by equation (12.1.5), we can write for the Fourier coefficients of the potential

$$\tilde{\phi}(k) = \tilde{g}(k)\tilde{\rho}_P(k)$$

with

$$\begin{aligned}\tilde{\rho}_P(k) &= \int_V d\mathbf{r}\sum_{i=1}^{N} q_i\delta(\mathbf{r}-\mathbf{r}_i)e^{-i\mathbf{k}\cdot\mathbf{r}} \\ &= \sum_{i=1}^{N} q_i e^{-i\mathbf{k}\cdot\mathbf{r}_i}.\end{aligned} \tag{12.1.12}$$

These equations show that in Fourier space the solution of Poisson's equation is simply obtained by multiplying $\tilde{\rho}(k)$ and $\tilde{g}(k)$ for all k vectors.

In the following we will also use another property of the Fourier transform. If we have a function $f_1(x)$, which is the convolution ($\star$) of two other functions $f_2(x)$ and $f_3(x)$,

$$f_1(x) = f_2(x) \star f_3(x) \equiv \int dx'\, f_2(x')f_3(x-x'),$$

then the Fourier coefficients of these functions are related by a simple multiplication:

$$\tilde{f}_1(k) = \tilde{f}_2(k)\tilde{f}_3(k).$$

For example, if we have a charge distribution that does not consist of simple point charges, but a more "smeared out" distribution,

$$\rho(\mathbf{r}) = \sum_i q_i \gamma(\mathbf{r} - \mathbf{r}_i) = \int d\mathbf{r}' \, \gamma(\mathbf{r}') \rho_p(\mathbf{r} - \mathbf{r}'), \tag{12.1.13}$$

where $\gamma(\mathbf{r})$ is the "shape" of the charge distribution of a single charge, then the Poisson equation for this system takes, in Fourier space, the form

$$\tilde{\phi}(k) = \tilde{g}(k)\tilde{\gamma}(k)\tilde{\rho}_P(k).$$

Fourier Part of Ewald Sum

We now apply the properties of the Poisson equation in Fourier form to compute the electrostatic potential at a point r_i due to a charge distribution $\rho_1(r)$ that consists of a periodic sum of Gaussians:

$$\rho_1(r) = \sum_{j=1}^{N} \sum_{\mathbf{n}} q_j (\alpha/\pi)^{\frac{3}{2}} \exp\left[-\alpha \left|\mathbf{r} - (\mathbf{r}_j + \mathbf{n}L)\right|^2\right].$$

To compute the electrostatic potential $\phi_1(r)$ due to this charge distribution, we use Poisson's equation:

$$-\nabla^2 \phi_1(r) = 4\pi\rho_1(r),$$

or in Fourier form,

$$k^2 \phi_1(k) = 4\pi\rho_1(k).$$

Fourier transforming the charge density ρ_1 yields

$$\begin{aligned}
\rho_1(\mathbf{k}) &= \int_V d\mathbf{r} \, \exp(-i\mathbf{k} \cdot \mathbf{r}) \rho_1(\mathbf{r}) \\
&= \int_V d\mathbf{r} \, \exp(-i\mathbf{k} \cdot \mathbf{r}) \sum_{j=1}^{N} \sum_{\mathbf{n}} q_j (\alpha/\pi)^{\frac{3}{2}} \exp\left[-\alpha \left|\mathbf{r} - (\mathbf{r}_j + \mathbf{n}L)\right|^2\right] \\
&= \int_{\text{all space}} d\mathbf{r} \, \exp(-i\mathbf{k} \cdot \mathbf{r}) \sum_{j=1}^{N} q_j (\alpha/\pi)^{\frac{3}{2}} \exp\left[-\alpha \left|\mathbf{r} - \mathbf{r}_j\right|^2\right] \\
&= \sum_{j=1}^{N} q_j \exp(-i\mathbf{k} \cdot \mathbf{r}_j) \exp(-k^2/4\alpha).
\end{aligned} \tag{12.1.14}$$

If we now insert this expression in Poisson's equation, we obtain

$$\phi_1(k) = \frac{4\pi}{k^2} \sum_{j=1}^{N} q_j \exp(-i\mathbf{k} \cdot \mathbf{r}_j) \exp(-k^2/4\alpha), \tag{12.1.15}$$

where it should be noted that this expression is defined only for $\mathbf{k} \neq 0$. This is a direct consequence of the conditional convergence of the Ewald sum. For the time being, we shall assume that the term with $\mathbf{k} = 0$ is equal to 0. As we shall see later, this assumption is consistent with a situation where the periodic system is embedded in a medium with infinite dielectric constant.

We now compute the contribution to the potential energy due to ϕ_1, using equation (12.1.1). To this end, we first compute $\phi_1(r)$:

$$\begin{aligned}\phi_1(r) &= \frac{1}{V}\sum_{\mathbf{k}\neq 0} \phi_1(\mathbf{k}) \exp(i\mathbf{k}\cdot\mathbf{r}) \qquad (12.1.16)\\ &= \sum_{\mathbf{k}\neq 0}\sum_{j=1}^{N} \frac{4\pi q_j}{k^2} \exp[i\mathbf{k}\cdot(\mathbf{r}-\mathbf{r}_j)]\exp(-k^2/4\alpha),\end{aligned}$$

and hence,

$$\begin{aligned}\mathcal{U}_1 &\equiv \frac{1}{2}\sum_i q_i\phi_1(r_i)\\ &= \frac{1}{2}\sum_{\mathbf{k}\neq 0}\sum_{i,j=1}^{N} \frac{4\pi q_i q_j}{Vk^2} \exp[i\mathbf{k}\cdot(\mathbf{r}_i-\mathbf{r}_j)]\exp(-k^2/4\alpha)\\ &= \frac{1}{2V}\sum_{\mathbf{k}\neq 0} \frac{4\pi}{k^2} |\rho(\mathbf{k})|^2 \exp(-k^2/4\alpha), \qquad (12.1.17)\end{aligned}$$

where we have used the definition

$$\rho(\mathbf{k}) \equiv \sum_{i=1}^{N} q_i \exp(i\mathbf{k}\cdot\mathbf{r}_i). \qquad (12.1.18)$$

Correction for Self-Interaction

The contribution to the potential energy given in equation (12.1.17) includes a term $(1/2)q_i\phi_{\text{self}}(r_i)$ due to the interaction between a continuous Gaussian charge cloud of charge q_i and a point charge q_i located at the center of the Gaussian. This term is spurious, and we should correct for it. We therefore must compute the potential energy at the origin of a Gaussian charge cloud. The charge distribution that we have overcounted is

$$\rho_{\text{Gauss}}(r) = q_i(\alpha/\pi)^{\frac{3}{2}} \exp(-\alpha r^2).$$

We can compute the electrostatic potential due to this charge distribution using Poisson's equation. Using the spherical symmetry of the Gaussian charge cloud, we can write Poisson's equation as

$$-\frac{1}{r}\frac{\partial^2 r\phi_{\text{Gauss}}(r)}{\partial r^2} = 4\pi\rho_{\text{Gauss}}(r)$$

or

$$-\frac{\partial^2 r\phi_{\text{Gauss}}(r)}{\partial r^2} = 4\pi r \rho_{\text{Gauss}}(r).$$

Partial integration yields

$$\begin{aligned} -\frac{\partial r\phi_{\text{Gauss}}(r)}{\partial r} &= \int_{\infty}^{r} dr\ 4\pi r \rho_{\text{Gauss}}(r) \\ &= -2\pi q_i (\alpha/\pi)^{\frac{3}{2}} \int_{r}^{\infty} dr^2\ \exp(-\alpha r^2) \\ &= -2 q_i (\alpha/\pi)^{\frac{1}{2}} \exp(-\alpha r^2). \end{aligned} \tag{12.1.19}$$

A second partial integration gives

$$\begin{aligned} r\phi_{\text{Gauss}}(r) &= 2 q_i (\alpha/\pi)^{\frac{1}{2}} \int_0^r dr\ \exp(-\alpha r^2) \\ &= q_i \operatorname{erf}\left(\sqrt{\alpha} r\right), \end{aligned} \tag{12.1.20}$$

where, in the last line, we have employed the definition of the error function: $\operatorname{erf}(x) \equiv (2/\sqrt{\pi}) \int_0^x \exp(-u^2)\, du$. Hence,

$$\phi_{\text{Gauss}}(r) = \frac{q_i}{r} \operatorname{erf}\left(\sqrt{\alpha} r\right). \tag{12.1.21}$$

To compute the spurious self term to the potential energy, we must compute $\phi_{\text{Gauss}}(r)$ at $r = 0$. It is easy to verify that

$$\phi_{\text{Gauss}}(r = 0) = 2 q_i (\alpha/\pi)^{\frac{1}{2}}.$$

Hence, the spurious contribution to the potential energy is

$$\begin{aligned} \mathcal{U}_{\text{self}} &= \frac{1}{2} \sum_{i=1}^{N} q_i \phi_{\text{self}}(r_i) \\ &= (\alpha/\pi)^{\frac{1}{2}} \sum_{i=1}^{N} q_i^2. \end{aligned} \tag{12.1.22}$$

The spurious self-interaction $\mathcal{U}_{\text{self}}$ should be *subtracted* from the sum of the real-space and Fourier contributions to the Coulomb energy. Note that equation (12.1.22) does not depend on the particle positions. Hence, during a simulation, this term is constant, provided that the values of all (partial) charges (and particles) in the system remain fixed.

Real-Space Sum

Finally, we must compute the electrostatic energy due to the point charges screened by oppositely charged Gaussians. Using the results of section 12.1.1, in particular equation (12.1.21), we can immediately write the (short-range) electrostatic potential due to a point charge q_i surrounded by a Gaussian with net charge $-q_i$:

$$\begin{aligned}\phi_{\text{short-range}}(r) &= \frac{q_i}{r} - \frac{q_i}{r}\text{erf}\left(\sqrt{\alpha}r\right) \\ &= \frac{q_i}{r}\text{erfc}\left(\sqrt{\alpha}r\right),\end{aligned} \tag{12.1.23}$$

where the last line defines the complementary error function $\text{erfc}(x) \equiv 1 - \text{erf}(x)$. The total contribution of the screened Coulomb interactions to the potential energy is then given by

$$\mathcal{U}_{\text{short-range}} = \frac{1}{2}\sum_{i \neq j}^{N} q_i q_j \text{erfc}\left(\sqrt{\alpha}r_{ij}\right)/r_{ij}. \tag{12.1.24}$$

The total electrostatic contribution to the potential energy now becomes the sum of equations (12.1.17), (12.1.22), and (12.1.24):

$$\begin{aligned}\mathcal{U}_{\text{Coul}} &= \frac{1}{2V}\sum_{\mathbf{k}\neq 0}\frac{4\pi}{k^2}|\rho(\mathbf{k})|^2 \exp(-k^2/4\alpha) \\ &\quad - (\alpha/\pi)^{\frac{1}{2}}\sum_{i=1}^{N} q_i^2 \\ &\quad + \frac{1}{2}\sum_{i\neq j}^{N}\frac{q_i q_j \text{erfc}\left(\sqrt{\alpha}r_{ij}\right)}{r_{ij}}.\end{aligned} \tag{12.1.25}$$

12.1.2 Dipolar Particles

It is straightforward to derive the corresponding expressions for the potential energy of a system containing dipolar molecules. The only modification is that we must everywhere replace q_i by $-\boldsymbol{\mu}_i \cdot \boldsymbol{\nabla}_i$. For example, the electro-

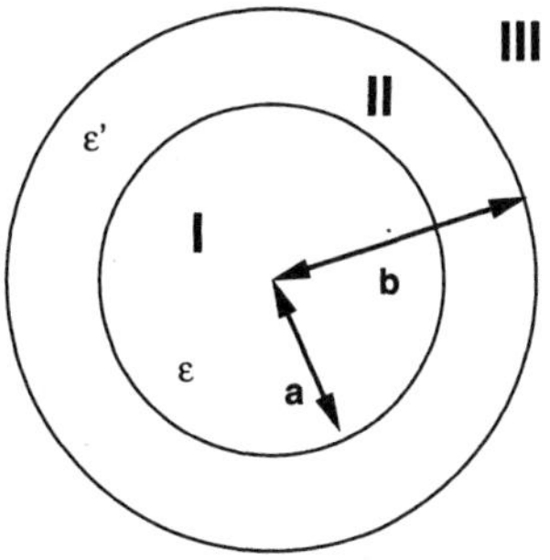

Figure 12.2: Spherical dielectric surrounded by a sphere.

static energy of a dipolar system becomes

$$\begin{aligned}\mathcal{U}_{\text{dipolar}} &= \frac{1}{2V}\sum_{\mathbf{k}\neq 0}\frac{4\pi}{k^2}|\mathbf{M}(k)|^2\exp(-k^2/4\alpha)\\ &\quad -\frac{2\pi}{3}(\alpha/\pi)^{\frac{3}{2}}\sum_{i=1}^{N}\mu_i^2\\ &\quad +\frac{1}{2}\sum_{i\neq j}^{N}\left[(\boldsymbol{\mu}_i\cdot\boldsymbol{\mu}_j)B(r_{ij})-(\boldsymbol{\mu}_i\cdot\mathbf{r}_{ij})(\boldsymbol{\mu}_j\cdot\mathbf{r}_{ij})C(r_{ij})\right],\end{aligned} \tag{12.1.26}$$

where

$$\begin{aligned}B(r) &\equiv \frac{\operatorname{erfc}(\sqrt{\alpha}r)}{r^3}+2(\alpha/\pi)^{\frac{1}{2}}\frac{\exp(-\alpha r^2)}{r^2},\\ C(r) &\equiv 3\frac{\operatorname{erfc}(\sqrt{\alpha}r)}{r^5}+2(\alpha/\pi)^{\frac{1}{2}}(2\alpha+3/r^2)\frac{\exp(-\alpha r^2)}{r^2},\end{aligned}$$

and

$$\mathbf{M}(k)\equiv\sum_{i=1}^{N} i\boldsymbol{\mu}_i\cdot\mathbf{k}\exp(i\mathbf{k}\cdot\mathbf{r}_i).$$

Again, this expression applies to a situation where the periodic system is embedded in a material with infinite dielectric constant.

12.1.3 Dielectric Constant

To derive an expression for the dielectric constant of a polar fluid, we consider the system shown in Figure 12.2: a large spherical dielectric with radius a and dielectric constant ϵ (region I) surrounded by a much larger sphere with radius b and dielectric constant ϵ' (region II). The entire system is placed in vacuum (region III), and an electric field **E** is applied. The potential at a given point in this system follows from the solution of the Poisson equation with the appropriate boundary conditions (continuity of the normal component of the displacement **D** and tangential component of the electric field **E**) at the two boundaries between regions I and II, and II and III. In the limit $a \to \infty$, $b \to \infty$, $a/b \to 0$, we can write, for the electric field in region I,

$$\mathbf{E}_{\mathrm{I}} = \frac{9\epsilon'}{(\epsilon'+2)(2\epsilon'+\epsilon)}\mathbf{E}, \tag{12.1.27}$$

which gives, for the polarization **P**,

$$\mathbf{P} \equiv \frac{\epsilon - 1}{4\pi}\mathbf{E}_{\mathrm{I}} = \frac{9\epsilon'(\epsilon-1)}{4\pi(\epsilon'+2)(2\epsilon'+\epsilon)}\mathbf{E}. \tag{12.1.28}$$

In order to make contact with linear response theory, we should compute the polarization of the system as a function of the applied *external* field, i.e., the electric field that would be present in the system in the absence of the particles. Using equation (12.1.27), it is easy to derive that the electrostatic field $\mathbf{E}'_{\mathrm{I}}$ that would be present in region I if it were empty is given by equation (12.1.27) with $\epsilon = 1$:

$$\mathbf{E}'_{\mathrm{I}} = \frac{9\epsilon'}{(\epsilon'+2)(2\epsilon'+1)}\mathbf{E}.$$

The field $\mathbf{E}'_{\mathrm{I}}$ is uniform throughout region I. If we assume that the system is isotropic and that linear response theory is sufficient, we can write for the polarization

$$\begin{aligned}\langle \mathbf{P} \rangle &= \frac{1}{VQ}\int d\mathbf{r}^N \sum_{i=1}^{N} \mu_i \exp\left[-\beta\left(\mathcal{H}_0 - \sum_{i=1}^{N} \mu_i \cdot \mathbf{E}'_{\mathrm{I}}\right)\right] \\ &= \frac{\beta}{3V}\left(\langle \mathbf{M}^2 \rangle - \langle \mathbf{M} \rangle^2\right)\mathbf{E}'_{\mathrm{I}}. \end{aligned} \tag{12.1.29}$$

Comparison of equations (12.1.29) and (12.1.28) yields

$$\langle \mathbf{P} \rangle = \frac{1}{3}\beta\rho g_k \mu^2 \mathbf{E}'_{\mathrm{I}}, \tag{12.1.30}$$

where the g_k is the Kirkwood factor, which is defined as

$$g_k \equiv \frac{1}{N\mu^2}\left(\langle \mathbf{M}^2 \rangle - \langle \mathbf{M} \rangle^2\right),$$

where **M** is the total dipole moment

$$\mathbf{M} = \sum_{i=1}^{N} \mu_i.$$

Combining equations (12.1.28) and (12.1.30) gives

$$\frac{(\epsilon - 1)(2\epsilon' + 1)}{(2\epsilon' + \epsilon)} = \frac{4}{3}\pi\beta\rho g_k \mu^2.$$

For a simulation with conducting boundary conditions ($\epsilon' \to \infty$), the expression for the dielectric constant becomes

$$\epsilon = 1 + \frac{4}{3}\pi\rho\beta g_k \mu^2. \tag{12.1.31}$$

This result shows that the fluctuations of the dipole moment depends on the dielectric constant of the *surrounding* medium. This, in turn, implies that, for a polar system, the Hamiltonian itself depends on the dielectric constant ϵ' of the surrounding medium.

12.1.4 Boundary Conditions

It may appear strange that the form for the potential energy of an infinite periodic system of ions or dipoles should depend on the nature of the boundary conditions at infinity. However, for systems of charges or dipoles, this is a very real effect that has a simple physical interpretation. To see this, consider the system shown in Figure 12.2. The fluctuating dipole moment of the unit cell **M** gives rise to a surface charge at the boundary of the sphere, which, in turn, is responsible for a homogeneous depolarizing field:

$$\mathbf{E} = -\frac{4\pi\mathbf{P}}{2\epsilon' + 1},$$

where $\mathbf{P} \equiv \mathbf{M}/V$. Now let us consider the reversible work per unit volume that must be performed against this depolarizing field to create the net polarization **P**. Using

$$dw = -\mathbf{E}d\mathbf{P} = \frac{4\pi}{2\epsilon' + 1}\mathbf{P}d\mathbf{P},$$

we find that the total work needed to polarize a system of volume V equals

$$\mathcal{U}_{\text{pol}} = \frac{2\pi}{2\epsilon' + 1}P^2 V = \frac{2\pi}{2\epsilon' + 1}M^2/V$$

or, using the explicit expression for the total dipole moment of the periodic box,

$$\mathcal{U}_{\text{pol}} = \frac{2\pi}{(2\epsilon' + 1)V}\left|\sum_{i=1}^{N} \mathbf{r}_i q_i\right|^2,$$

in the Coulomb case, and

$$\mathcal{U}_{\text{pol}} = \frac{2\pi}{(2\epsilon' + 1)V} \left| \sum_{i=1}^{N} \mu_i \right|^2 ,$$

in the dipolar case. This contribution to the potential energy corresponds to the $\mathbf{k} = \mathbf{0}$ term that we have neglected thus far. It is permissible to ignore this term if the depolarizing field vanishes. This is the case if our periodic system is embedded in a medium with infinite dielectric constant (a conductor, $\epsilon' \to \infty$), which is what we have assumed throughout.

For simulations of ionic systems, it is essential to use such "conducting" boundary conditions; for polar systems, it is merely advantageous. For a discussion of these subtle points, see [325].

12.1.5 Accuracy and Computational Complexity

In the Ewald summation, the calculation of the energy is performed in two parts: the real-space part (12.1.23) and the part in Fourier space (12.1.17). For a given implementation, we have to choose the parameter α that characterizes the shape of the Gaussian charge distributions, r_c the real-space cutoff distance, and k_c the cutoff in Fourier space. In fact, it is common to write k_c as $2\pi/Ln_c$, where n_c is a positive integer. The total number of Fourier components within this cutoff value is equal to $(4\pi/3)n_c^3$. The values of these parameters depend on the desired accuracy ϵ, that is, the root mean-squared difference between the exact Coulombic energy and the results from the Ewald summation. Expressions for the cutoff errors in the Ewald summation method[1] have been derived in [326,327]. For the energy, the standard deviation of the real-space cutoff error of the total energy is

$$\delta E_R \approx Q \left(\frac{r_c}{2L^3}\right)^{\frac{1}{2}} \frac{1}{(\alpha r_c)^2} \exp\left(-\alpha^2 r_c^2\right) \tag{12.1.32}$$

and for the Fourier part of the total energy

$$\delta E_F \approx Q \frac{n_c^{1/2}}{\alpha L^2} \frac{1}{(\pi n_c/\alpha L)^2} \exp\left[-\left(\pi n_c/\alpha L\right)^2\right] , \tag{12.1.33}$$

where

$$Q = \sum_i q_i^2 .$$

Note that for both the real-space part and the Fourier part, the strongest dependence of the estimated error on the parameters α, r_c, and n_c is through

[1]The accuracy is dependent on whether we focus on the energy (for Monte Carlo) or on the forces (for Molecular Dynamics).

a function of the form $\exp(-x^2)/x^2$. We now impose that these two functions have the same value ϵ. The value of x for which $\exp(-x^2)/x^2 = \epsilon$ we denote by s. Hence $\epsilon = \exp(-s^2)/s^2$. Then it follows from Equation (12.1.32) that

$$r_c = \frac{s}{\alpha} \tag{12.1.34}$$

and from equation (12.1.33) we obtain

$$n_c = \frac{sL\alpha}{\pi}. \tag{12.1.35}$$

If we insert these expressions for r_c and n_c back into the expressions (12.1.32) and (12.1.33), we find that both errors have the same functional form:

$$\delta E_R \approx Q \left(\frac{s}{\alpha L^3}\right)^{1/2} \frac{\exp(-s^2)}{s^2}$$

and

$$\delta E_F \approx Q \left(\frac{s}{2\alpha L^3}\right)^{1/2} \frac{\exp(-s^2)}{s^2}.$$

Hence, changing s affects both errors in the same way. We now estimate the computational effort involved in evaluating the Ewald sum. To this end, we write the total computational time as the sum of the total time in real space and the total time in Fourier space

$$\tau = \tau_R N_R + \tau_F N_F, \tag{12.1.36}$$

where τ_R is the time needed to evaluate the real part of the potential of a pair of particles and τ_F is the time needed to evaluate the Fourier part of the potential per particle and per k vector. N_R and N_F denote the number of times these terms need to be evaluated to determine the total energy or the force on the particles. If we assume a uniform distribution of particles, these two numbers follow from the estimates of r_c and n_c:

$$\begin{aligned} N_R &= \frac{4}{3}\pi \frac{s^3 N^2}{\alpha^3 L^3} \\ N_F &= \frac{4}{3}\pi \frac{s^3 \alpha^3 L^3 N}{\pi^3}. \end{aligned}$$

The value of α follows from minimization of equation (12.1.36)

$$\alpha = \left(\frac{\tau_R \pi^3 N}{\tau_F L^6}\right)^{\frac{1}{6}},$$

which yields for the time

$$\tau = \frac{8\sqrt{\tau_r \tau_f} N^{3/2} s^3}{3\sqrt{\pi}} = \mathcal{O}\left(N^{3/2}\right). \tag{12.1.37}$$

Note that, with the above expression for α, the parameters r_c and n_c follow from equations (12.1.34) and (12.1.35) respectively, once we have specified the desired accuracy. To optimize the Ewald summation one has to make an estimate of τ_R/τ_F. This ratio depends on the details of the particular implementation of the Ewald summation and can be obtained from a short simulation.[2]

We conclude this section with a few comments concerning the implementation. First of all, when using equation (12.1.34) to relate r_c to α, one should make sure that $r_c \leq L/2$; otherwise the real part of the energy cannot be restricted to the particles in the box $\mathbf{n} = \mathbf{0}$.

A second practical point is the following: in most simulations, there are short-range interactions between the particles, in addition to the Coulomb interaction. Usually, these short-range interactions also have a cutoff radius. Clearly, it is convenient if the same cutoff radius can be used for the short-range interactions and for the real-space part of the Ewald summation. However, if this is done, the parameters of the Ewald summation need not have their optimum values.

12.2 Fast Multipole Method

An algorithm that is of order $\mathcal{O}(N)$ is the fast multipole method. The multipole method is based on the idea that a group of particles at a large distance can be considered one big cluster, for which it is not necessary to calculate all particle-particle interactions individually. By clustering the system into bigger and bigger groups, the interactions can be approximated. This approach of Appel [329] leads to an order $\mathcal{O}(N)$ algorithm [330]. This algorithm was further refined by Barnes and Hut [331]. In the original algorithm of Appel, the clusters were approximated as a single charge. Greengard and Rokhlin [316] developed an algorithm in which the charge distribution in a cluster is approximated by a multipole expansion. Schmidt and Lee extended this method to systems with periodic boundary conditions [332].

Algorithm

Next, we give a schematic description of the Greengard and Rokhlin algorithm in three dimensions; a more detailed description can be found in [328, 332, 333]. Essential in this algorithm is the use of octal trees and multipole expansions, which are described before we discuss the algorithm.

An octal tree can be constructed in the following way (see Figure 12.3). The original system is defined to be the unique level-zero cell. Level-one

[2] A typical value of this ratio is $\tau_R/\tau_F = 3.6$ [328].

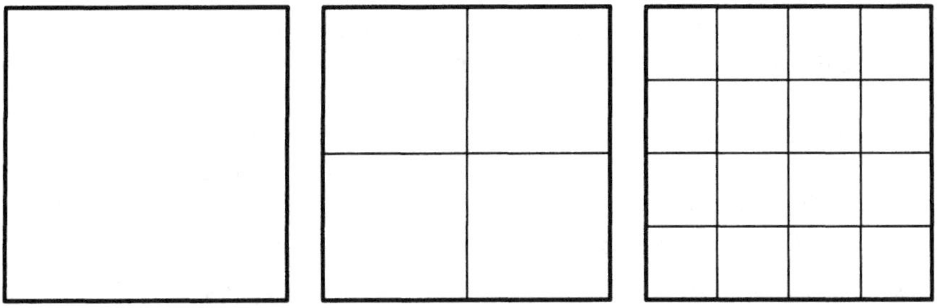

Figure 12.3: Octal tree in two dimensions: the left figure shows the original level-zero system, the middle figure the level-one cells, and the right figure the level-two cells.

cells are obtained by dividing the parent cell into eight (four in two dimensions) children. This subdivision can be continued till a maximum level, denoted level R, has been obtained. We refer to a cell as $C_i^{(l)}$, where l denotes its level and i the index that refers to its position.

In the Greengard and Rokhlin method various multipole expansions are used. The formula for these expansions are given, for example, in [328]; here we give a short description of the four formulas that are used. In the description of the algorithm we refer to the operation by the words in italics.

- *Multipole expansion* of the charges in a cell. Suppose we have a distribution of k charges at location $\mathbf{r}_i = (r_i, \theta_i, \phi_i)$ with charge q_i, then the potential at a point $\mathbf{r}' = (r', \theta', \phi')$ sufficiently far away is given by

$$\Phi(\mathbf{r}') = \sum_{n=0}^{\infty} \sum_{m=-n}^{n} \frac{M_n^m}{r^{n+1}} Y_n^m(\theta', \phi') \tag{12.2.1}$$

$$M_n^m = \sum_{i=1}^{k} q_i r_i^n Y_n^{-m}(\theta_i, \phi_i), \tag{12.2.2}$$

where Y_n^{-m} are the associated Legendre functions.

- *Multipole translation*. To calculate the multipole expansion of the parent cell, for example, one can use the multipole expansion of its children. However, for each of these children, the origin of this expansion (12.2.1) is different for each cell. For example, if multipole expansions have been made for the children, *multipole translations* are used to translate these expansions to a new, common origin, the center of the parent box. This translated multipole can be used only for sufficiently far-away positions.

- *Local expansion*. To calculate the potential energy of all particles in a cell one can determine the potential due to the multipoles in other cells. This involves a conversion of the multipole expansions into a local expansion that converges in the region of interest. Since, in these multipole expansions, the close-by cells are excluded, within a cell the potential energy does not change much; therefore it is advantageous to make a Taylor expansion of the local field in each cell. This field can then be used to calculate the energy for all atoms in this cell. This expansion can be used only for positions in the same cell.

- *Translation of the local expansion.* If we have calculated the local expansion with respect to a given origin, for example, the local expansion of the parent box, we can translate this local expansion to, for example, the center of its children cells.

The Greengard and Rokhlin algorithm consists of the following steps:

1. An octal tree of level R is constructed; the charge distribution in each of the 8^R cells is described with a multipole expansion (12.2.1) about the center of a box.
2. The multipole expansion of each of the children of a parent is translated to a multipole expansion around the center of the parent box. This procedure is repeated until level zero has been reached.
3. The calculation of the energy is done in two steps. Consider a particle in level-R cell $C_i^{(R)}$. The calculation of the interactions with the particles in the same cell and in the 26 neighboring level-R cells is performed directly:

$$u_{\text{clo}} = \sum_{\text{close}} \frac{q_i q_j}{r_{ij}}.$$

The interactions with the remainder of the particles are done by calculating the local field expansion in cell $C_i^{(R)}$. To calculate this local field, the following steps are used starting from level two (for levels zero and one, a local expansion does not exist, and therefore they are by definition zero, since there are no cells sufficiently far away).

(a) For each cell in level r, transform the local expansion of the parent cell to the center of the current cell.

(b) Add to the local potentials the transformation of the multipole potentials of the same level cells that fulfill the following conditions:

- The cell must be sufficiently far away for the multipole expansion to be valid; this excludes the neighboring cells.
- The cells must be sufficiently close that their contribution has not been included in the local field of the parent of the cell;

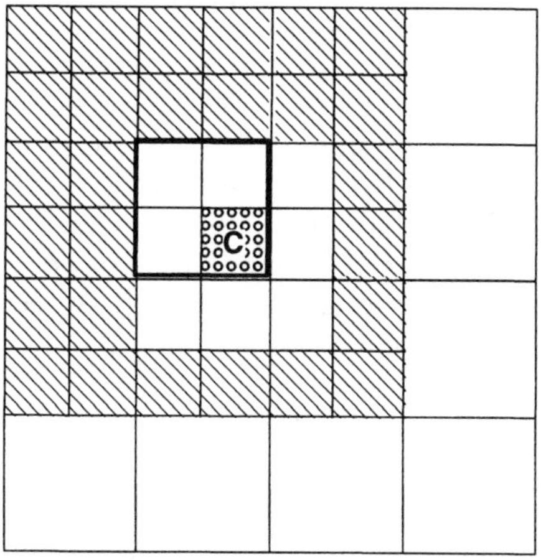

Figure 12.4: The same level cells (shaded) of which the multipole potentials are transferred to the current cell to give the local expansion. The current cell is denoted by C, the neighbors of C do not contribute because they touch C. The white cells of the parent level did not touch the parent cell of C (indicated with a thick line) and therefore did contribute to the local expansion of the parent cell.

this excludes the cells that contribute to local potential initialized in step 3a. In Figure 12.4 these cells are shown. In three dimensions, $8 \times 27 - 27 = 189$ of these cells fulfill this condition.

After level R has been reached the far-way contribution of the charges in cell $C_i^{(R)}$ is calculated from the local expansion $U^{Lo(R)}$:

$$u_{far} = \sum_i U^{Lo}(\mathbf{r}_i).$$

It is important to note that the local expansion $U^{Lo(R)}$ can be used for all particles in the same cell. Therefore in practice the calculation of this term is usually done after step 2 of this algorithm.

The preceding algorithm is for a finite system of charge. To apply this algorithm to a system with periodic boundary conditions, the following modifications have to be made. In a system with periodic boundary conditions the level-zero cell is surrounded by periodic images that have the same multipole expansion around their center. In the algorithm we have started step 3a, assuming that the local field of level zero is 0. For the periodic system, this has to be replaced by the local expansion of the potential from all images except the nearest 26 neighbors. If we have calculated this term, the remainder of the algorithm remains the same. The summation over all periodic images is done using an Ewald summation technique (see [332] for details). Since this involves the multipole expansion of the parent cell (which does not depend on the number of particles), the overhead in computer time is very small. However, the timings of Schmidt and Lee do show that to arrive at the same accuracy as for the nonperiodic system one needs to use a higher octal tree or more multipole moments.

12.3 Particle Mesh Approaches

The CPU time required for a fully optimized Ewald summation scales with the number of particles as $\mathcal{O}(N^{3/2})$. In many applications we not only have the long-range interactions but short-range interactions as well. For such systems it may be convenient to use the same cutoff radius for the real-space sum in the Ewald summation as for the short-range interactions. For a fixed cutoff, however, the calculation of the Fourier part of the Ewald summation scales as $\mathcal{O}(N^2)$, which makes the Ewald summation inefficient for large systems. Note that it is only the reciprocal-space part of the Ewald sum that suffers from this drawback. Clearly, it would be advantageous to have an approach that handles the Fourier part more efficiently. Several schemes for solving this problem have been proposed. They all exploit the fact that the Poisson equation can be solved much more efficiently if the charges are distributed on a mesh. The efficiency and accuracy of such mesh-based algorithms depend strongly on the way in which the charges are attributed to mesh points. Below, we briefly discuss the basics of the particle-mesh approach. However, a full description of the technical details is beyond the scope of this book.

The earliest particle-mesh scheme for molecular simulations was developed by Hockney and Eastwood [24]. The charges in the systems were interpolated on a grid to arrive at a discretized Poisson equation. For a regular grid this equation can be solved efficiently using the Fast Fourier Transform (FFT) [33]. The computer time associated with the FFT technique scales as $\mathcal{O}(N \log N)$, where N denotes the number of points of discrete Fourier transform. In its simplest implementation, the particle-mesh method is fast, but not very accurate. The technique was subsequently improved by splitting the calculation into a short-range and a long-range contribution. In the spirit of the Ewald method, the short-range part is then calculated directly from the particle-particle interactions while the particle-mesh technique is used for the long-range contribution.

Below, we briefly discuss the particle-mesh methods and their relation to the Ewald-sum approach. We will not attempt to present an exhaustive description of all existing particle-mesh methods. The reason is twofold: first of all, a systematic presentation of this subject—as given, for instance, in a paper by Deserno and Holm [334]—would require a chapter on its own. Secondly, Deserno and Holm [334] have shown that most of the "good" alternative methods, such as the Particle Mesh Ewald (PME) [317] and Smooth Particle Mesh Ewald (SPME) [335], are very similar in spirit and can be seen as variations of the original particle-particle/particle-mesh (PPPM) technique of Hockney and Eastwood [24]. The choice of the method to use, often depends on the application. For example, Monte Carlo simulations require an accurate estimate of the energy, while in Molecular Dynamics simulations we need to compute the forces accurately. Some particle-mesh schemes are better suited to do one, some to do the other.

The idea of the PPPM method is to split the Coulomb potential into two parts by using the following (trivial) identity:

$$\frac{1}{r} = \frac{f(r)}{r} + \frac{1 - f(r)}{r}. \tag{12.3.1}$$

The idea of using a switching function is similar to the splitting of the Ewald summation into a short-range and a long-range part. Pollock and Glosli [336] found that different choices for $f(r)$ yield comparable results, although the efficiency of the method does depend strongly on a careful choice of this function. Darden *et al.* [317] have shown that, if one uses the same Gaussian screening function as in the Ewald summation, the PPPM technique becomes indeed very similar to the Ewald method.

It is instructive to recall the Fourier-space contribution of the energy:

$$\mathcal{U}_1 = \frac{1}{2V} \sum_{k \neq 0} \frac{4\pi}{k^2} |\tilde{\rho}(\mathbf{k})|^2 \exp(-k^2/4\alpha).$$

Following Deserno and Holm [334], we write the Fourier-space contribution as

$$\begin{aligned} \mathcal{U}_1 &= \frac{1}{2} \sum_{i=1}^{N} q_i \left(\frac{1}{V} \sum_{k \neq 0} \tilde{g}(\mathbf{k}) \tilde{\gamma}(\mathbf{k}) \tilde{\rho}(\mathbf{k}) e^{i\mathbf{k}\cdot\mathbf{r}_i} \right) \\ &= \frac{1}{2} \sum_{i=1}^{N} q_i \phi^k(\mathbf{r}_i), \end{aligned} \tag{12.3.2}$$

where $\phi^k(\mathbf{r}_i)$ can be interpreted as the electrostatic potential due to the second term in equation (12.3.1):

$$\phi^k(\mathbf{r}_i) = \frac{1}{V} \sum_{\mathbf{k}} \tilde{g}(\mathbf{k}) \tilde{\gamma}(\mathbf{k} \neq 0) \tilde{\rho}(\mathbf{k}) \exp(i\mathbf{k} \cdot \mathbf{r}_i).$$

As a product in Fourier space corresponds to a convolution in real space, we see that the potential $\phi^k(\mathbf{r}_i)$ is due to the original charge distribution $\rho(x)$, convoluted by a smearing function $\gamma(r)$. The Ewald summation is recovered if we choose a Gaussian smearing function; in which case $f(r)$ is given by an error function.

In order to evaluate the above expression for the Fourier part of the electrostatic energy using a discrete fast Fourier transform, we have to perform the following steps [334, 337]:

1. Charge assignment: Up to this point, the charges in the system are not localized on lattice points. We now need a prescription to assign the charges to the grid points.
2. Solving Poisson's equation: Via a FFT technique the Poisson equation for our discrete charge distribution has to be solved (the Poisson equation on a lattice can also be solved efficiently, using a diffusion algorithm [338].
3. Force assignment (in the case of MD): Once the electrostatic energy has been obtained from the solution of the Poisson equation, the forces have to be calculated and assigned back to the particles in our system.

At every stage, there are several options to choose from. Deserno and Holm have made a careful study of the relative merits of the various options and their combinations [334]. Below we give a brief summary of their observations. For further details the reader is referred to the original article.

To assign the charges of the system to a grid, a charge assignment function, $W(\mathbf{r})$, is introduced. For example, in a one-dimensional system, the fraction of a unit charge at position x assigned to a grid point at position x_p is given by $W(x_p - x)$. Hence, if we have a charge distribution $\rho(x) = \sum_i q_i \delta(x - x_i)$, then the charges at a grid point x_p are given by

$$\rho_M(x_p) = \frac{1}{h} \int_0^L dx\, W(x_p - x)\rho(x), \tag{12.3.3}$$

where L is the box diameter and h is the mesh spacing. The number of mesh points in one dimension, M, is equal to L/h. The factor $1/h$ ensures that ρ_M is a density. Many choices for the function $W(x)$ are possible. Deserno and Holm [334] listed the properties that $W(x)$ should have. $W(x)$ should be an even function and the function should be normalized in such a way that the sum of the fractional charges equals the total charge of the system. Since the computational costs is proportional to the number of particles and the number of mesh points to which the single charge is distributed, a function with a small support decreases the computational cost. In addition, one would like to reduce the errors due to the discretization as much as possible. If a particle moves through the system and passes from one grid point to another, the function $W(x)$ should not yield abrupt changes in the fractional charges.

A particularly nice way to approach the charge assignment problem was described by Essmann *et al.* [335]. These authors argue that the problem of discretizing the Fourier transform can be viewed as an interpolation problem. Consider a single term in the (off-lattice) Fourier sum $q_i e^{-i\mathbf{k}\cdot\mathbf{r}_i}$. This term cannot be used in a discrete Fourier transform, because $\mathbf{r}$ does not, in general, correspond to a mesh point. However, we can *interpolate* $e^{-i\mathbf{k}\cdot\mathbf{r}_i}$ in terms of values of the complex exponential at mesh points. For convenience, consider a one-dimensional system. Moreover, let us assume that x varies between 0 and L and that there are M equidistant mesh points in this interval. Clearly, the particle coordinate x_i is located between mesh points $[Mx_i/L]$ and $[Mx_i/L]+1$, where [..] denotes the integer part of a real number. Let us denote the real number Mx_i/L by u_i. We can then write an order-2p interpolation of the exponential as

$$e^{-ik_x x_i} \approx \sum_{j=-\infty}^{\infty} W_{2p}(u_i - j)e^{-ik_x Lj/M},$$

where the W_{2p}'s denote the interpolation coefficients. Strictly speaking the sum over j contains only M terms. However, to account for the periodic boundary conditions, we have written it as if $-\infty < j < \infty$. For an interpolation of order 2p, only the 2p mesh point nearest to x_i contributes to the sum. For all other points, the weights W_{2p} vanish. We can now approximate the Fourier transform of the complete charge density as

$$\rho_k \approx \sum_{i=1}^{N} q_i \sum_{j=-\infty}^{\infty} W_{2p}(u_i - j)e^{-ik_x Lj/M}.$$

This can be rewritten as

$$\rho_k \approx \sum_{j} e^{-ik_x Lj/M} \sum_{i=1}^{N} q_i W_{2p}(u_i - j).$$

We can interpret the above expression as a *discrete* Fourier transform of a "meshed" charge density $\rho(j) = \sum_{i=1}^{N} q_i W_{2p}(u_i - j)$. This shows that the coefficients W_{2p} that were introduced to give a good interpolation of $e^{-ik_x x_i}$ end up as the charge-assignment coefficients that attribute off-lattice charges to a set of lattice points.

While the role of the coefficients W is now clear, there are still several choices possible. The most straightforward is to use the conventional Lagrange interpolation method to approximate the exponential (see Darden *et al.* [317] and Petersen [327]). The Lagrange interpolation scheme is useful for

Monte Carlo simulations, but less so for the Molecular Dynamics method. The reason is that although the Lagrangian coefficients are everywhere continuous, their derivative is not. This is problematic when we need to compute the forces acting on charged particles (the solution is that a separate interpolation must be used to compute the forces). To overcome this drawback of the Lagrangian interpolation scheme, Essmann *et al.* suggested the so-called SPME method [335]. The SPME scheme uses exponential Euler splines to interpolate complex exponentials. This approach results in weight functions W_{2p} that are $2p-2$ times continuously differentiable. It should be stressed that we cannot automatically use the *continuum* version of Poisson's equation in all interpolation schemes. In fact equation (12.1.11) is only consistent with the Lagrangian interpolation schemes. To minimize discretization errors, other schemes, such as the SPME method, require other forms of the Green's function $\tilde{g}(k)$ (see ref. [334]).

In section 12.1.5 we discussed how the parameter α in the conventional Ewald sum method can be chosen such that it minimizes the numerical error in the energy (or in the forces). Petersen [327] has derived similar expressions for the PME method. Expressions that apply to the PPPM method [24] and the SPME scheme are discussed by Deserno and Holm [339]. In the case of the force computation, matters are complicated by the fact that, in a particle-mesh scheme, there are several inequivalent ways to compute the electrostatic forces acting on the particles. Some such schemes do not conserve momentum, others do—but at a cost. The choice of what is the "best" method, depends largely on the application [334].

This concludes our discussion of particle-mesh schemes. While we have tried to convey the spirit of these algorithms, we realize that this description is not sufficiently detailed to be of any help in the actual implementation of such an algorithm. We refer readers who are considering implementation of one of the particle-mesh schemes to the articles of Essmann *et al.* [335], Deserno and Holm [334], and, of course, Hockney and Eastwood [24].

Example 14 (Algorithms to Calculate Long-Range Interactions)

A detailed comparison of the various algorithms for determining the long-range interactions has been performed by Esselink [328]. Esselink considered an ensemble of cubic systems with a density $\rho = 1$. Each system consists of N randomly distributed particles. The charges were assigned a random value between -1 and 1 in such a way that the total charge on the system was made 0. The algorithms compared were the naive approach, which is a summation of all pairs of particles (PP), the Appel algorithm (AP), and the Greengard and Rokhlin algorithm (GR); these three methods were applied to a nonperiodic system. The Ewald summation (EW) was tested on a periodic version of the system. All algorithms have been optimized for the accuracy required.

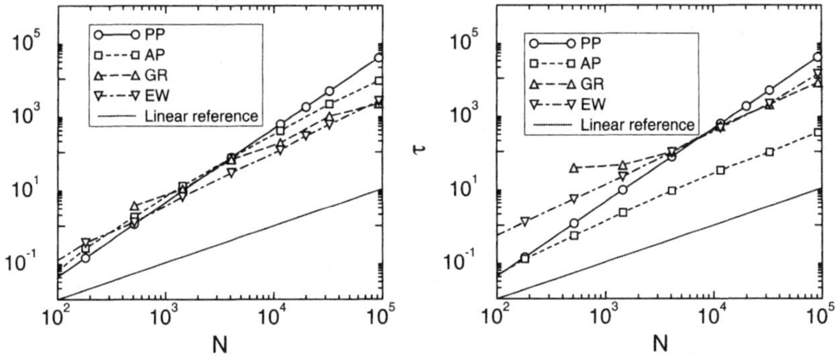

Figure 12.5: Comparison of algorithms for long-range interactions: the left figure shows CPU time τ as a function of the number of particles N for the force and the right figure shows that for the energy. PP denotes the summation over all particle pairs, AP denotes Appel's algorithm, GR the Greengard and Rokhlin algorithm, and EW the Ewald summation. The data are taken from [328].

In Figure 12.5 the efficiencies of the various algorithms for the energy and force calculations are compared.

For nonperiodic systems, for $N > 4000$ both the AP and GR algorithms outperform the naive PP algorithm. The GR algorithm for $N > 300$ is more efficient than the AP algorithm. However, if the energy alone is sufficient (as in the case of Monte Carlo simulations), the AP algorithm is more efficient than the GR algorithm and outperforms the PP algorithm for $N > 200$. Ding *et al.* [333] observed a slightly lower break-even point of the GR and PP algorithms for $N > 300$.

For periodic systems it is not possible to use the PP algorithm. For the force calculation, the EW method is more efficient than the GR algorithm for $N < 100,000$. It is important to note that Esselink used a nonperiodic version for the GR and AP algorithms; for a periodic version this break-even point will shift to a slightly larger number of particles. Ding *et al.* [340] observed a break-even point of the EW and GR methods for only 300 particles, while Schmidt and Lee [332] obtained a break-even point at several thousand particles. The results of Esselink are supported by Petersen [327]. The reason for this large difference is not clear, but may very much depend on different (more efficient) implementations of the Ewald summation.

Esselink did not include the particle-mesh-type methods in his comparison. Petersen has shown that in the range of $N = 10^4$–10^5 the PME technique is superior to the Ewald summation and fast multipole methods. Luty *et al.* [341] and Pollock and Glosli [336] obtained a conclusion result for the PPPM method. Pollock and Glosli even concluded that for any number of particles they have investigated ($\approx 10^6$) the PPPM method, despite the $\mathcal{O}(N \log N)$ complexity, is more efficient than the fast multipole methods, which have an $\mathcal{O}(N)$ complexity. If the fast multipole method is combined with multiple-time step integration (see section 15.3), a more favorable break-even point is obtained [342]. For a more detailed comparison see ref. [337].

12.4 Ewald Summation in a Slab Geometry

In the previous sections we discussed the treatment of long-range interactions in three-dimensional systems. For some applications one is interested in a system that is finite in one dimension and infinite in the other two dimensions. Examples of such systems are fluids adsorbed in slit-like pores or monolayers of surfactants.

Special techniques are required to compute long-range interactions in such inhomogeneous systems. The most straightforward solution would be to use the same approach as for the three-dimensional Ewald summation, but restrict the reciprocal-space sum to vectors in the x, y directions [343,344]. The energy we wish to calculate is

$$\mathcal{U}_{\text{Coul}} = \frac{1}{2} \sum_{i,j=1}^{N} \sum_{\mathbf{n}}{}' \frac{q_i q_j}{|\mathbf{r}_{ij} + \mathbf{n}|},$$

where the summation over $\mathbf{n} = (L_x n_x, L_y n_y, 0)$ indicates that periodicity is only imposed in the x, y directions. As in the ordinary Ewald summation the prime indicates that for cell $(0, 0, 0)$ the terms $i = j$ should be omitted. We have a two-dimensional periodicity in the x, y directions for which we can use the Fourier representation. The resulting expression for the energy is [345]

$$\begin{aligned} \mathcal{U}^{\text{Coul}} &= \frac{1}{2} \sum_{i,j=1}^{N} q_i q_j \left[\sum_{\mathbf{n}}{}' \frac{\text{erfc}(\alpha |\mathbf{r}_{ij} + \mathbf{n}|)}{|\mathbf{r}_{ij} + \mathbf{n}|} + \frac{\pi}{L^2} \sum_{h>0} \cos(\mathbf{h} \cdot \mathbf{r}_{ij}) F(h, z_{ij}, \alpha) \right. \\ &\quad \left. - g(z_{ij}, \alpha) \right] - \frac{\alpha}{\sqrt{\pi}} \sum_{i=1}^{M} q_i^2, \end{aligned} \tag{12.4.1}$$

where $\mathbf{h} \equiv (2\pi m_x / L_x, 2\pi m_y / L_y, 0)$ denotes a reciprocal lattice vector, z_{ij} is the distance between two particles in the z direction, and α is the screening parameter. The function $F(h, z_{ij}, \alpha)$

$$F(h, z, \alpha) = \frac{\exp(hz)\text{erfc}\left[\alpha z + h/(2\alpha)\right] + \exp(-hz)\text{erfc}\left[-\alpha z + h/(2\alpha)\right]}{2h} \tag{12.4.2}$$

corrects for the inhomogeneity in the nonperiodic direction. If the system is truly two-dimensional, this term takes a simpler form. The function $g(z, \alpha)$

$$g(z, \alpha) = z \operatorname{erf}(\alpha z) + \exp\left[-(z\alpha)^2\right] / (\alpha\sqrt{\pi}) \tag{12.4.3}$$

is an additional self-term of charge interactions in the central cell that must be subtracted from the reciprocal-space sum. In a neutral system with all particles in the plane $z = 0$ this term disappears. The last term in equation (12.4.1) is the same self-term that appears in the normal Ewald summation (12.1.22). The details of the derivation can be found in refs. [343–347].

From a computational point of view equation (12.4.1) is inconvenient. Unlike the three-dimensional case, the double sum over the particles in the Fourier part of equation (12.4.1) can, in general, not be expressed in terms of the square of a single sum. This makes the calculation much more expensive than its three-dimensional counterpart. Several methods have been developed to increase the efficiency of the evaluation of the Ewald sum for slab geometries. Spohr [348] showed that the calculation can be made more efficient by the use of a look-up table combined with an interpolation scheme and the long-distance limit given by equation (12.4.6).

Hautman and Klein [349] considered the case in which the deviation of the charge distribution from a purely two-dimensional system is small. For such a system one can introduce a Taylor expansion in z, to separate the in-plane contributions x, y in $1/\sqrt{x^2 + y^2 + z^2}$ from the out-of-plane contributions. Using this approach, Hautman and Klein derived an expression in which the Fourier contribution can again be expressed in terms of sums over single particles. However, unless the ratio $z/\sqrt{x^2 + y^2} \ll 1$, the Taylor expansion converges very poorly. Therefore the applicability of this method is limited to systems in which all charges are close to a single plane. An example of such a system would be a self-assembled monolayer in which only the head groups carry a charge [349].

An obvious idea would be to use the three-dimensional Ewald summation by placing a slab of vacuum in between the periodic images (see Figure 12.7). Spohr has shown [348], however, that even with a slab that is four times the distance between the charges one does not obtain the correct limiting behavior (see Example 15). The reason is that a periodically repeated slab behaves like a stack of parallel plate capacitors. If the slab has a net dipole moment, then there will be spurious electric fields between the periodic images of the slab. More importantly, the usual assumption that the system is embedded in a conducting sphere does not correctly account for the depolarizing field that prevails in a system with a (periodic) slab geometry. Yeh and Berkowitz [350] have shown that one can add a correction term to obtain the correct limiting behavior in the limit of an infinitely thin slab. In the limit of an infinitely thin slab in the z direction, the force on a charge

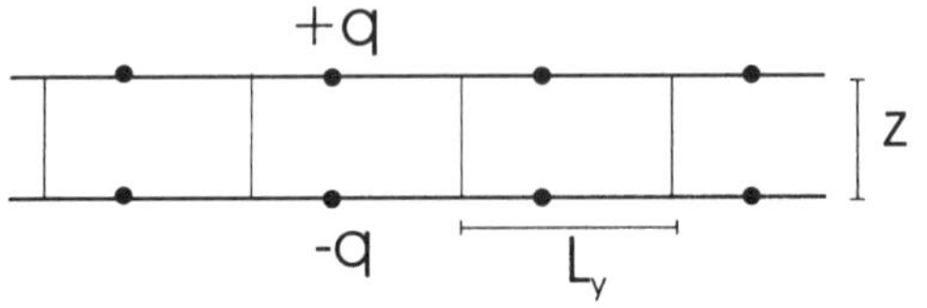

Figure 12.6: A system containing two point charges at positions $(0,0,0)$ and $(0,0,z)$; because of the periodic boundary conditions in the x and y directions, two oppositely charge "sheets" are formed. There are no periodic boundary conditions in the z direction.

q_i due to the depolarizing field is given by [351]

$$F_z = -\frac{4\pi q_i}{V} M_z, \tag{12.4.4}$$

and the total electrostatic energy due to this field is

$$U_c = -\frac{2\pi}{V} M_z^2, \tag{12.4.5}$$

where M_z is the net dipole moment of the simulation cell in the z direction

$$M_z = \sum_{i=1}^{N} q_i z_i.$$

If the slab is not infinitely thin compared to the box dimensions, higher-order correction terms have to be added. However, Yeh and Berkowitz [350] have shown that the lowest-order correction is sufficient if the spacing between the periodically repeated slabs is three to five times larger than the thickness of the slab (see also Crozier *et al.* [352]).

Example 15 (Ewald in Slab)

To illustrate the difficulties that arise when computing long-range forces in a slab geometry, Spohr and co-workers [348, 352] considered a simple example of two point charges: a charge $+q$ at $(0,0,z)$ and a charge $-q$ at $(0,0,0)$. The system is finite in the z direction and periodic in the x,y directions with box sizes L_x and L_z (see Figure 12.6). Because of the periodic boundary conditions the system forms two "sheets" of opposite charge.

In the limit $z \to \infty$, the distance between the periodic images of the charge is small compared to the distance between the sheets. We can therefore assume a uniform charge density $q/(L_x L_y)$ on each sheet. In this limit the force acting between the two particles is given by

$$F_z = \frac{2\pi q^2}{L_x L_y}. \tag{12.4.6}$$

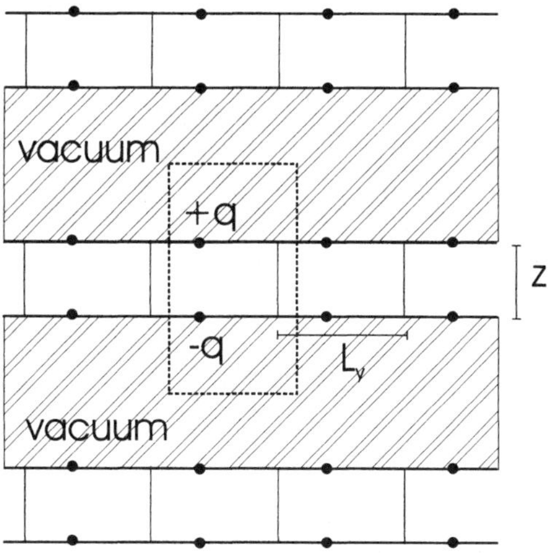

Figure 12.7: The system of Figure 12.6 artificially made periodic in the z direction by adding a slab of vacuum.

It is instructive to compare the various methods to compute the long-range interactions in this geometry. The true forces are given by the two-dimensional Ewald summation (12.4.1) and we can compare the following methods:

- Two-Dimensional Ewald summation, this solution is given by equation (12.4.1), which is the "exact" solution to this problem.
- Bare Coulomb Potential, we simple assume that the periodic images do not exist (or give a zero contribution). The resulting forces follow Coulomb law.
- Truncated and Shifted Coulomb Potential, in this method it is assumed that beyond $r_c = 9$ the potential is zero. To remove the discontinuity at $r = R_c$ the potential is shifted as well (see section 3.2.2).
- Three-Dimensional Ewald Summation, in this approximation a layer of vacuum is added. The total system (vacuum plus slab) is seen as a normal periodic three-dimensional system (see Figure 12.7) for which the three-dimensional Ewald summation (see equation (12.1.25)) is used. To study the effect of the thickness of the slab of vacuum, two systems are considered, one with $L_z = 3L_x$ and a larger one with $L_z = 3L_x$.
- 3-Dimensional Ewald Summation with Correction Term, this method is similar to the previous one; i.e., the normal three-dimensional Ewald summation is used with an additional slab of vacuum, except that now we correct for the spurious dipolar interactions, using equation (12.4.4).

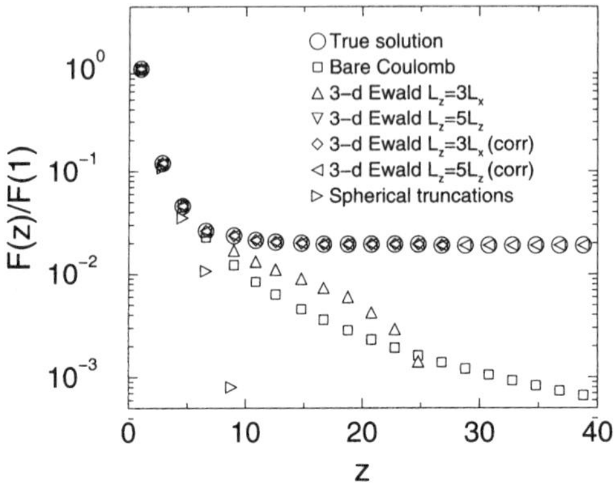

Figure 12.8: Comparison of various methods for approximating the long-range interaction for two charges of the slab geometry shown in Figure 12.6.

In Figure 12.8 we compare the various approximations with the true two-dimensional solution. The bare Coulomb potential and the shifted and truncated Coulomb potential both give a zero force in the limit $z \to \infty$ and therefore do not lead to the correct limiting behavior. Although the three-dimensional Ewald summation gives a better approximation of the correct solution, it still has the incorrect limiting behavior for both a small and a large added slab of vacuum. The corrected three-dimensional Ewald summation, however, does reproduce the correct solution, for both a slab of vacuum of $3L_x$ and that of $5L_x$.

Chapter 13

Biased Monte Carlo Schemes

Up to this point, we have not addressed a fairly obvious question: what is the point of using the Monte Carlo technique in simulations? After all, Molecular Dynamics simulations can be used to study the static properties of many-body systems and, in addition, MD provides information about their dynamical behavior. Moreover, a standard MD simulation is computationally no more expensive than the corresponding MC simulation. Hence, it would seem tempting to conclude that the MC method is an elegant but outdated scheme.

As the reader may have guessed, we believe that there are good reasons to use MC rather than MD in certain cases. But we stress the phrase *in certain cases*. All other things being equal, MD is clearly the method of choice. Hence, if we use the Monte Carlo technique, we should always be prepared to justify our choice. Of course, the reasons may differ from case to case. Sometimes it is simply a matter of ease of programming: in MC simulations there is no need to compute forces. This is irrelevant if we work with pair potentials, but for many-body potentials, the evaluation of the forces may be nontrivial. Another possible reason is that we are dealing with a system that has no natural dynamics. For instance, this is the case in models with discrete degrees of freedom (e.g., Ising spins). And, indeed, for simulations of lattice models, MC is almost always the technique of choice. But even in off-lattice models with continuous degrees of freedom, it is sometimes better, or even essential, to use Monte Carlo sampling. Usually, the reason to choose the MC technique is that it allows us to perform *unphysical* trial moves, that is, moves that cannot occur in nature (and, therefore, have no counterpart in Molecular Dynamics) but are essential for the equilibration of the system.

This introduction is meant to place our discussion of Monte Carlo techniques for simulating complex fluids in a proper perspective: in most published simulations of complex (often macromolecular) fluids, Molecular Dynamics is used, and rightly so. The Monte Carlo techniques that we discuss here have been developed for situations where either MD cannot be used at all or the natural dynamics of the system are too slow to allow the system to equilibrate on the time scale of a simulation.

Examples of such simulations are Gibbs ensemble and grand-canonical Monte Carlo simulations. Both techniques require the exchange of particles, either between a reservoir and the simulation box or between the two boxes. Such particle exchanges are not related to any real dynamics and therefore require the use of Monte Carlo techniques. But, in the case of complex fluids, in particular fluids consisting of chain molecules, the conventional Monte Carlo techniques for grand-canonical or Gibbs ensemble simulations fail. The reason is that, in the case of large molecules, the probability of acceptance of a random trial insertion in the simulation box is extremely small and hence the number of insertion attempts has to be made prohibitively large. For this reason, the conventional grand-canonical and Gibbs ensemble simulations were limited to the study of adsorption and liquid-vapor phase equilibria of small molecules.

13.1 Biased Sampling Techniques

In this chapter,[1] we discuss extensions of the standard Monte Carlo algorithm that allow us to overcome some of these limitations. The main feature of these more sophisticated Monte Carlo trial moves is that they are no longer completely random: the moves are biased in such a way that the molecule to be inserted has an enhanced probability to "fit" into the existing configuration. In contrast, no information about the present configuration of the system is used in the generation of normal (unbiased) MC trial moves: that information is used only to accept or reject the move (see Chapters 3 and 5). Biasing a Monte Carlo trial move means that we are no longer working with a symmetric *a priori* transition matrix. To satisfy detailed balance, we therefore also should change the acceptance rules. This point is discussed in some detail. Clearly, the price we pay for using configurationally biased MC trial moves is a greater complexity of our program. However, the reward is that, with the help of these techniques, we can sometimes speed up a calculation by many orders of magnitude. To illustrate this, we shall discuss examples of simulations that were made possible only through the use of bias sampling.

---

[1]Readers who are not familiar with the Rosenbluth scheme are advised to read section 11.2 first.

13.1.1 Beyond Metropolis

The general idea of biased sampling is best explained by considering a simple example. Let us assume that we have developed a Monte Carlo scheme that allows us to generate trial configurations with a probability that depends on the potential energy of that configuration:

$$\alpha(o \to n) = f[\mathcal{U}(n)].$$

For the reverse move, we have

$$\alpha(n \to o) = f[\mathcal{U}(o)].$$

Suppose we want to sample the N,V,T ensemble, which implies that we have to generate configurations with a Boltzmann distribution (5.2.2). Imposing detailed balance (see section 5.1) yields, as a condition for the acceptance rule,

$$\frac{\mathrm{acc}(o \to n)}{\mathrm{acc}(n \to o)} = \frac{f[\mathcal{U}(o)]}{f[\mathcal{U}(n)]} \exp\{-\beta[\mathcal{U}(n) - \mathcal{U}(o)]\}.$$

A possible acceptance rule that obeys this condition is

$$\mathrm{acc}(o \to n) = \min\left(1, \frac{f[\mathcal{U}(o)]}{f[\mathcal{U}(n)]} \exp\{-\beta[\mathcal{U}(n) - \mathcal{U}(o)]\}\right). \tag{13.1.1}$$

This derivation shows that we can introduce an arbitrary biasing function $f(\mathcal{U})$ in the sampling scheme and generate a Boltzmann distribution of configurations, provided that the acceptance rule is modified in such a way that the bias is removed from the sampling scheme. Ideally, by biasing the probability to generate a trial conformation in the right way, we could make the term on the right-hand side of equation (13.1.1) always equal to unity. In that case, every trial move will be accepted. In Chapter 14.3, we have seen that it is sometimes possible to achieve this ideal situation. However, in general, biased generation of trial moves is simply a technique for enhancing the acceptance of such moves without violating detailed balance.

We now give some examples of the use of non-Metropolis sampling techniques to demonstrate how they can be used to enhance the efficiency of a simulation.

13.1.2 Orientational Bias

To perform a Monte Carlo simulation of molecules with an intermolecular potential that depends strongly on the relative molecular orientation (e.g., polar molecules, hydrogen-bond formers, liquid-crystal forming molecules), it is important to find a position that not only does not overlap with the other molecule but also has an acceptable orientation. If the probability of finding a suitable orientation by chance is very low, we can use biased trial moves to enhance the acceptance.

Algorithm

Let us consider a Monte Carlo trial move in which a randomly selected particle has to be moved and reoriented. We denote the old configuration by o and the trial configuration by n. We use standard random displacement for the translational parts of the move, but we bias the generation of trial orientations, as follows:

1. Move the center of mass of the molecule over a (small) random distance and determine all those interactions that do not depend on the orientations. These interactions are denoted by $u^{pos}(n)$. In practice, there may be several ways to separate the potential into orientation-dependent and orientation-independent parts.
2. Generate k trial orientations $\{\mathbf{b}_1, \mathbf{b}_2, \cdots, \mathbf{b}_k\}$ and for each of these trial orientations, calculate the energy $u^{or}(\mathbf{b}_i)$.
3. We define the Rosenbluth[2] factor

$$W(n) = \sum_{j=1}^{k} \exp[-\beta u^{or}(\mathbf{b}_j)]. \tag{13.1.2}$$

   Out of these k orientations, we select one, say, n, with a probability

$$p(\mathbf{b}_n) = \frac{\exp[-\beta u^{or}(\mathbf{b}_n)]}{\sum_{j=1}^{k} \exp[-\beta u^{or}(\mathbf{b}_j)]}. \tag{13.1.3}$$

4. For the old configuration, o, the part of the energy that does not depend on the orientation of the molecules is denoted by $u^{pos}(o)$. The orientation of the molecule in the old position is denoted by $\mathbf{b}_o$, and we generate $k-1$ trial orientations denoted by $\mathbf{b}_2, \cdots, \mathbf{b}_k$. Using these k orientations, we determine

$$W(o) = \exp[-\beta u^{or}(\mathbf{b}_o)] + \sum_{j=2}^{k} \exp[-\beta u^{or}(\mathbf{b}_j)]. \tag{13.1.4}$$

5. The move is accepted with a probability

$$\text{acc}(o \to n) = \min\left(1, \frac{W(n)}{W(o)} \exp\{-\beta[u^{pos}(n) - u^{pos}(o)]\}\right). \tag{13.1.5}$$

It is clear that equation (13.1.3) ensures that energetically favorable configurations are more likely to be generated. An example implementation of this scheme is shown in Algorithm 22. Next, we should demonstrate that the sampling scheme is correct.

[2]Since this algorithm for biasing the orientation of the molecules is very similar to an algorithm developed by Rosenbluth and Rosenbluth in 1955 [295] for sampling configurations of polymers (see section 11.2), we refer to the factor W as the Rosenbluth factor.

Algorithm 22 (Orientational Bias)

```
PROGRAM orien_bias                      move a particle to a random
                                        position using an orient. bias
o=int(ranf()*npart)+1                   select a particle at random
xt=ranf()*box                           start: generate new configuration
call ener(xt,en)                        calculate u^pos
wn=exp(-beta*en)
sumw=0
do j=1,k                                generate k trial orientations
  call ranor(b(j))                      random vector on a sphere
  call enero(xt,b(j),eno)               calculate trial orientation j u^or(j)
  w(j)= exp(-beta*eno)                  calculate Rosenbluth factor (13.1.2)
  sumw=sumw+w(j)
enddo
call select(w,sum,n)                    select one of the orientations
bn=b(n)                                 n is the selected conformation
wn=wn*sumw                              Rosenbluth factor new configuration
                                        consider the old conformation
call ener(x(o),en)                      calculate u^pos
wo=exp(-beta*en)
sumw=0
do j=1,k                                consider k trial orientations
 if (j.eq.1) then
   b(j)=u(o)                            use actual orientation of particle o
 else
   call ranor(b(j))                     generate a random orientation
 endif
 call enero(x(o),b(j),eno)              calculate energy of trial orientation j
 sumw=sumw+exp(-beta*eno)               calculate Rosenbluth factor (13.1.4)
enddo
wo=wo*sumw                              Rosenbluth factor old configuration
if (ranf().lt.wn/wo)                    acceptance test (13.1.5)
+    call accept                        accepted: do bookkeeping
end
```

Comments to this algorithm:

1. *The subroutine* `ener` *calculates the energy associated with the position, the subroutine* `enero` *the energy associated with the orientations.*
2. *The subroutine* `ranor` *generates a random vector on a unit sphere (Algorithm 42), subroutine* `accept` *does the bookkeeping associated with the acceptance of a new configuration, and the subroutine* `select` *selects one of the orientations with probability* $p(i) = w(i)/\sum_j w(j)$ *(see, Algorithm 41).*

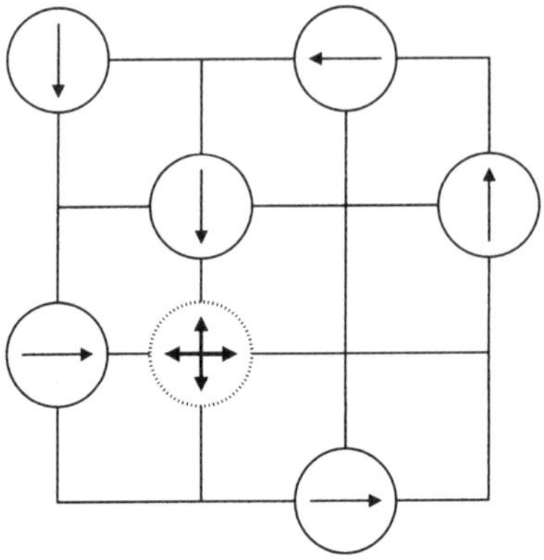

Figure 13.1: Lattice model in which the molecules can take four orientations (indicated by arrows, $k = 4$). The dotted circle indicates the trial position of the particle that we attempt to move.

Justification of Algorithm

To show that the orientational-bias Monte Carlo scheme just described is correct, that is, generates configurations according to the desired distribution, it is convenient to consider lattice models and continuum models separately. For both cases we assume that we work in the canonical ensemble, for which the distribution of configurations is given by equation (5.2.2)

$$\mathcal{N}(\mathbf{q}^N) \propto \exp[-\beta \mathcal{U}(\mathbf{q}^N)],$$

where $\mathcal{U}(\mathbf{q}^N)$ is the sum of orientational and nonorientational part of the energy:

$$\mathcal{U} = u^{or} + u^{pos}.$$

We first consider a lattice model.

Lattice Models

We assume that the molecules in our lattice model can have k discrete orientations (see Figure 13.1). We impose the condition of detailed balance (5.1.1):

$$K(o \to n) = K(n \to o).$$

The flow of configurations o to n is (equation (5.1.2))

$$K(o \to n) = \mathcal{N}(o) \times \alpha(o \to n) \times acc(o \to n). \tag{13.1.6}$$

In the orientational-bias scheme, the probability of selecting conformation n is (see equation (13.1.3))

$$\alpha(o \to n) = \frac{\exp[-\beta u^{or}(n)]}{W(n)}.$$

Imposing detailed balance and substitution of the desired distribution for $\mathcal{N}(n)$ and $\mathcal{N}(o)$ imposes the following condition on the acceptance rules:

$$\begin{aligned}\frac{\text{acc}(o \to n)}{\text{acc}(n \to o)} &= \frac{\exp[-\beta \mathcal{U}(n)]}{\exp[-\beta \mathcal{U}(o)]} \times \frac{\exp[-\beta u^{or}(o)]}{W(o)} \times \frac{W(n)}{\exp[-\beta u^{or}(n)]} \\ &= \frac{W(n)}{W(o)} \exp\{-\beta[u^{pos}(n) - u^{pos}(o)]\}. \end{aligned} \quad (13.1.7)$$

Acceptance rule (13.1.5) satisfies this condition. This demonstrates that for a lattice model detailed balance is fulfilled.

Continuum Model

If the orientation of a molecule is described by a continuous variable, then there is an essential difference with the previous case. In the lattice model all the possible orientations can be considered explicitly, and the corresponding Rosenbluth factor can be calculated exactly. For the continuum case, we can never hope to sample *all* possible orientations. It is impossible to determine the exact Rosenbluth factor since an infinite number of orientations are possible.[3] Hence, the scheme for lattice models, in which the Rosenbluth factor for all orientations is calculated, cannot be used for a continuum model. A possible solution would be to use a large but finite number of trial directions. Surprisingly, this is not necessary. It is possible to devise a *rigorous* algorithm using an *arbitrary subset* of all possible trial directions. The answer we get does *not* depend on the number of trial directions we choose but the statistical accuracy does.

Let us consider the case in which we use a set of k trial orientations; this set is denoted by

$$\{\mathbf{b}\}_k = \{\mathbf{b}_1, \mathbf{b}_2, \cdots, \mathbf{b}_k\}.$$

Conformation $\mathbf{b}_n$ can be selected only if it belongs to the set $\{\mathbf{b}\}_k$. The set of all sets $\{\mathbf{b}\}_k$ that includes conformation n is denoted by

$$\mathcal{B}_n = \{\{\mathbf{b}\}_k | \mathbf{b}_n \in \{\mathbf{b}\}_k\}.$$

Every element of $\mathcal{B}_n$ can be written as $(\mathbf{b}_n, b^*)$, where b^* is the set of $k-1$ additional trial orientations. In the flow of configuration o to n, we have to

[3] In Example 17 we discuss a special case for which the Rosenbluth factor *can* be calculated exactly.

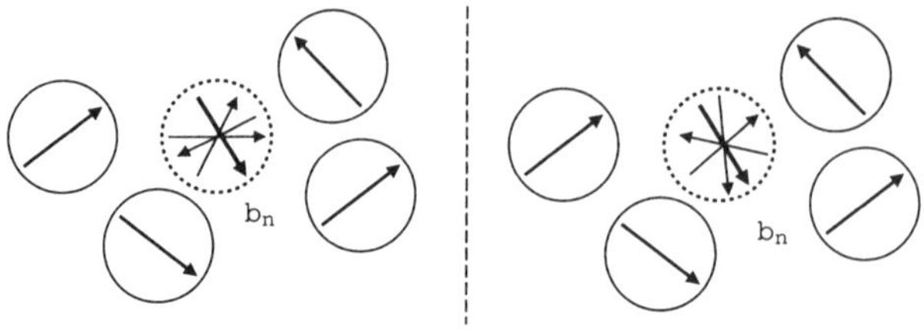

Figure 13.2: Continuum model in which the molecule can have an arbitrary orientation (indicated by arrows). The figure shows two different sets of four trial orientations that both include orientation $\mathbf{b}_n$.

consider the sum over all sets in $\mathcal{B}_n$

$$K(o \to n) = \mathcal{N}(o) \sum_{i \in \mathcal{B}_n} \alpha(o \to n, i) \times \text{acc}(o \to n, i), \tag{13.1.8}$$

in which the probability of generating configuration n and the acceptance depend on the particular set of trial orientations i.

Similarly, for the reverse move, we define the set $\mathcal{B}_o$

$$\mathcal{B}_0 = \{\{\mathbf{b}\}_k | \mathbf{b}_o \in \{\mathbf{b}\}_k\},$$

for which each element can be written as $(\mathbf{b}_o, b'^*)$. The expression for the reverse flow then becomes

$$K(n \to o) = \mathcal{N}(n) \sum_{j \in \mathcal{B}_o} \alpha(n \to o, j) \times \text{acc}(n \to o, j). \tag{13.1.9}$$

It should be stressed that infinitely many different sets of orientations include $\mathbf{b}_n$, and the same holds for sets that include $\mathbf{b}_o$. Moreover, the probability of selecting $\mathbf{b}_n$ from such a set depends on the remainder of the set b^* (see Figure 13.2). Hence, the acceptance probability must also depend on the sets b^* and b'^*.

Detailed balance is certainly obeyed if we impose a much stronger condition, "super-detailed balance," which states that for every particular choice of the sets b^* and b'^*, detailed balance should be obeyed,

$$\begin{aligned} K(o \to n, b^*, b'^*) &= K(n \to o, b'^*, b^*), \\ \mathcal{N}(o)\, \alpha(o \to n, b^*, b'^*)\, \text{acc}(o \to n, b^*, b'^*) & \\ &= \mathcal{N}(n)\, \alpha(n \to o, b'^*, b^*)\, \text{acc}(n \to o, b'^*, b^*), \end{aligned} \tag{13.1.10}$$

in which b^* and b'^* are two sets of $k-1$ arbitrary additional trial orientations. It may seem strange that the sets b^* and b'^* show up on *both* sides of the equations. However, bear in mind that, to decide on the acceptance of the forward move, one should generate both the set b^* that includes the new orientation *and* the set b'^* around the old orientation. Hence, the construction of a trial move includes both sets of trial orientations. As the probabilities of generating b^* and b'^* appear on both sides of the equations, they cancel each other. Moreover, the *a priori* probability of generating a random orientation $\mathbf{b}_n$ in the forward move is equal to the *a priori* probability of generating $\mathbf{b}_o$ in the reverse move. So these generation probabilities also cancel each other. This leads to a great simplification of the acceptance criterion. For the canonical ensemble, substitution of equations (13.1.2) and (13.1.3) yields

$$\begin{aligned}\frac{\mathrm{acc}(o \to n, b^*, b'^*)}{\mathrm{acc}(n \to o, b'^*, b^*)} &= \frac{\exp[-\beta \mathcal{U}(n)]}{\exp[-\beta \mathcal{U}(o)]} \frac{\exp[-\beta u^{or}(o)]}{W(\mathbf{b}_o, b'^*)} \frac{W(\mathbf{b}_n, b^*)}{\exp[-\beta u^{or}(n)]} \\ &= \frac{W(\mathbf{b}_n, b^*)}{W(\mathbf{b}_o, b'^*)} \exp\{-\beta [u^{pos}(n) - u^{pos}(o)]\}.\end{aligned} \tag{13.1.11}$$

As acceptance rule (13.1.5) satisfies this condition, detailed balance is indeed obeyed.

Note that, in this demonstration, we did not have to assume that the number of trial orientations k had to be large. In fact, the result is *independent* of the number of trial orientations.

Example 16 (Orientational Bias of Water)

Cracknell *et al.* [353] used an orientational-bias scheme to simulate liquid water. At ambient temperature, water has a relatively open structure, in which the water molecules form a network due to the hydrogen bonds. To insert a water molecule successfully, one has not only to place the molecule in an empty spot but also find a good orientation. The method used by Cracknell *et al.* to find this optimum orientation is similar to the one introduced in this section, in the sense that a bias in the orientation is introduced and is subsequently removed by adjusting the acceptance rules. Yet, the philosophy behind the approach of Cracknell *et al.* is fundamentally different.

In the scheme of Cracknell *et al.*, a random position of a water molecule r is generated and one trial orientation ω is drawn from a distribution $f(r, \omega)$. The problem is that the optimum distribution $f(r, \omega)$ is not known *a priori* and depends on the conformations of the other water molecules. However, as we have shown, any distribution can be used (as long as detailed balance and microscopic reversibility are obeyed). Since the construction of the true orientational distribution requires too much computer time, Cracknell *et al.* constructed a distribution that was meant to mimic the true distribution. To this end, one axis of the water molecule was given a random orientation and, for the other axis a biasing scheme was used. For this axis, n equidistant angles ψ_i were generated

$$\psi_i = 2\pi p/n, \quad p \in \{1, \cdots, n\}.$$

For each of these, the Boltzmann factor of the energy was calculated

$$f_i = C\exp(-\beta u_{\psi_i}).$$

Assuming that the Boltzmann weight varies linearly between test points, these n points span an approximate orientational distribution $f(\psi)$. For instance, for $\psi \in [2\pi p/n, 2\pi(p+1)/n]$, the distribution f is given by

$$f(\psi) = \frac{C}{2\pi/n}\{(2\pi(p+1)/n - \psi)f_p + (\psi - 2\pi p/n)f_{p+1}\}.$$

The constant C was fixed by the requirement that the orientational distribution be normalized. Using a standard rejection scheme, a trial orientation is generated according to the distribution specified by $f(\psi)$. For liquid water under ambient conditions, this method gives an improvement of a factor 2–3 over the conventional random insertion.

The main difference between the scheme of Cracknell *et al.* and the algorithm just discussed is that in Cracknell *et al.*'s scheme an attempt is made to construct a continuous distribution that approaches the true distribution in the limit of large n. In contrast, for the scheme of section 13.1.2, the shape of the true distribution does not matter. In particular, it is not necessary to reconstruct the distribution or to calculate a normalization factor.

Example 17 (Dipoles Embedded in Spherical Atoms)
In systems with dipoles, the energy depends on the mutual orientation of the molecules and a bias in the sampling of the orientation can be useful. For models of dipoles embedded in an otherwise spherical particle (e.g., the dipolar hard-sphere fluid) the scheme of section 13.1.2 can be implemented elegantly as pointed out by Caillol [225]. In equations (13.1.2) and (13.1.4), the Rosenbluth factor is calculated by sampling k trial orientations. For a dipolar hard sphere (or any point dipole), we can calculate the Rosenbluth factors exactly once the electric field $(\mathbf{E})$ at the position of the inserted particle and that at the position of the old configuration are known:

$$\begin{aligned} W(\mathbf{r}) &= \int d\mathbf{b}\, \exp[-\beta\boldsymbol{\mu}\cdot\mathbf{E}(\mathbf{r})] \\ &= \frac{\sinh[\beta|\boldsymbol{\mu}|\,|\mathbf{E}(\mathbf{r})|]}{\beta|\boldsymbol{\mu}|\,|\mathbf{E}(\mathbf{r})|}, \end{aligned}$$

where $\boldsymbol{\mu}$ is the dipole moment of the molecule.[4] A trial orientation can now

[4] In fact, there is a subtlety with this expression. It assumes that the component of the local electric field in the direction of the dipole does not depend on the absolute orientation of the dipole. This seems obvious. But, in the case of an Ewald summation, where the long-range interaction of a molecule with its periodic images is represented by a Fourier sum, this condition is not quite satisfied.

be drawn directly from the distribution

$$p(\mathbf{r}, \omega) = \frac{\exp[-\beta \boldsymbol{\mu} \cdot \mathbf{E}(\mathbf{r})]}{W(r)}.$$

13.2 Chain Molecules

The sampling of equilibrium conformations of polymers is usually time consuming. The main reason is that the natural dynamics of polymers are dominated by topological constraints (for example, chains cannot cross) and hence any algorithm based on the real motion of macromolecules will suffer from the same problem. For this reason, many "unphysical" Monte Carlo trial moves have been proposed to speed up the sampling of polymer conformations (see, e.g., [299]). In this section we introduce the configurational-bias Monte Carlo scheme [293, 297, 354, 355]. This simulation technique can be used for systems where it is not possible to change the conformation of a macromolecule by successive small steps.

13.2.1 Configurational-Bias Monte Carlo

The starting point for the configurational-bias Monte Carlo technique is the scheme introduced by Rosenbluth and Rosenbluth in 1955 [295]. The Rosenbluth scheme itself also was designed as a method to sample polymer conformations.[5] A drawback of the Rosenbluth scheme is, however, that it generates an unrepresentative sample of all polymer conformations; that is, the probability of generating a particular conformation using this scheme is *not* proportional to its Boltzmann weight. Rosenbluth and Rosenbluth corrected for this bias in the sampling of polymer conformations by introducing a conformation-dependent weight factor W. However, as was shown in detail by Batoulis and Kremer [300], this correction procedure, although correct in principle, in practice works only for relatively short chains (see Example 13).

The solution of this problem is to bias the Rosenbluth sampling in such a way that the correct (Boltzmann) distribution of chain conformations is recovered in a Monte Carlo sequence. In the configurational-bias scheme to be discussed next, the Rosenbluth weight is used to bias the *acceptance* of trial conformations generated by the Rosenbluth procedure. As we shall show, this guarantees that all chain conformations are generated *with the correct Boltzmann weight*.

[5] The Rosenbluth scheme is discussed in some detail in the context of a free energy calculation of a chain molecule in Chapter 11.

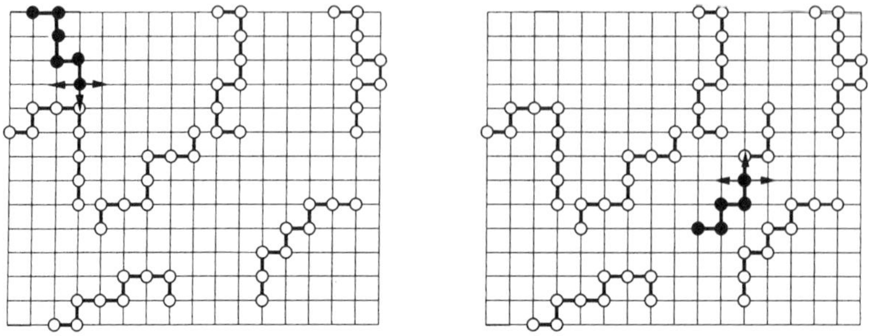

Figure 13.3: Sketch of the configurational-bias Monte Carlo scheme. The left figure shows the generation of a new configuration and the right figure shows the retracing of the old conformation. The arrows indicate the three trial positions.

13.2.2 Lattice Models

Algorithm

The configurational-bias Monte Carlo algorithm consists of the following steps:

1. Generate a trial conformation using the Rosenbluth scheme (see Figure 13.3, left) to grow the entire molecule, or part thereof, and compute its Rosenbluth weight $W(n)$.
2. "Retrace" the old conformation (see Figure 13.3, right) and determine its Rosenbluth factor.
3. Accept the trial move with a probability

$$\text{acc}(o \to n) = \min[1, W(n)/W(o)]. \tag{13.2.1}$$

The generation of a trial conformation n of a polymer consisting of ℓ monomers is generated using an algorithm based on the method of Rosenbluth and Rosenbluth (see Figure 13.3):

1. The first atom is inserted at random, and its energy is denoted by $u_1(n)$, and[6] $w_1(n) = k\exp[-\beta u_1(n)]$, where k is the coordination number of the lattice, for example, $k = 6$ for a simple cubic lattice.
2. For the next segment, with index i, there are k possible trial directions. The energy of trial direction j is denoted by $u_i(j)$. From the k possible

[6]The factor k in the definition of the Rosenbluth weight of the first segment, strictly speaking, is unnecessary. We introduce it only here to make the subsequent notation more compact.

directions, we select one, say, n, with a probability

$$p_i(n) = \frac{\exp[-\beta u_i(n)]}{w_i(n)}, \tag{13.2.2}$$

where $w_i(n)$ is defined as

$$w_i(n) = \sum_{j=1}^{k} \exp[-\beta u_i(j)]. \tag{13.2.3}$$

The interaction energy $u_i(j)$ includes all interactions of segment i with other molecules in the system and with segments 1 through $i-1$ of the same molecule. It does not include the interactions with segments $i+1$ to ℓ. Hence, the total energy of the chain is given by $\mathcal{U}(n) = \sum_{i=1}^{\ell} u_i(n)$.

3. Step 2 is repeated until the entire chain is grown and we can determine the Rosenbluth factor of configuration n:

$$W(n) = \prod_{i=1}^{\ell} w_i(n). \tag{13.2.4}$$

Similarly, to determine the Rosenbluth factor of the old configuration, o, we use the following steps (see Figure 13.3).

1. One of the chains is selected at random. This chain is denoted by o.
2. We measure the energy of the first monomer $u_1(o)$ and compute $w_1(o) = k\exp[-\beta u_1(o)]$.
3. To compute the Rosenbluth weight for the remainder of the chain, we determine the energy of monomer i at its actual position, and also the energy it would have had had it been placed in any of the other $k-1$ sites neighboring the actual position of monomer $i-1$ (see Figure 13.3). These energies are used to calculate

$$w_i(o) = \exp[-\beta u_i(o)] + \sum_{j=2}^{k} \exp[-\beta u_i(j)].$$

4. Once the entire chain has been retraced, we determine its Rosenbluth factor:

$$W(o) = \prod_{i=1}^{\ell} w_i(o). \tag{13.2.5}$$

Algorithm 23 (Basic Configurational-Bias Monte Carlo)

```
   PROGRAM CBMC                    configurational-bias Monte Carlo

   new_conf=.false.                first retrace (part of) the old conf.
   call grow(new_conf,wo)          to calculate its Rosenbluth factor

   new_conf=.true.                 next consider the new configuration
   call grow(new_conf,wn)          grow (part of) a chain and calculate
                                   the Rosenbluth factor of the new conf.
   if (ranf().lt.wn/wo)            acceptance test (13.2.6)
  +   call accept                  accept and do bookkeeping
   end
```

Comments to this algorithm:

1. *This algorithm shows the basic structure of the configurational-bias Monte Carlo method. The details of the model are considered in the subroutine* `grow` *(see Algorithm 24 for a polymer on a lattice).*
2. *The subroutine* `accept` *takes care of the bookkeeping of the new configuration.*

Finally the trial move from o to n is accepted with a probability given by

$$\mathrm{acc}(o \to n) = \min[1, W(n)/W(o)]. \tag{13.2.6}$$

A schematic example of the implementation of this scheme is given in Algorithms 23 and 24. We now have to demonstrate that the acceptance rule (13.2.6) correctly removes the bias of generating new segments in the chain introduced by using equation (13.2.2).

Justification of the Algorithm

The demonstration that this algorithm samples a *Boltzmann* distribution is similar to the one for the orientational-bias algorithm for lattice models (section 13.1.2).

The probability of generating a particular conformation n follows from the repetitive use of equation (13.2.2):

$$\alpha(o \to n) = \prod_{i=1}^{\ell} \frac{\exp[-\beta u_i(n)]}{w_i(n)} = \frac{\exp[-\beta \mathcal{U}(n)]}{W(n)}. \tag{13.2.7}$$

Algorithm 24 (Growing a Chain on a Lattice)

```
SUBROUTINE grow(new_conf,w)       grow an ℓ bead polymer on a lattice
                                  with coordination number k and
                                  calculate its Rosenbluth factor w
if (new_conf) then
  xn(1)=ranf()*box                insert the first monomer
else
  o=ranf()*npart+1                select old chain at random
  xn(1)=x(o,1)
endif
call ener(xn(1),en)               calculate energy
w=k*exp(-beta*en)                 Rosenbluth factor first monomer
do i=2,ell
  sumw=0
  do j=1,k                        consider the k trial directions
    xt(j)=xn(i-1)+b(j)            determine trial position
    call ener(xt(j),en)           determine energy trial position j
    w(j)=exp(-beta*en)
    sumw=sumw+w(j)
  enddo
  if (new_conf) then
    call select(w,sumw,n)         select one of the trial position
    xn(i)=xt(n)                   direction n is selected
  else
    xn(i)=x(o,i)
  endif
  w=w*sumw                        update Rosenbluth factor
enddo
return
end
```

Comments to this algorithm:

1. *If* `new_conf=.true.` *generate a new configuration, if* `new_conf = .false.` *retrace an old one.*
2. *In a lattice model we consider all possible trial positions, denoted by* `b(j)`, *therefore, for the old configuration, the actual position is automatically included.*
3. *The subroutine* `select` *(Algorithm 41) selects one of the trial positions with probability* $p(i) = w(i)/\sum_j w(j)$. *The subroutine* `ener` *calculates the energy of the monomer at the given position with the other polymers and the monomers of the chain that already have been grown.*

Similarly, for the reverse move,

$$\alpha(n \to o) = \frac{\exp[-\beta \mathcal{U}(o)]}{W(o)}. \tag{13.2.8}$$

The requirement of detailed balance (5.1.1) imposes the following condition on the acceptance criterion:

$$\frac{\text{acc}(o \to n)}{\text{acc}(n \to o)} = \frac{W(n)}{W(o)}. \tag{13.2.9}$$

Clearly, the proposed acceptance criterion (13.2.6) satisfies this condition.

It should be stressed that the value of factor $W(o)$ depends on the direction in which the old configuration is retraced: if we start from monomer 1, we find a different numerical value for $W(o)$ than if we start from monomer ℓ. As a consequence the probability of such a move depends on the way the factor $W(o)$ has been calculated. Although such a dependence is at first sight counterintuitive, both ways of retracing the old conformation—starting with monomer 1 or with monomer ℓ—result in the correct distribution of states, as long as both ways occur with equal probability during the simulation. This is automatically satisfied in the case of linear chains of identical segments where the labeling of the terminal groups is completely arbitrary.

13.2.3 Off-lattice Case

Next we consider configurational-bias Monte Carlo for off-lattice systems. As with the orientational moves described in section 13.1.2, some aspects in a continuum version of configurational-bias Monte Carlo require special attention. In section 13.1.2 we already showed that it may be possible to develop a configurational-bias sampling scheme even when it is impossible to calculate the Rosenbluth factor exactly. For chain molecules, we can follow basically the same approach.

The other important point that we have to consider is the way in which trial conformations of a chain molecule are generated. In a lattice model, the number of trial conformations is dictated by the lattice. In an off-lattice system, one could generate trial segments with orientations distributed uniformly on a unit sphere. However, for many models of interest this procedure is not very efficient, in particular when there are strong intramolecular interactions (e.g., bending and torsion potentials). The efficiency of a configurational-bias Monte Carlo algorithm depends to a large extent on the method used of generating the trial orientations. For example, an isotropic distribution of trial directions is well suited for completely flexible chains. In contrast, for a stiff chain (e.g., liquid-crystal forming polymer), such a trial position will almost always be rejected because of the intramolecular interactions.

Algorithm

From the preceding discussion, it follows that the intramolecular interactions should be taken into account in generating the set of trial conformations. Here, we consider the case of a flexible molecule *with* contributions to the internal energy due to bond bending and torsion. The fully flexible case then follows trivially. Consider a chain of ℓ linear segments, the potential energy of a given conformation $\mathcal{U}$ has two contributions:

1. The *bonded potential energy* $\mathcal{U}^{\text{bond}}$ is equal to the sum of the contributions of the individual joints. A joint between segments i and $i+1$ (say) has a potential energy u_i^{bond} that depends on the angle θ between the successive segments. For instance, $u_i^{\text{bond}}(\theta)$ could be of the form $u_i^{\text{bond}}(\theta)=k_\theta(\theta-\theta_0)^2$. For realistic models for polyatomic molecules, u_i^{bond} includes all local bonded potential energy changes due to the bending and torsion of the bond from atom $i-1$ to atom i.
2. The *external potential energy* $\mathcal{U}^{\text{ext}}$ accounts for all interactions with other molecules and for all the nonbonded intramolecular interactions. In addition, interactions with any external field that may be present are also included in $\mathcal{U}^{\text{ext}}$.

In what follows we shall denote a chain in the absence of the external interactions as the *ideal* chain. Note that this is a purely fictitious concept, as real chains always have nonbonded intramolecular interactions.

To perform a configurational-bias Monte Carlo move, we apply the following "recipe" to construct a conformation of a chain of ℓ segments. The construction of chain conformations proceeds segment by segment. Let us consider the addition of one such segment. To be specific, let us assume that we have already grown $i-1$ segments and are trying to add segment i. This is done in two steps. First we generate a trial conformation n, next we consider the old conformation o. A trial conformation is generated as follows:

1. Generate a fixed number (say k) trial segments. The orientations of the trial segments are distributed according to the Boltzmann weight associated with the bonded interactions of monomer i (u_i^{bond}). We denote this set of k different trial segments by

$$\{\mathbf{b}\}_k = \{\mathbf{b}_1, \cdots, \mathbf{b}_k\},$$

where the probability of generating a trial segment $\mathbf{b}$ is given by

$$p_i^{\text{bond}}(\mathbf{b})d\mathbf{b} = \frac{\exp[-\beta u_i^{\text{bond}}(\mathbf{b})]d\mathbf{b}}{\int d\mathbf{b}\exp[-\beta u_i^{\text{bond}}(\mathbf{b})]} = C\exp[-\beta u_i^{\text{bond}}(\mathbf{b})]d\mathbf{b}. \tag{13.2.10}$$

2. For all k trial segments, we compute the external Boltzmann factors $\exp[-\beta u_i^{ext}(\mathbf{b}_i)]$, and out of these, we select one, denoted by n, with a probability

$$p_i^{ext}(\mathbf{b}_n) = \frac{\exp[-\beta u_i^{ext}(\mathbf{b}_n)]}{w_i^{ext}(n)}, \tag{13.2.11}$$

where we have defined

$$w_i^{ext}(n) = \sum_{j=1}^{k} \exp[-\beta u_i^{ext}(\mathbf{b}_j)]. \tag{13.2.12}$$

3. The selected segment n becomes the ith segment of the trial conformation of the chain.
4. When the entire chain is grown, we calculate the Rosenbluth factor of the chain:

$$W^{ext}(n) = \prod_{i=1}^{\ell} w_i^{ext}(n), \tag{13.2.13}$$

where Rosenbluth factor of the first monomer is defined by

$$w_1^{ext}(n) = k \exp[-\beta u_1^{ext}(\mathbf{r}_1)], \tag{13.2.14}$$

where $\mathbf{r}_1$ is the position of the first monomer.

For the old configuration, a similar procedure to calculate its Rosenbluth factor is used.

1. One of the chains is selected at random. This chain is denoted o.
2. The external energy of the first monomer is calculated. This energy involves only the external interactions. The Rosenbluth weight of this first monomer is given by

$$w_1^{ext}(o) = k \exp[-\beta u_1^{ext}(o)]. \tag{13.2.15}$$

3. The Rosenbluth factors of the other $\ell - 1$ segments are calculated as follows. We consider the calculation of the Rosenbluth factor of segment i. We generate a set of $k-1$ orientations with a distribution prescribed by the bonded interactions (13.2.10). These orientations, together with the actual bond between segment $i-1$ and i, form the set of k orientations $(\mathbf{b}_o, \mathbf{b}'^*)$. These orientations are used to calculate the external Rosenbluth factor:

$$w_i^{ext}(o) = \sum_{j=1}^{k} \exp[-\beta u_i^{ext}(\mathbf{b}_j)]. \tag{13.2.16}$$

4. For the entire chain the Rosenbluth factor of the old conformation is defined by

$$W^{ext}(o) = \prod_{i=1}^{\ell} w_i^{ext}(o). \tag{13.2.17}$$

After the new configuration has been generated and the Rosenbluth factor of the old configuration has been calculated, the move is accepted with a probability

$$\text{acc}(o \to n) = \min[1, W^{ext}(n)/W^{ext}(o)]. \tag{13.2.18}$$

We still have to show that this sampling scheme is correct.

Justification of Algorithm

Comparison with the lattice version shows that for the off-lattice case, two aspects are different. First, for a model with continuous degrees of freedom, we cannot calculate the Rosenbluth factor exactly. This point has been discussed in detail in section 13.1.2 for the orientational-bias scheme. As in section 13.1.2, we impose super-detailed balance. Second, the way in which we generate trial conformations is different for off-lattice than for lattice models. In a lattice model there is no need to separate the interactions in bonded and external ones. We have to show that the way in which we treat bonded interactions does not perturb the sampling.

The probability of generating a chain of length ℓ is the product of the probability of generating a trial orientation (13.2.10) and the probability of selecting this orientation (13.2.11); for all monomers this gives, as a probability of generating conformation n,

$$\alpha(o \to n) = \prod_{i=1}^{\ell} p_i(o \to n) = \prod_{i=1}^{\ell} p_i^{bond}(n) p_i^{ext}(n). \tag{13.2.19}$$

In the following, we consider the expressions for one of the ℓ segments, to keep the equations simple. A given set of k trial orientations, which includes orientation n, is denoted by $(\mathbf{b}_n, b^*)$ (see section 13.1.2). As before, we stress that the generation of the additional trial orientations (b'^*) around the old segment $(\mathbf{b}_o)$ is an essential part of the *generation* of the trial move. We denote the probability of generating the combined set b^*, b'^* by

$$\mathcal{P}^{bond}(b^*, b'^*).$$

Hence, the flow of configurations is given by

$$\begin{aligned} K(o \to n, b^*, b'^*) &= \mathcal{N}(o) \times \alpha(o \to n, b^*, b'^*) \times \text{acc}(o \to n, b^*, b'^*) \\ &= \exp[-\beta u(o)] \times C \exp[-\beta u^{\text{bond}}(n)] \times \frac{\exp[-\beta u^{\text{ext}}(n)]}{\mathcal{W}^{\text{ext}}(\mathbf{b}_n, b^*)} \\ &\quad \times \text{acc}(o \to n, b^*, b'^*) \mathcal{P}^{\text{bond}}(b^*, b'^*). \end{aligned} \tag{13.2.20}$$

For the reverse move, we have

$$\begin{aligned} K(n \to o, b'^*, b^*) &= \mathcal{N}(n) \times \alpha(n \to o, b'^*, b^*) \times \text{acc}(n \to o, b'^*, b^*) \\ &= \exp[-\beta u(n)] \times C \exp[-\beta u^{\text{bond}}(o)] \times \frac{\exp[-\beta u^{\text{ext}}(o)]}{\mathcal{W}^{\text{ext}}(\mathbf{b}_o, b'^*)} \\ &\quad \times \text{acc}(n \to o, b'^*, b^*) \mathcal{P}^{\text{bond}}(b^*, b'^*). \end{aligned} \tag{13.2.21}$$

Recall that the total energy of a monomer is the sum of the bonded and external contributions:

$$u(n) = u^{\text{bond}}(n) + u^{\text{ext}}(n).$$

We now impose super-detailed balance (13.1.10). The factors $\mathcal{P}^{\text{bond}}(b^*, b'^*)$ on both sides of the equation cancel each other, and we get the following simple criterion for the acceptance rule:

$$\frac{\text{acc}(o \to n, b^*, b'^*)}{\text{acc}(n \to o, b'^*, b^*)} = \frac{\mathcal{W}^{\text{ext}}(\mathbf{b}_n, b^*)}{\mathcal{W}^{\text{ext}}(\mathbf{b}_o, b'^*)}. \tag{13.2.22}$$

This demonstration was only for a single segment in a chain. For the entire chain, the corresponding acceptance criterion is obtained analogously. It is simply the product of the terms for all segments:

$$\frac{\text{acc}[o \to n, (b_1^*, \cdots, b_\ell^*)]}{\text{acc}[n \to o, (b'^*_1, \cdots, b'^*_\ell)]} = \frac{\prod_{i=1}^{\ell} w_i^{\text{ext}}(\mathbf{b}_n, b^*)}{\prod_{i=1}^{\ell} w_i^{\text{ext}}(\mathbf{b}_o, b'^*)} = \frac{W[n, (b^*{}_1, \cdots, b^*{}_\ell)]}{W[o, (b'^*_1, \cdots, b'^*_\ell)]}. \tag{13.2.23}$$

And, indeed, our acceptance rule (13.2.18) satisfies this condition. The equation shows that, because the trial orientations are generated with a probability (13.2.10) prescribed by the bonded energy, this energy does *not* appear in the acceptance rules. In Case Study 19, a detailed discussion is given on the advantages of this approach. It is important to note that we do not need to know the normalization constant C of equation (13.2.10).

The basic structure of an algorithm for configurational-bias Monte Carlo for continuum models is very similar to the lattice version (Algorithm 23); the main difference is the way in which configurations are generated.

Case Study 18 (Equation of State of Lennard-Jones Chains)
To illustrate the configurational-bias Monte Carlo technique described in this section, we determine the equation of state of a system consisting of eight-bead chains of Lennard-Jones particles. The nonbonded interactions are described by a truncated and shifted Lennard-Jones potential. The potential is truncated at $R_c = 2.5\sigma$. The bonded interactions are described with a harmonic spring

$$u^{vib}(l) = \begin{cases} 0.5k_{vib}(l-1)^2 & 0.5 \leq l \leq 1.5 \\ \infty & \text{otherwise} \end{cases},$$

where l is the bond length, the equilibrium bond length has been set to 1, and $k_{vib} = 400$.

The simulations are performed in cycles. In each cycle, we perform on average N_{dis} attempts to displace a particle, N_{cbmc} attempts to (partly) re-grow a chain, and N_{vol} attempts to change the volume (only in the case of N,P,T simulations). If we regrow a chain, the configurational-bias Monte Carlo scheme is used. In this move we select at random the monomer from which we start to regrow. If this happens to be the first monomer, the entire molecule is regrown at a random position. For all the simulations, we used eight trial orientations. The lengths of trial bonds are generated with a probability prescribed by the bond-stretching potential (see Case Study 19).

In Figure 13.4 the equation of state as obtained from N,V,T simulations is compared with one obtained from N,P,T simulations. This isotherm is well above the critical temperature of the corresponding monomeric fluid ($T_c = 1.085$, see Figure 3.3), but the critical temperature of the chain molecules is appreciably higher [356].

13.3 Generation of Trial Orientations

The efficient generation of good trial conformations is an essential aspect of the configurational-bias Monte Carlo scheme for continuum models with strong intramolecular interactions. For some models (for example, Gaussian chains) it is possible to generate this distribution directly. For an arbitrary model we can use the acceptance-rejection technique [33] of generating the trial orientations.

Here, we show how a rejection technique can be used to generate trial positions efficiently. The number of trial directions in the CBMC scheme can be chosen at will. Often, the optimal number of trial directions is determined empirically. However, more systematic techniques exist to compute this optimal number [357].

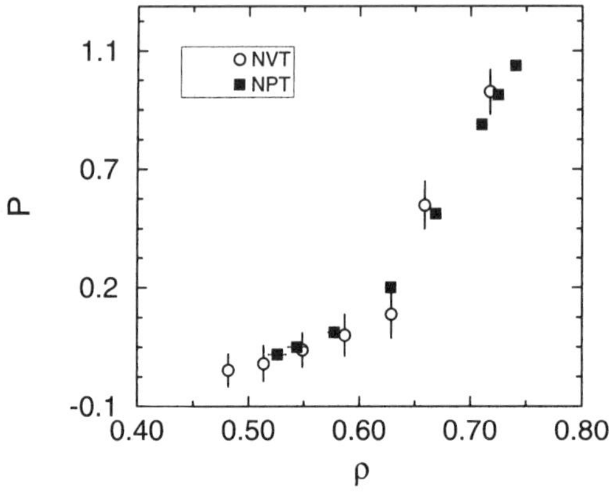

Figure 13.4: Equation of state of an eight-bead Lennard-Jones chain as obtained from N,V,T and N,P,T simulations using the configurational-bias Monte Carlo scheme. The simulations are performed with 50 chains at a temperature $T = 1.9$.

13.3.1 Strong Intramolecular Interactions

Let us consider as an example a model of a molecule in which the bonded interactions include bond stretching, bond bending, and torsion. The external interactions are the nonbonded interactions. A united atom model of an alkane is a typical example of such a molecule.

The probability that we generate a trial configuration **b** is given by, (see equation (13.2.10))

$$P(\mathbf{b})d\mathbf{b} = C\exp[-\beta u^{\text{bond}}(\mathbf{b})]d\mathbf{b}. \tag{13.3.1}$$

It is convenient to represent the position of an atom using the bond length r, bond angle θ, and torsional angle ϕ (see Figure 13.5). With these coordinates the volume element $d\mathbf{b}$ is given by

$$d\mathbf{b} = r^2\, dr d\cos\theta d\phi. \tag{13.3.2}$$

The bonded energy is the sum of the bond-stretching potential, the bond-bending potential, and the torsion potential:

$$u^{\text{bond}}(r,\theta,\phi) = u_{\text{vib}}(r) + u_{\text{bend}}(\theta) + u_{\text{tors}}(\phi). \tag{13.3.3}$$

Substitution of equations (13.3.3) and (13.3.2) into equation (13.3.1) gives

$$\begin{aligned} P(\mathbf{b})\, d\mathbf{b} &= P(r,\theta,\phi) r^2\, dr d\cos\theta d\phi \\ &= C\exp[-\beta u_{\text{vib}}(r)] r^2 dr \times \exp[-\beta u_{\text{bend}}(\theta)] d\cos\theta \\ &\quad \times \exp[-\beta u_{\text{tors}}(\phi)] d\phi. \end{aligned} \tag{13.3.4}$$

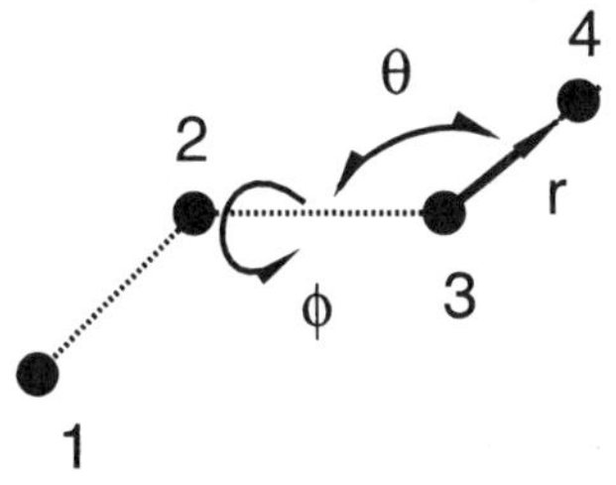

Figure 13.5: Schematic sketch of a part of a molecule.

Many models use a fixed bond length, in which case the first term in equation (13.3.4) is a constant.

Let us consider the molecule shown in Figure 13.5. The first atom is placed at a random position and we now have to add the second atom. For convenience, it is assumed that the model has a fixed bond length. The second atom has no bonded interactions other than the constraints on the bond length. The distribution of trial orientations, equation (13.3.4), reduces to

$$P_2(\mathbf{b})d\mathbf{b} \propto d\cos\theta d\phi. \tag{13.3.5}$$

Hence, the trial orientations are randomly distributed on the surface of a sphere (such a distribution can be generated with Algorithm 42 in Appendix J).

For the third atom, the bonded energy contains the bond-bending energy as well. This gives, for the distribution of trial orientations,

$$P_3(\mathbf{b})d\mathbf{b} \propto \exp[-\beta u_{\text{bend}}(\theta)]\, d\cos\theta d\phi. \tag{13.3.6}$$

To generate k trial orientations distributed according to equation (13.3.6), we again generate a random vector on a unit sphere and determine the angle θ. This vector is accepted with a probability $\exp[-\beta u_{\text{bend}}(\theta)]$. If rejected, this procedure is repeated until a value of θ has been accepted. In [33], this acceptance-rejection method is shown to indeed give the desired distribution of trial orientations. In this way, k (or $k-1$, for the old conformation) trial orientations are generated.

An alternative scheme would be to generate angle θ uniformly ($\theta \in [0, \pi]$) and to determine the bond-bending energy corresponding to this angle. This angle θ is accepted with a probability $\sin(\theta)\exp[-\beta u_{\text{bend}}(\theta)]$. If rejected, this procedure is repeated until a value of θ has been accepted. The selected value of θ is supplemented with a randomly selected angle ϕ. These two angles determine a new trial orientation.

Algorithm 25 (Growing an Alkane)

```
SUBROUTINE grow(new_conf,w)             grow or retrace an alkane and
                                        calculate its Rosenbluth factor w
if (new_conf) then                      new_conf =.true.: new conf.
   ib=int(ranf()*ell)+1                 start to grow from position ib
   ibnewconf=ib                         store starting position
else                                    new_conf =.false.: old conf.
   ib=ibnewconf                         same starting position to regrow
endif                                   as used for the new configuration
do i=1,ib-1
  xn(i)=x(i)                            store positions that are not regrown
enddo
w=1
do i=ib,ell
 if (ib.eq.1) then                      first atom
   if (new_conf) then
    xt(1)=ranf()*box                    generate random position
   else
    xt(1)=xn(1)                         use old position
   endif
   call enerex(xt(1),eni)               calculate (external) energy
   w=k*exp(-beta*eni)                   and Rosenbluth factor
 else                                   second and higher atoms
   sumw=0
   do j=1,k
    if (.not.new_conf
+      .and. j.eq.1) then
     xt(1)=x(i)                         actual position as trial orientation
    else
     call next_ci(xt(j),xn,i)           generate trial position
    endif
    call enerex(xt(j),eni)              (external) energy of this position
    wt(j)= exp(-beta*eni)
    sumw=sumw+wt(j)
   enddo
   w=w*sumw                             update Rosenbluth factor
   if (new_conf) then
     call select(wt,sumw,n)             select one of the trial orientations
     xn(i)=xt(n)
     xstore(i)=xt(n)                    store selected configuration
   else                                 for bookkeeping
     xn(i)=x(i)
   endif
 endif
enddo
return
end
```

Comments to this algorithm:

1. *Subroutine* `enerex` *calculates the external energy of an atom at the given position, and subroutine* `select` *selects one of the trial positions with probability* $p(i) = w(i)/\sum_j w(j)$ *(Algorithm 41).*
2. *Subroutine* `next_ci` *adds the next atom to the chain as prescribed by the bonded interactions (Algorithms 26, 27, and 28 are examples for ethane, propane, and higher alkanes, respectively).*

For the fourth and higher atoms, the bonded energy includes both bond-bending and torsion energy. This gives, for equation (13.3.4),

$$p_l^{\text{bond}}(\mathbf{b})d\mathbf{b} \propto \exp[-\beta u_{\text{bend}}(\theta)] \exp[-\beta u_{\text{tors}}(\phi)] \, d\cos\theta d\phi. \tag{13.3.7}$$

We again generate a random vector on a sphere and calculate the bond-bending angle θ and torsion ϕ. These angles are accepted with a probability $\exp\{-\beta[u_{\text{bend}}(\theta) + u_{\text{tors}}(\phi)]\}$. If these angles are rejected, new vectors are generated until one gets accepted.

Again an alternative scheme is to determine first a bond-bending angle θ by generating θ uniformly on $[0, \pi]$ and calculating the bond-bending energy corresponding to this angle. This angle θ is then accepted with a probability $\sin(\theta) \exp[-\beta u_{\text{bend}}(\theta)]$. This procedure is continued until we have accepted an angle. Next we generate a torsion angle randomly on $[0, 2\pi]$ and accept this angle with a probability $\exp[-\beta u_{\text{tors}}(\phi)]$, again repeating this until a value has been accepted. In this scheme the bond angle and torsion are generated independently, which can be an advantage in cases where the corresponding potentials are sharply peaked.

The acceptance-rejection technique is illustrated in Algorithms 25–28 for different n-alkanes. For all-atom or explicit-hydrogen models of hydrocarbons, a different strategy is needed for which we refer the reader to the relevant literature [358,359].

Case Study 19 (Generation of Trial Configurations of Ideal Chains)

In section 13.2.3, we emphasized the importance of efficiently generating trial segments for molecules with strong intramolecular interactions. In this case study, we quantify this. We consider the following bead-spring model of a polymer. The nonbonded interactions are described with a Lennard-Jones potential and the bonded interactions with a harmonic spring:

$$u^{\text{vib}}(l) = \begin{cases} 0.5k_{\text{vib}}(l-1)^2 & 0.5 \le l \le 1.5 \\ \infty & \text{otherwise} \end{cases},$$

where l is the bond length, the equilibrium bond length has been set to 1, and $k_{\text{vib}} = 400$. The bonded interaction is only the bond stretching. The external (nonbonded) interactions are the Lennard-Jones interactions. We consider the following two schemes of generating a set of trial positions:

Algorithm 26 (Growing Ethane)

```
SUBROUTINE next_c2(xn,xt,i)        generate a trial position for ethane
                                   position of the first atom is known
call bondl(l)                      generate bond length
call ranor(b)                      generate vector on unit sphere
xt(i)=xn(i-1)+l*b
return
end
```

Comment to this algorithm:

1. *The subroutine* ranor *generates a random vector on a unit sphere (Algorithm 42), and the subroutine* bondl *(Algorithm 43) generates the bond length prescribed by the bonded interactions.*

Algorithm 27 (Growing Propane)

```
 SUBROUTINE next_c3(xn,         generate a trial position for ith atom
+           xt,i)               position of the (i − 1)th atom is known
 call bondl(l)                  generate bond length
 if (i.eq.2) then               second atom
   call next_c2(xn,xt,i)        use Algorithm 26
 else if (i.eq.3) then          third atom
   call bonda(xn,b,i)           generate orientation of the
   xt=xn(2)+l*b                 new position with desired bond angle
 else
   STOP 'error'
 endif
 return
 end
```

Comment to this algorithm:

1. *The subroutine* ranor *generates a random vector on a unit sphere (Algorithm 42), the subroutine* bondl *(Algorithm 43) generates the bond length prescribed by the bonded interactions (for the second atom, only bond stretching), and the subroutine* bonda *generates a vector on a unit sphere with bond angle prescribed by the bond-bending potential (Algorithm 45).*

Algorithm 28 (Generating a Trial Position for an Alkane)

```
SUBROUTINE next_cn(xn,xt,i)      generate a trial position for ith atom
                                 position of atoms (i − 1) are known
call bondl(l)                    generate bond length
if (i.eq.2) then                 second atom
  call next_c2(xn,xt,i)          use Algorithm 26
else if (i.eq.3) then            third atom
  call next_c3(xn,xt,i)          use Algorithm 27
else if (i.ge.4) then            fourth and higher atoms
  call tors_bonda(xn,b,i)        generate vector with prescribed
  xt=xn(i-1)+l*b                 bond and torsional angles
endif
return
end
```

Comment to this algorithm:

1. *The subroutine* `tors_bonda` *(Algorithm 46) generates bond bending and a torsional angle prescribed by the corresponding potentials.*

1. Generate a random orientation with bond length uniformly distributed in the spherical shell between limits chosen such that they bracket all acceptable bond lengths. For instance, we could consider limits that correspond to a 50% stretching or compression of the bond. In that case, the probability of generating bond length l is given by

$$p_1(l)\begin{cases} \propto C\mathrm{d}l \propto l^2\mathrm{d}l & 0.5 \leq l \leq 1.5 \\ 0 & \text{otherwise} \end{cases} .$$

2. Generate a random orientation and the bond length prescribed by the bond-stretching potential (as described in Algorithm 26). The probability of generating bond length l with this scheme is

$$p_2(l)\begin{cases} \propto C\exp[-\beta u^{vib}(l)]\mathrm{d}l = C\exp[-\beta u^{vib}(l)]l^2\mathrm{d}l & 0.5 \leq l \leq 1.5 \\ 0 & \text{otherwise} \end{cases} .$$

Let us consider a case in which the system consists of ideal chains. Ideal chains are defined (see section 13.2.3) as chains having only *bonded* interactions.

Suppose we use method 1 to generate the set of k trial orientations with bond lengths $l_1, \cdots, l_k$, then the Rosenbluth factor for atom i is given by

$$w_i(n) = \sum_{j=1}^{k} \exp[-\beta u^{\text{vib}}(l_j)].$$

The Rosenbluth factor of the entire chain is

$$W(n) = \prod_{i=1}^{\ell} w_i(n).$$

For the old conformation a similar procedure is used to calculate its Rosenbluth factor:

$$W(o) = \prod_{i=1}^{\ell} w_i(o).$$

In absence of external interactions the Rosenbluth factor of the first atom is defined to be $w_1 = k$.

In the second scheme, we generate the set of k trial orientations with a bond length distribution $p_2(l)$. If we use this scheme, we have to consider only the external interaction. Since, for an ideal chain, the external interactions are by definition 0, the Rosenbluth factor for each atom is given by

$$w_i^{\text{ext}}(n) = \sum_{j=1}^{k} \exp[-\beta u^{\text{ext}}(l_j)] = k,$$

and similarly, for the old conformation

$$w_i^{\text{ext}}(o) = k.$$

Hence, the Rosenbluth weight is the same for the new and the old conformations:

$$W^{\text{ext}}(n) = \prod_{i=1}^{\ell} w_i^{\text{ext}}(n) = k^{\ell}$$

and

$$W^{\text{ext}}(o) = \prod_{i=1}^{\ell} w_i^{\text{ext}}(o) = k^{\ell}.$$

The acceptance rule for the first scheme is

$$\text{acc}(o \to n) = \min[1, W(n)/W(o)]$$

and for the second scheme is

$$\text{acc}(o \to n) = \min[1, W^{\text{ext}}(n)/W^{\text{ext}}(o)] = 1.$$

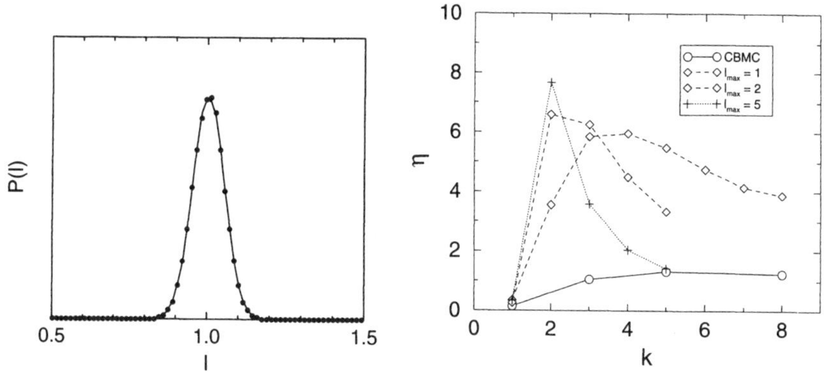

Figure 13.6: Comparison of methods 1 and 2 for the distribution of bond lengths l (left) and the distribution of the radius of gyration R_g (right). The solid lines represent the results for method 1, the dots for method 2 ($\ell = 5$ and $k = 5$).

Inspection of these acceptance rules shows that, in the second scheme, *all* configurations generated are accepted, whereas in the first scheme this probability depends on the bond-stretching energy and therefore will be less than 1. Hence, it is clearly useful to employ the second scheme.

To show that the results of schemes 1 and 2 are indeed equivalent, we compare the distribution of the bond length of the chain and the distribution of the radius of gyration in Figure 13.6. The figure shows that the results for the two methods are indeed indistinguishable. The efficiency of the two methods, however, is very different. In Table 13.1, the difference in acceptance probability is given for some values of the bond-stretching force constant and various chain lengths. The table shows that if we use method 1 and generate a uniformly distributed bond length, we need to use at least 10 trial orientations to have a reasonable acceptance for chains longer than 20 monomers. Note that the corresponding table for the second method has a 100% acceptance for all values of k independent of the chain length.

Most of the simulations, however, do not involve ideal chains but chains with external interactions. For chains with external interactions, the first method performs even worse. First of all, we generate the chains the same way as in the case of the ideal chains. The bonded interactions are the same and we need to generate at least the same number of trial directions to get a reasonable acceptance. In addition, if there are external interactions, we have to calculate the nonbonded interactions for *all* of those trial positions. The calculation of the nonbonded interactions takes most of the CPU time; yet, in the first method, most of the trial orientations are doomed to be re-

k	$\ell = 5$	$\ell = 10$	$\ell = 20$	$\ell = 40$	$\ell = 80$	$\ell = 160$
1	0.6	$\ll$0.01	$\ll$0.01	$\ll$0.01	$\ll$0.01	$\ll$0.01
5	50	50	10	$\ll$0.01	$\ll$0.01	$\ll$0.01
10	64	58	53	42	$\ll$0.01	$\ll$0.01
20	72	66	60	56	44	$\ll$0.01
40	80	72	67	62	57	40
80	83	78	72	68	62	60

Table 13.1: Probability of acceptance (%) for ideal chains using uniformly distributed bond lengths (method 1), where ℓ is the chain length, and k is the number of trial orientations. The value for the spring constant is $k_{vib} = 400$ (see [289]). For method 2, the acceptance would have been 100% for all values of k and ℓ.

jected solely on the basis of the bonded energy. These two reasons make the second scheme much more attractive than the first.

13.3.2 Generation of Branched Molecules

The generation of trial configurations for branched alkanes requires some care. Naively, one might think that it is easiest to grow a branched alkane atom by atom. However, at the branchpoint we have to be careful. Suppose we have grown the backbone shown in Figure 13.7 and we now have to add the branches b_A and b_B. The total bond-bending potential has three contributions, given by

$$u_{bend} = u_{bend}(c_1, c_2, b_A) + u_{bend}(c_1, c_2, b_B) + u_{bend}(b_A, c_2, b_B).$$

Vlugt [360] pointed out that, because of the term $u_{bend}(b_A, c_2, b_B)$, it is better not to generate the positions of b_A and b_B independently. Suppose that we would try to do this anyway. We would then generate the first trial position, b_A, according to

$$P(b_A) \propto \exp\left[-\beta u_{bend}(c_1, c_2, b_A)\right],$$

next we would generate the second trial position, b_B, using

$$P(b_B|b_A) \propto \exp\{-\beta\left[u_{bend}(c_1, c_2, b_B) + u_{bend}(b_A, c_2, b_B)\right]\},$$

where $P(b_B|b_A)$ denotes the probability of generating b_B for a given position of segment b_A. However, if we would generate both positions at the same time, then the probability is given by

$$P(b_A, b_B) \propto \exp\{-\beta\left[u_{bend}(c_1, c_2, b_A) + u_{bend}(c_1, c_2, b_B) + u_{bend}(b_A, c_2, b_B)\right]\}.$$

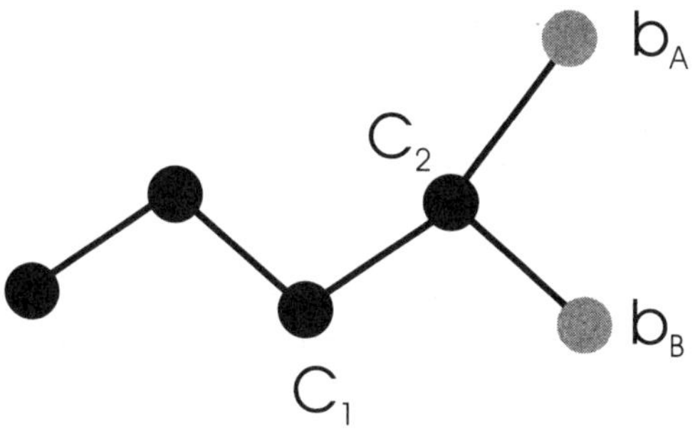

Figure 13.7: Growth of a banched alkane.

The two schemes are only equivalent if

$$P(b_A, b_B) = P(b_B|b_A)P(b_A).$$

In general this equality does not hold. To see this, compare the probability of generating configuration b_A for the two schemes. This probability is obtained by integrating over all orientations b_B. If both chains are inserted at the same time, we find that

$$\begin{aligned} P(b_A) &= \int db_B P(b_A, b_B) \\ &\propto \exp\left[-\beta u_{bend}(c_1, c_2, b_A)\right] \\ &\quad \times \int db_B \exp\{-\beta\left[u_{bend}(c_1, c_2, b_A) + u_{bend}(b_A, c_2, b_B)\right]\}. \end{aligned}$$

For the sequential scheme, we would have obtained

$$\begin{aligned} P(b_A) &= \int db_B P(b_B|b_A)P(b_A) \\ &= P(b_A) \\ &\propto \exp\left[-\beta u_{bend}(c_1, c_2, b_A)\right] \end{aligned}$$

as, in this scheme, segment b_A is inserted before segment b_B. Therefore the probability $P(b_A)$ cannot depend on b_B.

We can now easily see that if we use a model in which the two branches are equivalent, for example, isobutane, the sequential scheme does not generate equivalent *a priori* distributions for the two branches. Of course, the generation of trial segments is but one step in the CBMC scheme. Any bias introduced at this stage can be removed by incorporating the ratio of the true and the biased distributions in the acceptance criterion. However, the resulting algorithm may be inefficient. Vlugt *et al.* [361] have shown that simply ignoring the bias introduced by the "sequential" scheme will result in small, but noticeable, errors in the distribution of the bond angles.

As the insertion of two segments at the same time is less efficient than sequential insertion, several strategies have been proposed to increase the efficiency of the simultaneous generation of branches.

For molecules in which the bond-bending potential has three contributions (as in the example above), the simplest scheme is to generate two random vectors on a sphere and use the conventional rejection scheme to generate configurations with a probability proportional to their Boltzmann weight [362]. One can also use this approach for more complex potentials that include torsion. If the random generation of trial directions becomes inefficient, it may be replaced by a simple Monte Carlo scheme [361].

For some intramolecular potential it may even be necessary to add more than two atoms at the same time to ensure a proper *a priori* distribution of added segments. In fact, for some molecules that have multiple torsional angles, such as 2,3-dimethylbutane, this approach would imply that all atoms have to be added at the same time. To avoid such many-particle insertions, Martin and Siepmann [363] developed a scheme similar to the multiple-first-bead algorithm (see section 13.5).

The idea is to use a random insertion to generate several trial positions and to use a CBMC scheme to select acceptable candidates using the internal energies only. These configurations that are distributed according to the correct intramolecular Boltzmann weight will subsequently be used in another CBMC scheme that involves the more expensive external energy calculations.

To see how this approach works, assume that we have a model with internal interactions given by u^{int}. A single segment is added using the following steps:

1. First generate a set of n_{int} *random* trial positions and for each position compute the internal energy, $U^{int}(i)$, and calculate the Rosenbluth factor associated with this *internal* energy

$$W^{int}(n) = \sum_{j=1}^{n_{int}} \exp\left[-\beta U^{int}(j)\right].$$

A possible orientation is then selected using

$$p^{int}(j) = \frac{\exp\left[-\beta U^{int}(j)\right]}{W^{int}(n)}.$$

2. Step 1 is repeated to generate k trial positions which are then fed into the conventional CBMC scheme to compute the Rosenbluth factor using the external potential $W^{ext}(n)$.

3. A similar method is used for the old configuration, giving $W^{\text{int}}(o)$ and $W^{\text{ext}}(o)$.
4. A move is accepted using

$$\text{acc}(o \to n) = \min\left(1, \frac{W^{\text{int}}(n)W^{\text{ext}}(n)}{W^{\text{int}}(o)W^{\text{ext}}(o)}\right).$$

Depending on the details of the potential, further refinements are possible. One can, for instance, separate the bond-bending potential and the torsion potential. This would imply three nested CBMC steps giving three different Rosenbluth factors. For more details see [363].

13.4 Fixed Endpoints

A drawback of the conventional configurational-bias Monte Carlo scheme is that it regrows a chain molecule, either partly or completely, starting from one of the endpoints. For dense systems, where only relatively short segments of the molecule can be regrown successfully, the configurational-bias Monte Carlo scheme reduces to the reptation scheme. This implies that the equilibration of the middle segments of a chain proceeds very slowly—for heteropolymers, where reptation moves are forbidden, the situation is even worse. The same restriction applies to chain molecules that have either end rigidly anchored to a surface. Finally, conventional configurational-bias Monte Carlo cannot be applied at all to ring polymers.

In the present section, we discuss how the configurational-bias Monte Carlo scheme can be extended to include sampling of chain conformations with fixed endpoints. With such a scheme it is possible to relax the interior of a chain as efficiently as the endpoints. Ring polymers can be considered special examples of chain molecules with fixed endpoints. Another interesting example that can be treated in the same way is the sampling of path integrals [364], but this falls outside the scope of this book. In addition, we discuss some alternative Monte Carlo techniques, such as concerted rotations and end-bridging Monte Carlo, which have been developed by Theodorou and co-workers [365].

13.4.1 Lattice Models

Let us first consider configurational-bias Monte Carlo between fixed endpoints for a chain molecule on a simple cubic lattice. If we remove n segments of the molecule between two fixed endpoints $\mathbf{r}_1$ and $\mathbf{r}_2$, we cannot simply regrow the molecule by the normal Rosenbluth scheme, because this does not ensure that a trial conformation starting at $\mathbf{r}_1$ will end at $\mathbf{r}_2$. Clearly, we must bias our regrowth scheme in such a way that the trial conformation is forced to terminate at $\mathbf{r}_2$. To achieve this, we use the following scheme. Suppose that we start our regrowth at position $\mathbf{r}_1$. On a three-dimensional lattice, this coordinate is represented by three integer coordinates $\{k_1, l_1, m_1\}$. The final position is denoted by $\{k_2, l_2, m_2\}$. The total number of ideal (i.e., nonself-avoiding) random walks of length n between $\mathbf{r}_1$ and n $\mathbf{r}_2$ is denoted by $\Omega(\mathbf{r}_1, \mathbf{r}_2; n)$. We can always compute the number of ideal random walks between fixed endpoints analytically as it is simply a finite sum of multinomial coefficients [366, 367]. Let us next consider the growth of one segment, starting at $\mathbf{r}_1$. In the original configurational-bias Monte Carlo scheme, we would consider all k possible trial directions. And we would select one of these directions, say direction j, with a probability

$$P(j) = \frac{\exp[-\beta u^{ext}(j)]}{\sum_{j'=1}^{k} \exp[-\beta u^{ext}(j')]},$$

where $u^{ext}(j)$ denotes the potential energy of trial segment j due to all other particles already in the system. In the present case, we use a different weight factor to select the trial segment, namely,

$$P(j) = \frac{\exp[-\beta u^{ext}(j)]\Omega(\mathbf{r}_1 + \Delta\mathbf{r}(j), \mathbf{r}_2; n-1)}{\sum_{j'=1}^{k} \exp[-\beta u^{ext}(j')]\Omega(\mathbf{r}_1 + \Delta\mathbf{r}(j'), \mathbf{r}_2; n-1)}. \qquad (13.4.1)$$

In other words, the probability of selecting a given trial direction is proportional to the number of ideal random walks of length $n-1$ that start at the position of the trial segment and terminate at $\mathbf{r}_2$. In this way, we *guarantee* that we generate only conformations that start at $\mathbf{r}_1$ and terminate at $\mathbf{r}_2$. However, as before, we must correct for the bias that we have introduced. We do this by constructing a modified Rosenbluth weight W: $W = \prod_{i=1}^{n} w_i$ with

$$\begin{aligned} w_i &\equiv \frac{\sum_{j'=1}^{k} \exp[-\beta u^{ext}(j')]\Omega[\mathbf{r}_i + \Delta\mathbf{r}(j'), \mathbf{r}_2; n-i]}{\sum_{j'=1}^{k} \Omega[\mathbf{r}_i + \Delta\mathbf{r}(j'), \mathbf{r}_2; n-i]} \\ &= \frac{\sum_{j'=1}^{k} \exp[-\beta u^{ext}(j')]\Omega[\mathbf{r}_i + \Delta\mathbf{r}(j'), \mathbf{r}_2; n-i]}{\Omega[\mathbf{r}_i, \mathbf{r}_2; n-i+1]}. \end{aligned} \qquad (13.4.2)$$

If we now multiply the probability of generating a given trial conformation Γ with the Rosenbluth weight of that conformation, we find that

$$\begin{aligned} P_{gen}(\Gamma) \times W(\Gamma) &= \prod_{i=1}^{n} \left\{ \frac{\exp[-\beta u^{ext}(j)]\Omega[\mathbf{r}_i + \Delta\mathbf{r}(j), \mathbf{r}_2; n-i]}{\sum_{j'=1}^{k} \exp[-\beta u^{ext}(j')]\Omega[\mathbf{r}_i + \Delta\mathbf{r}(j'), \mathbf{r}_2; n-i]} \right. \\ &\quad \times \left. \frac{\sum_{j'=1}^{k} \exp[-\beta u^{ext}(j')]\Omega[\mathbf{r}_i + \Delta\mathbf{r}(j'), \mathbf{r}_2; n-i]}{\Omega[\mathbf{r}_i, \mathbf{r}_2; n-i+1]} \right\} \\ &= \prod_{i=1}^{n} \left\{ \frac{\exp[-\beta u^{ext}(j)]\Omega[\mathbf{r}_i + \Delta\mathbf{r}(j), \mathbf{r}_2; n-i]}{\Omega[\mathbf{r}_i, \mathbf{r}_2; n-i+1]} \right\}. \end{aligned} \qquad (13.4.3)$$

The modified Rosenbluth weight has been chosen such that all but one of the factors involving the number of ideal conformations cancel each other:

$$
\begin{aligned}
P_{gen}(\Gamma) \times \mathcal{W}(\Gamma) &= \prod_{i=1}^{n} \frac{\exp[-\beta u^{ext}(i)]}{\Omega(\mathbf{r}_1, \mathbf{r}_2; n)} \\
&= \frac{\exp[-\beta \mathcal{U}^{ext}(\Gamma)]}{\Omega(\mathbf{r}_1, \mathbf{r}_2; n)}. \qquad (13.4.4)
\end{aligned}
$$

The remaining factor Ω is the same for all conformations of length n that start at $\mathbf{r}_1$ and terminate at $\mathbf{r}_2$; hence, it drops out when we compute the relative probabilities of the old and new conformations. As before, the actual Monte Carlo scheme involves generating the trial conformation using the scheme indicated in equation (13.4.1) and accepting the new conformation with a probability given by

$$
\text{acc}(o \to n) = \min[1, \mathcal{W}(n)/\mathcal{W}(o)]. \qquad (13.4.5)
$$

A total regrowth of a ring polymer of length ℓ can be accomplished by choosing $\mathbf{r}_1 = \mathbf{r}_2$ and $n = \ell$.

13.4.2 Fully Flexible Chain

Again, it is possible to extend configurational-bias Monte Carlo to sample chain conformations between fixed endpoints, using our knowledge of the exact expression for the number (or, more precisely, the probability density) of ideal (nonself-avoiding) conformations of n segments between fixed endpoints $\mathbf{r}_1$ and $\mathbf{r}_2$. If we denote the probability density to find segment $i + 1$ at a distance r from segment i by $p_1(r)$, then we have the following recursion relation between the probability density of the end-to-end separation of chains of length n and $n + 1$:

$$
P(\mathbf{r}_{12}; n+1) = \int d\Delta P(\mathbf{r}_{12} - \Delta; n) p_1(\Delta). \qquad (13.4.6)
$$

From equation (13.4.6) and the fact that $p_1(r)$ is normalized, we immediately deduce the inverse relation:

$$
P(\mathbf{r}_{12}; n) = \int d\Delta P(\mathbf{r}_{12} + \Delta; n+1). \qquad (13.4.7)
$$

In the special case that all segments are of fixed length a, the expression for this probability density is [368]

$$
P(\mathbf{r}_{12}; n) = \frac{\sum_{k=0}^{k \le (n - r_{12}/a)/2} (-1)^k \binom{n}{k} (n - 2k - r_{12}/a)^{n-2}}{2^{n+1}(n-2)! \pi a^2 r_{12}}, \qquad (13.4.8)
$$

where $r_{12} \equiv |\mathbf{r}_1 - \mathbf{r}_2|$. This expression is valid for all $n > 1$. As before, we wish to modify the configurational-bias Monte Carlo sampling of conformations of a fully flexible chain in such a way that the chain is forced to terminate at $\mathbf{r}_2$. There are two ways to do this. In one approach, we include the bias in the probability with which we generate trial directions; in the second, the bias is in the acceptance probability. In either case, our approach does not depend on the specific form of $p_1(r)$, but only on the existence of the recurrence relation (13.4.7).

In the first approach, we use the following scheme of generating the ith segment out of ℓ segments to be regrown. We generate k trial segments, all starting at the current trial position $\mathbf{r}$, such that the *a priori* probability of generating a given trial direction (say, Γ_j) is proportional to the probability of having an ideal chain conformation of length $\ell - i$ between this trial segment and the final position $\mathbf{r}_2$. Let us denote this *a priori* probability by $p_{\text{bond}}(\Gamma_j)$. By construction, $p_{\text{bond}}(\Gamma_j)$ is normalized. Using equation (13.4.7) we can easily derive an explicit expression for p_{bond}:

$$\begin{aligned} p_{\text{bond}}(\Gamma) &= \frac{p_1(\Gamma)P(\mathbf{r}+\Gamma-\mathbf{r}_2;\ell-i)}{\int d\Gamma' p_1(\Gamma')P(\mathbf{r}+\Gamma'-\mathbf{r}_2;\ell-i)} \\ &= \frac{p_1(\Gamma)P(\mathbf{r}+\Gamma-\mathbf{r}_2;\ell-i)}{P(\mathbf{r}-\mathbf{r}_2;\ell-i+1)}. \end{aligned} \tag{13.4.9}$$

From here on, we treat the problem just like the sampling of a continuously deformable chain, described in section 13.2.3. That is, we select one of the k trial directions with a probability

$$P_{\text{sel}}(j) = \frac{\exp[-\beta u^{\text{ext}}(\Gamma_j)]}{\sum_{j'=1}^{k} \exp[-\beta u^{\text{ext}}(\Gamma_{j'})]}.$$

The contribution to the total Rosenbluth weight of the set of k trial directions generated in step i is

$$w_i \equiv \frac{\sum_{j'=1}^{k} \exp[-\beta u^{\text{ext}}(\Gamma_{j'})]}{k}.$$

The overall probability of moving from the old conformation Γ_{old} to a new conformation Γ_{new} is proportional to the product of the probability of generating the new conformation and the ratio of the new to the old Rosenbluth weights. The condition of (super-)detailed balance requires that the product of the probability of generating the new conformation times the Rosenbluth weight of that conformation is (but for a factor that is the same for the old and new conformations) equal to the product of the Boltzmann weight of that conformation and the properly normalized probability of generating the corresponding ideal (i.e., noninteracting) conformation. If we write the expression for this product, we find that

$$\begin{aligned}
&\prod_{i=1}^{\ell} P_{\text{gen}}[\Gamma_j(i)]w_i \\
&\quad = \left(\frac{p_1(\mathbf{r}_i - \mathbf{r}_{i-1})P(\mathbf{r}_i - \mathbf{r}_2; \ell - i)}{P(\mathbf{r}_{i-1} - \mathbf{r}_2; \ell - i + 1)}\right)\left(\frac{\exp\{-\beta u^{\text{ext}}[\Gamma_j(i)]\}}{\sum_{j'=1}^{k}\exp\{-\beta u^{\text{ext}}[\Gamma_{j'}(i)]\}}\right) \\
&\quad\quad \times \left(\frac{\sum_{j'=1}^{k}\exp\{-\beta u^{\text{ext}}[\Gamma_{j'}(i)]\}}{k}\right) \\
&\quad = \frac{\exp[-\beta\mathcal{U}^{\text{ext}}(\Gamma_{\text{total}})]\prod_{i=1}^{\ell} p_1(\mathbf{r}_i - \mathbf{r}_{i-1})}{k^{\ell}P(\mathbf{r}_{12}; \ell)}.
\end{aligned} \tag{13.4.10}$$

As the last line of this equation shows, the conformations are indeed generated with the correct statistical weight. In ref. [369] this scheme has been applied to simulate model homopolymers, random heteropolymers, and random copolymers consisting of up to 1000 Lennard-Jones beads. For molecules with strong intramolecular interactions, the present scheme will not work and other approaches are needed.

13.4.3 Strong Intramolecular Interactions

In the previous section we have shown that we can use the configurational-bias Monte Carlo scheme to grow a chain of length n between two fixed endpoints $\mathbf{r}_1$ and $\mathbf{r}_2$ if we know the probability density of conformations of length n between these points. For the special case of a fully flexible chain this probability distribution is known analytically. For chains with strong intramolecular interactions such an analytical distribution is not known. Wick and Siepmann [370] and Chen and Escobedo [371] have show that one can use an approximated distribution. Chen and Escobedo [371] estimate this distribution using a simulation of an isolated chain with bonded interactions only. Wick and Siepmann [370] proposed a scheme in which this estimated probability distribution is further refined during the simulation.

13.4.4 Rebridging Monte Carlo

If we model a realistic polymer or peptide we have to include bond-bending and torsional potentials. Suppose that we rotate in the interior of a polymer a randomly selected torsional angle by an amount $\Delta\phi$. If we would keep all other torsional angles of the remainder of the chain fixed, a tiny change of this torsional angle would lead to a large displacement of the last atom of the chain. If, on the other hand, one would only displace the neighboring

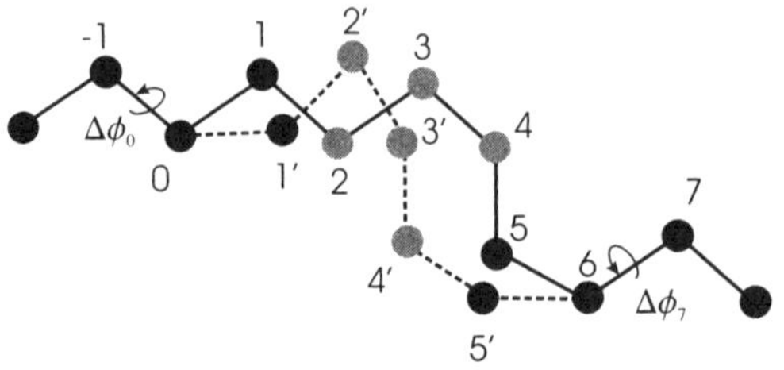

Figure 13.8: Schematic drawing of the rebridging Monte Carlo scheme. Suppose we give the atoms 1 and 5 a new position, for example, by rotating around the -1,0 and 6,7 bonds by angles $\Delta\phi_0$ and $\Delta\phi_7$, respectively. If we would not change the positions of the trimer, consisting of the gray atoms 2, 3, and 4, the intramolecular energy would increase significantly. The rebridging problem is to find a new conformation of the trimer with the same bond length and bond angle as the old conformation that bridges the new positions $1'$ and $5'$.

atoms, the intramolecular interactions of the chain would increase significantly, again limiting the maximum rotation. We would like to ensure that the rotation affects only a small part of the interior of the chain and that it results in a redistribution of atoms that does not result in a large increase in the intramolecular energy. *Concerted rotation* and *rebridging* and *end-bridging* Monte Carlo are schemes that have been developed by Theodorou and co-workers [60,365,372] to perform such Monte Carlo moves.

In Figure 13.8 the rebridging problem is sketched schematically. Suppose we give the atoms 1 and 5 a new position by a random rotation of the driver angles ϕ_0 and ϕ_7. Assume that all bond lengths and bond angles have a prescribed value, for example, their equilibrium value or any other specified value. The rebridging problem is to find all possible conformations of the trimer consisting of the atoms 2, 3, and 4 that rebridge the new positions of atoms 1 and 5 given the constraints of the prescribed bond lengths and angles. Wu and Deem [373] have shown that for the rebridging problem an analytical solution exists and that the maximum number of solutions is strictly limited to 16. Alternatively, in refs. [365, 372] it is shown how to numerically locate *all* solutions of the rebridging problem.

Suppose that we have all solutions of the rebridging problem, either by the analytical solution of Wu and Deem or by the numerical scheme of Theodorou and co-workers. The next step is to use this in a Monte Carlo scheme. The scheme that we discuss here is only valid for the interior segments of a polymer. For the ends of the chains, a slightly different scheme has to be used [60]:

1. The present conformation of the polymer is denoted by o. We gener-

ate the new configuration of the polymer, n, by selecting an atom and a direction (forward or backward) at random. This defines the atoms pair 1 and 5. These atoms are given new positions 1′ and 5′ by performing a random rotation around bonds −1,0 and 6,7, respectively (see Figure 13.8).

2. Solve the rebridging problem to locate all possible conformations of the trimer that bridge the new positions of atoms 1′ and 5′. The total number of conformations is denoted by N_n and out of these we select one conformation, say n, at random.[7] If no such conformation is found the move is rejected.

3. For the old conformation, we also locate all possible conformation, i.e., we solve the rebridging problem to locate the conformations of the trimer that bridge the old positions of atoms 1 and 5. This number of conformation is denoted by N_o.

4. In the rebridging scheme, we use a dihedral angle ϕ to generate a new configuration. This implies a temporary change of coordinate system; a Jacobian is associated with this change. In general, this Jacobian is not equal to 1. This Jacobian has to be taken into account in the acceptance rules [60]. The equations for this Jacobian can be found in refs. [60,372,373]. Here, we assume that these determinants for the old and new conformation have been calculated and are denoted by $J(o)$ and $J(n)$, respectively.

5. Of the new and old conformations the energies are calculated, $U(o)$ and $U(n)$, respectively.

6. The new conformation is accepted with a probability

$$\text{acc}(o \to n) = \min\left(1, \frac{\exp[-\beta U(n)]J(n)/N_n}{\exp[-\beta U(o)]J(o)/N_o}\right).$$

In refs. [60,372] the proof is given that this rebridging scheme obeys detailed balance and samples a Boltzmann distribution.

The reason it is important to find *all* solutions of the rebridging problem is to ensure detailed balance. Suppose that the determinants of the Jacobians are one and the energiegs are zero, then without the terms $1/N_n$ and $1/N_o$ the acceptance probability would be one for all possible new conformations. Suppose that we have a single solution for the new conformation, $N_n = 1$, and for the old conformation $N_o = 2$. Without the correction we would violate detailed balance since the *a priori* probability of the reverse move is only a half. Pant and Theodorou [372] have developed an alternative scheme

[7] One could use the configurational-bias Monte Carlo scheme as an alternative for the random selection.

in which one has to find only a single rebridging conformation, which is the first solution of their numerical scheme. To ensure detailed balance one has to check that the old conformation should also be the first solution to which the numerical scheme converges.

One can also use the rebridging scheme to connect atoms of *different* chains. The idea of end-bridging Monte Carlo is to alter the connectivity of the chain by bridging atoms from different chains. The simplest form is to rebridge a chain end to an interior segment of another chain with a trimer. Such an end-bridging Monte Carlo move induces a very large jump in configuration space. An important aspect of such an end-bridging move is, however, that it alters the chain lengths of the two chains. Therefore, such a move cannot be used if it is important to keep the chain length fixed. However, in most practical applications of polymers one does not have a single chain length but a distribution of chain lengths. Pant and Theodorou [372] have shown that the resulting chain length distribution resembles a truncated Gaussian distribution.

One can envision a chain length distribution as a mixture of a very large number of components, each component characterized by its chain length l. Imposing the chain length distribution is equivalent to imposing the chemical potentials of the various components. This suggests that we could combine these end-bridging moves with the *semigrand ensemble* simulation technique (see section 9.1) to determine whether a change of the polymer length should be accepted.

In principle one can use two rebridging moves for the interior segments of two chains. This would allow us to perform moves in which the total chain length remains constant. Whether in practice such a scheme will work depends on the probability that two segments of different chains with the *same* number of end segments connected to it are sufficiently close to each other.

Tests show that the rebridging method is very efficient for polymer melts with chain length up to C_{30}. For chains up to C_{70} rebridging Monte Carlo still samples the local structure efficiently, but fails to sample chain characteristics at larger length scales such as the end-to-end vector. End-bridging Monte Carlo effectively relaxes chains up to C_{500} [365]. Another important application of rebridging Monte Carlo is the possibility of simulating cyclic molecules. This application is illustrated by Wu and Deem in their study of cis/trans isomerisation of proline-containing cyclic peptides [373,374].

13.5 Beyond Polymers

Thus far, the configurational-bias scheme has been presented exclusively as a method of generating polymer conformations. The method is more general than that. It can be used as a scheme to perform collective rearrangements of any set of labeled coordinates. In fact, the scheme can be used to carry out Monte Carlo moves to swap n small particles within a volume ΔV with one large particle that occupies the same (excluded) volume. This application of the CBMC scheme has been exploited by Biben *et al.* [375, 376] to study mixtures of large and small hard spheres. Gibbs ensemble simulations of mixtures of spherical colloids and rodlike polymers were performed by Bolhuis and Frenkel [377] (see Example 18), using CBMC-style particle swaps and a closely related approach was employed by Dijkstra and co-workers to study phase separation [366, 367] in mixtures of large and small hard-core particles on a lattice. An application of CBMC for improving the sampling of ionic solutions has been proposed by Shelley and Patey [378].

A different application of the CBMC ideas is used by Esselink *et al.* [379] to develop an algorithm to perform Monte Carlo moves in parallel. Parallel Monte Carlo appears to be a contradiction in terms, since the Monte Carlo procedure is an intrinsically sequential process. One has to know whether the current move is accepted or rejected before one can continue with the next move. The conventional way of introducing parallelism is to distribute the energy calculation over various processors or to farm out the calculation by performing separate simulations over various processors. Although the last algorithm is extremely efficient and requires minimum skills to use a parallel computer, it is not a truly parallel algorithm. For example, farming out a calculation is not very efficient if the equilibration of the system takes a significant amount of CPU time. In the algorithm of Esselink *et al.* several trial positions are generated in parallel, and out of these trial positions the one with the highest probability of being accepted is selected. This selection step introduces a bias that is removed by adjusting the acceptance rules. The generation of each trial move, which includes the calculation of the energy (or Rosenbluth factor in the case of chain molecules), is distributed over the various processors. Loyens *et al.* have used this approach to perform phase equilibrium calculations in parallel using the Gibbs ensemble technique [380].

An interesting application of this parallel scheme is the multiple-first-bead algorithm. In a conventional CBMC simulation one would have to grow an entire chain before one can reject a configuration that is "doomed" from the start because the very first bead has an unfavorable energy. If the chains are long this can be inefficient and it becomes advantageous to use a multiple-first-bead scheme [379]. Instead of generating a single trial position for the first bead, k trial positions are generated. The energy of these beads, $u_1(j)$ with $j = 1, \ldots, k$, is calculated, and one of these beads, say j, is selected using the Rosenbluth criterion:

$$P_{1st}(j) = \frac{\exp[-\beta u_1(j)]}{w_1}$$

where

$$w_1(n) = \sum_{i=1}^{k} \exp[-\beta u_1(i)].$$

Also for the old configuration one should use a similar scheme to compute $w_1(o)$. For some moves the same set of first beads used for the new configuration can be used to compute the Rosenbluth factor for the old configuration [381]. To ensure detailed balance the Rosenbluth factors associated with the multiple-first beads should be taken into account in the acceptance rules:

$$\text{acc}(o \to n) = \min\left(1, \frac{w_1(n)W(n)}{w_1(o)W(o)}\right),$$

where $W(n)$ and $W(o)$ are the (conventional) Rosenbluth factors of the new and the old configuration of the chain, respectively, excluding the contribution of the first segment. Vlugt *et al.* [382] have shown that a multiple-first-bead move can increase the efficiency of simulations of n-alkanes up to a factor of 3.

Another extension of the CBMC principles is the use of a dual-cutoff radius [382]. The idea is that usually in a particular trial conformation is accepted not because it is energetically very favorable, but because its competitors are so unfavorable. This suggests that one can use a much cheaper potential to perform a prescreening of acceptable trial configurations in a CBMC move. Let us split the potential into a contribution that is cheap to compute and the expensive remainder:

$$U(r) = U^{\text{cheap}}(r) + \Delta U(r).$$

This can be done, for example, by splitting the potential into a long-range and short-range part. We can now use the cheap part in our CBMC scheme to generate trial configurations. The probability of generating a given configuration is then

$$P^{\text{cheap}}(n) = \frac{\exp[-\beta U^{\text{cheap}}(n)]}{W^{\text{cheap}}(n)}$$

and the move is accepted using

$$\text{acc}(o \to n) = \min\left(1, \frac{W^{\text{cheap}}(n)}{W^{\text{cheap}}(o)} \exp\{-\beta[\Delta U(n) - \Delta U(o)]\}\right).$$

In ref. [382] it is shown that this scheme obeys detailed balance. The advantage of this algorithm is that the expensive part of the energy calculation has to be performed only once and not for every trial segment. A typical application would be to include the Fourier part of an Ewald summation in ΔU. Many variations on this theme are possible.

Example 18 (Mixtures of Colloids and Polymers)
Configurational-bias Monte Carlo (CBMC) was presented as a scheme to sample conformations of chain molecules. In fact, the method is more general than that. It can be used to perform collective rearrangements of any set of labeled coordinates. For instance, the scheme can be used to carry out Monte Carlo moves to swap n small particles within a volume ΔV with one large particle that occupies the same (excluded) volume. This application of the CBMC scheme has been exploited by Biben [375] to study mixtures of large and small hard spheres. Gibbs ensemble simulations of mixtures of spherical colloids and rodlike polymers were performed in ref. [377] using CBMC-style particle swaps, and a closely related approach was employed by Dijkstra *et al.* [366, 367] to study phase separation of mixtures of large and small hard-core particles on a lattice.

Below, we briefly discuss an example of such a CBMC scheme, related to the phase behavior of colloidal suspensions [377]. Examples of colloidal solutions are milk, paint, and mayonnaise. Since a single colloidal particle may contain more than 10^9 atoms, it is not practical to model such a particle as a collection of atoms. It is better to describe colloidal solutions using coarse-grained models. For example, a suspension of sterically stabilized silica spheres in a nonpolar solvent can be described surprisingly accurately with a hard-sphere potential. Similar to the hard-sphere fluid, such a colloidal suspension has a "fluid-solid" transition, but not a "liquid-gas" transition. To be more precise, the colloidal particles undergo a transition from a liquid-like arrangement to a crystalline structure. But in either case, the solvent remains liquid. In what follows, we use the terms "crystal," "liquid," and "gas" to refer to the state of the colloidal particles in suspension. Experimentally, it is observed that a liquid-gas transition can be induced in a suspension of hard-sphere colloids by adding nonadsorbing polymers.

The addition of polymers induces an effective attraction between the colloidal particles. This attraction is not related to any change in the internal energy of the system, but to an increase in the entropy. It is not difficult to understand the origin of such entropic attractions. Let us assume that the polymers in solution do not interact with each other. This is never rigorously true, but for dilute solutions of long, thin molecules, it is a good first approximation. The translational entropy of N polymers in a volume V is then equal to that of N ideal-gas molecules occupying the same volume: $S^{(0)}_{trans} = \text{constant} + N k_B \ln V$, where the constant accounts for all those contributions that do not depend on the volume V. In the absence of colloids, the volume accessible to the polymers is equal to V_0, the volume of the container. Now suppose that we add one hard colloidal particle with radius R_c. As the polymers cannot penetrate the colloidal particle, such a colloid excludes the polymers from a spherical volume with radius $R_{excl} \equiv R_c + R_p$, where R_p is the effective radius of the polymer (for flexible polymers, R_p is on the order of the radius of gyration, and for rigid polymers, R_p is of order $\mathcal{O}(L)$, where L is the length of the polymer). Let us denote the volume excluded by one colloid by ν^c_{excl}. Clearly, the entropy of N polymers in the system that contains one colloid is $S^{(1)}_{trans} = \text{constant} + Nk_B \ln(V_0 - \nu^c_{excl})$. Now consider what happens if we have *two* colloidal spheres in the solution. Naively, one might think that the entropy of the polymer solution is now equal to $S^{(2)}_{trans} = \text{constant} + Nk_B \ln(V_0 - 2\nu^c_{excl})$. However, this is only true if the two colloids are far apart. If they are touching, their exclusion zones overlap, and the total excluded volume ν^{pair}_{excl} is *less than* $2\nu^c_{excl}$. This implies that the entropy of the polymers is *larger* when the colloids are touching than when they are far apart. Therefore, we can lower the free energy of the polymer solution by bringing the colloids close together. And this is the origin of entropic attraction. The strength of the attraction can be tuned by changing the polymer concentration and, for sufficiently high polymer concentrations, the colloidal suspensions may undergo a "liquid-vapor" phase separation.

In the present example, we consider the phase behavior of a mixture of colloidal hard spheres and thin hard rods [377]. In principle, we can use Gibbs ensemble simulations to study the "vapor-liquid" coexistence in this mixture. However, a conventional Gibbs ensemble simulation is likely to fail as the transfer of a colloidal sphere from one simulation box to the other will, almost certainly, result in an overlap of the sphere with some of the rodlike polymers. We can now use the CBMC approach to perform such a trial move with a higher chance of success. In this scheme, we perform the following steps:

1. Randomly select a sphere in one of the boxes, and insert this sphere at a random position in the other box.
2. Remove all the rods that overlap with this sphere. These rods are inserted in the other box. The positions and orientations of the rods are chosen such that they intersect with the volume vacated by the colloid —but apart from that, they are random. Even though we have thus ensured that the rods are in, or near, the "cavity" left by the colloidal sphere, they are very likely to overlap with one or more of the remaining spheres. However, if one tries several orientations and positions of the rods and selects an acceptable configuration using the configurational-bias Monte Carlo scheme, one can strongly enhance the acceptance probability of such particle swaps.

The results of these Gibbs ensemble simulations are presented in Figure 13.9. This figure shows that if one increases the fugacity (and thereby the concentration) of the rods, a demixing into a phase with a low density of spheres and a phase with a high density of spheres occurs. The longer the rods, the lower the concentration at which this demixing occurs. We

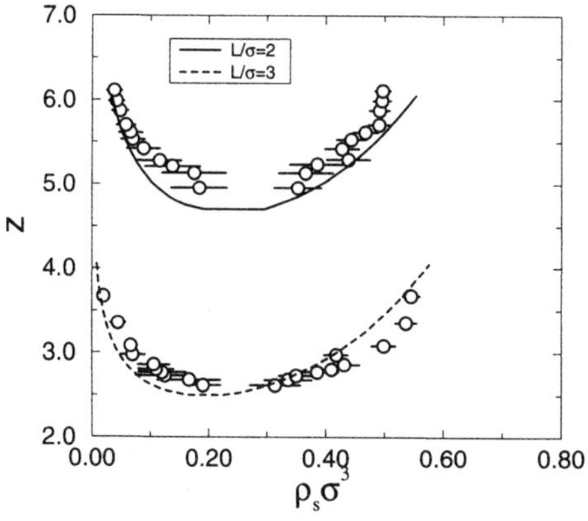

Figure 13.9: Coexistence curves for a mixture of hard spheres and thin rods [377]. The horizontal axis measures the density and the vertical axis the fugacity (= $\exp(\beta\mu)$). L/σ is the ratio of the length of the rods to the diameter of the hard spheres.

stress once again that, in this system, only hard-core interactions between the particles exist. Therefore this demixing is driven by entropy alone.

13.6 Other Ensembles

13.6.1 Grand-Canonical Ensemble

In Chapter 5, we introduced the grand-canonical ensemble in the context of simulations of systems in open contact with a reservoir. An essential ingredient of Monte Carlo simulations in this ensemble is the random insertion or removal of particles. Clearly, such simulations will be efficient only if there is a reasonable acceptance probability of particle-insertion moves. In particular for polyatomic molecules, this is usually a problem. Let us consider the system mentioned in Example 2, a grand-canonical ensemble simulation of the adsorption of molecules in the pores of a microporous material such as a zeolite. For single atoms, the probability that we find an arbitrary position that does not overlap with one of the atoms in the zeolite lattice is on the order 1 in 10^3. For dimers, we have to find two positions that do not overlap, and if we assume that these positions are independent, the probability of success will be 1 in 10^6. Clearly, for the long-chain molecules, the probability of a successful insertion is so low that to obtain a reasonable number of accepted insertions, the number of attempts needs to be prohibitively large. In the present section, we demonstrate how configurational-bias Monte Carlo technique can be used in the grand-canonical ensemble to make the exchange step of chain molecules more probable.

Algorithm

As in the general scheme of the configurational-bias Monte Carlo technique for off-lattice systems, we divide the potential energy of a given conformation into a bonded potential energy ($\mathcal{U}^{\text{bond}}$), which includes the local intramolecular interactions, and an *external* potential energy ($\mathcal{U}^{\text{ext}}$), which includes the intermolecular interactions and the nonbonded intramolecular interactions (see section 13.2.3). A chain that has only bonded interactions is defined as an ideal chain. Let us now consider the Monte Carlo trial moves for the insertion and removal of particles.

Particle Insertion To insert a particle into the system, we use the following steps:

1. For the first monomer, a random position is selected, and the energy of this monomer is calculated. This energy is denoted by $u_1^{\text{ext}}(n)$ and we define $w_1^{\text{ext}}(n) = k \exp[-\beta u_1^{\text{ext}}(n)]$ (as before, the factor k is introduced only to simplify the subsequent notation).
2. For the following monomers, a set of k trial positions is generated. We denote these positions by $\{\mathbf{b}\}_k = (\mathbf{b}_1, \mathbf{b}_2, \cdots, \mathbf{b}_k)$. This set of trial orientations is generated using the bonded part of the potential, which results in the following distribution for the ith monomer:

$$p_i^{\text{bond}}(\mathbf{b})d\mathbf{b} = C \exp[-\beta u_i^{\text{bond}}(\mathbf{b})]d\mathbf{b} \tag{13.6.1}$$

with

$$C^{-1} \equiv \int d\mathbf{b} \exp[-\beta u_i^{\text{bond}}(\mathbf{b})]. \tag{13.6.2}$$

Note that the way the trial orientations are generated depends on the type of monomer being added (see section 13.3). For each of these trial positions the external energy, $u_i^{\text{ext}}(\mathbf{b}_j)$, is calculated, and one of these positions is selected with a probability

$$p_i^{\text{ext}}(\mathbf{b}_n) = \frac{\exp[-\beta u_i^{\text{ext}}(\mathbf{b}_n)]}{w_i^{\text{ext}}(n)}, \tag{13.6.3}$$

in which

$$w_i^{\text{ext}}(n) = \sum_{j=1}^{k} \exp[-\beta u_i^{\text{ext}}(\mathbf{b}_j)].$$

3. Step 2 is repeated until the entire alkane of length ℓ has been grown, and the normalized Rosenbluth factor can be calculated:

$$\mathcal{W}^{\text{ext}}(n) \equiv \frac{W^{\text{ext}}(n)}{k^{\ell}} = \prod_{i=1}^{\ell} \frac{w_i^{\text{ext}}(n)}{k}. \tag{13.6.4}$$

4. The new molecule is accepted with a probability

$$\text{acc}(N \to N+1) = \min\left(1, \frac{q(T)\exp(\beta\mu^B)V}{(N+1)} \mathcal{W}^{\text{ext}}(n)\right), \tag{13.6.5}$$

where μ^B is the chemical potential of a reservoir consisting of *ideal chain* molecules and $q(T)$ is the kinetic contribution to the molecular partition function (for atoms, $q(T) = 1/\Lambda^3$).

Particle Removal To remove a particle from the system, we use the following algorithm:

1. A particle, say, o, is selected at random, the energy of the first monomer is calculated and is denoted by $u_1^{\text{ext}}(o)$, and we determine $w_1^{\text{ext}}(o) = k\exp[-\beta u_1^{\text{ext}}(o)]$.
2. For the following segments of the chain, the external energy $u_i^{\text{ext}}(o)$ is calculated and a set of $k-1$ trial orientations is generated with a probability given by equation (13.6.1). Using this set of orientations and the actual position, we calculate for monomer i:

$$w_i^{\text{ext}}(o) = \exp[-\beta u_i^{\text{ext}}(o)] + \sum_{j=2}^{k} \exp[-\beta u_i^{\text{ext}}(\mathbf{b}_j)].$$

3. After step 2 is repeated for all ℓ monomers and we compute for the entire molecule:

$$\mathcal{W}^{\text{ext}}(o) \equiv \frac{W^{\text{ext}}(o)}{k^{\ell}} = \prod_{i=1}^{M} \frac{w_i^{\text{ext}}(o)}{k}. \tag{13.6.6}$$

4. The selected molecule is removed with a probability

$$\text{acc}(N \to N-1) = \min\left(1, \frac{N}{q(T)V\exp(\beta\mu^B)} \frac{1}{\mathcal{W}^{\text{ext}}(o)}\right). \tag{13.6.7}$$

We have defined μ^B as the chemical potential of a reservoir consisting of ideal chains. It is often convenient to use as a reference state the ideal gas of nonideal chains (i.e., chains that have both bonded and nonbonded intramolecular interactions). This results in a simple, temperature-dependent shift of the chemical potential:

$$\beta\mu^B \equiv \beta\mu_{\text{id.chain}} = \beta\mu_{\text{nonid.chain}} + \ln\left\langle \mathcal{W}^{\text{nonbonded}} \right\rangle, \tag{13.6.8}$$

where $\langle \mathcal{W}^{\text{nonbonded}} \rangle$ is the average Rosenbluth factor due to the nonbonded intramolecular interactions. This Rosenbluth factor has to be determined in a separate simulation of a single chain molecule. For more details about reference states, see Appendix G. In the same appendix, we also discuss the relation between the chemical potential and the imposed pressure (the latter quantity is needed when comparing with real experimental data). To show that the preceding algorithm does indeed yield the correct distribution, we have to demonstrate, as before, that detailed balance is satisfied. As the proof is very similar to those shown before, we will not reproduce it here. For more details, the reader is referred to [304].

Example 19 (Adsorption of Alkanes in Zeolites)
In Example 2 grand-canonical simulations were used to determine the adsorption of methane in the zeolite silicalite. Using the scheme described in the present section, Smit and Maesen computed adsorption isotherms of the longer alkanes [383]. Adsorption isotherms are of interest since they may signal phase transitions, such as capillary condensation or wetting, of the fluid inside the pores [384]. Capillary condensation usually shows up as a step or rapid variation in the adsorption isotherm. It is often accompanied by hysteresis, but not always; for instance, experiments on flat substrates [385] found evidence for steps in the adsorption isotherm without noticeable hysteresis.

Since the pores of most zeolites are of molecular dimensions, adsorbed alkane molecules behave like a one-dimensional fluid. In a true one-dimensional system, phase transitions are not expected to occur. To the extent that zeolites behave as a one-dimensional medium, one therefore might expect that the adsorption isotherms of alkanes in zeolites exhibit no steps. If steps occur, they are usually attributed to capillary condensation in the exterior secondary pore system formed by the space between different crystals. For silicalite, adsorption isotherms have been determined for various n-alkanes, and, indeed, for the short-chain alkanes (methane–pentane) the isotherms exhibit no steps. The same holds for decane. For hexane and heptane, however, steplike features are observed (for experimental details, see [383]).

In the simulations of Smit and Maesen [383] the alkane molecules are modeled with a united atom model; that is, CH_3 and CH_2 groups are considered as single interaction centers [386]. The zeolite is modeled as a rigid

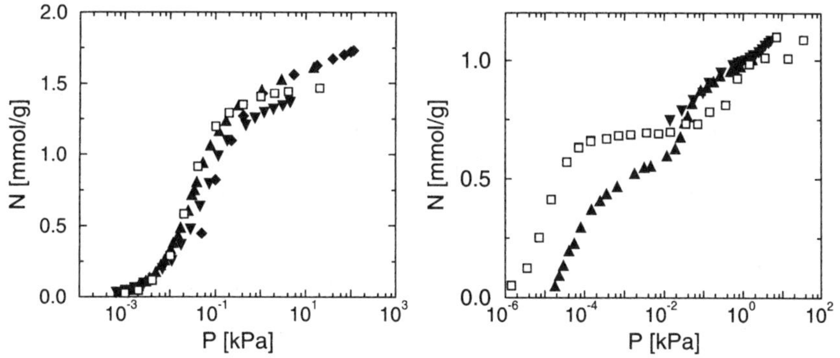

Figure 13.10: Adsorption isotherms of butane (left) and heptane (right); the closed symbols are experimental data and the open symbols the results from simulations at T = 298 K.

crystal and the zeolite-alkane interactions are assumed to be dominated by the interaction with the oxygen atoms and are described by a Lennard-Jones potential.

Figure 13.10 compares the simulated adsorption isotherms of various alkanes in silicalite with experimental data. For butane, a smooth isotherm is observed and the agreement between experiments and simulation is good. For hexane and heptane, the agreement is good at high pressures but at low pressures deviations indicate that the zeolite-alkane model may need to be refined. It is interesting to note that, for heptane, both the experiments and the simulations show a step at approximately half the loading. Since the simulations are performed on a perfect single crystal, this behavior must be due to a transition of the fluid inside the pores and cannot be attributed to the secondary pore system.

Silicalite has two types of channels, straight and zigzag, which are connected via intersections. It so happens that the length of a hexane molecule is on the order of the length of the period of the zigzag channel. The simulations show that, at low chemical potential, the hexane molecules move freely in these channels and the molecules will spend part of their time at the intersections. If a fraction of the intersections is occupied, other molecules cannot reside in the straight channels at the same time. At high pressures, almost all hexane molecules fit exactly into the zigzag channel. They no longer move freely and keep their noses and tails out of the intersection. In such a configuration the entire straight channel can now be tightly packed with hexane molecules. This may explain the plateau in the adsorption isotherm; to fill the entire zeolite structure neatly, the hexane molecules located in zigzag channels first have to be "frozen" in these channels. This "freezing" of the positions of the hexane molecules implies a loss of entropy and therefore will occur only if the pressure (or chemical potential) is sufficiently high to compensate for this loss. This also makes it clear why we do not observe a step for molecules shorter or longer than hexane or heptane. If the molecules are longer, they will always be partly in the intersection and nothing can be gained by a collective freezing in the zigzag channels. If the molecules are shorter than one period of the zigzag channel, a single molecule will not occupy an entire period and a second molecule will enter, which results in a different type of packing. The interesting aspect is that after the simulations were published this observation has been confirmed by experiments [387].

Also the adsorption behavior of mixtures of hydrocarbons has many surprising effects [361, 388].

13.6.2 Gibbs Ensemble Simulations

In Chapter 8, the Gibbs ensemble technique was introduced as an efficient tool for simulating vapor-liquid phase equilibria. One of the Monte Carlo steps in the Gibbs ensemble technique is the transfer of molecules between the liquid phase and gas phase. For long-chain molecules, this step, if carried out completely randomly, results in a prohibitively low acceptance of particle exchanges. Therefore, the Gibbs ensemble technique used to be limited to systems containing atoms or small molecules. However, by combining the Gibbs ensemble method with configurational-bias Monte Carlo, the method can be made to work for much longer chain molecules.

Algorithm

Let us consider a continuum system with strong intramolecular interactions. In section 13.2.3 it is shown that for such a system it is convenient to separate the potential energy into two contributions: the bonded intramolecular energy ($\mathcal{U}^{\text{bond}}$) and the "external" energy ($\mathcal{U}^{\text{ext}}$) that contains the intermolecular interactions and the nonbonded intramolecular interactions. As in the original implementation of the Gibbs ensemble, we attempt to exchange a molecule between the two boxes. However, while in section 8.1 the molecules were inserted at random, we now use the following procedure to grow a molecule atom by atom in a randomly selected box. Let us assume this is box 1 with volume V_1. The number of particles in this box is denoted by n_1.

1. The first atom is inserted at a random position, and the (external) energy $u_1^{\text{ext}}(n)$ is calculated together with

$$w_1^{\text{ext}}(n) = k \exp[-\beta u_1^{\text{ext}}(n)]. \tag{13.6.9}$$

2. To insert the next atom i, k trial orientations are generated. The set of k trial orientations is denoted by $\{\mathbf{b}\}_k = \mathbf{b}_1, \mathbf{b}_2, \cdots, \mathbf{b}_k$. These orientations are not generated at random but with a probability that is a function of the bonded part of the intramolecular energy:

$$p_i^{\text{bond}}(\mathbf{b}_n) = C \exp[-\beta u_i^{\text{bond}}(\mathbf{b}_n)]. \tag{13.6.10}$$

   Of each of these trial orientations the external energy is calculated $u_i^{\text{ext}}(\mathbf{b}_j)$ together with the factor

$$w_i^{\text{ext}}(n) = \sum_{j=1}^{k} \exp[-\beta u_i^{\text{ext}}(\mathbf{b}_j)]. \tag{13.6.11}$$

   Out of these k trial positions, we select one with probability

$$p_i^{\text{ext}}(\mathbf{b}_n) = \frac{\exp[-\beta u_i^{\text{ext}}(\mathbf{b}_i)]}{w_i^{\text{ext}}(n)}. \tag{13.6.12}$$

3. Step 2 is repeated $\ell - 1$ times until the entire molecule is grown and the Rosenbluth factor of the molecule can be calculated:

$$\mathcal{W}^{\text{ext}}(n) = \prod_{i=1}^{\ell} w_i^{\text{ext}}(n). \tag{13.6.13}$$

For the other box, we select a molecule at random and we determine its Rosenbluth factor, using the following procedure:

1. A particle is selected at random.
2. The (external) energy of the first atom is determined $u_1^{\text{ext}}(o)$ together with

$$w_1^{\text{ext}}(o) = k \exp[-\beta u_1^{\text{ext}}(o)]. \tag{13.6.14}$$

3. For the next atom, $k - 1$ trial orientations are generated with a probability given by equation (13.6.10). These trial orientations, together with the actual position of atom i ($\mathbf{b}_o$), form the set $\{\mathbf{b}'\}_k$ for which we determine the factor

$$w_i^{\text{ext}}(o) = \exp[-\beta u_i^{\text{ext}}(o)] + \sum_{j=2}^{k} \exp[-\beta u_i^{\text{ext}}(\mathbf{b}'_j)]. \tag{13.6.15}$$

4. Step 2 is repeated $\ell - 1$ times until we have retraced the entire chain and its Rosenbluth factor can be calculated:

$$\mathcal{W}^{\text{ext}}(o) = \prod_{l=1}^{\ell} w_i^{\text{ext}}(o). \tag{13.6.16}$$

We then accept this move with probability

$$\text{acc}(o \to n) = \min\left(1, \frac{V_1(N - n_1)}{(V - V_1)(n_1 + 1)} \frac{W^{\text{ext}}(n)}{W^{\text{ext}}(o)}\right). \qquad (13.6.17)$$

The proof of the validity of this algorithm, again, is very similar to those shown earlier in this chapter. We therefore refer the interested reader to [356, 389, 390]. The combination of the Gibbs ensemble technique with the configurational-bias Monte Carlo method has been used to determine the vapor-liquid coexistence curve of chains of Lennard-Jones beads [356, 389] and alkanes [386, 390–392]. In Example 20, an application of this method is described.

Example 20 (Critical Properties of Alkanes)
Alkanes are thermally unstable above approximately 650 K, which makes experimental determination of the critical point of alkanes longer than decane (C_{10}) extremely difficult. The longer alkanes, however, are present in mixtures of practical importance for the petrochemical industry. In these mixtures, the number of components can be so large that it is not practical to determine all phase diagrams experimentally. One therefore has to rely on predictions made by equations of state. The parameters of these equations of state are directly related to the critical properties of the pure components. Therefore, the critical properties of the long-chain alkanes are used in the design of petrochemical processes, even if they are unstable close to the critical point [393]. Unfortunately, experimental data are scarce and contradictory, and one has to rely on semi-empirical methods to estimate the critical properties [393].

Siepmann *et al.* [386, 390] have used the combination of the Gibbs ensemble technique and configurational-bias Monte Carlo to simulate vapor-liquid equilibria of the n-alkanes under conditions where experiments are not (yet) feasible. Phase diagrams are very sensitive to the choice of interaction potentials. Most available models for alkanes have been obtained by fitting simulation data to experimental properties of the liquid under standard conditions. In Figure 13.11 the vapor-liquid curve of octane as predicted by some of these models is compared with experimental data. This figure shows that the models of [394, 395], which give nearly identical liquid properties, yield estimates of the critical temperature of octane that differ by 100 K. Siepmann *et al.* [386, 390] used these vapor-liquid equilibrium data to improve the existing models.

In Figure 13.12 the critical temperatures and densities as predicted by the model of Siepmann *et al.* are plotted versus the carbon number. The simulations reproduce the experimental critical points very well. There is considerable disagreement, however, between the various experimental estimates of the critical densities. Much of our current knowledge of the critical

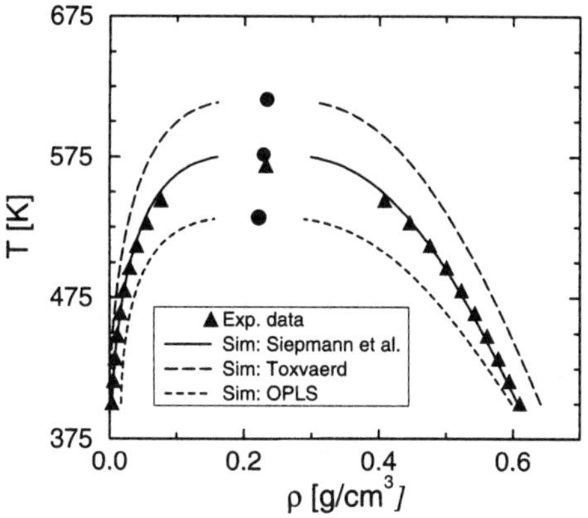

Figure 13.11: Vapor-liquid curve of octane: comparison of Gibbs ensemble simulations using the so-called OPLS model of Jorgensen and co-workers [394], the model of Toxvaerd [395], and the model of Siepmann *et al.* [386, 390].

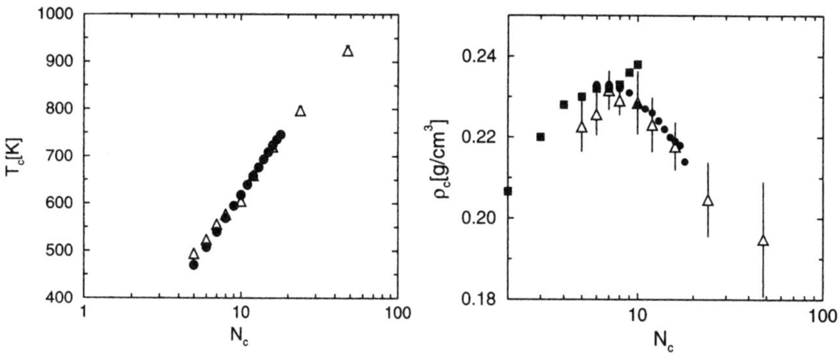

Figure 13.12: Critical temperature T_c (left) and density ρ_c (right) as a function of carbon number N_c. The open symbols are the simulation data and the closed symbols are experimental data.

properties of the higher alkanes is based on extrapolations of fits of the experimental data up to C_8. The most commonly used extrapolations assume that the critical density is a a monotonically increasing function of the carbon number, approaching a limiting value for the very long alkanes [393, 396]. In contrast to these expectations, the experimental data of Anselme *et al.* [397] indicate that the critical density has a maximum for C_8 and then decreases monotonically. The data of Steele (as reported in [396]), however, do not give any evidence for such a maximum (see Figure 13.12). The simulations indicate the same trend as that observed by Anselme *et al.* In this context, it is interesting to note that Mooij *et al.* [356], Sheng *et al.* [398], and Escobedo and de Pablo [399] used Monte Carlo simulations to study the vapor-liquid curve of a polymeric bead-spring model for various chain lengths. These studies also show a decrease of the critical density as a function of chain length. Such a decrease of the critical density with chain length is a general feature of long-chain molecules, as was already pointed out by Flory.

The Gibbs ensemble technique makes it possible to compute efficiently the liquid-vapor coexistence curve of realistic models for molecular fluids. This makes it possible to optimize the parameters of the model to yield an accurate description of the entire coexistence curve, rather than of a single state point. It is likely, but not inevitable, that a model that describes the phase behavior correctly, will also yield reasonable estimates of other properties, such as viscosity or diffusivity. Mondello and Grest have shown that this is indeed true for the diffusion coefficient of linear hydrocarbons [400, 401], while Cochran, Cummings, and co-workers [402, 403] found the same for the viscosity. The hydrocarbon model that was used in these studies had been optimized to reproduce experimental vapor-liquid coexistence data [386, 390]. Improved force fields have since been proposed for linear alkanes [358, 404, 405], branched alkanes [363], alkenes [406, 407], alkylbenzenes [406], and alcohols [408, 409].

13.7 Recoil Growth

To find numerical schemes that are more efficient than conformational-bias Monte Carlo (CBMC), we should first understand why CBMC works better than a scheme that employs random trial moves. Suppose that we have a system with hard-core interactions and the probability of successfully inserting a monomer is a. If we assume that the insertion of an m-mer is equivalent to inserting m independent monomers, then the probability of a successful random insertion of an n-mer is

$$p_m^{\text{random}} \approx a^m.$$

For a dense system, $a \ll 1$, and therefore random insertion only works for very short chains. With the CBMC scheme we try k trial orientations and our growing scheme fails if *all* of the k trial orientations result in an overlap. The probability that we grow a chain successfully is therefore

$$p_m^{\text{CBMC}} \approx a\left[1-(1-a)^k\right]^{m-1} = ab^{m-1}.$$

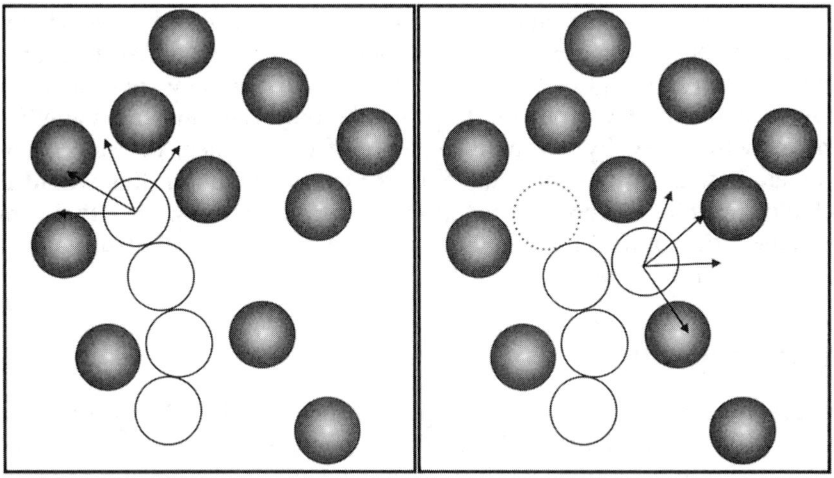

Figure 13.13: The conformational-bias Monte Carlo scheme fails if the molecule is trapped in a dead alley (left); irrespective of the number of trial orientations the CBMC scheme will never generate an acceptable conformation. In the recoil growth scheme (right) the algorithm "recoils" back to a previous monomer and attempts to regrow from there.

This crude estimate suggests that by increasing k, the number of trial orientations, we can make b arbitrarily close to 1 and hence obtain a reasonable insertion probability for any chain length and at any density. In practice, simply increasing k will not solve the problem. First of all, there is a practical limitation: increasing k increases the computational cost. More importantly, the assumption that the probability of a successful insertion of a monomer is equal and independent for each trial position is not correct. For instance, if we have grown into a "dead alley" where there is simply no space for an additional monomer (see Figure 13.13), then no matter how often we try, the insertion will not be accepted. At high densities such dead alleys are the main reason the CBMC method becomes inefficient. This suggests that we need a computational scheme that allows us to escape from these dead alleys.

The recoil growth (RG) scheme is a dynamic Monte Carlo algorithm that was developed with the dead-alley problem in mind [410, 411]. The algorithm is related to earlier static MC schemes due to Meirovitch [412] and Alexandrowicz and Wilding [413]. The basic strategy of the method is that it allows us to escape from a trap by "recoiling back" a few monomers and retrying the growth process using another trial orientation. In contrast, the CBMC scheme looks only one step ahead. Once a trial orientation has been selected, we cannot "deselect" it, even if it turns out to lead into a dead alley. The recoil growth scheme looks several monomers ahead to see whether traps are to be expected before a monomer is irrevocably added to the trial conformation (see Figure 13.13). In this way we can alleviate (but not remove) the dead-alley problem. In principle, one could also do something similar with CBMC by adding a sequence of l monomers per step. However, as there are k possible directions for every monomer, this would involve computing k^l energies per group. Even though many of these trial monomers do not lead to acceptable conformations, we would still have to compute all interaction energies.

13.7.1 Algorithm

In order to explain the practical implementation of the RG algorithm, let us first consider a totally impractical, but conceptually simple scheme that will turn out to have the same net effect. Consider a chain of l monomers. We place the first monomer at a random position. Next, we generate k trial positions for the second monomer. From each of these trial positions, we generate k trial positions for the third monomer. At this stage, we have generated k^2 "trimer" chains. We continue in the same manner until we have grown k^{l-1} chains of length l. Obviously, most of the conformations thus generated have a vanishing Boltzmann factor and are, therefore, irrelevant. However, some may have a reasonable Boltzmann weight and it is these conformations that we should like to find. To simplify this search, we introduce a concept that plays an important role in the RG algorithm: we shall distinguish between trial directions that are "open" and thoase that are "closed." To decide whether a given trial direction, say b, for monomer j is open, we compute its energy $u_j(b)$. The probability[8] that trial position b is open is given by

$$p_j^{open}(b) = \min(1, \exp[-\beta u_j(b)]), \tag{13.7.1}$$

For hard-core interactions, the decision whether a trial direction is open or closed is unambiguous, as $p_j^{open}(b)$ is either zero or one. For continuous interactions we compare $p_j^{open}(b)$ with a random number between 0 and 1. If the random number is less than $p_j^{open}(b)$, the direction is open; otherwise it is closed. We now have a tree with k^{l-1} branches but many of these branches are "dead," in the sense that they emerge from a "closed" monomer. Clearly, there is little point in exploring the remainder of a branch if it does not correspond to an "open" direction. This is where the RG algorithm comes in. Rather than generating a host of useless conformations, it generates them "on the fly." In addition, the algorithm uses a cheap test to check if a given

[8]This probability can be chosen in many alternative ways and may be used to optimize a simulation. However, the particular choice discussed here appears to work well for Lennard-Jones and hard-core potentials.

branch will "die" within a specified number of steps (this number is denoted by l_{max}). The algorithm then randomly chooses among the available open branches. As we have only looked a distance l_{max} ahead, it may still happen that we have picked a branch that is doomed. But the probability of ending up in such a dead alley is much lower than that in the CBMC scheme.

In practice, the recoil growth algorithm consists of two steps. The first step is to grow a new chain conformation using only "open" directions. The next step is to compute the weights of the new and the old conformations.

The following steps are involved in the generation of a new conformation:

1. The first monomer of a chain is placed at a random position. The energy of this monomer is calculated (u_1). The probability that this position is "open" is given by equation (13.7.1). If the position is closed we cannot continue growing the chain and we reject the trial conformation. If the first position is open, we continue with the next step.

2. A trial position b_{i+1} for monomer $i + 1$ is generated starting from monomer i. We compute the energy of this trial monomer $u_{i+1}(b)$ and, using equation (13.7.1), we decide whether this position is open or closed. If this direction is closed, we try another trial position, up to a maximum[9] of k trial orientations. As soon as we find an open position we continue with step 3.

   If not a single open trial position is found, we make a recoil step. The chain retracts one step to monomer $i - 1$ (if this monomer exists), and the unused directions (if any) from step 2, for $i - 1$, are explored. If all directions at level $i - 1$ are exhausted, we attempt to recoil to $i - 2$. The chain is allowed to recoil a total of l_{max} steps, i.e., down to length $i - l_{max} + 1$.

   If, at the maximum recoil length, all trial directions are closed, the trial conformation is discarded.

3. We have now found an "open" trial position for monomer $i+1$. At this point monomer $i - l_{max}$ is permanently added in the new conformation; i.e., a recoil step will not reach this monomer anymore.

4. Steps 2 and 3 are repeated until the entire chain has been grown.

In the naive version of the algorithm sketched above, we can consider the above steps as a procedure for searching for an open branch on the existing tree. However, the RG procedure does this by generating the absolute minimum of trial directions compatible with the chosen recoil distance l_{max}.

[9]The maximum number of trial orientation should be chosen in advance—and may depend on the index i— but is otherwise arbitrary.

Once we have successfully generated a trial conformation, we have to decide on its acceptance. To this end, we have to compute the weights, $W(n)$ and $W(o)$, of the new and the old conformations, respectively. This part of the algorithm is more expensive. However, we only carry it out once we know for sure that we have successfully generated a trial conformation. In contrast, in CBMC it may happen that we spend much of our time computing the weight factor for a conformation that terminates in a dead alley.

In the RG scheme, the following algorithm is used to compute the weight of the new conformation:

1. Consider that we are at monomer position i (initially, of course, $i = 1$). In the previous stage of the algorithm, we have already found that at least one trial direction is available (namely, the one that is included in our new conformation). In addition, we may have found that a certain number of directions (say k_c) are closed—these are the ones that we tried but that died within l_{max} steps. We still have to test the remaining $k_{rest} \equiv k - 1 - k_c$ directions. We randomly generate k_{rest} trial positions for monomer $i+1$ and use the recoil growth algorithm to test whether at least one "feeler" of length l_{max} can be grown in this direction grown (unless $i+l_{max} > l$; in that case we only continue until we have reached the end of the chain). Note that, again, we do *not* explore all possible branches. We only check if there is at least *one* open branch of length l_{max} in each of the k_{rest} directions. If this is the case, we call that direction "available." We denote the total number of available directions (including the one that corresponds to the direction that we had found in the first stage of the algorithm) by m_i. In the next section we shall derive that monomer i contributes a factor $w_i(n)$ to the weight of the chain, where $w_i(n)$ is given by

$$w_i(n) = \frac{m_i(n)}{p_i^{open}(n)}$$

and $p_i^{open}(n)$ is given by equation (13.7.1).

2. Repeat the previous step for all i from 1 to $l-1$. The expression for the partial weight of the final monomer seems ambiguous, as $m_l(n)$ is not defined. An easy (and correct) solution is to choose $m_l(n) = 1$.

3. Next compute the weight for the entire chain:

$$W(n) = \prod_{i=1}^{\ell} w_i(n) = \prod_{i=1}^{\ell} \frac{m_i(n)}{p_i^{open}(n)}. \qquad (13.7.2)$$

For the calculation of the weight of the old conformation, we use almost the same procedure. The difference is that, for the old conformation, we have to generate $k-1$ additional directions for every monomer i. The weight is again related to the total number of directions that start from monomer i and that are "available," i.e., that contain at least one open feeler of length l_{max}:

$$W(o) = \prod_{i=1}^{\ell} w_i(o) = \prod_{i=1}^{\ell} \frac{m_i(o)}{p_i^{open}(o)}.$$

Finally, the new conformation is accepted with a probability:

$$acc(o \to n) = \min(1, \exp[-\beta U(n)]W(n)/\exp[-\beta U(o)]W(o)), \qquad (13.7.3)$$

where $U(n)$ and $U(o)$ are the energies of the new and old conformations, respectively. In the next section, we demonstrate that this scheme generates a Boltzmann distribution of conformations.

13.7.2 Justification of the Method

The best way to arrive at the acceptance rule for the recoil growth scheme is to pretend that we actually carry out the naive brute-force calculation where we first generate the tree of *all* k^{l-1} trial conformations. We denote this tree by T_n and the *a priori* probability for generating this tree by $P_T(T_n)$. *Next* we test which links are "open" or "closed." The decision whether a monomer direction is "open" or "closed" is made on the basis of the probabilities equation (13.7.1) and we denote the probability that we have a particular set O_n of "open" monomers (and all others "closed") by $P_O(O_n|T_n)$. Let us note the number of "open" monomers in this set by $N(O_n)$ and the number of "closed" monomers by $N(C_n)$. It is easy to see that the probability of generating this particular set is given by

$$P_O(O_n|T_n) = \prod_{j=1}^{N(O_n)} p_j^{open}(b) \prod_{k=1}^{N(C_n)} (1 - p_k^{open}(b)).$$

Finally we try to select one completely open conformation by randomly selecting, at every step, one of the "available" trial directions, i.e., a direction that is connected to (at least) one feeler that does not "die" within l_{max} steps. At every step, there are $m_i(n)$ such directions. Hence the probability of selecting a given direction is simply $1/m_i(n)$ and the total probability that a specific conformation will be selected on the previously generated tree of all possible conformations is

$$P_S(n|O_n) = \prod_{i=1}^{l-1} \frac{1}{m_i(n)}$$

if all m_i are nonzero, and

$$P_S(n|O_n) = 0$$

otherwise. The fact that the algorithm leaves out many redundant steps (*viz.* generating the "doomed" branches or checking if there is more than one open feeler in a given direction) is irrelevant for the acceptance rule. The overall probability that we generate a trial conformation n on the set O_n, $P_S(n|O_n)$, in a tree T_n is

$$P_T(T_n) \times P_O(O_n|T_n) \times P_S(n|O_n). \tag{13.7.4}$$

In order to compute acceptance probability of a trial move, we should consider the reverse move where the *old* configuration is generated. By analogy to the forward case, this probability is given by

$$P_T(T_o) \times P_O(O_o|T_o) \times P_S(o|O_o). \tag{13.7.5}$$

We wish our MC scheme to obey detailed balance. However, just as in the CBMC case, it is easier to impose the stronger condition of *super-detailed balance*. This implies that, in the forward move, we also should consider the probability of generating a complete tree of possible conformations around the "old" conformation and the probability that a subset of all monomers on this tree is "open." We denote the probability of generating this tree by $P'_T(T'_o)$, where the prime indicates that this is the probability of generating all branches of the old tree, except the already existing old conformation. Clearly

$$P_T(T_o) = P'_T(T'_o) \times P_{gen}(o), \tag{13.7.6}$$

where $P_{gen}(o)$ denotes the probability of generating the old conformation. As in the CBMC scheme, we can include strong intramolecular interactions in the generation of these trial monomers (see section 13.3). $P_{gen}(o)$ will then be of the form (see section 13.3)

$$P_{gen}(o) = \left(\prod_{i=1}^{l} p_i^{bond}(b_o) \right). \tag{13.7.7}$$

Similarly, we have to consider the probability $P'_O(O'_o|T'_o)$ that a set O'_o on this tree is "open." Again, the prime indicates that we should not include the old conformation itself. Again, it is easy to see that

$$P_O(O_o|T_o) = \left(\prod_{i=1}^{l} p_i^{open}(b_o) \right) P'_O(O'_o|T'_o). \tag{13.7.8}$$

The *a priori* probability of generating a trial move from o to n is then given by

$$\alpha(o \to n|T_n, O_n, T_o, O_o) = P_T(T_n) \times P_O(O_n|T_n) \times P_S(n|O_n) \times P'_T(T'_o) \times P'_O(O'_o|T'_o). \tag{13.7.9}$$

For the reverse move $n \to o$, we can derive a similar expression:

$$\alpha(n \to o|T_n, O_n, T_o, O_o) = P_T(T_o) \times P_O(O_o|T_o) \times P_G(o|O_o) \times P'_T(T'_n) \times P'_O(O'_n|T'_n). \tag{13.7.10}$$

In these equations we have used the notation $(o \to n|T_n, O_n, T_o, O_o)$ to indicate that we consider a transition from o to n (or vice versa) for a *given* set of "embedding" conformations. Clearly, there are many different trees and sets of open orientations that include the same conformations n and o.

Our super-detailed balance condition now becomes

$$\mathcal{N}(o) \times \alpha(o \to n|T_n, O_n, T_o, O_o)\mathrm{acc}(o \to n|T_n, O_n, T_o, O_o) = \mathcal{N}(n) \times \alpha(n \to o|T_n, O_n, T_o, O_o)\mathrm{acc}(n \to o|T_n, O_n, T_o, O_o). \tag{13.7.11}$$

All terms in this equation are known, except the acceptance probabilities. We now derive an expression for the ratio $\mathrm{acc}(o \to n|T_n, O_n, T_o, O_o)/\mathrm{acc}(n \to o|T_n, O_n, T_o, O_o)$. To this end, we insert equations (13.7.6) and (13.7.8) (and the corresponding expressions for $P'_T(T'_n)$ and $P'_O(O'_n|T'_n)$) into our super-detailed balance condition equation (13.7.11). This leads to a huge simplification as there is a complete cancellation of all probabilities for generating "open" or "closed" monomers that do *not* belong to the new (or the old) conformation. What remains is

$$\mathcal{N}(o) \times P_{gen}(n) \left(\prod_{i=1}^{l} \frac{p_i^{open}(b_n)}{m_i(n)} \right) \mathrm{acc}(o \to n|T_n, O_n, T_o, O_o) = \mathcal{N}(n) \times P_{gen}(o) \left(\prod_{i=1}^{l} \frac{p_i^{open}(b_o)}{m_i(o)} \right) \mathrm{acc}(n \to o|T_n, O_n, T_o, O_o). \tag{13.7.12}$$

In order to simplify the notation, we shall assume that the trial directions are uniformly distributed, i.e., see equation (13.7.7), p^{bond}= constant. From equation (13.7.7) it then follows that $P_{gen}(n)$ and $P_{gen}(o)$ are identical constants.

Our expression for the ratio of the acceptance probabilities then becomes

$$\frac{\text{acc}(o \to n)}{\text{acc}(n \to o)} = \frac{\mathcal{N}(n)\prod_{i=1}^{\ell} p_i^{\text{open}}(o)/m_i(o)}{\mathcal{N}(o)\prod_{i=1}^{\ell} p_i^{\text{open}}(n)/m_i(n)}, \tag{13.7.13}$$

where we have dropped the indices $T_n, O_n, \ldots$. Using the definitions of $W(n)$ and $W(o)$ (equation (13.7.2) and below),

$$\frac{\text{acc}(o \to n)}{\text{acc}(n \to o)} = \frac{\mathcal{N}(n)W(n)}{\mathcal{N}(o)W(o)}. \tag{13.7.14}$$

This is precisely the acceptance rule given by equation (13.7.3). This concludes our "derivation" of the recoil growth scheme. The obvious question is: how well does it perform? A comparison between CBMC and the RG algorithm was made by Consta *et al.* [411], who studied the behavior of Lennard-Jones chains in solution. The simulations showed that for relatively short chains ($\ell = 10$) at a density of $\rho = 0.2$, the recoil growth scheme was a factor of 1.5 faster than CBMC. For higher densities $\rho = 0.4$ and longer chains $N = 40$ the gain could be as large as a factor 25. This illustrates the fact that the recoil scheme is still efficient, under conditions where CBMC is likely to fail. For still higher densities or still longer chains, the relative advantage of RG would be even larger. However, the bad news is that, under those conditions, *both* schemes become very inefficient.

While the recoil growth scheme is a powerful alternative to CBMC, the RG strategy is not very useful for computing chemical potentials (see [411]). More efficient schemes for computing the chemical potential are the recursive sampling scheme and the Pruning-Enriched Rosenbluth Method (PERM) (see Chapter 11).

Case Study 20 (Recoil Growth Simulation of Lennard-Jones Chains)
To illustrate the recoil growth (RG) method, we make a comparison between this method and conformational-bias Monte Carlo (CBMC). Consider 20 Lennard-Jones chains of length 15. The monomer density is $\rho = 0.3$ at temperature $T = 6.0$. Two bonded monomers have a constant bond length of 1.0, while three successive particles have a constant bond angle of 2.0 radians.

In Figure 13.14 the distribution of the end-to-end vector, R_E, of the chain is plotted. In this figure we compare the results from a CBMC and a RG. Since both methods generate a Boltzmann distribution of conformations, the results are identical (as they should be).

For this speficic example, we have compared the efficiency, η, of the two methods. The efficiency is defined as the number of accepted trial moves per amount of CPU time. For CBMC we see that the efficiency increases as we increase k, the number of trial orientations, from 1 to 4. From 4 to 8 the

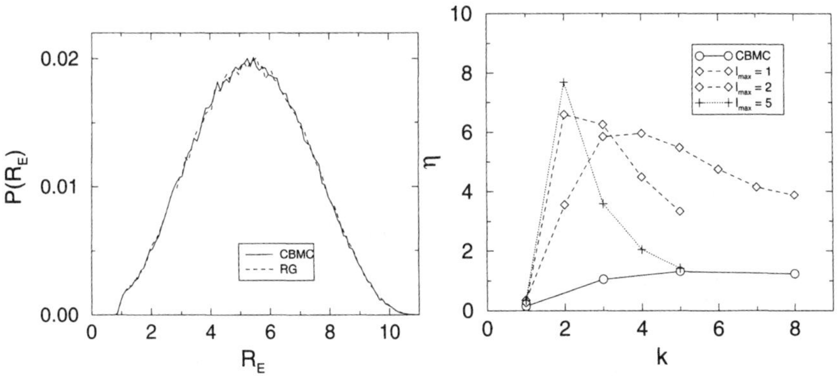

Figure 13.14: Comparison of configurational-bias Monte Carlo (CBMC) with recoil growth for the simulation of Lennard-Jones chains of length 15. The left figure gives the distribution of the end-to-end distance (R_E). In the right figure the efficiency (η) is a function of the number of trial directions (k) for different recoil lengths (l_{max}) as well as for CBMC.

efficiency is more or less constant, and above 8 a decrease in the efficiency is observed.

In the RG scheme we have two parameters to optimize: the number of trial orientations k and the recoil length l_{max}. If we use only one trial orientation, recoiling is impossible, since there are no other trial orientations. If we use a recoil length of 1, the optimum number of trial orientations is 4 and for larger recoil lengths the optimum is reached with less trial orientations. Interestingly, the global optimum is 2 trial orientations and a recoil length of 3–5. In this regime, the increase in CPU time associated with a larger recoil length is compensated by a higher acceptance. In the present study, optimal RG was a factor 8 more efficient than optimal CBMC.

13.8 Questions and Exercises

Question 20 (Biased CBMC) *In a configurational-bias Monte Carlo simulation, trial positions are selected with a probability that is proportional to the Boltzmann factor of each trial segment. However, in principle one can use another probability function [382] to select a trial segment. Suppose that the probability of selecting a trial segment* i *is proportional to*

$$p_i \propto \exp\left[-\beta^\star u_i\right]$$

in which $\beta^\star \neq \beta$.

1. *Derive the correct acceptance/rejection rule for this situation.*
2. *Derive an expression for the excess chemical potential when this modified CBMC method is used to generate configurations of test particles.*
3. *What happens if* $\beta^\star \to \infty$ *and if* $\beta^\star \to 0$?

Exercise 15 (CBMC of a Single Chain)
In this exercise, we will look at the properties of a single chain molecule. We will compare various sampling schemes. Suppose that we have a chain molecule of length n in which there are the following interactions between beads:

- Two successive beads have a fixed bond length l. We will use $l = 1$.
- Three successive beads have a bond-bending interaction

$$U = \frac{1}{2} k_t (\theta - \theta_0)^2,$$

  in which θ is the bond angle, θ_0 is the equilibrium bond angle, and k_t is a constant. We will use $\theta_0 = 2.0$ rad ($\approx 114.6°$) and $k_t = 2.0$.
- Every pair of beads that is separated by more than two bonds has a soft repulsive interaction

$$U(r) = \begin{cases} \frac{A(r - r_{rcut})^2}{r_{cut}^2} & r \leq r_{cut} \\ 0 & r > r_{cut} \end{cases},$$

  in which r_{cut} is the cutoff radius (we will use $r_{cut} = 1.0$ and $A > 0$).

An interesting property of a chain molecule is the distribution of the end-to-end distance, which is the distance between the first and the last segments of the chain. There are several possible schemes for studying this property:

Dynamic Schemes In a dynamic scheme, a Markov chain of states is generated. The average of a property B is the average of B over the elements of the Markov chain

$$\langle B \rangle \approx \frac{\sum_{i=1}^{i=N} B_i}{N}.$$

In the limit $N \to \infty$ this expression becomes exact. Every new configuration is accepted or rejected using an acceptance citerion:

- When unbiased chains are generated:

$$\text{acc}(o \to n) = \min(1, \exp\{-\beta [U(n) - U(o)]\}),$$

  in which U is the total energy (soft repulsion and bond bending) of a chain.

- When configurational-bias Monte Carlo is used:

$$\mathrm{acc}\,(o \rightarrow n) = \min\left(1, \frac{W(n)}{W(o)}\right),$$

in which

$$W = \frac{\prod_{i=2}^{i=n} \sum_{j=1}^{j=k} \exp\left[-\beta U(i,j)\right]}{k^{n-1}}.$$

In this equation, k is the number of trial positions and $U(i,j)$ is the energy of the jth trial position of the ith chain segment. The term $U(i,j)$ does not contain the bond-bending potential, because that potential has already been used to generate the trial positions.

Static Schemes In a static scheme, all configurations are generated independently. To obtain a canonical average, every configuration is weighted with a factor R

$$\langle B \rangle = \frac{\sum_{i=1}^{i=N} B_i \times R_i}{\sum_{i=1}^{i=N} R_i}.$$

For R_i we can write:

- When random chains are generated:

$$R_i = \exp\left[-\beta U_i\right].$$

  Here, U_i is the total energy of the chain.

- When CBMC is used:

$$R_i = W. \tag{13.8.1}$$

1. On the book's website you can find a program for calculating chain properties using these four methods. However, some additional programming has to be done in the file *grow.f*, which is a subroutine for growing a new chain using either CBMC or random insertion.
2. Compare the end-to-end distance distributions of the four methods. Which method has the best performance? Investigate how the efficiency of CBMC depends on the number of trial directions (k).
3. Investigate the influence of chain length on the end-to-end distance distribution. For which chain lengths do the four methods start to fail?
4. For high temperatures (and for low k_t and A), the end-to-end distance distribution looks like the distribution of a nonself-avoiding random walk. This means that the chain segments are randomly oriented

and the segments are allowed to overlap. For the mean square end-to-end distance, we can write

$$\frac{\langle r^2 \rangle}{l^2} = \left\langle \left(\sum_{i=1}^{i=n} x_i^2 \right) + \left(\sum_{i=1}^{i=n} y_i^2 \right) + \left(\sum_{i=1}^{i=n} z_i^2 \right) \right\rangle,$$

in which (x_i, y_i, z_i) are the projections of each segment on the (x, y, z) axes

$$\begin{aligned} x_i &= \sin(\theta_i)\cos(\phi_i) \\ y_i &= \sin(\theta_i)\sin(\phi_i) \\ z_i &= \cos(\theta_i). \end{aligned}$$

This set of equations can be reduced to

$$\frac{\langle r^2 \rangle}{l^2} = n. \tag{13.8.2}$$

- Derive equation (13.8.2). Hint: the following equations will be very useful:

  $$\begin{aligned} \cos^2(\theta_i) + \sin^2(\theta_i) &= 1 \\ \cos(\theta_i - \theta_j) &= \cos(\theta_i)\cos(\theta_j) + \sin(\theta_i)\sin(\theta_j) \\ \langle \cos(\theta_i - \theta_j) \rangle &= 0. \end{aligned}$$

  The last equation holds because $\theta_i - \theta_j$ is uniformly distributed.
- Modify the program in such a way that $\langle r^2 \rangle$ is calculated for a nonself-avoiding random walk. Compare your results with the analytical solution.
- Does

  $$\langle r^2 \rangle \propto n$$

  hold for a chain with a potential energy function described in this exercise? Investigate the influence of A on the end-to-end distance distribution.

Exercise 16 (CBMC of a Simple System)
Consider a system with three coordinates (x_1, x_2, x_3) and phase space density

$$\rho(x_1, x_2, x_3) = \exp\left[-\left(x_1^2 + x_2^2 + x_3^2\right)\right] = \exp\left[-r^2\right].$$

We wish to calculate the average $\langle r^2 \rangle$,

$$\langle r^2 \rangle = \frac{\iiint dx_1 dx_2 dx_3 r^2 \rho}{\iiint dx_1 dx_2 dx_3 \rho},$$

using the CBMC algorithm of Falcioni and Deem [414]:

- Generate k sets of new coordinates $B_1, \cdots, B_k$ by adding random vectors to the old configuration (A_1).
- Select one set (i) with a probability proportional to its Boltzmann factor,

$$p_i = \exp\left[-r_{B_i}\right].$$

The corresponding Rosenbluth factor weight is

$$W(n) = \sum_{j=1}^{j=k} \exp\left[-r_{B_j}^2\right].$$

- Starting from the selected configuration B_i, $k-1$ configurations (A_2, $\cdots, A_k$) are generated by adding a uniform vector to B_i. A_1 is the old configuration. This leads to the Rosenbluth factor of the old configuration

$$W(o) = \sum_{j=1}^{j=k} \exp\left[-r_{A_j}^2\right].$$

- The new configuration B_i is accepted with a probability

$$\text{acc}(o \to n) = \min\left(1, \frac{W(n)}{W(o)}\right).$$

1. Make a sketch of the configurations $A_1, \cdots, A_k$ and $B_1, \cdots, B_k$. Show that for $k = 1$, this algorithm reduces to the standard Metropolis algorithm for particle displacements.
2. Prove that this algorithm obeys detailed balance.
3. Calculate $\langle r^2 \rangle$ analytically using

$$\int_0^\infty \exp\left[-a^2 x^2\right] dx = \frac{\sqrt{\pi}}{2|a|}.$$

4. On the book's website you can find a computer program for this CBMC sampling scheme. This program, however, has to be completed by you (see the file *cbmc.f*)! Make sure that your estimate of $\langle r^2 \rangle$ is independent of the number of trial directions (k).
5. What happens with the fraction of accepted trial moves when the number of trial directions (k) is increased? Make a plot of the fraction of accepted trial moves as a function of k for various maximum displacements. Explain your results.
6. Why is this CBMC method useful when the system is initially far from equilibrium?

Chapter 14

Accelerating Monte Carlo Sampling

In this chapter we discuss various advanced Monte Carlo techniques. These methods illustrate the flexibility the Monte Carlo method gives us to develop novel ways to sample a system efficiently.

14.1 Parallel Tempering

The method of parallel tempering [415–417] is a Monte Carlo scheme that has been derived to achieve good sampling of systems that have a free-energy landscape with many local minima. It resembles the technique of simulating annealing [418] and is related to several other schemes such as the extended-ensemble method [415], simulated tempering [416, 419], and J-walking [420]. Closely related schemes had been proposed by Nezbeda and Kolafa [421] in the context of the Widom particle insertion method, while Shing and Azadipour and later Vega *et al.* proposed a (mildly approximate) version of this method for grand-canonical Monte Carlo [422] and Molecular Dynamics [423] simulations.

In parallel tempering we consider n-systems. In each of these systems we perform a simulation in the canonical (NVT) ensemble, but each system is in a different thermodynamic state. Usually, but not necessarily, these states differ in temperature. In what follows we assume—for convenience—that this is the case. Systems with a sufficiently high temperature pass over the potential. The low-temperature systems, on the other hand, mainly probe the local free energy minima. The idea of parallel tempering is to include MC trial moves that attempt to "swap" systems that belong to different thermodynamic states, e.g., to swap a high-temperature system with a low-temperature system. If the temperature difference between the two systems is very large, such a swap has a very low probability of being accepted (see below). This is very similar to particle displacement in ordinary Monte Carlo, were if one uses a very large maximum displacement a move has a very low probability of being accepted. The solution to this problem is to use many small steps. In parallel tempering we use intermediate temperatures in a similar way. Instead of making attempts to swap between a low and a high temperature, we swap between ensembles with a small temperature difference.

Let us call the temperature of system i, T_i, and the n systems are numbered according to an increasing temperature scale, $T_1 < T_2 < \cdots < T_n$. We define an extended ensemble that is the combination of all n subsystems. The partition function of this extended ensemble is the product of all individual NVT_i ensembles:

$$Q_{\text{extended}} = \prod_{i=1}^{n} Q_{NVT_i} = \prod_{i=1}^{n} \frac{1}{\Lambda_i^{3N} N!} \int d\mathbf{r}_i^N \exp[-\beta_i \mathcal{U}(\mathbf{r}_i^N)],$$

where $\mathbf{r}_i^N$ denotes the positions of the N particles in system i. For the canonical ensemble the temperature dependence of Λ does not play a role since the total number of particles is constant. However, for extensions to grand-canonical ensemble simulation this factor should be taken into account. To sample this extended ensemble it is in principle sufficient to perform NVT simulations of all individual ensembles. But we can also introduce a Monte Carlo move, which consists of a swap between two ensembles. The acceptance rule of a swap between ensembles i and j follows from the condition of detailed balance. If we denote the configuration of system i by $\mathbf{i} = \mathbf{r}_i^N$, this condition reads

$$\begin{aligned} &\mathcal{N}(\mathbf{i}, \beta_i)\mathcal{N}(\mathbf{j}, \beta_j) \times \alpha\left[(\mathbf{i}, \beta_i), (\mathbf{j}, \beta_j) \to (\mathbf{j}, \beta_i), (\mathbf{i}, \beta_j)\right] \\ &\qquad \times \text{acc}\left[(\mathbf{i}, \beta_i), (\mathbf{j}, \beta_j) \to (\mathbf{j}, \beta_i), (\mathbf{i}, \beta_j)\right] \\ &= \mathcal{N}(\mathbf{i}, \beta_j)\mathcal{N}(\mathbf{j}, \beta_i) \times \alpha\left[(\mathbf{i}, \beta_j), (\mathbf{j}, \beta_i) \to (\mathbf{i}, \beta_i), (\mathbf{j}, \beta_j)\right] \\ &\qquad \times \text{acc}\left[(\mathbf{i}, \beta_j), (\mathbf{j}, \beta_i) \to (\mathbf{i}, \beta_i), (\mathbf{j}, \beta_j)\right]. \end{aligned}$$

If we perform the simulations in such a way that the *a priori* probability, α, of performing a particular swap move is equal for all conditions, we obtain as acceptance rules

$$\begin{aligned} &\frac{\text{acc}[(\mathbf{i}, \beta_i), (\mathbf{j}, \beta_j) \to (\mathbf{j}, \beta_i), (\mathbf{i}, \beta_j)]}{\text{acc}[(\mathbf{i}, \beta_j), (\mathbf{j}, \beta_i) \to (\mathbf{i}, \beta_i), (\mathbf{j}, \beta_j)]} \\ &\quad = \frac{\exp\left[-\beta_i U(\mathbf{j}) - \beta_j U(\mathbf{i})\right]}{\exp\left[-\beta_i U(\mathbf{i}) - \beta_j U(\mathbf{j})\right]} \\ &\quad = \exp\{(\beta_j - \beta_i)\left[U(\mathbf{i}) - U(\mathbf{j})\right]\}. \end{aligned} \qquad (14.1.1)$$

It is important to note that, as we know the total energy of a configuration anyway, these swap moves are very inexpensive since they do not involve additional calculations.

It should be stressed that the swap moves do not disturb the Boltzmann distribution corresponding to a particular ensemble. Therefore one can determine ensemble averages from every individual ensemble just as we do for an ordinary Monte Carlo simulations. This is an important improvement over simulating annealing, since in simulating annealing ensemble averages are not defined. Parallel tempering is a true equilibrium Monte Carlo scheme.

To see the power of parallel tempering consider a single particle moving in one dimension. This particle moves in the external potential shown in Figure 14.1. Depending on the temperature this particle will be able to cross some of the barriers. A typical probability distribution of an NVT_i simulation of three ensembles is shown in Figure 14.1. These simulations are done using particle displacements only. As one can expect, only at very high temperatures is the entire phase space sampled. At low temperatures the molecule is trapped in a pocket of phase space.

Let us next consider the same three systems and use the extended ensemble and allow for swaps between the ensembles. These swaps are illustrated in Figure 14.2. In this figure we can trace the history of the various ensembles and the figure illustrates that exchanges between low and high temperatures occur via the intermediate temperatures. If we now determine the probability distribution at the various temperatures we see a striking improvement for the low temperatures. For this system the entire phase space is being explored. As can be expected, for the high temperatures there are no differences. In Case Study 21 the details of these simulations are discussed. Parallel tempering is not restricted to the NVT ensemble and can be extended to other ensembles as well.

Case Study 21 (Parallel Tempering of a Single Particle)
To demonstrate the parallel tempering technique, we have simulated a single particle on a simple one-dimensional potential:

$$U(x) = \begin{cases} \infty & x < -2 \\ 1 \times (1 + \sin(2\pi x)) & -2 \leq x \leq -1.25 \\ 2 \times (1 + \sin(2\pi x)) & -1.25 \leq x \leq -0.25 \\ 3 \times (1 + \sin(2\pi x)) & -0.25 \leq x \leq 0.75. \\ 4 \times (1 + \sin(2\pi x)) & 0.75 \leq x \leq 1.75 \\ 5 \times (1 + \sin(2\pi x)) & 1.75 \leq x \leq 2 \\ \infty & x > 2 \end{cases}$$

This function in plotted in Figure 14.1 (left). This potential is caricature of a glassy material with energy barriers of varying height. This potential is

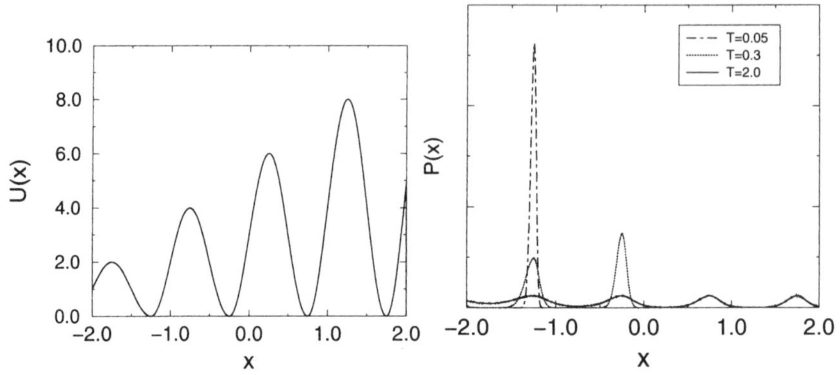

Figure 14.1: (left) Potential energy ($U(x)$) as a function of the position x. (right) Probability ($P(x)$) of finding a particle at position x for various temperatures (T) as obtained from ordinary Monte Carlo simulations. The lower-temperature systems are not able to cross the highest barrier.

constructed in such a way that the equilibrium density at the very minimum of the potential is very nearly constant.

Clearly, if we perform an ordinary NVT Monte Carlo simulation a particle can only pass a barrier if the temperature is sufficiently high. This is illustrated in Figure 14.1 (right) in which we have plotted the probability distributions $P(x)$ for $T = 0.05$, $T = 0.3$, and $T = 2.0$ as obtained from a Monte Carlo simulation. Apparently, at $T = 0.05$ and $T = 0.3$ the particle is not able to cross all energy barriers, while for $T = 2.0$ the whole system is sampled.

We now apply the parallel tempering scheme to see whether the sampling can be improved. In this Monte Carlo algorithm, we have performed two types of trial moves:

1. Particle displacement. The particle was given a uniformly distributed random displacement between -0.1 and 0.1. This displacement is accepted with the conventional acceptance rule for a simulation in the NVT ensemble

$$\text{acc}(o \to n) = \min\{1, \exp[-\beta(U(n) - U(o))]\}.$$

2. Hamiltonian swapping. The Hamiltonians of two randomly selected neighboring temperatures (i and j) were swapped. This trial move is accepted with a probability given by equation (14.1.1)

$$\text{acc}(o \to n) = \min(1, \exp[(\beta_i - \beta_j) \times (U_j - U_i)]).$$

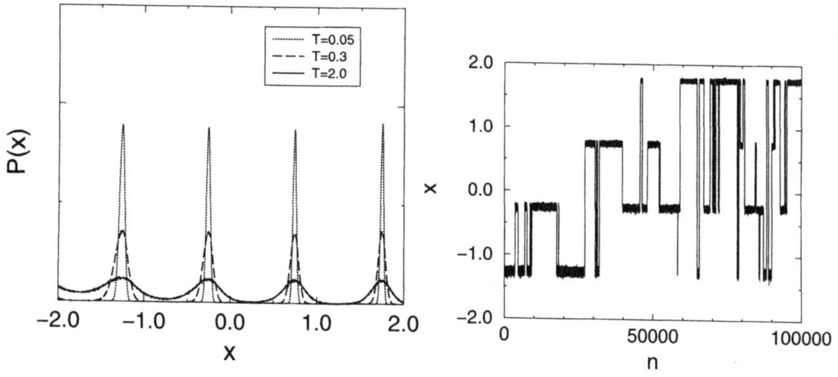

Figure 14.2: (left) Probability (P (x)) of finding a particle at position x for various temperatures (T) using parallel tempering. (right) Position (x) as a function of the number of Monte Carlo trial moves (n) for T = 0.05. For graphical purposes, we have used 0.05% swap moves here.

During the simulation, it was first decided which system to consider. Next, it was decided whether to do a displacement or a Hamiltonian swap move. Following Falcioni and Deem [414], we have used 10% swap moves and 90% particle displacements.

The results of the parallel tempering simulations are shown in Figure 14.2. We find a dramatic improvement in the sampling. We now have a constant equal peak height of the particle to be in any of the wells. The effect of the swapping moves for the system at the lowest temperature is illustrated in Figure 14.2 (left). These swapping moves help the particle to cross barriers during the simulation.

Example 21 (Zeolite Structure Solution)
Falcioni and Deem [414] used simulating tempering as a method to find zeolite structures. Zeolites are inorganic crystals with a known chemical structure. The interesting aspect of these materials is that there are about 118 different known zeolite structures. But many more zeolite structures are theoretically possible. If one makes a new zeolite, one often has a powder and it is much more time consuming, if possible at all, to make a sufficiently large crystal from which the crystal structure can be determined.

The problem that Falcioni and Deem were interested in was finding the zeolite structure that corresponds best to an experimental powder diffraction pattern. Experimentally, it is well known that in zeolites the angles between three successive T-sites (positions of the Si atoms in an all silica zeolite) are clustered around 109.5°. Falcioni and Deem assumed that any new zeolite structure should have angles with a similar distribution. In a similar way rules were deduced for the average bond distance and other properties of the structures. These rules were translated into a pseudo-Hamiltonian; this Hamiltonian has a high value for those crystal structures that do not resemble the characteristics of a zeolite and a low value for those crystals that have zeolite-like characteristics. An additional term was added to this structural pseudo-Hamiltonian that indicated the difference between the experimental and the calculated X-ray powder pattern of the new crystal structure. Falcioni and Deem then assumed that the unknown zeolite structure corresponds to the structure that minimizes this combined Hamiltonian. Various Monte Carlo rules were invented to change the crystal structure, and a new structure was accepted or rejected by using this pseudo-Hamiltonian. Since this pseudo-Hamiltonian has many local minima, parallel tempering was used to find the optimal structure.

To test their method Falcioni and Deem used the information of 32 zeolite structures to generate the pseudo-Hamiltonian. This information was sufficient to solve the structure of the remainder of all known zeolite structures.

Thus far, we have considered examples of the parallel tempering method in which the systems that are being simulated in parallel differ only in a single control variable. It is, of course, possible to use more than one variable. For example, Yan and de Pablo [424] performed a parallel tempering study of a number of systems that differed both in temperature and in chemical potential. But there is no need to limit the choice of control variable to the usual "thermodynamic" ones. For example, one may perform parallel simulations of systems that differ in the parameters characterizing the intermolecular potential. An example is the parallel tempering simulation that Yan and de Pablo [425] performed on a number of polymer systems that differed in the polymer chain length. Bunker and Dünweg [426] applied parallel tempering using systems consisting of polymers that have different excluded volume interactions. In Examples 22 and 23 we discuss these applications.

Another application is to combine parallel tempering with multihistogram umbrella sampling, to map out a free energy landscape (see, e.g., [427]). As the multihistogram technique requires several simulations anyway, the sampling efficiency can be increased at no extra cost by integrating these calculations with the parallel tempering scheme.

Example 22 (Parallel Tempering and Phase Equilibria)
One way to compute the coexistence curve of a Lennard-Jones fluid is to use grand-canonical Monte Carlo. During such a simulation the density probability function is monitored. At coexistence two peaks in this distribution should appear: one corresponding to the gas phase and the other to the liquid phase. However, in practice, such a simulation is limited to conditions close to the critical point. The reason is that one needs to sample both the liquid and the gas phases from a single simulation. Only close to the critical

point are the density fluctuations sufficiently large that one can observe both the liquid and the gas. Wilding and Bruce [100, 428] used this method to accurately locate the critical point of a Lennard-Jones fluid using finite-size scaling techniques.

Far below the critical point, however, the probability of observing a density fluctuation that would spontaneously take the systems from the liquid to the gas phase or vice versa is so low that the method would become impractical. Yan and de Pablo [424] circumvented this problem by using parallel tempering to ensure that both the liquid and the vapor phases are sampled in a single simulation. The idea is that fluctuations between the liquid and the gas phases do occur regularly close to the critical point. The parallel tempering scheme then ensures that these fluctuations are "transmitted" to temperatures well below the critical point. Yan and de Pablo call this method hyper-parallel tempering and simulated 18 different systems with temperatures between the triple point and the critical point and chemical potentials ranging from $\beta\mu = -6,97$ at the triple point to $-\beta\mu = -2.74$ at the critical temperature.

Suppose that we have a system i, with temperature T_i, excess chemical potential μ_i^{ex}, energy U_i, and number of particle N_i, and a system j with T_j, μ_j^{ex}, U_j, and N_j. The probability of finding these configurations is

$$P(\alpha) \propto \exp(\beta_\alpha \mu_\alpha^{ex} N_\alpha - \beta_\alpha U_\alpha).$$

In a parallel tempering move, we exchange the chemical potential and the temperature of the two systems without changing the configuration of either system. The new configuration of system i is then characterized by i $(U_i, N_i, \beta_j, \mu_j^{ex})$, while for system j we have $(U_j, N_j, \beta_i, \mu_i^{ex})$. Such a trial move is accepted with probability

$$\mathrm{acc}(o \to n) = \min\left(1, \exp\left[(\beta_j \mu_j^{ex} - \beta_i \mu_i^{ex})(N_i - N_j) - (\beta_j - \beta_i)(U_i - U_j)\right]\right).$$

Such parallel tempering simulations generate accurate density probability functions that will contain both the liquid and gas peaks close to coexistence.

The next step is to determine the coexistence properties from these density probability functions. For a given temperature, the value of the chemical potential at coexistence is fixed. In order to observe liquid-vapor coexistence, one should therefore tune the chemical potential such that the density histogram has peaks of equal area around the liquid and the vapor densities. Yan and de Pablo showed that one can avoid performing many simulations to locate the coexistence point in this way. To this end, they used the histogram reweighting technique as developed by Ferrenberg and Swendsen [185,429] (see section 7.3.1).

Suppose we perform a simulation at temperature T and chemical potential μ. The idea of the histogram reweighting technique is that we can use the information of this simulation to estimate the properties at a different temperature T' and chemical potential μ'. In a parallel tempering simulation one should choose the temperatures and chemical potentials of the various systems in such a way that there is a sufficient overlap of the energy and density distributions; otherwise the parallel tempering moves would never be accepted. Such overlap of the different histograms is also required for a successful application of the histogram reweighting technique. Hence parallel tempering as used by Yan and de Pablo is ideally suited to be combined with histogram reweighting (see also ref. [427]).

Yan and de Pablo showed that, provided one has a good set of temperatures and chemical potentials, the combined parallel temperature and histogram reweighting technique is very efficient in determining the coexistence curve of the Lennard-Jones fluid. However, finding a good set of temperatures and chemical potentials may require some trial and error.

Example 23 (Parallel Tempering and Polymers)
In a parallel tempering simulation one can also perform swaps between systems that have slightly different Hamiltonians. In some cases, this can be very useful. Suppose that system i has a temperature T and energy $U_i^{(i)} = \sum_{k>l}^N u^{(i)}(r_{kl})$ and system j a temperature T and energy $U_j^{(j)} = \sum_{k>l}^N u^{(j)}(r_{kl})$, where $u^{(i)}$ and $u^{(j)}$ are *different* potentials. As a parallel tempering move we can swap systems i and j. We take the positions of the particles in system i and recompute the energy using the intermolecular potential of system j; this energy is denoted by $U_i^{(j)}$. In a similar way we compute for system j the energy using the intermolecular potential of system i, $U_j^{(i)}$. The acceptance rule for such a move reads

$$\text{acc}(o \to n) = \min\left(1, \exp\left[-\beta(U_i^{(j)} - U_i^{(i)}) + (U_j^{(i)} + U_j^{(j)})\right]\right).$$

This type of parallel tempering move can be combined with others involving the temperature, pressure, or chemical potential.

Bunker and Dünweg used "Hamiltonian" parallel tempering to simulate a long-chain polymer. The polymer was modeled using a bead-spring model with a purely repulsive Lennard-Jones interaction between the beads:

$$U^{\text{pol}}(r) = \begin{cases} A - Br^2 & r \le r_{PT} \\ 4\epsilon\left[\left(\frac{\sigma}{r}\right)^{12} - \left(\frac{\sigma}{r}\right)^{6}\right] & r_{PT} < r \le 2^{1/6}\sigma, \\ 0 & r > 2^{1/6}\sigma \end{cases}$$

where A and B were chosen such that, for a given value of r_{PT}, the potential and its first derivative were continuous. We are interested in the properties of the model system with $r_{PT} = 0$. The other systems are simply added to facilitate equilibration. For instance, for $r_{PT} = 2^{1/6}\sigma$, the core repulsion vanishes and polymer chains can pass through each other. In their parallel tempering scheme, Bunker and Dünweg simulated a number of systems with different values of r_{PT}. A drawback of this "Hamiltonian" parallel tempering scheme is that most of the simulation time is spent on systems that are *not* of physical interest. After all, we are only interested in the thermodynamic behavior of the system with $r_{PT} = 0$. However, as is discussed in ref. [426], the gain in sampling efficiency due to the use of the parallel tempering scheme is still sufficient to make the scheme competitive.

Yan and de Pablo [425] used a parallel tempering scheme to study phase coexistence in a polymeric system. In their simulations, each of the subsystems contained a tagged chain that has a different chain length. For example, in box i this chain has a length M_i and in box j a length M_j. In the last box, k, this chain has the same chain length as the other chains M_k. This method resembles the expanded ensemble scheme of Escobedo and de Pablo [399], but in the expanded ensemble scheme only one box is simulated that switches between states, while in the implementation of de Pablo all systems are simulated in parallel. To compute the vapor-liquid coexistence curve of their polymers, Yang and de Pablo performed in addition parallel tempering moves in the grand-canonical ensemble in which both the temperature and the chemical potential of each of these subsystems changes (see Example 22). Without the parallel tempering moves in which the chain length is changed, one would have to insert or remove an entire chain at once. With the parallel tempering this insertion or deletion can be done very efficiently in those systems for which the tagged chain is short.

14.2 Hybrid Monte Carlo

In Molecular Dynamics simulations, all particle coordinates are updated simultaneously. In conventional MC simulations, only a few coordinates are changed in a trial move. As as consequence, collective molecular motions are not well represented by Monte Carlo, and this may adversely affect the rate of equilibration. The advantage of Monte Carlo is that, unlike MD, we can carry out unphysical moves. Moreover, in MC the system is not constrained to move on a hypersurface where some Hamiltonian is conserved. The time step in Molecular Dynamics is limited by the need to conserve energy. Clearly, no such constraint applies to Monte Carlo. For this reason, many authors have attempted to combine the natural dynamics of MD with the large jumps in configuration space possible in MC. The book by Allen and Tildesley [19] describes a number of such techniques (force bias MC, Langevin Dynamics, smart Monte Carlo) that basically work by including some or all information about the intermolecular forces in the construction of a collective MC trial move.

A technique that has achieved much attention during the past few years is the hybrid Monte Carlo scheme [430]. The basic idea behind this scheme is that one can use MD to generate Monte Carlo trial moves. At first sight, there is no advantage in doing so. However, the criteria for what constitutes a good Monte Carlo trial move are more tolerant than the specifications of a good Molecular Dynamics time step. In particular, one can take a time step that is too long for MD. Energy will not be conserved in such a trial move. However, as long as one uses an algorithm that is time reversible and area preserving (i.e., that conserves volume in phase space), such collective moves can be used as a Monte Carlo trial move. Fortunately, a systematic way now exists to construct time-reversible, area-preserving MD algorithms, using the multiple-time-step MD scheme of Tuckerman *et al.* [71]. The usual Metropolis algorithm can then be used to decide on the acceptance or rejection of the move (see, e.g., [431, 432]). For every trial move, the particle velocities are chosen at random from a Maxwell distribution. In fact, it is often advantageous to construct a trial move that consists of a sequence of MD steps. The reason is that, due to the randomization of the velocities, the diffusion constant of the system becomes quite low if the velocities are randomized well before the natural decay of the velocity autocorrelation function.

Yet one cannot make the time step for a single hybrid MC move too long, because then the acceptance would become very small. As a consequence, the performance of hybrid MC is not dramatically better than that of the corresponding Molecular Dynamics. Moreover, the acceptance probability of hybrid MC moves of constant length decreases with the system size, because the root-mean-square error in the energy increases with $N^{1/2}$. MD does not suffer from a similar problem. That is to say, the noise in the total energy increases with N, but the stability of the MD algorithm does not deteriorate. Hence, for very large systems, MD will always win. For more normal system sizes, hybrid MC may be advantageous.

It is also interesting to use hybrid MC on models that have an expensive (many-body) potential energy function that may, to a first approximation, be modeled using a cheap (pair) potential. We could then perform a sequence of MD steps, using the cheap potential. At the end of this collective (MD) trial move, we would accept or reject the resulting configuration by applying the Metropolis criterion with the *true* potential energy function. Many variations of this scheme exist. In any event, the hybrid Monte Carlo method is a scheme that requires fine tuning [432].

Forrest and Suter [433] have devised a hybrid MC scheme that samples polymer conformations, using fictitious dynamics of the generalized coordinates. This scheme leads to an improved sampling of the polymer conformations compared to normal MD. The interesting feature of the dynamics used in ref. [433] is that it uses a Hamiltonian that has the same potential energy function as the original polymer model. In contrast, the kinetic part of the Hamiltonian is adjusted to speed up conformational changes.

14.3 Cluster Moves

One of the main problems in the simulation of complex liquids is that the "natural" dynamics of these systems may be quite slow. As a consequence, normal Molecular Dynamics or Monte Carlo methods may not be adequate to achieve a good sampling of configuration space. For this reason, various schemes for speeding up the sampling rate have been proposed. Many of these techniques rely on the fact that, in Monte Carlo simulations, one has the freedom to carry out unphysical moves. To give an example: in the simulation of large (bio)molecules, it is often difficult to explore the different possible conformations of such molecules, because probable conformations are often separated by high free energy barriers. A technique for alleviating this problem is to allow the molecule to explore pathways from one conformation to another that do not *cross* potential energy barriers but bypass them. This is achieved by allowing the molecule to make excursions into a higher "embedding" dimension [434].

14.3.1 Clusters

As explained at the beginning of section 13.2.3, the crucial step in configurational-bias Monte Carlo is that the "bias" in the generation of trial conformations results in an enhanced acceptance of these trials moves. Ideally, we would like to bias the generation of trial moves in such a way that *every* move is always accepted. Surprisingly, this is sometimes possible. Swendsen and Wang [435] (for a review, see [37]) have shown that, at least for certain classes of lattice problems, it is possible to perform cluster moves that have an acceptance probability of 100%. In this context, a "cluster" is a group of particle coordinates (or spins) that are changed collectively in a Monte Carlo move. That is, the interactions within a cluster do not change in a cluster move, but the interaction of the cluster with the remainder of the system may change.

The central idea behind the Swendsen-Wang (SW) scheme and subsequent extensions and modifications is to generate trial configurations with a probability that is proportional to the Boltzmann weight of that configuration. As a result, the subsequent trial moves can be accepted with 100% probability. We use a somewhat simplified derivation of the Swendsen-Wang for cluster moves based, again, on the condition for detailed balance. Consider an "old" configuration (labeled by a superscript o) and a "new" configuration (denoted by a superscript n). Detailed balance is satisfied if the following equality holds:

$$\begin{aligned}
&\mathcal{N}(\text{o})P^{\text{o}}_{\text{Gen}}(\{\text{cluster}\})P^{\{\text{cl}\}}(\text{o} \to \text{n})\text{acc}(\text{o} \to \text{n}) \\
&= \mathcal{N}(\text{n})P^{\text{n}}_{\text{Gen}}(\{\text{cluster}\})P^{\{\text{cl}\}}(\text{n} \to \text{o})\text{acc}(\text{n} \to \text{o}),
\end{aligned} \tag{14.3.1}$$

where $\mathcal{N}(\text{o})$ is the Boltzmann weight of the old configuration, $P^{\text{o}}_{\text{Gen}}(\{\text{cluster}\})$ denotes the probability of generating a specific cluster, starting from the old configuration of the system. The term $P^{\{\text{cl}\}}(\text{o} \to \text{n})$ is the probability of transforming the generated cluster from the old to the new situation. Finally, $\text{acc}(\text{o} \to \text{n})$ is that acceptance probability of a given trial move. We can simplify equation (14.3.1) in two ways. First of all, we require that the *a priori* probability $P^{\{\text{cl}\}}(\text{o} \to \text{n})$ be the same for the forward and reverse moves. Moreover, we wish to impose $P_{\text{acc}} = 1$ for both forward and reverse moves. This may not always be feasible. However, for the simple case that we discuss next, this is indeed possible. The detailed balance equation then becomes

$$\mathcal{N}(\text{o})P^{\text{o}}_{\text{Gen}}(\{\text{cluster}\}) = \mathcal{N}(\text{n})P^{\text{n}}_{\text{Gen}}(\{\text{cluster}\}) \tag{14.3.2}$$

or

$$\frac{P^{\text{o}}_{\text{Gen}}(\{\text{cluster}\})}{P^{\text{n}}_{\text{Gen}}(\{\text{cluster}\})} = \frac{\mathcal{N}(\text{n})}{\mathcal{N}(\text{o})} = \exp(-\beta\Delta\mathcal{U}), \tag{14.3.3}$$

where $\Delta\mathcal{U}$ is the difference in energy between the new and the old configurations. The trick is then to find a recipe for cluster generation that will satisfy equation (14.3.3). To illustrate how this works, consider the Ising model. The extension to many other models is straightforward.

Ising Model

For the construction of the Ising SW algorithm, the dimensionality of the model is irrelevant. Consider a given configuration of the spin system, with N_p spin pairs parallel and N_a spin pairs antiparallel. The total energy of that configuration is

$$\mathcal{U} = (N_a - N_p)J,$$

where J denotes the strength of the nearest-neighbor interaction. The Boltzmann weight of that configuration is

$$\mathcal{N}(\text{o}) = \exp[-\beta J(N_a - N_p)]/Z,$$

where Z is the partition function of the system. In general, Z is unknown, but that is unimportant. We are concerned only that Z is a constant. Next, we construct clusters by creating bonds between spin pairs according to the following recipe:

- If nearest neighbors are *antiparallel*, they are not connected.
- If nearest neighbors are *parallel*, they are connected with probability p and disconnected with probability $(1 - p)$.

Here, it is assumed that J is positive. If J is negative (antiferromagnetic interaction), parallel spins are not connected, while antiparallel spins are connected with a probability p.

In the case that we consider, there are N_p parallel spin pairs. The probability that n_c of these are connected and $n_b = N_p - n_c$ are "broken" is

$$P^o_{\mathrm{Gen}}(\{\mathrm{cluster}\}) = p^{n_c}(1-p)^{n_b}.$$

Note that this is the probability to connect (or break) a *specified* subset of all links between parallel spins. Once the connected bonds have been selected, we can define the clusters in the system. A cluster is a set of spins that is at least singly connected by bonds. We now choose our subset of clusters to flip. After the cluster flipping, the number of parallel and antiparallel spin pairs will have changed, for example,

$$N_p(n) = N_p(o) + \Delta$$

and (hence)

$$N_a(n) = N_a(o) - \Delta.$$

Therefore, the total energy of the system will have changed by an amount $-2J\Delta$:

$$\mathcal{U}(n) = \mathcal{U}(o) - 2J\Delta.$$

Let us now consider the probability of making the *reverse* move. To do this, we should generate the *same* cluster structure, but now starting from a situation where there are $N_p + \Delta$ parallel spin pairs and $N_a - \Delta$ antiparallel pairs. As before, the bonds between antiparallel pairs are assumed to be broken (this is compatible with the same cluster structure). We also know that the new number of connected bonds, n'_c, must be equal to n_c, because the same number of connected bonds is required to generate the same cluster structure. The difference appears when we consider how many of the bonds between parallel spins in the new configuration should be broken (n'_b). Using

$$N_p(n) = n'_c + n'_b = n_c + n'_b = N_p(o) + \Delta = n_c + n_b + \Delta,$$

we see that

$$n'_b = n_b + \Delta.$$

If we insert this in equation (14.3.3), we obtain

$$\begin{aligned}\frac{P^{o}_{\mathrm{Gen}}(\{\mathrm{cluster}\})}{P^{n}_{\mathrm{Gen}}(\{\mathrm{cluster}\})} &= \\ \frac{p^{n_c}(1-p)^{n_b}}{p^{n_c}(1-p)^{n_b+\Delta}} &= \\ (1-p)^{-\Delta} &= \frac{\mathcal{N}(n)}{\mathcal{N}(o)} = \exp(2\beta J\Delta). \end{aligned} \tag{14.3.4}$$

To satisfy this equation, we must have

$$1-p = \exp(-2\beta J)$$

or

$$p = 1-\exp(-2\beta J),$$

which is the Swendsen-Wang rule.

General Cluster Moves

In general, it is not possible to design clusters such that trial moves are always accepted. However, it is often convenient to perform clustering to enhance the acceptance of trial moves. For instance, in molecular systems with very strong short-range attractions, trial moves that pull apart two neighboring particles are very likely to be rejected. It is preferable therefore to include trial moves that attempt to displace the tightly bound particles as a single cluster. To do this, we have to specify a rule for generating clusters. Let us assume that we have such a rule that tells us that particles i and j belong to a single cluster with probability $p(i,j)$ and are disconnected with a probability $1-p(i,j)$. Here, $p(i,j)$ depends on the state (relative distance, orientation, spin, etc.) of particles i and j. Moreover, we require that $p(i,j)$ be unchanged in a cluster move if both i and j belong to the cluster, and also if neither particle belongs to the cluster. For instance, $p(i,j)$ could depend on the current distance of i and j only. If we denote the potential energy of the old (new) configuration by $\mathcal{U}_0$ ($\mathcal{U}_1$), the detailed balance condition requires that

$$\begin{aligned}&\exp(-\beta\mathcal{U}_o)\prod_{kl}[1-p^{f}(k,l)]\mathrm{acc}(o\to n)\\ &= \exp(-\beta\mathcal{U}_n)\prod_{kl}[1-p^{r}(k,l)]\mathrm{acc}(n\to o),\end{aligned} \tag{14.3.5}$$

where k denotes a particle *in* the cluster and l a particle outside it. The superscripts f and r denote forward and reverse moves. In writing equation (14.3.5), we have assumed that the probability of forming bonds completely within, or completely outside, the cluster is the same for forward and reverse moves. From equation (14.3.5), we derive an expression for the ratio of the acceptance probabilities:

$$\frac{\mathrm{acc}(o \to n)}{\mathrm{acc}(n \to o)} = \exp\{-\beta[\mathcal{U}(n) - \mathcal{U}(o)]\} \prod_{kl} \frac{1 - p^r(k,l)}{1 - p^f(k,l)}. \tag{14.3.6}$$

Clearly, many choices for p_{kl} are possible. A particularly simple form was chosen by Wu *et al.* [436], who assumed that $p(i,j) = 1$ for r_{ij} less than a critical distance r_c and $p(i,j) = 0$ beyond that distance (see Example 24). Note that the acceptance rule in equation (14.3.5) guarantees that two particles that did not belong to the same cluster in the old configuration will not end up at a distance less than r_c.

Example 24 (Micelle Formation)
Surfactants are amphiphilic molecules that consist of two chemically distinct parts. If these two parts cannot be connected, then one part will preferentially dissolve in another solvent compared to the other part. The most common case is that one part of the molecule is water soluble and the other oil soluble. But, as the two parts of the molecule are connected, the dissolution of surfactant molecules in a pure solvent (say, water) causes frustration, as the oil soluble part of the molecule is dragged along into the water phase. Beyond a certain critical concentration, the molecules resolve this frustration by self-assembling into micelles. Micelles are (often spherical) aggregates of surfactant molecules in which the surfactants are organized in such a way that the hydrophilic heads point toward the water phase and the hydrophobic tails toward the interior of the micelle. It is of considerable interest to study the equilibrium properties of such a micellar solution. However, Molecular Dynamics simulations on model micelles have shown that the micelles move on a time scale that is long compared to the time it takes individual surfactant molecules to move [437]. Using conventional MC, rather than MD, does not improve the situation: it is relatively easy to move a single surfactant, but it takes many displacements of single surfactants to achieve an appreciable rearrangement of the micelles in the system. Yet, to sample the equilibrium properties of micellar solutions, the micelles must be able to move over distances that are long compared to their own diameter, they must be able to exchange surfactant molecules, and they must even be able to merge or break up. As a consequence, standard simulations of micellar self-assembly are very slow.

Wu *et al.* [436] have used cluster moves to speed up the simulation of micellar solutions. They use a cluster MC scheme that makes it possible to displace entire micelles with respect to each other. The specific model for a surfactant solution that Wu *et al.* studied was based on a lattice model proposed by Stillinger [438]. In this model, the description of surfactants is highly simplified: the hydrophilic and hydrophobic groups are considered

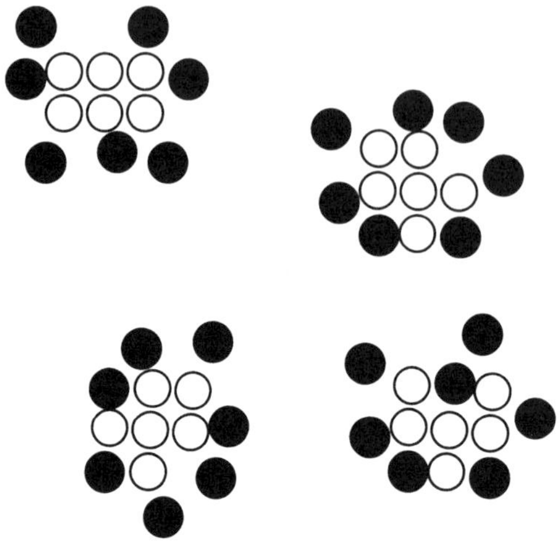

Figure 14.3: Snapshot of a typical configuration of the surfactant model of Wu *et al.* [436], in which the hydrophilic and hydrophobic parts of surfactant molecules are represented by charges (black head and white tail). Under appropriate conditions, the surfactants self-assemble into micelles.

independent (unbonded) particles. The constraint that the head and tail of a surfactant be physically linked is translated into an electrostatic attraction between these groups. The magnitude of these effective charges depends on density and temperature. A typical configuration is shown in Figure 14.3. In such a system it is natural to cluster the molecules in a micelle and subsequently move entire micelles. The cluster criterion used by Wu *et al.* is

$$p(i,j) = \begin{cases} = 1 & \text{if} \quad r_{ij} < r_c \\ = 0 & \text{if} \quad r_{ij} > r_c \end{cases} .$$

Wu *et al.* used $r_c = 1$, which implies that two particles belong to the same cluster if they are on neighboring lattice sites.

In the first step of the algorithm, the clusters are constructed using the preceding criterion. Subsequently, a cluster is selected at random and given a random displacement. It is instructive to consider the case in which we would use the ordinary acceptance rules to move the cluster; that is,

$$\text{acc}(o \to n) = \min[1, \exp(-\beta\Delta U)].$$

Figure 14.4 shows such a cluster move. The first step (top) is the construction of the cluster, followed by a displacement of one of the clusters (middle). If we accept this move with the probability given by the preceding equation, we would violate microscopic reversibility. Since we have moved the clusters in such a way that they touch each other, in the next step, these two clusters would be considered a single cluster. It then will be impossible to separate them to retrieve the initial configuration.

If we use the correct acceptance rule, equation (14.3.6),

$$\mathrm{acc}(o \to n) = \min\left[1, \exp(-\beta\Delta U) \prod_{kl} \frac{1 - p^{\mathrm{new}}(k,l)}{1 - p^{\mathrm{old}}(k,l)}\right],$$

then this move would be rejected because $p^{\mathrm{new}}(k,l) = 1$. Since these cluster moves do not change the configuration of the particles in a cluster and do not change the total number of particles in a cluster, it is important to combine these cluster moves with single particle moves, or use a cluster criterion $p(i,j)$ that allows the number of particles in a cluster to change.

Orkoulas and Panagiotopoulos have used such cluster moves to simulate the vapor-liquid coexistence curve of the restricted primitive model of an ionic fluid [439, 440].

14.3.2 Early Rejection Scheme

One of the differences of simulations of models with continuous interactions compared to those of models with hard-core potentials is the way Monte Carlo moves are optimized. For example, if hard-core interactions are present, one can reject a move as soon as a single overlap is detected. For continuous potentials, all interactions must be computed before a trial move can be accepted or rejected. As a consequence, on average, it is cheaper to perform a trial move of a hard-core particle that results in rejection than in acceptance. This leads to the strange situation where it is appreciably cheaper to perform a MC simulation of a true hard-core model than of a corresponding model that has very steep but continuous repulsive interactions. Intuitively, one would expect that, also for continuous potentials, it should be possible to reject at an early stage those trial moves that are almost certainly "doomed" because they will result in a large increase of the potential energy. In practice this is often achieved by assuming that, beyond a certain critical distance, the continuous potential can be replaced by a hard-core potential. In this section we show that it is not really necessary to make such an assumption. Rather, we can use the idea behind the Swendsen-Wang scheme to formulate criteria that allow us to reject at an early stage "doomed" trial moves for particles with a continuous intermolecular potential [441].

To see how this approach works, consider a trial displacement of particle i to a new position. We now construct "bonds" between this particle and a neighboring particle j with a probability

$$p_{\mathrm{bond}}(i,j) = \max\left[0, 1 - \exp(-\beta\Delta u_{i,j})\right], \qquad (14.3.7)$$

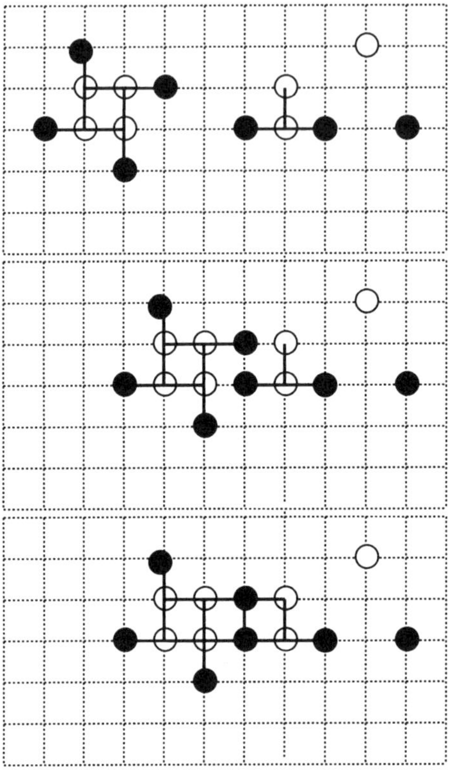

Figure 14.4: Violation of detailed balance in a cluster move; the top figure shows the four clusters in the system; in the middle figure one of the clusters is given a random displacement, which brings this cluster into contact with another cluster; and the bottom figure shows that new configuration has only three clusters if the moves have been accepted.

where $\Delta u_{i,j} = u^n(i,j) - u^o(i,j)$ is change in the interaction energy of particles i and j caused by the trial displacement of particle i. If j is not connected to i, we proceed with the next neighbor k, and so on. But as soon as a bond is found between i and any of its neighbors, we reject the trial move. Only if particle i is not bonded to any of its neighbors do we accept the trial move.

Before discussing the advantages and disadvantages of this scheme, we first demonstrate that this scheme satisfies detailed balance (see section 5.2). The probability that we accept a move from the old to the new position is given by

$$\text{acc}(o \to n) = \prod_{j \neq i} [1 - p_{\text{bond}}(i,j)] = \exp\left[-\beta \sum_{j \neq i}^{+} \Delta u_{i,j}(o \to n)\right],$$

where the summation is over particles j for which the $\Delta u(i,j)$ is positive. For the reverse move $n \to o$, we have

$$\text{acc}(n \to o) = \prod_{j \neq i} [1 - p_{\text{bond}}(i,j)] = \exp\left[-\beta \sum_{j \neq i}^{+} \Delta u_{i,j}(n \to o)\right].$$

The summation is over all particles j for which the reverse move causes an increase in energy. In addition, we can write

$$\Delta u_{i,j}(n \to o) = -\Delta u_{i,j}(o \to n),$$

which gives for the probability of accepting the reverse move $n \to o$

$$\text{acc}(n \to o) = \exp\left[+\beta \sum_{j \neq i}^{-} \Delta u_{i,j}(o \to n)\right],$$

where the summation is over all particle j for which the energy $u_{i,j}(o \to n)$ decreases. Substitution of these two expressions for the acceptance probabilities under the condition of detailed balance (5.1.1), for the ratio of the acceptance probabilities, gives

$$\begin{aligned}
\frac{\text{acc}(o \to n)}{\text{acc}(n \to o)} &= \frac{\exp\left[-\beta \sum_{j \neq i}^{+} \Delta u_{i,j}(o \to n)\right]}{\exp\left[+\beta \sum_{j \neq i}^{-} \Delta u_{i,j}(o \to n)\right]} \\
&= \exp\left[-\beta \sum_{j \neq i} \Delta u_{i,j}(o \to n)\right] = \frac{\mathcal{N}(n)}{\mathcal{N}(o)},
\end{aligned}$$

which demonstrates that detailed balance is indeed obeyed.

It is interesting to compare this scheme with the original Metropolis algorithm. In the bond formation scheme it is possible that a move is rejected even when the total energy decreases. Hence, although this scheme yields a valid Monte Carlo algorithm, it is *not* equivalent to the Metropolis method. Equation (14.3.7) ensures that, if a trial displacement puts particle i in a very unfavorable position where $\Delta u_{ij} \gg 0$, then it is very likely that a bond will form between these particles and hence the trial move can be rejected. This advantage of the early-rejection scheme, at least partly, is offset because for all pair energies that increase due to the trial displacement, a random number must be drawn to test whether a bond will be formed. If a particle interacts with many other particles, then this scheme requires the calculation of a large number of random numbers. This can be avoided if we use the early rejection scheme only for nearest neighbors (i.e., those that are most likely to contribute to strong repulsive interactions when particle i is moved).[1] If this stage of the trial move is passed, the interaction with the remaining particles is then computed in the usual way, and accepted or rejected with the conventional Metropolis rule. The early-rejection scheme is not limited to single particle moves. In fact, it is probably most useful when applied to complex many-particle moves. For example, in the configurational-bias Monte Carlo scheme (see Chapter 13), one has to grow an entire molecule to calculate its Rosenbluth factor before one can reject or accept such a move. If one of the first segments has been placed at an unfavorable position, such that the new configuration is "doomed," then the early-rejection scheme could be used to avoid having to complete the growth of a new polymer configuration.

[1]Here, too, we must be careful: nearest neighbors are only those particles within a certain distance before and after the trial displacement of particle i.

Chapter 15

Tackling Time-Scale Problems

One might argue that it would be sufficient to limit our discussion of Molecular Dynamics simulations of atomic systems, because, after all, molecules are made up of atoms. Hence, if we know how to simulate the dynamics of nonbonded atoms, we also know how to simulate atoms that belong to a molecule. Although this statement is correct, as long as it is legitimate to ignore the quantum nature of intramolecular motions, it is usually not advisable to employ the same simulation techniques to atoms that belong to a molecule as to free atoms. The reason is that the characteristic time scales associated with intramolecular motions are typically a factor 10–50 shorter than the time over which the translational velocity of a molecule changes appreciably. In a Molecular Dynamics simulation, the time step should be chosen such that it is appreciably shorter than the shortest relevant time scale in the simulation. If we simulate the intramolecular dynamics of molecules explicitly, this implies that our time step should be shorter than the period of the highest-frequency intramolecular vibration. This would make the simulation of molecular substances very expensive. Therefore, techniques for tackling this problem have been developed. Here, we will discuss three approaches: constraints, extended Lagrangians, and multiple-time-step simulations.

Multiple-time-scale Molecular Dynamics [71], is based on the observation that forces associated with a high-frequency intramolecular vibration can be integrated efficiently with a time step different than the time step used for the integration of the intermolecular forces.

An alternative is to treat the bonds (and, sometimes, bond angles) in molecules as rigid. The Molecular Dynamics equations of motion are then solved under the constraint that the rigid bonds and bond angles do not change during our simulation. The motion associated with the remaining degrees of freedom is presumably slower, and hence we can again use a long time step in our simulations. Here, we briefly explain how such constraints are implemented in a Molecular Dynamics simulation.

In addition, we illustrate how extended Lagrangians can be used in "on-the-fly" optimization problems. The first and foremost example of a technique that uses an extended Lagrangian for this purpose is the Car-Parrinello "ab-initio" MD method [442]. We shall not discuss this technique because quantum simulations fall outside the scope of this book. Rather, we shall illustrate the idea of the Car-Parrinello approach to optimization using a purely classical example.

15.1 Constraints

To get a feel for the way in which constrained dynamics works, let us first consider a simple example, namely, a single particle that is constrained to move on a surface (e.g., a sphere) in space, $f(x, y, z) = 0$. The (Lagrangian) equations of motion for the unconstrained particle are (see Appendix A)

$$\frac{\partial}{\partial t}\frac{\partial \mathcal{L}}{\partial \dot{\mathbf{q}}} = \frac{\partial \mathcal{L}}{\partial q}. \qquad (15.1.1)$$

As the Lagrangian, $\mathcal{L}$, is equal to $\mathcal{K}_{\text{kin}} - \mathcal{U}_{\text{pot}}$, the equation of motion for the unconstrained particle is

$$m\ddot{q} = -\frac{\partial \mathcal{U}}{\partial q}.$$

Now, suppose that we have the particle initially on the surface $f(x, y, z) = 0$. Moreover, we impose that, initially, the particle moves tangential to the constraint plane:

$$\dot{f} = \dot{q} \cdot \nabla f = 0.$$

But as the particle moves, its velocity changes such that it is no longer tangent to the constraint surface. To keep the particle on the constraint surface, we now apply a fictitious force (the constraint force) in such a way that the new velocity is again perpendicular to ∇f.

In the general case, the dynamics should satisfy many constraints simultaneously (e.g., many bond lengths). Let us denote the functions describing these constraints by $\sigma_1, \sigma_2, \cdots$. For instance, σ_1 may be a function that is equal to 0 when atoms i and j are at a fixed distance d_{ij}:

$$\sigma_1(\mathbf{r}_i, \mathbf{r}_j) = r_{ij}^2 - d_{ij}^2.$$

We now introduce a new Lagrangian $\mathcal{L}'$ that contains all the constraints:

$$\mathcal{L}' = \mathcal{L} - \sum_{\alpha} \lambda_\alpha \sigma_\alpha(\mathbf{r}^N),$$

where α denotes the set of constraints and λ_α denotes a set of (as yet undetermined) Lagrange multipliers. The equations of motion that correspond to this new Lagrangian are

$$\frac{\partial}{\partial t}\frac{\partial \mathcal{L}'}{\partial \dot{q}} = \frac{\partial \mathcal{L}'}{\partial q} \tag{15.1.2}$$

or

$$\begin{aligned} m_i \ddot{q}_i &= -\frac{\partial \mathcal{U}}{\partial q_i} - \sum_\alpha \lambda_\alpha \frac{\partial \sigma_\alpha}{\partial q_i} \\ &\equiv F_i + \sum_\alpha G_i(\alpha). \end{aligned} \tag{15.1.3}$$

The last line of this equation defines the constraint force $\mathbf{G}_\alpha$. To solve for the set λ_α, we require that the second derivatives of all σ_α vanish (our initial conditions were chosen such that the first derivatives vanished):

$$\begin{aligned} \frac{\partial \dot{\sigma}_\alpha}{\partial t} &= \frac{\partial \dot{q}\nabla\sigma_\alpha}{\partial t} \\ &= \ddot{q}\nabla\sigma_\alpha + \dot{q}\dot{q} : \nabla\nabla\sigma_\alpha \\ &= 0. \end{aligned} \tag{15.1.4}$$

Using equation (15.1.3), we can rewrite this equation as

$$\begin{aligned} \frac{\partial \dot{\sigma}_\alpha}{\partial t} &= \sum_i \frac{1}{m_i}\left[F_i + \sum_\beta G_i(\beta)\right] \nabla_i\sigma_\alpha + \sum_{i,j} \dot{q}_i\dot{q}_j\nabla_i\nabla_j\sigma_\alpha \\ &= \sum_i \frac{1}{m_i} F_i\nabla_i\sigma_\alpha - \sum_i \frac{1}{m_i}\sum_\beta \lambda_\beta\nabla_i\sigma_\beta\nabla_i\sigma_\alpha + \sum_{i,j}\dot{q}_i\dot{q}_j\nabla_i\nabla_j\sigma_\alpha \\ &\equiv \mathcal{F}_\alpha - \Lambda_\beta M_{\alpha\beta} + \mathcal{T}_\alpha \\ &= 0. \end{aligned} \tag{15.1.5}$$

In the last line of equation (15.1.5), we have written the equation on the previous line in matrix notation. The formal solution of this equation is

$$\Lambda = \mathbf{M}^{-1}(\mathcal{F} + \mathcal{T}). \tag{15.1.6}$$

This formal solution of the Lagrangian equations of motion in the presence of constraints, unfortunately, is of little practical use. The reason is that, in a simulation, we do not solve differential equations but difference equations. Hence there is little point in going through the (time-consuming) matrix inversion needed for the exact solution of the differential equation, because this procedure does not guarantee that the constraints also will be accurately satisfied in the solution of the difference equation.

Before we proceed, let us consider a simple example of constrained dynamics, namely, a particle moving on the surface of a sphere of radius d. In that case, we can write our constraint function σ as

$$\sigma = \frac{1}{2}\left(r^2 - d^2\right).$$

The constraint force **G** is equal to

$$\mathbf{G} = -\lambda \boldsymbol{\nabla}\sigma = -\lambda\mathbf{r}.$$

To solve for λ, we impose $\ddot{\sigma} = 0$:

$$\begin{aligned} \partial_t \dot{\sigma} &= \partial_t(\dot{\mathbf{r}} \cdot \mathbf{r}) \\ &= (\ddot{\mathbf{r}} \cdot \mathbf{r}) + \dot{r}^2 = 0. \end{aligned} \qquad (15.1.7)$$

The Lagrangian equation of motion is

$$\begin{aligned} \ddot{\mathbf{r}} &= \frac{1}{m}(\mathbf{F} + \mathbf{G}) \\ &= \frac{1}{m}(\mathbf{F} - \lambda\mathbf{r}). \end{aligned}$$

For convenience, we assume that no external forces are acting on the particle ($\mathbf{F} = \mathbf{0}$). Combining equations (15.1.7) and (15.1.8), we obtain

$$-\frac{\lambda}{m} r^2 + \dot{r}^2 = 0. \qquad (15.1.8)$$

Hence

$$\lambda = \frac{m\dot{r}^2}{r^2},$$

and the constraint force **G** is equal to

$$\mathbf{G} = -\lambda\mathbf{r} = -\frac{m\dot{r}^2}{r^2}\mathbf{r}.$$

Recall that, on the surface of a sphere, the velocity $\dot{\mathbf{r}}$ is simply equal to ωr. Hence we can also write the constraint force as

$$\mathbf{G} = -m\omega^2\mathbf{r},$$

which is the well-known expression for the centripetal force.

This simple example will help us to understand what goes wrong when we insert the preceding expression for the constraint force into an MD algorithm, e.g., the Verlet scheme. In the absence of external forces, we would get the following algorithm for a particle on the surface of a sphere:

$$\mathbf{r}(t + \Delta t) = 2\mathbf{r}(t) - \mathbf{r}(t - \Delta t) - \omega^2 \Delta t^2 \mathbf{r}(t).$$

How well is the constraint $r^2 = d^2$ satisfied? To get an impression, we work out the expression for r^2 after one time step. Assuming that the constraint was satisfied at $t = 0$ and at $t = -\Delta t$, we find that, at $t = \Delta t$,

$$\begin{aligned} r^2(t+\Delta t) &= d^2\left\{5 + (\omega\Delta t)^4 - 4(\omega\Delta t)^2 + \cos(\omega\Delta t)[2(\omega\Delta t)^2 - 4]\right\} \\ &\approx d^2\left[1 - \frac{(\omega\Delta t)^4}{6} + \mathcal{O}(\Delta t^6)\right]. \end{aligned}$$

At first sight, this looks reasonable, and the constraint violation is of order $\Delta^4 t$, as is to be expected for the Verlet scheme. However, whereas for center-of-mass motion we do not worry too much about errors of this order in the trajectories, we should worry in the case of constraints. In the case of translational motion, we argued that two trajectories that are initially close but subsequently diverge exponentially still may both be representative of the true trajectories of the particles in the system. However, if we find that, due to small errors in the integration of the equations of motion, the numerical trajectories diverge exponentially from the constraint surface, then we are in deep trouble. The conclusion is that we should not rely on our algorithm to satisfy the constraints (although, in fact, for the particle on a sphere, the Verlet algorithm performs remarkably well). We should construct our algorithm such that the constraints are rigorously obeyed.

The most straightforward solution to this problem is not to fix the Lagrange multiplier λ by the condition that the second derivative of the constraint vanishes but by the condition that the constraint is exactly satisfied after one time step. In the case of the particle on a sphere, this approach would work as follows. The equation for the position at time $t + \Delta t$ in the presence of the constraint force is given by

$$\begin{aligned} \mathbf{r}(t+\Delta t) &= 2\mathbf{r}(t) - \mathbf{r}(t-\Delta t) - \frac{\lambda}{m}\mathbf{r}(t) \\ &= \mathbf{r}_u(t+\Delta t) - \frac{\lambda}{m}\mathbf{r}(t), \end{aligned}$$

where $r_u(t+\Delta t)$ denotes the new position of the particle in the absence of the constraint force. We now impose that the constraint $r^2 = d^2$ is satisfied at $t + \Delta t$:

$$\begin{aligned} d^2 &= \left[\mathbf{r}_u(t+\Delta t) - \frac{\lambda}{m}\mathbf{r}(t)\right]^2 \\ &= r_u^2(t+\Delta t) - \frac{2\lambda}{m}\mathbf{r}(t)\cdot\mathbf{r}_u(t+\Delta t) + \left[\frac{\lambda}{m}r(t)\right]^2. \end{aligned}$$

This expression is a quadratic equation in λ,

$$\left(\frac{\lambda}{m}d\right)^2 - \frac{2\lambda}{m}\mathbf{r}(t)\cdot\mathbf{r}_u(t+\Delta t) + \mathbf{r}_u^2(t+\Delta t) - d^2 = 0,$$

and the solution is

$$\lambda = \frac{\mathbf{r}(t) \cdot \mathbf{r}_u(t+\Delta t) - \sqrt{[\mathbf{r}(t) \cdot \mathbf{r}_u(t+\Delta t)]^2 - d^2[r_u^2(t+\Delta t) - d^2]}}{d^2/m}.$$

For the trivial case of a particle on a spherical surface, this approach clearly will work. However, for a large number of constraints, it will become difficult, or even impossible, to solve the quadratic constraint equations analytically. Why this is so can be seen by considering the form of the Verlet algorithm in the presence of ℓ constraints:

$$\mathbf{r}_i^{\text{constrained}}(t+\Delta t) = \mathbf{r}_i^{\text{unconstrained}}(t) - \frac{\Delta t^2}{m_i} \sum_{k=1}^{\ell} \lambda_k \boldsymbol{\nabla}_i \sigma_k(t). \tag{15.1.9}$$

If we satisfy the constraints at time $t + \Delta t$, then $\sigma_k^c(t+\Delta t) = 0$. But if the system would move along the unconstrained trajectory, the constraints would not be satisfied at $t + \Delta t$. We assume that we can perform a Taylor expansion of the constraints:

$$\begin{aligned} \sigma_k^c(t+\Delta t) &= \sigma_k^u(t+\Delta t) + \sum_{i=1}^{N} \left(\frac{\partial \sigma_k}{\partial \mathbf{r}_i}\right)_{r_i^u(t+\Delta t)} \cdot [\mathbf{r}_i^c(t+\Delta t) - \mathbf{r}_i^u(t+\Delta t)] \\ &\quad + \mathcal{O}(\Delta t^4). \end{aligned} \tag{15.1.10}$$

If we insert equation (15.1.9) for $r_i^u - r_i^c$ in equation (15.1.10), we get

$$\sigma_k^u(t+\Delta t) = \sum_{i=1}^{N} \frac{\Delta t^2}{m_i} \sum_{k'=1}^{\ell} \boldsymbol{\nabla}_i \sigma_k(t+\Delta t) \boldsymbol{\nabla}_i \sigma_{k'}(t) \lambda_{k'}. \tag{15.1.11}$$

We note that equation (15.1.11) has the structure of a matrix equation:

$$\boldsymbol{\sigma}^u(t+\Delta t) = \Delta t^2 \mathbf{M} \boldsymbol{\Lambda}. \tag{15.1.12}$$

By inverting the matrix, we can solve for the vector $\boldsymbol{\Lambda}$. However, as we had truncated the Taylor expansion in equation (15.1.10), we should then compute the σ's at the corrected positions, and iterate the preceding equations until convergence is reached.

Although the approach sketched here will work, it is not computationally cheap because it requires a matrix inversion at every iteration. In practice, therefore, often one uses a simpler iterative scheme to satisfy the constraints. In this scheme, called *SHAKE* [443], the iterative scheme just sketched is not applied to all constraints simultaneously but to each constraint in succession. To be more precise, we use the Taylor expansion of equation (15.1.10) for σ_k, but then we approximate $r_i^c - r_i^u$ as

$$\mathbf{r}_i^c(t+\Delta t) - \mathbf{r}_i^u(t) \approx -\frac{\Delta t^2 \lambda_k}{m_i} \boldsymbol{\nabla}_i \sigma_k(t). \tag{15.1.13}$$

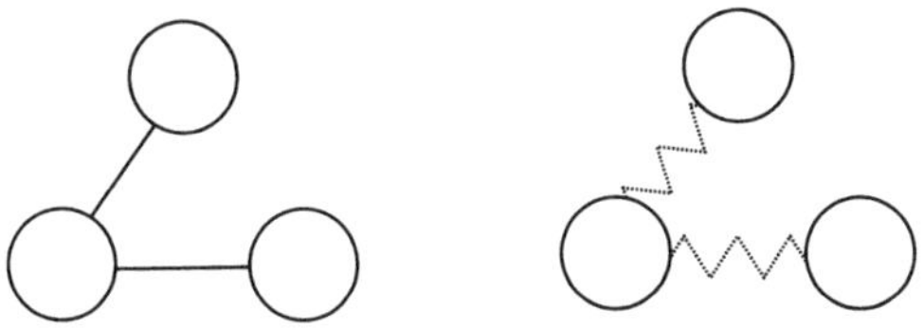

Figure 15.1: Symmetric trimer with bond length d and internal bond angle ψ. (left) Bonds are represented by an infinitely stiff spring; (right) bonds are represented by hard constraints in the equation of motion of the dimer.

If we insert equation (15.1.13) in equation (15.1.10), we get

$$\sigma_k^u(t+\Delta t) = \Delta t^2 \lambda_k \sum_{i=1}^{N} \frac{1}{m_i} \nabla_i \sigma_k(t+\Delta t) \nabla_i \sigma_k(t), \tag{15.1.14}$$

and hence our estimate for λ_k is

$$\lambda_k \Delta t^2 = \frac{\sigma_k^u(t+\Delta t)}{\sum_{i=1}^{N} \frac{1}{m_i} \nabla_i \sigma_k(t+\Delta t) \nabla_i \sigma_k(t)}. \tag{15.1.15}$$

In a simulation, we treat all constraints in succession during one cycle of the iteration and then repeat the process until all constraints have converged to within the desired accuracy. De Leeuw *et al.* have shown how one may cast the problem of constrained dynamics in a Hamiltonian form [444]. The available evidence seems to indicate that this formalism results in algorithms used to solve the equations of motion that lead to less drift from the constraint surface than the algorithms used to solve the corresponding Lagrangian equations of motion. Why this should be so is not obvious.

15.1.1 Constrained and Unconstrained Averages

Thus far, we have presented constrained dynamics as a convenient scheme for modeling the motion of molecules with stiff internal bonds. The advantage of using constrained dynamics was that we could use a longer time step in our Molecular Dynamics algorithm when the high-frequency vibrations associated with the stiff degrees of freedom were eliminated. However, a hidden danger lies in using constraints: the averages computed in a system with hard constraints and in a system with arbitrarily stiff but nonrigid bonds are not the same. To give a well-known specific example, consider a fully flexible trimer (see Figure 15.1). We wish to fix the bond lengths r_{12} and r_{23}. We can do this in two ways. One is to impose the constraints $r_{12}^2 = d^2$ and $r_{23}^2 = d^2$ in the equations of motion of the trimer. The other is to link the atoms in the trimer by harmonic springs, such that

$$\mathcal{U}_{\text{Harmonic}} = \frac{\alpha}{2}\left[(r_{12} - d)^2 + (r_{23} - d)^2\right].$$

Intuitively, one might expect that the limit $\alpha \to \infty$ would be equivalent to dynamics with hard constraints, but this is not so. In fact, if we look at $P(\psi)$, the distribution of the internal angle ψ, we find that

$$\begin{array}{ll} P(\psi) = c\sin\psi & \text{(Harmonic forces)} \\ P(\psi) = c\sin\psi\sqrt{1-(\cos\psi)^2/4} & \text{(Hard constraints).} \end{array} \tag{15.1.16}$$

We shall try to explain the origin for this difference in behavior of "hard" and "soft" constraints. A convenient way of discussing constraints is to start with the Lagrangian of the system, $\mathcal{L} = \mathcal{K} - \mathcal{U}$. Thus far, we had expressed the kinetic ($\mathcal{K}$) and potential ($\mathcal{U}$) energy of the system in terms of the Cartesian velocities and coordinates of the atoms. However, when we talk about bonds and bond angles or, for that matter, any other function of the coordinates that has to be kept constant, it is more convenient to use generalized coordinates, denoted by q. We choose our generalized coordinates such that every quantity we wish to constrain corresponds to a single generalized coordinate. We denote by q_H the set of generalized coordinates that describes the quantities that are effectively, or rigorously, fixed. The remaining soft coordinates are denoted by q_S. The potential energy function $\mathcal{U}$ is a function of both q_H and q_S:

$$\mathcal{U}(q) = \mathcal{U}(q_H, q_S).$$

If we rigorously fix the hard coordinates such that $q_H = \sigma$, then the potential energy is a function of q_S, while it depends parametrically on σ:

$$\mathcal{U}_{\text{hard}}(q_S) = \mathcal{U}_{\text{soft}}(\sigma, q_S).$$

Let us now express the Lagrangian in terms of these generalized coordinates:

$$\begin{aligned} \mathcal{L} &= \frac{1}{2}\sum_{i=1}^{N} m_i \dot{r}_i^2 - \mathcal{U} \\ &= \frac{1}{2}\sum_{i=1}^{N} m_i \dot{q}_\alpha \frac{\partial \mathbf{r}_i}{\partial q_\alpha} \cdot \frac{\partial \mathbf{r}_i}{\partial q_\beta} \dot{q}_\beta - \mathcal{U} \\ &\equiv \frac{1}{2}\dot{\mathbf{q}} \cdot \mathsf{G} \cdot \dot{\mathbf{q}} - \mathcal{U}, \end{aligned} \tag{15.1.17}$$

where the last line of equation (15.1.17) defines the mass-weighted metric tensor G. We can now write the expression for the generalized momentum:

$$p_\alpha \equiv \frac{\partial \mathcal{L}}{\partial \dot{q}_\alpha} = G_{\alpha\beta}\dot{q}_\beta, \tag{15.1.18}$$

where summation of the repeated index β is implied. Next, we can write the Hamiltonian as a function of generalized coordinates and momenta:

$$\mathcal{H} = \frac{1}{2}\mathbf{p} \cdot \mathsf{G}^{-1} \cdot \mathbf{p} + \mathcal{U}(\mathbf{q}).$$

Once we have the Hamiltonian, we can write expressions for the equilibrium phase space density that determines all thermal averages. Although one could write expressions for all averages in the microcanonical ensemble (constant N, V, E), this is, in fact, not very convenient. Hence, we shall consider canonical averages (constant N, V, T). It is straightforward to write the expression for the canonical distribution function in terms of the generalized coordinates and momenta:

$$\rho(\mathbf{p}, \mathbf{q}) = \frac{\exp[-\beta\mathcal{H}(\mathbf{p}, \mathbf{q})]}{Q_{NVT}} \tag{15.1.19}$$

with

$$Q_{NVT} = \int \mathbf{dpdq} \, \exp[-\beta\mathcal{H}(\mathbf{p}, \mathbf{q})]. \tag{15.1.20}$$

The reason we can write equation (15.1.19) in this simple form is that the Jacobian of the transformation from Cartesian coordinates to generalized coordinates is 1. Let us now look at the canonical probability distribution function as a function of $\mathbf{q}$ only:

$$\begin{aligned} \rho(\mathbf{q}) &= c \int \mathbf{dp} \, \exp\{-\beta[\mathbf{p} \cdot \mathsf{G}^{-1} \cdot \mathbf{p}/2 + \mathcal{U}(\mathbf{q})]\} \\ &= c' \exp[-\beta\mathcal{U}(\mathbf{q})] \sqrt{|\mathsf{G}|}, \end{aligned} \tag{15.1.21}$$

where $|\mathsf{G}|$ denotes the (absolute value of the) determinant of G and c and c' are normalizing constants.

Thus far, we have not mentioned constraints. We have simply transformed the canonical distribution function from one set of phase space coordinates to another. Clearly, the answers will not depend on our choice of these coordinates. But now we introduce constraints. That is, in our Lagrangian (15.1.17) we remove the contribution to the kinetic energy due to the change in the hard coordinates; that is, we set $\dot{\mathbf{q}}_H = \mathbf{0}$, and in the potential energy function, we replace the coordinates $\mathbf{q}_H$ by the parameters $\boldsymbol{\sigma}$. The Lagrangian for the system with constraints is

$$\begin{aligned} \mathcal{L}^H &= \frac{1}{2}\sum_{i=1}^{N} m_i \dot{r}_i^2 - \mathcal{U} \\ &= \frac{1}{2}\sum_{i=1}^{N} m_i \dot{q}_\alpha^S \frac{\partial \mathbf{r}_i}{\partial q_\alpha^S} \cdot \frac{\partial \mathbf{r}_i}{\partial q_\beta^S} \dot{q}_\beta^S - \mathcal{U}(\mathbf{q}^S, \boldsymbol{\sigma}) \\ &= \frac{1}{2}\dot{\mathbf{q}}^S \cdot \mathsf{G}_S \cdot \dot{\mathbf{q}}^S - \mathcal{U}(\mathbf{q}^S, \boldsymbol{\sigma}). \end{aligned} \tag{15.1.22}$$

Note that the number of variables has decreased from 3N to $3N - \ell$, where ℓ is the number of constraints. The Hamiltonian of the constrained system is

$$\mathcal{H}^{H} = \frac{1}{2}\mathbf{p}^{S} \cdot \mathsf{G}_{S}^{-1} \cdot \mathbf{p}^{S} + \mathcal{U}(\mathbf{q}^{S}, \sigma),$$

where

$$p_{\alpha}^{S} \equiv \frac{\partial \mathcal{L}}{\partial \dot{q}_{\alpha}^{S}}. \qquad (15.1.23)$$

As before, we can write the phase space density. In this case, it is most convenient to write this density directly as a function of the generalized coordinates and momenta:

$$\rho(\mathbf{p}^{S}, \mathbf{q}^{S}) = \frac{\exp[-\beta \mathcal{H}(\mathbf{p}^{S}, \mathbf{q}^{S})]}{Q_{NVT}^{S}}. \qquad (15.1.24)$$

Let us now write the probability density in coordinate space:

$$\begin{aligned} \rho(\mathbf{q}^{S}) &= a \int d\mathbf{p}^{S} \exp\{-\beta[\mathbf{p}^{S} \cdot \mathsf{G}_{S} \cdot \mathbf{p}^{S}/2 + \mathcal{U}(\mathbf{q}^{S}, \sigma)]\} \\ &= a' \exp[-\beta \mathcal{U}(\mathbf{q}^{S}, \sigma)] \sqrt{|\mathsf{G}_{S}|}, \end{aligned} \qquad (15.1.25)$$

where a and a' are normalizing constants. Now compare this expression with the result that we would have obtained if we had applied very stiff springs to impose the constraints. In that case, we would have to use equation (15.1.21). For $\mathbf{q}^{H}{=}\sigma$, equation (15.1.21) predicts

$$\rho(\mathbf{q}^{S}) = c' \exp[-\beta \mathcal{U}(\mathbf{q}^{S}, \sigma)] \sqrt{|\mathsf{G}|}, \qquad (15.1.26)$$

and this is not the same result as given by equation (15.1.25). Ignoring constant factors, the ratio of the probabilities in the constrained and unconstrained system is given by

$$\frac{\rho(\mathbf{q}^{S})}{\rho(\mathbf{q}^{S}, \mathbf{q}_{H} = \sigma)} = \sqrt{\frac{|\mathsf{G}_{S}|}{|\mathsf{G}|}}.$$

This implies that, if we do a simulation in a system with hard constraints and we wish to predict average properties for the system with "stiff-spring" constraints, then we must compute a weighted average with a weight factor $\sqrt{|\mathsf{G}|/|\mathsf{G}_{S}|}$ to compensate for the bias in the distribution function of the constrained system. Fortunately, it is usually easier to compute the ratio $|\mathsf{G}|/|\mathsf{G}_{S}|$ than to compute $|\mathsf{G}|$ and $|\mathsf{G}_{S}|$ individually. To see this, consider the inverse of G

$$\mathsf{G}_{\alpha\beta}^{-1} = \sum_{i=1}^{N} m_{i}^{-1} \frac{\partial q_{\alpha}}{\partial \mathbf{r}_{i}} \cdot \frac{\partial q_{\beta}}{\partial \mathbf{r}_{i}}.$$

It is easy to verify that this is indeed the inverse of G:

$$\begin{aligned}
G_{\alpha\beta}G^{-1}_{\beta\gamma} &= \sum_{i,j=1}^{N} m_i \frac{\partial \mathbf{r}_i}{\partial q_\alpha}\frac{\partial \mathbf{r}_i}{\partial q_\beta}\frac{\partial q_\beta}{\partial \mathbf{r}_j}\frac{\partial q_\gamma}{\partial \mathbf{r}_j} m_j^{-1} \\
&= \sum_{i=1}^{N} \frac{\partial \mathbf{r}_i}{\partial q_\alpha}\frac{\partial q_\gamma}{\partial \mathbf{r}_i} \\
&= \delta_{\alpha\gamma}.
\end{aligned} \tag{15.1.27}$$

Now, let us write both the matrices G and G^{-1} in block form

$$G = \left(\begin{array}{c|c} G_S & A_{SH} \\ \hline A_{HS} & A_{HH} \end{array}\right) \tag{15.1.28}$$

and

$$G^{-1} = \left(\begin{array}{c|c} B_{SS} & B_{SH} \\ \hline B_{HS} & H \end{array}\right), \tag{15.1.29}$$

where the subscripts S and H denote soft and hard coordinates, respectively. The submatrix H is simply that part of G^{-1} that is quadratic in the derivatives of the constraints:

$$H_{\alpha\beta} = \sum_{i=1}^{N} m_i^{-1} \frac{\partial \sigma_\alpha}{\partial \mathbf{r}_i}\frac{\partial \sigma_\beta}{\partial \mathbf{r}_i}.$$

Now we construct a matrix X as follows. We take the first $3N - \ell$ columns of G and we complete it with the last ℓ columns of the unit matrix:

$$X = \left(\begin{array}{c|c} G_S & 0 \\ \hline A_{HS} & I \end{array}\right). \tag{15.1.30}$$

From the block structure of X, it is obvious that the determinant of X is equal to the determinant of G_S. Next, we multiply X with GG^{-1}, that is, with the unit matrix. Straightforward matrix multiplication shows that

$$GG^{-1}X = G\left(\begin{array}{c|c} I & B_{SH} \\ \hline 0 & H \end{array}\right). \tag{15.1.31}$$

Hence,

$$\begin{aligned}
|X| &= |G_S| \\
&= |GG^{-1}X| \\
&= |G||H|.
\end{aligned} \tag{15.1.32}$$

The final result is that

$$\frac{|G|}{|G_S|} = |H|. \tag{15.1.33}$$

We therefore can write the following relation between the coordinate space densities of the constrained and unconstrained systems:

$$\rho_{\text{flex}}(\mathbf{q}) = |\mathsf{H}|^{-\frac{1}{2}} \rho_{\text{hard}}(\mathbf{q}). \tag{15.1.34}$$

The advantage of this expression is that we have expressed the ratio of the determinants of a $3N \times 3N$ matrix and a $3N - \ell \times 3N - \ell$ matrix, by the determinant of an $\ell \times \ell$ matrix. In many cases, this simplifies the calculation of the weight factor considerably.

As a practical example, let us consider the case of the flexible trimer, discussed at the beginning of this section. We have two constraints:

$$\begin{aligned} \sigma_1 &= r_{12}^2 - d^2 = 0 \\ \sigma_2 &= r_{23}^2 - d^2 = 0. \end{aligned}$$

If all three atoms have the same mass m, we can write $|\mathsf{H}|$ as

$$|\mathsf{H}| = \frac{1}{m} \begin{vmatrix} \sum_i \frac{\partial \sigma_1}{\partial \mathbf{r}_i} \frac{\partial \sigma_1}{\partial \mathbf{r}_i} & \sum_i \frac{\partial \sigma_1}{\partial \mathbf{r}_i} \frac{\partial \sigma_2}{\partial \mathbf{r}_i} \\ \sum_i \frac{\partial \sigma_2}{\partial \mathbf{r}_i} \frac{\partial \sigma_1}{\partial \mathbf{r}_i} & \sum_i \frac{\partial \sigma_2}{\partial \mathbf{r}_i} \frac{\partial \sigma_2}{\partial \mathbf{r}_i} \end{vmatrix}.$$

Inserting the expressions for σ_1 and σ_2, we find that

$$|\mathsf{H}| = \frac{2}{m} \begin{vmatrix} 2r_{12}^2 & -\mathbf{r}_{12} \cdot \mathbf{r}_{23} \\ -\mathbf{r}_{12} \cdot \mathbf{r}_{23} & 2r_{23}^2 \end{vmatrix}.$$

Using the fact that $r_{12}^2 = r_{23}^2 = d^2$, we get

$$\begin{aligned} |\mathsf{H}| &= \frac{8}{m} \left(r_{12}^2 r_{23}^2 - (\mathbf{r}_{12} \cdot \mathbf{r}_{23})^2 \right) \\ &= \frac{8d^4}{m} \left(1 - \frac{\cos^2 \psi}{4} \right). \end{aligned} \tag{15.1.35}$$

Finally, we recover equation (15.1.16) for the ratio of the probability densities for the constrained and unconstrained systems:

$$\frac{\rho_{\text{flex}}}{\rho_{\text{hard}}} = |\mathsf{H}|^{\frac{1}{2}} = c\sqrt{1 - \frac{\cos^2 \psi}{4}}. \tag{15.1.36}$$

This ratio varies between 1 and 0.866, that is, at most some 15%. It should be noted that, in general, the ratio depends on the masses of the particles that participate in the constraints. For instance, if the middle atom of our trimer is much lighter than the two end atoms, then $\sqrt{|\mathsf{H}|}$ becomes $\sqrt{1 - \cos^2 \psi}$ $= |\sin \psi|$ and the correction due to the presence of hard constraints is not small. However, to put things in perspective, we should add that, at least for bond length constraints of the type most often used in Molecular Dynamics simulations, the effect of the hard constraints on the distribution functions appears to be relatively small.

Finally, we stress that, although constrained dynamics is discussed in the context of the simulation of polyatomic molecules, the same technique has many other applications. For instance, in the Car-Parrinello scheme for "ab initio" Molecular Dynamics (see, e.g., the articles by Galli in [40, 41]), constraints are imposed to keep the Kohn-Sham orbitals orthonormal. In the simulation of reaction rates (see, for example, Chapter 16 and the article by Ciccotti [445]), constrained dynamics is used to simulate the system at the top of the free energy barrier that separates reactants and products. In addition, it is often possible to compute the free energy difference between two realizations of a system, by computing the reversible work needed to transform one system into the other. Again, constrained dynamics simulations can be used to compute this reversible work.

15.2 On-the-Fly Optimization: Car-Parrinello Approach

Thus far, we have considered the use of extended Lagrangians only to perform simulations in ensembles other than the microcanonical. However, another important application of extended Lagrangians is to perform simulations subject to an expensive variational constraint. Car and Parrinello [442] pioneered this use of extended Lagrangians to maintain constraints in the context of the "ab initio" Molecular Dynamics method. The Car-Parrinello method is a Molecular Dynamics technique in which electronic density-functional theory (in the local density approximation) is used to compute the energies and densities of the valence electrons "on the fly" [442]. It is assumed that the system is in its electronic ground state and that electrons follow the nuclear motion adiabatically. The condition that the system is always in its electronic ground state seems to imply that an electronic energy minimization should be performed at every MD time step. Typically, this requires an iterative procedure. Moreover, it is important that this energy minimization is carried out to high accuracy, because a partially converged electronic density will lag behind the nuclear motion and therefore exert a systematic drag force on the nuclei. What Car and Parrinello showed is an alternative to this minimization scheme, based on the extended Lagrangian approach. In the Car-Parrinello method, the electronic density fluctuates around its optimal (adiabatic) value but, and this is the crucial point, even though at every step the system is not exactly in its electronic ground state, the electrons do not exert a systematic drag force on the nuclei. We discuss in more detail this aspect of the Car-Parrinello method, but not in the context of electronic structure calculations, as that topic is outside the scope of this book. For more details on the Car-Parrinello method for electronic structure calculations, we refer the reader to the excellent reviews of Galli and Pasquarello [446] and Remler and Madden [447]. The closest classical analogue of "ab initio" Molecular Dynamics is the method developed by Löwen *et al.* [448,449] to simulate counterion screening in colloidal suspensions of polyelectrolytes. In the approach of [448], the counterions are described by classical density-functional theory and an extended Lagrangian method is used to keep the free energy of the counterions close to its minimum.

Here we consider a somewhat simpler application of the Car-Parrinello approach to a classical system. As before, the aim of the method is to replace an expensive iterative optimization procedure by a cheap extended dynamical scheme. As a specific example, we consider a fluid of point-polarizable molecules. The molecules have a static charge distribution that we leave unspecified (for instance, we could be dealing with ions, dipoles, or quadrupoles). We denote the polarizability of the molecules by α. The total energy of this system is given by

$$\mathcal{U} = U_0 + U_{\text{pol}},$$

where U_0 is the part of the potential energy that does not involve polarization. The induction energy, U_{pol}, is given by [450]

$$U_{\text{pol}} = -\sum_i \mathbf{E}_i \cdot \boldsymbol{\mu}_i + \frac{1}{2\alpha}\sum_i (\boldsymbol{\mu}_i)^2,$$

where $\mathbf{E}_i$ is the local electric field acting on particle i and μ_i is the dipole induced on particle i by this electric field. Of course, the local field depends on the values of all other charges in the system. For instance, in the case of dipolar molecules,

$$\mathbf{E}_i = \mathsf{T}_{ij} \cdot \boldsymbol{\mu}_j^{\text{tot}},$$

where T_{ij} is the dipole-dipole tensor and μ_j^{tot} is the total (i.e., permanent plus induced) dipole moment of molecule j. We assume that the induced dipoles follow the nuclear motion adiabatically and that U_{pol} is always at its minimum. Minimizing U_{pol} with respect to the $\boldsymbol{\mu}_i$ yields

$$\boldsymbol{\mu}_i = \alpha \mathbf{E}_i. \tag{15.2.1}$$

Hence, to properly account for the molecular polarizability of an N-particle system, we would have to solve a set of 3N linear equations at every time step. If we solve this set of equations iteratively, we must make sure that the solution has fully converged, because otherwise the local field will exert a systematic drag force on the induced dipoles and the system will fail to conserve energy.

Now let us consider the Car-Parrinello approach to this optimization problem. The application of this extended Lagrangian method to polarizable molecules has been proposed by Rahman and co-workers [451] and by Sprik and Klein [452]. A closely related approach was subsequently advocated by Wilson and Madden [453]. The basic idea is to treat the induced dipoles as additional dynamical variables that are included in the Lagrangian:

$$\mathcal{L}(\mathbf{r}^N, \boldsymbol{\mu}^N) = \frac{1}{2}\sum_{i=1}^{N} m\dot{\mathbf{r}}_i^2 + \frac{1}{2}\sum_{i=1}^{N} M\dot{\boldsymbol{\mu}}_i^2 - \mathcal{U}, \tag{15.2.2}$$

where M is the mass associated with the motion of the dipoles. This Lagrangian yields the following equations of motion for the dipole moments:

$$M\ddot{\boldsymbol{\mu}}_i \equiv \frac{\partial \mathcal{L}}{\partial \boldsymbol{\mu}_i} = -\frac{\boldsymbol{\mu}_i}{\alpha} + \mathbf{E}_i.$$

The right-hand side of this equation can be considered as a generalized force that acts on the dipoles. In the limit that this force is exactly zero the iterative scheme is recovered. If the temperature associated with the kinetic energy of the dipoles is sufficiently low, the dipoles will fluctuate around their lowest-energy configuration. More important, there will be no systematic drag force on the dipoles, and hence the energy of the system will not drift.

To make sure that the induced dipoles are indeed close to their ground-state configuration, we should keep the temperature of the induced-dipole degrees of freedom low. Yet, at the same time, the dipoles should be able to adopt rapidly (adiabatically) to changes in the nuclear coordinates to ensure that the condition of minimum energy is maintained during the simulation. This implies that the masses associated with the induced dipoles should be small. In summary, we require that

$$T_\mu \ll T_r$$
$$M \ll m,$$

where the temperature of the induced dipoles is defined as

$$T_\mu = \frac{1}{2}\sum_{i=1}^{N} M\dot{\boldsymbol{\mu}}_i^2,$$

while the translational temperature is related in the usual way to the kinetic energy

$$T_r = \frac{1}{2}\sum_{i=1}^{N} m\dot{\mathbf{r}}_i^2.$$

The condition that the temperature of the induced dipoles should be much lower than the translational temperature seems to create a problem because, in an ordinary simulation, the coupling between induced-dipole moments and translational motion leads to heat exchange. This heat exchange will continue until the temperature of the induced dipoles equals the translational temperature. Hence, it would seem that we cannot fix the temperature of the induced dipoles independent of the translational temperature. However, here we can again make use of the Nosé-Hoover thermostats. Sprik and Klein [452] have shown that one can use two separate Nosé-Hoover thermostats to impose the temperature of the positions and to impose the (low) temperature of the polarization [454]. The mass M associated with the induced dipoles should be chosen such that the relaxation time of the polarization is on the same order of magnitude as the fastest relaxation in the liquid.

15.3 Multiple Time Steps

An alternative scheme for dealing with the high-frequency vibrational modes of polyatomic molecules is based on the Liouville formulation of the classical equations of motion. We separate the force on a particle into two parts:

$$\mathbf{F} = \mathbf{F}_{\text{short}} + \mathbf{F}_{\text{long}}.$$

Although this division is arbitrary, for our diatomic molecule we would divide the potential into the short-range interactions that are responisble for the bond vibration and the long-range attractive forces between the atoms. The idea is that on the time scale of the vibrations of the atoms, the long-range part of the potential hardly changes and therefore this "expensive potential" does not need to be updated as often as the "cheap" short-range part of the potential. This suggests using multiple time steps, a short time step for the vibration and a much longer one for the remainder of the interactions.

Martyna *et al.* [85] have used the Liouville formalism to solve the equations of motion using multiple time steps. Here, we consider the NVE ensemble. For details on how to use multiple time steps in other ensembles, we refer to [85]. Let us start with the simple case and derive the equations of motions for a single particle with force F. The Liouville operator for this system is equation (4.3.9):

$$\begin{aligned} iL &= iL_r + iL_p \\ &= \mathbf{v}\frac{\partial}{\partial \mathbf{r}} + \frac{\mathbf{F}}{m}\frac{\partial}{\partial \mathbf{v}}. \end{aligned}$$

The equations of motion follow from applying the Trotter formula (4.3.15) with time step Δt:

$$e^{iL\Delta t} \approx e^{iL_p \Delta t/2} e^{iL_r \Delta t} e^{iL_p \Delta t/2}.$$

The position and the velocity at time Δt follow from applying the Liouville operator under the initial condition $(\mathbf{r}(0), \mathbf{v}(0))$. As shown in section (4.3.3), $iL_r t$ corresponds to a shift in coordinates and $iL_p t$ to a shift in momenta. If we perform these operations in three steps, we obtain

$$\begin{aligned} e^{iL\Delta t} f[\dot{\mathbf{r}}(0), \mathbf{r}(0)] &= e^{iL_p\Delta t/2} e^{iL_r\Delta t} e^{iL_p\Delta t/2} f[\dot{\mathbf{r}}(0), \mathbf{r}(0)] \\ &= e^{iL_p\Delta t/2} e^{iL_r\Delta t} f[\dot{\mathbf{r}}(0) + \mathbf{F}(0)\Delta t/2m, \mathbf{r}(0)] \\ &= e^{iL_p\Delta t/2} f[\dot{\mathbf{r}}(0) + \mathbf{F}(0)\Delta t/2m, \mathbf{r}(0) + \dot{\mathbf{r}}(\Delta t/2)\Delta t] \\ &= f[\dot{\mathbf{r}}(0) + \mathbf{F}(0)\Delta t/2m + \mathbf{F}(\Delta t)\Delta t/2m, \mathbf{r}(0) + \dot{\mathbf{r}}(\Delta/2)\Delta t]. \end{aligned}$$

The equations of motion that follow are

$$\begin{aligned} \dot{\mathbf{r}}(\Delta t) &= \dot{\mathbf{r}}(0) + \frac{\Delta t}{2m}[\mathbf{F}(0) + \mathbf{F}(\Delta t)] \\ \mathbf{r}(\Delta t) &= \mathbf{r}(0) + \dot{\mathbf{r}}(\Delta/2)\Delta t, \end{aligned}$$

which the reader wil recognize as the velocity Verlet equations (see section 4.3.3).

Let us now separate the Liouville operator iL_p into two parts:

$$\begin{aligned} iL_{\text{short}} &= \frac{\mathbf{F}_{\text{short}}}{m}\frac{\partial}{\partial \mathbf{v}}. \\ iL_{\text{long}} &= \frac{\mathbf{F} - \mathbf{F}_{\text{short}}}{m}\frac{\partial}{\partial \mathbf{v}} = \frac{\mathbf{F}_{\text{long}}}{m}\frac{\partial}{\partial \mathbf{v}}. \end{aligned}$$

We use a Trotter expansion with two time steps: a long time step, Δt, and a short one, $\delta t = \Delta t/n$. The total Liouville operator then reads

$$\begin{aligned} e^{iL\Delta t} &= e^{i(L_{\text{short}} + L_{\text{long}} + L_r)\Delta t} \\ &\approx e^{iL_{\text{long}}\Delta t/2} e^{i(L_{\text{short}} + L_r)\Delta t} e^{iL_{\text{long}}\Delta t/2}. \end{aligned}$$

We can again apply a Trotter expansion for the terms iL_{long} and iL_r:

$$e^{iL\Delta t} = e^{iL_{\text{long}}\Delta t/2}\left[e^{iL_{\text{short}}\delta t/2n} e^{iL_r\delta t/n} e^{iL_{\text{short}}\delta t/2n}\right]^n e^{iL_{\text{long}}\Delta t/2}.$$

We apply this Liouville operator on the initial position and velocity. We first make a step using the expensive $\mathbf{F}_{\text{long}}$

$$e^{iL_{\text{long}}\Delta t/2} f[\dot{\mathbf{r}}(0), \mathbf{r}(0)] = f\left[\dot{\mathbf{r}}(0) + \mathbf{F}_{\text{long}}(0)\Delta t/2m, \mathbf{r}(0)\right],$$

followed by n small steps using the cheap $\mathbf{F}_{\text{short}}$ with the smaller time step, δt, or

$$\left[e^{iL_r\delta t/n} e^{iL_{\text{short}}\delta t/2n}\right]^n f\left[\dot{\mathbf{r}}(0) + \mathbf{F}_{\text{long}}(0)\Delta t/2m, \mathbf{r}(0)\right],$$

Algorithm 29 (Multiple Time Step)

```
   subroutine                            Multiple time step, f_long is
+     multi(f_long,f_short)              the long-range part and f_short
                                         the short-range part of the force
   vx=vx+0.5*delt*f_long                 velocity Verlet with time step Δt
   do it=1,n                             loop for the small time step
      vx=vx+0.5*(delt/n)*f_short         velocity Verlet with timestep Δt/n
      x=x+(delt/n)2*vx
      call force_short(f_short)          short-range forces
      vx=vx+0.5*(delt/n)*f_short
   enddo
   call force_all(f_long,f_short)        all forces
   vx=vx+0.5*delt*f_long
   return
   end
```

Comments to this algorithm:

1. *In the argument list of the subroutine call we have added* `f_long, f_short` *to indicate that in the velocity Verlet algorithm the force has to be known from the previous time step.*
2. *Subroutines* `force_short` *determines to short-range forces. Since this involves a small number of particles, the calculation of these forces is much faster than* `force_all` *in which all interactions are computed.*

which corresponds to solving the equations of motion using the velocity Verlet scheme using the force $\mathbf{F}_{\text{short}}$ with time step δt and initial conditions $\dot{\mathbf{r}}(0) + \mathbf{F}_{\text{long}}(0)\Delta t/2m, \mathbf{r}(0)$. It should be emphasized that this algorithm, by construction, is time reversible. In Algorithm 29 we illustrate how this multiple-time-step method (MTS) can be implemented.

Two applications of this algorithm are particularly important. One is the use of MTS algorithms to simulate the dynamics of molecules with stiff internal bonds. In Case Study 22 it is shown that this application of the MTS method is attractive, since for the case we consider, it is at least as efficient as constrained dynamics (see section 15.1). The second important area of application is as a time-saving device in the simulation of systems with computationally "expensive" potential-energy functions. Here the MTS method offers the possibility of carrying out many time steps with a "cheap" potential energy (e.g., an effective pair potential) and then performing the expensive correction every nth step. Procacci and co-workers have used this approach to reduce the computational costs associated with the long-range interactions of Coulombic systems [455, 456]. They have used the MTS ideas in combination with the Ewald summation (see Chapter 12.1) to reduce the CPU time for the calculation of long-range interactions.

Case Study 22 (Multiple Time Step versus Constraints)
In this Case Study we consider a system of diatomic Lennard-Jones molecules. We compare two models: the first model uses a fixed bond length l_0 between the two atoms of a molecule. In the second model, we use a bond-stretching potential given by

$$u_{\text{bond}}(l) = \frac{1}{2} k_b (l - l_0)^2,$$

where l is the distance between the two atoms in a molecule. In the simulations we used $k_b = 50000$ and $l_0 = 1$. In addition to the bond-stretching potential, all nonbonded atoms interact via a Lennard-Jones potential. The total number of diatomics was 125 and the box length 7.0 (in the usual reduced units). The Lennard-Jones potential was truncated at $r_c = 3.0$, while $T = 3.0$. The equations of motion are solved using bond constraints for the first model, while multiple time steps were used for the second model. All simulations were performed in the NVE ensemble.

It is interesting to compare the maximum time steps that can be used to solve the equations of motion for these two methods. As a measure of the accuracy with which the equations of motion are solved, we compute the average deviation of the initial energy, which is defined by Martyna *et al.* [138] as

$$E = \frac{1}{N_{\text{step}}} \sum_{i=1}^{N_{\text{step}}} \left| \frac{E(i\Delta t) - E(0)}{E(0)} \right|,$$

in which $E(i)$ is the total energy at time i.

For the bond constraints we use the SHAKE algorithm [443] (see also section 15.1). In the SHAKE algorithm, the bond lengths are exactly fixed at l_0 using an iterative scheme. In Figure 15.2 (left) the energy fluctuations are shown as a function of the time step. Normally one tolerates a noise level in E of $\mathcal{O}(10^{-5})$, which would correspond to a time step of 2×10^{-4} for the first model. This should be compared with a single-time-step Molecular Dynamics simulation using the second model. A similar energy noise level can be obtained with a time step of 9×10^{-5}, which is a factor 2 smaller.

To apply the multiple-time-step algorithm, we have to separate the intermolecular force into a short-range and a long-range part. In the short-range part we include the bond-stretching potential and the short-range part of the Lennard-Jones potential. To make a split in the Lennard-Jones potential we

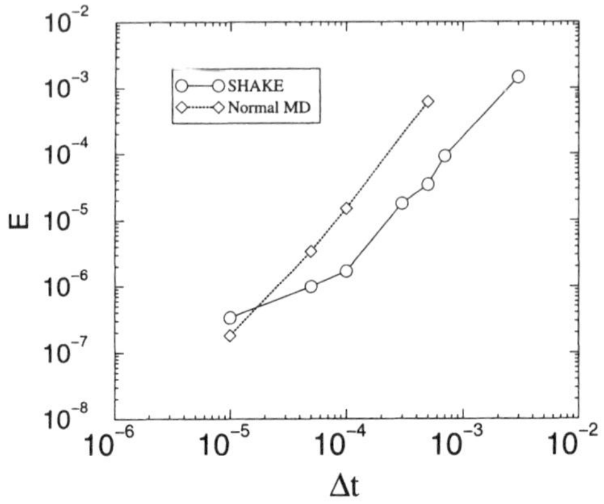

Figure 15.2: Comparison of the energy fluctuations as a function of the time step for a normal MD simulation with a harmonic bond potential and a constrained MD simulation with the SHAKE algorithm.

use a simple switching function $S(r)$:

$$\begin{aligned} u_{LJ}(r) &= u_{short}(r) + u_{long}(r) \\ u^{short}(r) &= S(r) \times u_{LJ}(r) \\ u^{long}(r) &= [1 - S(r)]\, u_{LJ}(r), \end{aligned}$$

where

$$S(r) = \begin{cases} 1 & 0 < r < r_c - \lambda \\ 1 + \gamma^2 (2\gamma - 3) & r_m - \lambda < r < r_m \\ 0 & r_m < r < r_c \end{cases}$$

and

$$\gamma = \frac{r - r_m + \lambda}{\lambda}. \tag{15.3.1}$$

In fact, there are other ways to split the total potential function [455,456]. We have chosen $\lambda = 0.3$ and $r_m = 1.7$. To save CPU time a list is made of all the atoms that are close to each other (see Appendix F for details); therefore the calculation of the short-range forces can be done very efficiently. For a noise level of 10^{-5}, one is able to use $\delta t = 10^{-4}$ and $n = 10$, giving $\Delta t = 10^{-3}$.

To compare the different algorithms in a consistent way, we compare in Figure 15.3 the efficiency of the various techniques. The efficiency η is defined as the length of the simulation (time step times the number of integration steps) divided by the amount of CPU time that was used. In the figure, we have plotted η for all simulations from Figure 15.2. For an energy noise level of 10^{-5}, the SHAKE algorithm is twice as efficient than normal MD

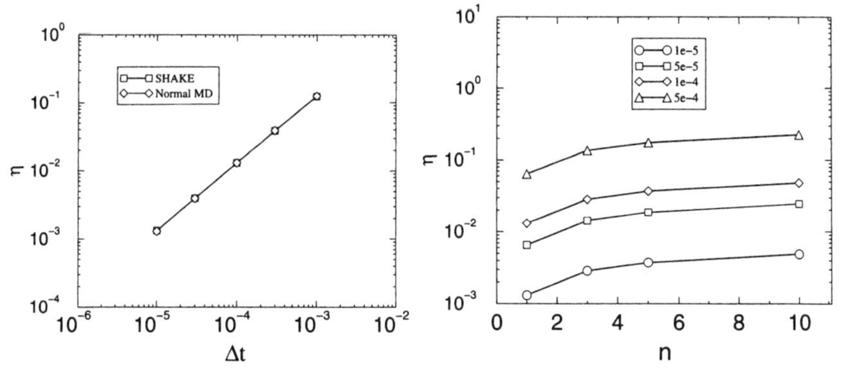

Figure 15.3: Comparison of the efficiency η for bond constraints (SHAKE) with normal molecular dynamics (left), and multiple times steps (right). The left figure gives the efficiency as a function of the time step and the right figure as a function of the number of small time steps n, $\Delta t = n\delta t$, where the value of δt is given in the symbol legend.

($n = 1$). This means that hardly any CPU time is spent in the SHAKE routine. However, the MTS algorithm is still two times faster ($n = 10$, $\delta t = 10^{-4}$) at the same efficiency.

Chapter 16

Rare Events

Molecular Dynamics simulations can be used to probe the natural time evolution of classical many-body systems on a time scale of 10^{-14} to 10^{-8} s (this upper limit depends of course on the computing power that is at our disposal). This time window is adequate for studying many structural and dynamical properties, provided that the relevant fluctuations decay on a time scale that it is appreciably shorter than 10^{-8}s. This is true for most equilibrium properties of simple liquids. It is also usually true for the dynamics associated with nonhydrodynamic modes. For hydrodynamic modes (typically, the modes that describe the diffusion or propagation of quantities that satisfy a conservation law, such as mass, momentum, or energy), the time scales can be much longer. But we still can use MD simulations to compute the transport coefficients that govern the hydrodynamic behavior by making use of the appropriate Green-Kubo relation. As explained in Appendix C, Green-Kubo relations allow us to express the hydrodynamic transport coefficients in terms of a time integral of a correlation function of a dynamical quantity that fluctuates on a microscopic time scale (e.g., the self-diffusion coefficient is equal to the integral of the velocity autocorrelation function).

Nevertheless, there are many dynamical phenomena that cannot be studied in this way. We discuss one particularly important example, namely, activated processes. Conventional MD simulations cannot be used to study activated processes. The reason is not that the relevant dynamics is slow, but rather that it involves a step that involves a rare event. However, if this rare event does take place, it usually happens quite quickly, i.e., on a time scale that can be followed by MD simulation. An example is the trans-to-gauche transition in an alkane: this process is infrequent if the barrier separating the two conformations is large compared to k_BT. Yet, once an unlikely fluctuation has driven the system to the top of the barrier, the actual barrier crossing is quick.

It turns out that in many cases, MD simulations can be used to compute the rate of such activated processes. Such calculations were first performed by Bennett in the context of diffusion in solids [457]. Subsequently, Chandler extended and generalized the approach to the calculation of reaction rates [187,458]. The basic idea behind these MD calculations is that the rate at which a barrier crossing proceeds is determined by the product of a static term, namely the probability of finding the system at the top of the barrier, and a dynamic term that describes the rate at which systems at the top of the barrier move to the other valley.

16.1 Theoretical Background

As a prototypical example, we consider a unimolecular reaction $A \rightleftharpoons B$, in which species A is transformed into species B. If the rate-limiting step of this reaction is a (classical) barrier crossing, then Molecular Dynamics simulations can be used to compute the rate constant of such a reaction (the best explanation of this approach is still to be found in Chandler's original 1978 paper [458]). In contrast, if the rate-limiting step is a tunneling event or the hopping from one potential-energy surface to another, the classical approach breaks down, and we should turn to quantum dynamical schemes that fall outside the scope of this book (see references in [459]). Let us first look at the phenomenological description of unimolecular reactions. We denote the number density of species A and B by c_A and c_B, respectively. The phenomenological rate equations are

$$\frac{dc_A(t)}{dt} = -k_{A\to B}c_A(t) + k_{B\to A}c_B(t) \tag{16.1.1}$$

$$\frac{dc_B(t)}{dt} = +k_{A\to B}c_A(t) - k_{B\to A}c_B(t). \tag{16.1.2}$$

Clearly, as the number of molecules is constant in this conversion reaction, the total number density is conserved:

$$\frac{d\left[c_A(t) + c_B(t)\right]}{dt} = 0. \tag{16.1.3}$$

In equilibrium, all concentrations are time independent, i.e., $\dot{c}_A = \dot{c}_B = 0$. This implies that

$$K \equiv \frac{\langle c_A \rangle}{\langle c_B \rangle} = \frac{k_{B\to A}}{k_{A\to B}}, \tag{16.1.4}$$

where K is the equilibrium constant of the reaction. Let us now consider what happens if we take a system at equilibrium, and apply a small perturbation, Δc_A, to the concentration of species A (and thereby of species B). We can write the rate equation that determines the decay of this perturbation as

$$\frac{d\Delta c_A(t)}{dt} = -k_{A\to B}\Delta c_A(t) - k_{B\to A}\Delta c_A(t),$$

where we have used equations (16.1.3) and (16.1.4). The solution to this equation is

$$\Delta c_A(t) = \Delta c_A(0)\exp[-(k_{A\to B} + k_{B\to A})t] \equiv \Delta c_A(0)\exp(-t/\tau_R), \quad (16.1.5)$$

where we have defined the reaction time constant

$$\tau_R = (k_{B\to A} + k_{A\to B})^{-1} = k_{A\to B}^{-1}\left(1 + \frac{\langle c_A\rangle}{\langle c_B\rangle}\right)^{-1} = \frac{c_B}{k_{A\to B}}, \quad (16.1.6)$$

where we have assumed that the total concentration $c_A + c_B = 1$. With this normalization, c_A is simply the probability that a given molecule is in state A.

Thus far, we have discussed the reaction from a macroscopic, phenomenological point of view. Let us now look at the microscopics. We do this in the framework of linear response theory. First of all, we must have a microscopic description of the reaction. This means that we need a recipe that allows us to measure how far the reaction has progressed. In the case of diffusion over a barrier from one free energy minimum to another, we could use the fraction of the distance traveled as a reaction coordinate. In general, reaction coordinates may be complicated, nonlinear functions of the coordinates of all particles. It is convenient to think of the reaction coordinate q simply as a generalized coordinate, of the type discussed in the previous section. In Figure 16.1, we show a schematic drawing of the free energy surface of the system as a function of the reaction coordinate q. If we wish to change the equilibrium concentration of species; A, we should apply an external perturbation that favors all states with $q < q^*$ relative to those with $q > q^*$.

By analogy to the discussion in Appendix C, we consider an external perturbation that changes the relative probabilities of finding species A and B. To achieve this, we add to the Hamiltonian a term that lowers the potential energy for $q < q^*$:

$$\mathcal{H} = \mathcal{H}_0 - \epsilon g_A(q - q^*), \quad (16.1.7)$$

where ϵ is a parameter that measures the strength of the perturbation. As we are interested in the linear response, we shall consider the limit $\epsilon \to 0$. The function $g_A(q - q^*)$ is chosen such that it is equal to 1 if the reaction coordinate q is in the range that corresponds to an equilibrium configuration of the "reactant," while $g_A(q - q^*)$ should be equal to 0 for a typical "product" configuration. The traditional choice for g_A is a Heaviside θ-function:

$$g_A(q - q^*) = 1 - \theta(q - q^*) = \theta(q^* - q),$$

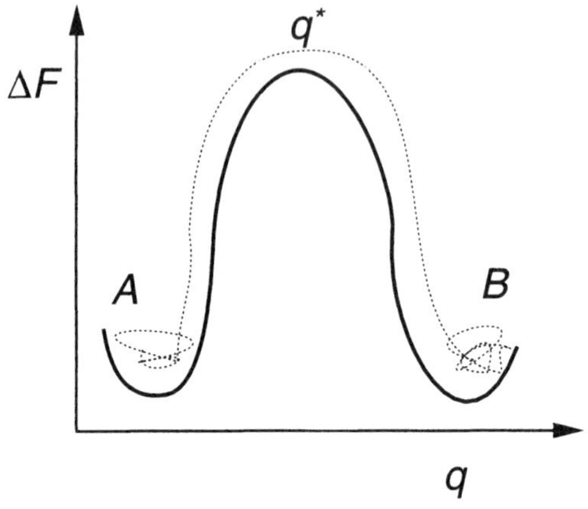

Figure 16.1: Schematic drawing of the free energy surface of a many-body system, as a function of the reaction coordinate q. For $q < q^*$, we have the reactant species A, for $q < q^*$, we have the product B. As will be discussed below, the choice of q^* is, to some extent, arbitrary. However, it is convenient to identify the value of the reaction coordinate at the top of the barrier with q^*.

where $\theta(x) = 1$ for $x > 0$ and $\theta(x) = 0$ otherwise. In what follows, we shall consider the more general case that g_A is equal to the θ-function in the reactant and product domains. However, unlike θ, g_A varies smoothly from 1 to 0 in the region of the free energy barrier. For the sake of simplicity, we refer to the states A and B as "reactants" and "products" in the chemical sense of the word. However, in general A and B can designate any pair of initial and final states that can interconvert by a barrier-crossing process.

Let us first consider the effect of a static perturbation of this type on the probability of finding the system in state A. We note that

$$\Delta c_A = \langle c_A \rangle_\epsilon - \langle c_A \rangle_0 = \langle g_A \rangle_\epsilon - \langle g_A \rangle_0 .$$

Here we have used the fact that g_A is equal to 1 in the reactant basin. Hence, the average value of g_A is simply equal to the probability of finding the system in state A. From equation (C.1.1) of section C.1, we find immediately that

$$\frac{\partial \Delta c_A}{\partial \epsilon} = \beta \left(\langle g_A^2 \rangle_0 - \langle g_A \rangle_0^2 \right) .$$

This equation can be simplified by noting that, outside the barrier region, g_A is either 1 or 0 and hence, $g_A^2(x) = g_A(x)$. In the barrier region, this equality need not hold—but those configurations hardly contribute to the equilibrium average. Hence,

$$\frac{\partial \Delta c_A}{\partial \epsilon} = \beta\left[\langle g_A \rangle_0 \left(1 - \langle g_A \rangle_0\right)\right] = \beta \langle c_A \rangle \langle c_B \rangle . \tag{16.1.8}$$

For what follows, it is convenient to define the function $g_B = 1 - g_A$. Clearly,

$$\langle g_B \rangle_0 = \langle (1 - g_A) \rangle_0 = \langle c_B \rangle_0 .$$

Next, consider what happens if we suddenly switch off the perturbation at time $t = 0$. The concentration of A will relax to its equilibrium value as described in equation (C.2.1) and we find that, to first order in ϵ,

$$\begin{aligned} \Delta c_A(t) &= \beta\epsilon \frac{\int d\Gamma \exp(-\beta \mathcal{H}_0)\left(g_A(0) - \langle g_A \rangle\right) \exp(iL_0 t)\left(g_A(0) - \langle g_A \rangle\right)}{\int d\Gamma \exp(-\beta \mathcal{H}_0)} \\ &= \beta\epsilon \langle \Delta g_A(0) \Delta g_A(t) \rangle . \end{aligned} \tag{16.1.9}$$

Finally, we can use equation (16.1.8) to eliminate ϵ from the above equation, and we find the following expression for the relaxation of an initial perturbation in the concentration of species A:

$$\Delta c_A(t) = \Delta c_A(0) \frac{\langle \Delta g_A(0) \Delta g_A(t) \rangle}{\langle c_A \rangle \langle c_B \rangle} . \tag{16.1.10}$$

If we compare this with the phenomenological expression, equation (16.1.5), we see that

$$\exp(-t/\tau_R) = \frac{\langle \Delta g_A(0) \Delta g_A(t) \rangle}{\langle c_A \rangle \langle c_B \rangle} . \tag{16.1.11}$$

Actually, we should be cautious with this identification. For very short times (i.e., times comparable to the average time that the system spends in the region of the barrier), we should not expect the autocorrelation function of the concentration fluctuations to decay exponentially. Only at times that are long compared to the typical barrier-crossing time should we expect equation (16.1.11) to hold. Let us assume that we are in this regime. Then we can obtain an expression for τ_R by differentiating equation (16.1.11):

$$-\tau_R^{-1} \exp(-t/\tau_R) = \frac{\langle g_A(0) \dot{g}_A(t) \rangle}{\langle c_A \rangle \langle c_B \rangle} = -\frac{\langle \dot{g}_A(0) g_A(t) \rangle}{\langle c_A \rangle \langle c_B \rangle} , \tag{16.1.12}$$

where we have dropped the Δ's, because the time derivative of the equilibrium concentration vanishes. Hence, for times that are long compared to molecular times, but still very much shorter than τ_R, we can write

$$\tau_R^{-1} = \frac{\langle \dot{g}_A(0) g_A(t) \rangle}{\langle c_A \rangle \langle c_B \rangle} \tag{16.1.13}$$

or, if we recall equation (16.1.6) for the relation between $k_{A\rightarrow B}$ and τ_R, we find

$$k_{A\rightarrow B}(t) = \frac{\langle \dot{g}_A(0) g_A(t)\rangle}{\langle c_A\rangle}. \tag{16.1.14}$$

In this equation, the time dependence of $k_{A\rightarrow B}(t)$ is indicated explicitly. However, we recall that it is only the long-time plateau value of $k_{B\rightarrow A}(t)$ that enters into the phenomenological rate equation. Finally, we can reexpress the correlation function in equation (16.1.14) by noting that

$$\dot{g}_A(q-q^*) = \dot{q}\frac{\partial g_A(q-q^*)}{\partial q} = -\dot{q}\frac{\partial g_B(q-q^*)}{\partial q} = -\dot{q}(\partial_q g_B),$$

where the last equality defines

$$\partial_q \equiv \frac{\partial}{\partial q}.$$

Hence,

$$k_{A\rightarrow B}(t) = \frac{\langle \dot{q}(\partial_q g_B)(0) g_B(t)\rangle}{\langle c_A\rangle}, \tag{16.1.15}$$

where we have used the fact that $\langle \dot{q}\rangle = 0$. A particularly convenient form of equation (16.1.15) that we shall use in section 16.3 is

$$k_{A\rightarrow B}(t) = \int_0^\infty dt\, \frac{\langle \dot{q}(0)(\partial_q) g_B(0)\dot{q}(t)(\partial_q g_B)(t)\rangle}{\langle c_A\rangle}. \tag{16.1.16}$$

But first we establish contact with the conventional "Bennett-Chandler" expression for the rate constant.

16.2 Bennett-Chandler Approach

If we choose $g_A = \theta(q^* - q)$—and hence $g_B = \theta(q - q^*)$—then we can rewrite equation (16.1.15) in the following way:

$$\begin{aligned} k_{A\rightarrow B}(t) &= \frac{\langle \dot{q}\delta(q(0)-q^*)\theta(q(t)-q^*)\rangle}{\langle c_A\rangle} \\ &= \frac{\langle \dot{q}\delta(q^*-q(0))\theta(q(t)-q^*)\rangle}{\langle \theta(q^*-q)\rangle}. \end{aligned} \tag{16.2.1}$$

In this way, we have expressed the rate constant $k_{A\rightarrow B}$ exclusively in microscopic quantities that can be measured in a simulation. Next, we shall see how this can be done. First, however, we establish the connection between equation (16.2.1) and the expression for the rate constant that follows from

Eyring's transition state theory. To this end, consider $k_{A\to B}(t)$ in the limit $t \to 0+$:

$$\begin{aligned}\lim_{t\to 0+} k_{A\to B}(t) &= \frac{\langle \dot{q}\theta(q(0+) - q^*)\delta(q^* - q(0))\rangle}{\langle \theta(q^* - q)\rangle} \\ &= \frac{\langle \dot{q}\theta(\dot{q})\delta(q^* - q(0))\rangle}{\langle \theta(q^* - q)\rangle}, \qquad (16.2.2)\end{aligned}$$

where we have used the fact that $\theta(q(0+) - q^*) = 1$, if $\dot{q} < 0$, and 0 otherwise. In other words $\theta(q(0+) - q^*) = \theta(\dot{q})$. The expression on the last line of equation (16.2.2) is the transition state theory prediction for the rate constant, $k^{TST}_{A\to B}$.

Now that we have a microscopic expression for the rate constant $k_{A\to B}$ (equation (16.2.1)), we should consider how to measure it by simulation. It will turn out that we cannot use conventional Molecular Dynamics simulations to measure the quantity on the right-hand side of equation (16.2.1). The reason is the following: what we need to compute is the product of the probability that the system is at point q^* at $t = 0$, multiplied by the generalized velocity $\dot{q}$ at $t = 0$ and this, in turn, multiplied by 1 if the system ends up on the product side and by 0 if the system returns to the reactant side. The problem is that the barrier is usually so high that the probability that this system will cross the barrier spontaneously is very low (in fact, if this were not the case, the whole idea of a time-scale separation between molecular times and times on the order τ_R would not make sense). Hence, if we sample equation (16.2.1) by normal Molecular Dynamics, we would get very poor statistics on $k_{A\to B}$. The solution of this problem is to constrain the system such that it is initially always at the top of the barrier. To be more precise, we equilibrate the system under the constraint $q = q^*$ and then we start a large number of runs in which the system is allowed to cross the barrier (i.e., without any constraints). But now we must be careful because in the previous section we argued that constraining the system, such that $q = q^*$, would affect the equilibrium distribution:

$$\rho_{\text{un-constrained}}(\mathbf{q}) = |H|^{-\frac{1}{2}}\rho_{\text{constrained}}(\mathbf{q}) \qquad (16.2.3)$$

with

$$H_{\alpha\beta} = \sum_{i=1}^{N} m_i^{-1}\frac{\partial\sigma_\alpha}{\partial \mathbf{r}_i}\frac{\partial\sigma_\beta}{\partial \mathbf{r}_i}.$$

In the present case, we have one constraint.[1] Our constraint is

$$\sigma \equiv q^* - q = 0.$$

[1]We assume, for the moment, that there are no other constraints in the system. For a discussion of the latter case, see e.g., ref. [445].

If q is a linear function of the Cartesian coordinates, there is no need to worry about the effect of the constraints on the distribution function, because $|H|$ is a constant. However, in general, q is a nonlinear function of all other coordinates, and we should consider the effect of $|H|$ on $\rho(\mathbf{q})$.

16.2.1 Computational Aspects

In practice, the computation of a rate constant consists of two steps. We write

$$k_{A\to B} = \frac{\langle \dot{q}(0)\delta(q^* - q(0))\theta(q(t) - q^*)\rangle}{\langle \delta(q^* - q(0))\rangle} \times \frac{\langle \delta(q^* - q)\rangle}{\langle \theta(q^* - q)\rangle}. \tag{16.2.4}$$

The first part on the left-hand side of equation (16.2.4) is a conditional average, namely the average of the product $\dot{q}(0)\theta(q(t) - q^*)$, given that $q(0) = q^*$. It is convenient to compute this conditional average using constraint Molecular Dynamics. But then we should correct for the fact that the hard constraint(s) will bias the initial distribution function:

$$\frac{\langle \dot{q}(0)\delta(q^* - q(0))\theta(q(t) - q^*)\rangle}{\langle \delta(q^* - q(0))\rangle} = \frac{\left\langle |H|^{-\frac{1}{2}}\dot{q}(0)\theta(q(t) - q^*)\right\rangle_c}{\left\langle |H|^{-\frac{1}{2}}\right\rangle_c}, \tag{16.2.5}$$

where the subscript c indicates a constrained average over initial configurations. Note, however, that although the initial coordinates from which the reaction trajectories are started have been generated from a constrained ensemble, the subsequent time evolution is not in the constrained ensemble.

We still must compute the second term on the right-hand side of equation (16.2.4), i.e., $\langle \delta(q^* - q)\rangle / \langle \theta(q^* - q)\rangle$. It is the probability density of finding the system at the top of the barrier, divided by the probability that the system is on the reactant side of the barrier. This ratio, which we shall denote by $P(q^*)$, cannot be sampled directly with constrained dynamics, because it is not in the form of a conditional average. We can, however, compute this probability density indirectly. Let us first look at the statistical mechanical expression for $P(q^*)$. In fact, it is convenient to look at a more general quantity $P(q')$, i.e., the probability density of finding the system at q':

$$P(q') \equiv \frac{\int d\mathbf{r}\, \exp(-\beta\mathcal{U})\delta(q' - q)}{\int d\mathbf{r}\, \exp(-\beta\mathcal{U})\theta(q^* - q)}. \tag{16.2.6}$$

From a direct equilibrium simulation of species A, we can measure $P(q')$ near the bottom of the free energy, at the reactant side. Below, we derive an explicit expression for the variation of $P(q')$ with q'. This allows us to compute the variation of $P(q')$ between the bottom and the top of the barrier, and in this way we can compute $P(q^*)$. In fact, rather than computing the derivative of $P(q')$ with respect to q', we differentiate $\ln P(q')$:

$$\frac{\partial \ln P(q')}{\partial q'} = \frac{\int d\mathbf{r} \exp(-\beta\mathcal{U})\partial\delta(q'-q)/\partial q'}{\int d\mathbf{r} \exp(-\beta\mathcal{U})\delta(q'-q)}. \tag{16.2.7}$$

We can reexpress the integral in the numerator by partial integration. To do this, we should first transform from the Cartesian coordinates $\mathbf{r}$ to a set of generalized coordinates $\{\mathbf{Q}, q\}$ that includes the reaction coordinate q. We denote the Jacobian of the transformation from $\mathbf{r}$ to $\{\mathbf{Q}, q\}$ by $|J|$. Now we carry out the partial integration

$$\begin{aligned}\frac{\partial \ln P(q')}{\partial q'} &= \frac{\int d\mathbf{Q}dq\, |J| \exp(-\beta\mathcal{U})\partial\delta(q'-q)/\partial q'}{\int d\mathbf{r}\, \exp(-\beta\mathcal{U})\delta(q'-q)} \\ &= \frac{\int d\mathbf{Q}dq\, \partial|J| \exp(-\beta\mathcal{U})/\partial q \delta(q'-q)}{\int d\mathbf{r}\, \exp(-\beta\mathcal{U})\delta(q'-q)} \\ &= \frac{\int d\mathbf{r}\, \partial(\ln(|J|) - \beta\mathcal{U})/\partial q \exp[-\beta\mathcal{U}]\delta(q'-q)}{\int d\mathbf{r}\, \exp(-\beta\mathcal{U})\delta(q'-q)} \\ &= \frac{\langle \partial(\ln(|J|) - \beta\mathcal{U})/\partial q \delta(q'-q)\rangle}{\langle \delta(q'-q)\rangle},\end{aligned} \tag{16.2.8}$$

where, in the third line, we have transformed back to the original Cartesian coordinates. It should be noted that the computation of the Jacobian $|J|$ can be greatly simplified [460].

As the averages both in the numerator and in the denominator contain $\delta(q-q')$, it is natural to express equation (16.2.8) in terms of constrained averages that can be computed conveniently in a constrained Molecular Dynamics simulation. Just as in equation (16.2.5), we must correct for the bias introduced by the hard constraint:

$$\frac{\partial \ln P(q')}{\partial q'} = \frac{\left\langle |H|^{-\frac{1}{2}} \partial(\ln(|J|) - \beta\mathcal{U})/\partial q \right\rangle_c}{\left\langle |H|^{-\frac{1}{2}} \right\rangle_c}, \tag{16.2.9}$$

where the subscript c denotes averaging in an ensemble where q is constrained to be equal to q'. If we integrate equation (16.2.9) from the bottom to the top of the barrier, we get

$$\ln\left(\frac{P(q^*)}{P(q=q_A)}\right) = \int_{q_A}^{q^*} dq' \frac{\left\langle |H|^{-\frac{1}{2}} \partial(\ln(|J|) - \beta\mathcal{U})/\partial q' \right\rangle_c}{\left\langle |H|^{-\frac{1}{2}} \right\rangle_c} \tag{16.2.10}$$

In practice, this integration has to be carried out numerically.

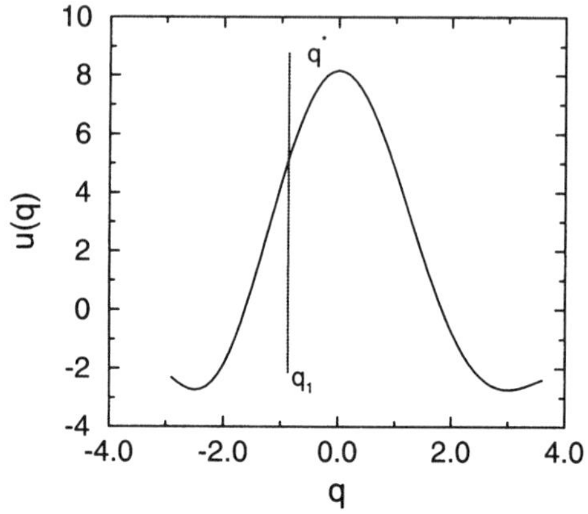

Figure 16.2: Potential-energy barrier for an ideal gas particle; if the particle has a position to the left of the dividing surface q_1 the particle is in state A (reactant). The region to the right of the barrier is designated as product B. The top of the barrier is denoted by q^* ($q^* = 0$).

By combining equations (16.2.5), (16.2.6), and (16.2.10), we finally have an expression for the rate constant $k_{A\to B}$ that can be computed numerically.

It should be noted that, in the above expression, we have assumed that the reaction coordinate is the only quantity that will be constrained in the simulation. If there are more constraints, *e.g.* if we simulate a reaction in a polyatomic fluid, then the expression for $k_{A\to B}$ becomes a bit more complicated (see the article by Ciccotti [445], and references therein).

The expression derived above for a unimolecular rate constant is, in no way, limited to chemical reactions. In fact, the same approach can be used to study any activated classical process, such as diffusion in solids, crystal nucleation, or transport through membranes.

Case Study 23 (Ideal Gas Particle over a Barrier)

To illustrate the "Bennett-Chandler" approach for calculating crossing rates, we consider an ideal gas particle moving in an external field. This particle is constrained to move on the dimensional potential surface shown in Figure 16.2. This example is rather unphysical because the moving particle cannot dissipate its energy. As a consequence, the motion of the particle is purely ballistic. We assume that, far away on either side of the barrier, the particle can exchange energy with a thermal reservoir. Transition state theory predicts a crossing rate given by equation (16.2.2):

$$k_{A\to B}^{TST} = \frac{1}{2}|\dot{q}|\frac{\exp[-\beta u(q^*)]}{\int_{-\infty}^{q^*} dq \exp[-\beta u(q)]} = \sqrt{\frac{k_B T}{2\pi m}}\frac{\exp[-\beta u(q^*)]}{\int_{-\infty}^{q^*} dq \exp[-\beta u(q)]}. \quad (a)$$

If we choose the dividing surface q_1 (see Figure 16.2) at the top of the barrier ($q_1 = q^*$) none of the particles that start off with a positive velocity will return to the reactant state. Hence, there is no recrossing of the barrier and transition state theory is exact for this system.

Note that transition state theory (equation (a)) predicts a rate constant that depends on the location of the dividing surface. In contrast, the Bennett-Chandler expression for the crossing rate is independent of location of the dividing surface (as it should be). To see this, consider the situation that the dividing surface is chosen to be the left of the top of the barrier (i.e., at $q_1 < q^*$). The calculation of the crossing rate according to equation (16.2.4) proceeds in two steps. First we calculate the relative probability of finding a particle at the dividing surface. And then we need to compute the probability that a particle that starts with an initial velocity $\dot{q}$ from this dividing surface will, in fact, cross the barrier. The advantage of the present example is that this probability can be computed explicitly. According to equation (16.2.6), the relative probability of finding a particle at q_1 is given by

$$\frac{\langle \delta(q - q_1) \rangle}{\langle \theta(q_1 - q) \rangle} = \frac{\exp[-\beta u(q_1)]}{\int_{-\infty}^{q_1} dq \, \exp[-\beta u(q)]}. \tag{b}$$

If the dividing surface is not at the top of the barrier, then the probability of finding a particle will be higher at q_1 than at q^*, but the fraction of the number of particles that actually cross the barrier will be less then predicted by transition state theory. It is convenient to introduce the time-dependent transmission coefficient $\kappa(t)$, defined as the ratio

$$\kappa(t) \equiv \frac{k_{A\to B}(t)}{k_{A\to B}^{TST}} = \frac{\langle \dot{q}(0)\delta(q(0) - q_1)\theta(q(t) - q_1) \rangle}{0.5 \, \langle |\dot{q}(0)| \rangle}. \tag{c}$$

The behavior of $\kappa(t)$ is shown in Figure 16.3 for various choices of q_1. The figure shows that for $t \to 0$ $\kappa(t) = 1$, and that for different values of q_1 we get different plateau values. The reason $\kappa(t)$ decays from its initial value is that particles that start off with too little kinetic energy cannot cross the barrier and recross the dividing surface (q_1). The plateau value of $\kappa(t)$ provides us with the correction that has to be applied to the crossing rate predicted by transition state theory. Hence, we see that as we change q_1, the probability of finding a particle at q_1 goes up, and the transmission coefficient goes down. But, as can be seen from Figure 16.3, the actual crossing rate (which is proportional to the product of these two terms) is independent of q_1, as it should be. Now consider the case that $q_1 > q^*$. In that case, all particles starting with positive $\dot{q}$ will continue to the product side. But now there is also a fraction of the particles with negative $\dot{q}$ that will proceed to the product side. These events will give a negative contribution to κ. And the net result is that the transmission coefficient will again be less than predicted by transition

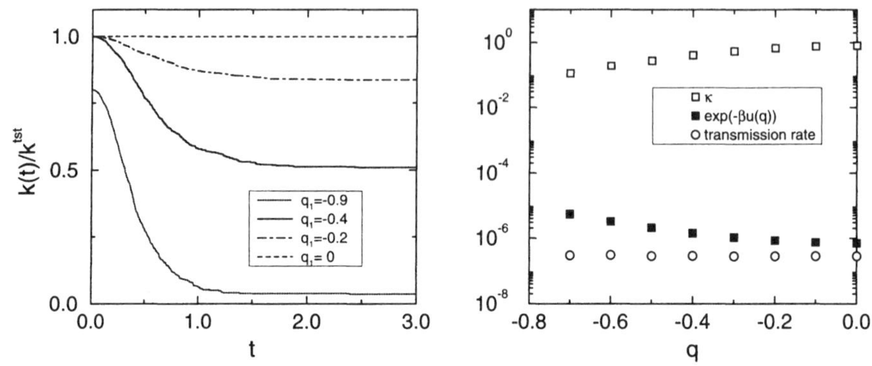

Figure 16.3: Barrier recrossing: the left figure gives the transmission coefficient as a function of time for different values of q_1. The right-hand figure shows, in a single plot, the probability density of finding the system at $q = q_1$ (solid squares), the transmission coefficient κ (open squares), and the overall crossing rate (open circles), all plotted as a function of the location of the dividing surface. Note that the overall crossing rate is independent of the choice of the dividing surface.

state theory. Hence, the important thing is not if a trajectory ends up on the product side, but if it starts on the reactant side and proceeds to the product side. In a simulation, it is therefore convenient always to compute trajectories in pairs: for every trajectory starting from a given initial configuration with a velocity $\dot{q}$, we also compute the time-reversed trajectory, i.e., the one starting from the same configuration with a velocity $-\dot{q}$. If both trajectories end up on the same side of the barrier then their total contribution to the transmission coefficient is clearly zero. Only if the forward and time-reversed trajectories end up on different sides of the barrier, do we get a contribution to κ. In the present (ballistic) case, this contribution is always positive. But in general, this contribution can also be negative (namely, if the initial velocity at the top of the barrier is not in the direction where the particle ends up).

We chose this simple ballistic barrier-crossing problem because we can easily show explicitly that the transmission rate is independent[2] of the location of q_1. We start with the observation that the sum of the kinetic and potential energies of a particle that crosses the dividing surface q_1 is constant. Only those particles that have sufficient kinetic energy can cross the

[2]The general proof that the long-time limit of the crossing rate is independent of the location of the dividing surface was given by Miller [461].

barrier. We can easily compute the long-time limit of $\langle \dot{q}(0)\theta(q(t) - q_1)\rangle$:

$$\begin{aligned}\langle \dot{q}(0)\theta(q(\infty) - q_1)\rangle &= \sqrt{\frac{m\beta}{2\pi}} \int_{v_\epsilon}^{\infty} dv\, v \exp(-\beta m v^2/2) \\ &= \sqrt{\frac{1}{2\pi m\beta}} \exp(-\frac{1}{2}\beta m v_\epsilon^2),\end{aligned}$$

where v_ϵ is the minimum velocity needed to achieve a successful crossing. v_ϵ is given by

$$\frac{1}{2} m v_\epsilon^2 + u(q_1) = u(q^*).$$

It then follows that

$$\langle \dot{q}(0)\theta(q(\infty) - q_1)\rangle = \sqrt{\frac{1}{2\pi m\beta}} \exp\{-\beta[u(q^*) - u(q_1)]\}.$$

This term exactly compensates the Boltzmann factor, $\exp(-\beta u(q_1))$, associated with the probability of finding a particle at q_1. Hence, we have shown that the overall crossing rate is given by equation (a), independent of the choice of q_1.

The reader may wonder why it is so important to have an expression for the rate constant that is independent of the precise location of the dividing surface. The reason is that, although it is straightforward to find the top of the barrier in a one-dimensional system, the precise location of the saddle point in a reaction pathway of a many-dimensional system is usually difficult to determine. With the Bennett-Chandler approach it is not necessary to know the exact location of the saddle point. Still, it is worth trying to get a reasonable estimate, as the statistical accuracy of the results is best if the dividing surface is chosen close to the true saddle point.

The nice feature of the Bennett-Chandler expression for barrier-crossing rates is that it allows us to compute rate constants under conditions where barrier recrossings are important, for instance, if the motion over the top of the barrier is more diffusive than ballistic. Examples of such systems are the cyclohexane interconversion in a solvent [462] and the diffusion of nitrogen in an argon crystal [463].

16.3 Diffusive Barrier Crossing

In the previous section we described the Bennett-Chandler expression for the rate of activated processes. This expression is widely used in numerical simulation. However, although the expression is correct for arbitrary barrier crossings (provided that the barrier is much larger than k_BT), it is not

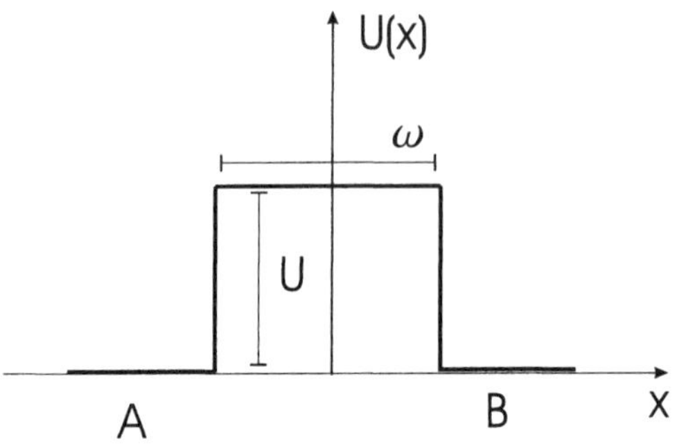

Figure 16.4: Simple model for a diffuse barrier crossing: a square barrier of height U and width ω that separates two macroscopic states, A and B.

always computationally efficient. To see this, consider the expression for the transmission coefficient

$$\kappa(t) \equiv \frac{k_{A\to B}(t)}{k^{TST}_{A\to B}} = \frac{\langle \dot{q}(0)\delta(q(0)-q^*)\theta(q(t)-q^*)\rangle}{0.5\,\langle|\dot{q}(0)|\rangle}. \tag{16.3.1}$$

Clearly, if $\kappa \to 1$, we can use transition state theory (TST) to compute the crossing rate, once we know the barrier height. Hence, the only regime where equation (16.3.1) is of interest is when there are appreciable corrections to TST, i.e., when $\kappa \ll 1$. However, precisely in this regime, the numerical calculation of κ, using equation (16.3.1), is plagued by slow transient behavior and large statistical errors. To illustrate this, let us consider a simple example: a square barrier of height U and width ω that separates two macroscopic states, A and B (see Figure 16.4). For simplicity, we assume that, in equilibrium, the two states have the same probability, $P_{eq} \approx 0.5$ (the population of the barrier region is negligible). Moreover, we assume that the motion in the barrier region is diffusive. For this simple geometry, it is easy to write down the diffusion equation. This equation follows from the continuity equation

$$\frac{\partial\rho(q,t)}{\partial t} = -\frac{\partial}{\partial q}J(q,t), \tag{16.3.2}$$

which relates the local density $\rho(q,t)$ at point q and time t to the flux density $J(q,t)$. In addition, we have the constitutive equation for the diffusional flux in an external field

$$J(q,t) = -D\left[\beta\frac{\partial U(q)}{\partial q}\rho(q,t) + \frac{\partial\rho(q,t)}{\partial q}\right], \tag{16.3.3}$$

where $U(q)$ is the external potential. Combining this with the continuity equation, we obtain the Kramers equation [464],

$$\frac{\partial \rho(q,t)}{\partial t} = \frac{\partial}{\partial q}[D\beta \frac{\partial U}{\partial q}(q)\rho(q,t)] + D\frac{\partial^2 \rho(q,t)}{\partial q^2}, \tag{16.3.4}$$

where D is the diffusion constant of the system. In steady state ($\dot{\rho} = 0$), the flux density J is constant and hence it follows from equation (16.3.3) that

$$J^{st} = -aD, \tag{16.3.5}$$

where we have used the fact that $\partial U/\partial q = 0$ for $-\omega/2 < q < \omega/2$. It then follows from equation (16.3.3) that the probability distribution at the top of the barrier is a linear function of the reaction coordinate, q:

$$\rho^{st}(q) = aq + b \quad \text{for } -\omega/2 < q < \omega/2. \tag{16.3.6}$$

The constants a and b have to be determined from the boundary conditions. In equilibrium, $a = 0$ and $b = \rho_{eq}\exp(-\beta U)$, where ρ_{eq} is the density in states A and B. Let us suppose now that we increase initially the probability density in state A from its equilibrium value by an amount $\delta\rho_{eq}/2$, and decrease the probability density of state B by the same amount. If the barrier is high enough, the flux will be very small and the probabilities of states A and B will not change. In this case, the stationary probability distribution at the top of the barrier is

$$\rho^{st}(q) = e^{-\beta U}\rho_{eq}\left[1 - q\frac{\delta}{\omega}\right], \tag{16.3.7}$$

and the flux

$$J^{st} = D\frac{\delta\rho_{eq}}{\omega}e^{-\beta U}. \tag{16.3.8}$$

As expected, the flux decreases exponentially with the barrier height. The probability density at the top of the barrier is given by equation (16.3.7) if, and only if, the flux has reached its stationary value.

Now consider expression for the rate. We rewrite equation (16.3.1) as

$$\kappa(t) \equiv \frac{\langle \theta(q^* - q(0))\dot{q}(t)\delta(q(t) - q^*)\rangle}{0.5\,\langle|\dot{q}(0)|\rangle}. \tag{16.3.9}$$

Apart from a constant factor, $\kappa(t)$ is the flux through the transition state, q^* $(= 0)$, due to a step function probability profile at $t = 0$. As this step function differs from the linear steady state profile, the resulting flux will depend on time. We are interested in the plateau value of $\kappa(t)$ after the initial transient regime. The usual assumption is that this transient regime extends over typical "molecular" time scales. However, in the present case it is easy to show that the approach of $\kappa(t)$ to its plateau value can be quite slow. For times $t \ll \omega^2/D$, we can combine equations (16.3.4) and (16.3.2) to yield

$$\frac{\partial J(q,t)}{\partial t} \approx D \frac{\partial^2 J(q,t)}{\partial q^2}, \tag{16.3.10}$$

with the solution

$$J(q,t) \approx \epsilon D e^{-\beta u} \rho_{eq} \sqrt{\frac{1}{2\pi D t}} \exp\left(-\frac{(q-q^*)^2}{2Dt}\right). \tag{16.3.11}$$

We then find that $J(q^*,t)$ decays as $1/\sqrt{t}$ for times $t \ll \omega^2/D$. This means that the approach to the stationary state is very slow. But, more importantly, in the case of diffusive barrier crossings, the transmission coefficient κ is typically quite small. And small values of κ cannot be determined accurately using equation (16.3.9). To see this, consider the expression for the transmission coefficient:

$$\kappa = \frac{2}{\langle |\dot{q}| \rangle_{eq}} \langle \dot{q}(0) \theta(q(t) - q^*) \rangle_{q(0)=q^*} . \tag{16.3.12}$$

In a computer simulation, we put the system initially at q^* and let it evolve. We then compute $\theta(q(t) - q^*)$ for times that are long enough for equation (16.3.9) to have reached a plateau value. We repeat this procedure for n independent trajectories, and then estimate κ as

$$\kappa_{est} = \frac{2}{n \langle |\dot{q}| \rangle} \sum_{i=1}^{n} [\dot{q}(0) \theta(q(t) - q^*)]_i . \tag{16.3.13}$$

The statistical error in κ_{est} is given by

$$\sigma_\kappa^2 = \langle (\kappa_{est} - \langle \kappa \rangle)^2 \rangle. \tag{16.3.14}$$

Taking into account that the trajectories are uncorrelated and assuming that $\dot{q}$ and $\theta(q(t) - q^*)$ are Gaussian variables, we get

$$\sigma_\kappa^2 = \frac{4}{n \langle |\dot{q}| \rangle^2} \langle \dot{q}^2 \rangle \langle \theta^2 \rangle + \frac{1}{n} \kappa^2 . \tag{16.3.15}$$

If the transmission coefficient is very small, the second contribution in this expression is negligible, and

$$\sigma_\kappa^2 \sim \frac{4}{n \langle |\dot{q}| \rangle^2} \langle \dot{q}^2 \rangle \langle \theta^2 \rangle . \tag{16.3.16}$$

Moreover,

$$4\frac{\langle \dot{q}^2 \rangle}{\langle |\dot{q}| \rangle^2} \sim \mathcal{O}(1). \tag{16.3.17}$$

Finally, $\langle \theta^2 \rangle = 0.5$. Hence,

$$\sigma_\kappa^2 \sim \frac{1}{n}, \tag{16.3.18}$$

and the relative error is

$$\frac{\sigma_\kappa}{\kappa} \sim \frac{1}{\kappa\sqrt{n}}. \tag{16.3.19}$$

This shows that, even for a transmission coefficient as large as 0.1, we would need to follow about 10^4 trajectories in order to get an accuracy of only 10%. The reason the statistical error is so large is that we use the θ-function to detect transitions from A to B. In a diffusive barrier crossing process, where recrossings of the transition state are frequent, the time evolution of this θ-function resembles a random telegraph signal.

In summary, the Bennett-Chandler approach becomes inefficient for systems with low transmission coefficients because the scheme prepares the system in a state that is not close to the steady-state situation. In addition, this scheme employs the "noisy" θ-function to detect whether the system is in state B.

The obvious question is whether we can do better. Below, we show that this is indeed possible. First of all, we shall go back to equation (16.1.7) and try to devise a perturbation that prepares the system immediately close to the steady state. Secondly, we shall construct a more continuous "detector" function for measuring the concentration of state B. Below, we shall not discuss the general case, but explain the basic ideas in the context of our simple square-barrier model. We refer the reader to the literature [460] for a more general discussion.

As discussed above, the steady-state probability profile at the top of the barrier is a linear function of the reaction coordinate. Hence, if we set up a perturbation that has this shape, rather than a step function, we would eliminate the problem of the slow, diffusive approach to the steady-state crossing rate. Let us therefore replace the θ-function perturbation by a function $g(q)$ chosen such that $g(q) = \theta(q^* - q)$ outside the barrier region, while inside the barrier region[3] $g(q) = 1/2 - q/\omega$.

The change in the equilibrium concentration profile due to this perturbation is

$$\Delta\rho(q) = -e^{-\beta u}\rho_{eq}\frac{q\beta\epsilon}{\omega}. \tag{16.3.20}$$

[3] Note that a perturbation that is everywhere constant does not change the equilibrium distribution. Hence to compute the change in the concentration profile, we can focus on $g(q) = 1/2 - q/\omega$.

But, with the identification $\delta = \beta\epsilon$, this is precisely the (linear) concentration profile that corresponds to the steady state. Hence, with this perturbation, we have suppressed the initial transient. However, if we still use a θ-function to detect whether the system is in state B, the numerical results will still be noisy. So the second step is to replace the "detector" function for state B by $1 - g(q)$. Note that outside the barrier region $g(q) = \theta(q^* - q)$. Hence, replacing θ with g makes a negligible difference for our estimate of the concentration of B. Let us next consider the effect of this choice of the perturbation g on the statistical accuracy for the transmission coefficient κ. We start from equation (16.1.16) for the crossing rate

$$k_{A\to B} = \int_0^\infty dt\, \frac{\langle \dot{q}(0)(\partial_q g_B(0))\dot{q}(t)(\partial_q g_B(t))\rangle}{\langle c_A\rangle}.$$

Now, $g_B = 1 - g_A = \frac{1}{2} + q/\omega$ inside the barrier, and zero elsewhere. Inside the barrier region, we have at all times

$$\partial_q g_B = \frac{1}{\omega}$$

and hence

$$k_{A\to B} = \frac{1}{\omega^2}\int_0^\infty dt\, \frac{\langle \dot{q}(0)\dot{q}(t)\rangle^*}{\langle c_A\rangle},$$

where the asterisk indicates the condition that both $q(0)$ and $q(t)$ should be within the barrier region. If the velocity correlations decay on a time scale that is much shorter than the time it takes to diffuse across the barrier, then we can write

$$\langle \dot{q}(0)\dot{q}(t)\rangle^* \approx \langle \dot{q}(0)\dot{q}(t)\rangle\, \omega \exp(-\beta U)\rho_{eq}.$$

The transition state theory expression for $k_{A\to B}$ is

$$k_{A\to B}^{TST} = 0.5\,\langle|\dot{q}|\rangle\, \frac{\exp(-\beta U)\rho_{eq}}{\langle c_A\rangle}.$$

We then obtain the following expression for the transmission coefficient κ:

$$\kappa = \frac{2}{\omega\,\langle|\dot{q}|\rangle}\int_0^\infty dt\, \langle \dot{q}(0)\dot{q}(t)\rangle\,.$$

Making use of the Green-Kubo relation

$$D = \int_0^\infty dt\, \langle \dot{q}(0)\dot{q}(t)\rangle\,, \tag{16.3.21}$$

we obtain

$$\kappa = \frac{2D}{\omega\,\langle|\dot{q}|\rangle}.$$

As D is of order $\mathcal{O}(\langle|\dot{q}|\rangle\,\lambda)$, where λ is the mean-free path, we immediately see that

$$\kappa \sim \frac{\lambda}{\omega};$$

i.e., the transmission coefficient is approximately equal to the ratio of the mean-free path to the barrier width.

Next, we consider the statistical accuracy of our estimate for κ

$$\kappa_{est} = \frac{2}{\omega\langle|\dot{q}|\rangle n}\sum_{i=1}^{n}\int_0^t dt'\langle\dot{q}(0)\dot{q}(t')\rangle_i, \tag{16.3.22}$$

where we must remember that in all of the n trajectories considered the system is initially at the top of the barrier. Following essentially the same reasoning that led to equation (16.3.15) we now get

$$\begin{aligned}\langle(\Delta\kappa_{est})^2\rangle &= \frac{4}{\omega^2\langle|\dot{q}|\rangle^2 n}\left\{\int_0^t dt'\int_0^t dt''\langle\dot{q}(0)\dot{q}(t')\dot{q}(0)\dot{q}(t'')\rangle\right.\\ &\quad\left.-\left[\int_0^t dt'\langle\dot{q}(0)\dot{q}(t')\rangle\right]^2\right\}.\end{aligned}$$

If we assume, as before, that $\dot{q}$ is a Gaussian variable, then

$$\langle(\Delta\kappa_{est})^2\rangle = \frac{4}{\omega^2\langle|\dot{q}|\rangle^2 n}\left[\langle\dot{q}^2\rangle t\int_0^t dt'\langle\dot{q}(0)\dot{q}(t')\rangle + D^2\right]. \tag{16.3.23}$$

We consider the limit $t \to \infty$. In that limit $D \ll \langle\dot{q}^2\rangle t$ and hence

$$\langle(\Delta\kappa_{est})^2\rangle \sim \frac{4}{\omega^2\langle|\dot{q}|\rangle^2 n}\langle\dot{q}^2\rangle Dt. \tag{16.3.24}$$

The relative error in the computation of the transmission coefficient is now

$$\frac{\langle(\Delta\kappa_{est})^2\rangle^{1/2}}{\kappa} \sim \sqrt{\frac{\langle\dot{q}^2\rangle t}{Dn}}. \tag{16.3.25}$$

From the Green-Kubo relation equation (16.3.21) we see that the diffusion constant D is equal to $\langle\dot{q}^2\rangle\tau_c$, where τ_c is the decay time for velocity fluctuations. Hence,

$$\frac{\langle(\Delta\kappa_{est})^2\rangle^{1/2}}{\kappa} \sim \sqrt{\frac{t}{n\tau_c}}. \tag{16.3.26}$$

Typically, there is not much point in computing the correlation function $\langle\dot{q}(0)\dot{q}(t)\rangle$ for times much larger than τ_c. Hence, the relative error in κ is simply $1/\sqrt{n}$. If we compare this expression for the statistical accuracy in κ with that obtained in the Bennett-Chandler scheme equation (16.3.19)

$$\left[\frac{\langle(\Delta\kappa_{\text{est}})^2\rangle^{1/2}}{\kappa}\right]_{\text{Bennett-Chandler}} = \frac{1}{\kappa\sqrt{n}},$$

we conclude that, by a judicious choice of the scheme to compute κ, we have decreased the statistical error—for a given number of trajectories—by a factor κ. This implies that the present scheme is also applicable in the diffusive regime where $\kappa \ll 1$. Moreover, by suppressing the transient behavior, we have substantially reduced the time to compute a single barrier-crossing trajectory. The additional gain due to the suppression of transients is of order

$$\frac{\tau_{\text{diff}}}{\tau_c} = \frac{\omega^2}{D\tau_c} = \left(\frac{\omega}{\lambda}\right)^2 \approx \frac{1}{\kappa^2}.$$

Hence, the overall gain in speed is of order $1/\kappa^4$. Of course, the present analysis is based on a highly simplified example. A discussion of the application of the present method to more realistic diffusive barrier-crossing problems is discussed in detail in ref. [460].

16.4 Transition Path Ensemble

In the previous sections we have introduced a reaction coordinate as a variable for characterizing the state of the system. In the discussion we have implicitly assumed that it is relatively easy to define a proper order parameter. As we have seen in Case Study 23, the result of a simulation should be independent of this choice of order parameter, but an optimal choice can make a simulation much more efficient. However, in some practical cases the choice of reaction coordinate is not as obvious. For example, if the order parameter involves the complex reorganization of the solvent, the bottleneck may be quite different from the bottleneck in the system without solvent. In such a case, if one uses the experience with the system in vaccum one may end up with a very poor order parameter. As a consequence one has to do a very time-consuming calculation of the crossing rate. Chandler and coworkers [465–468] have developed the transition path ensemble, which is based on earlier ideas of Pratt [469], in which a Monte Carlo method is used to find the transition paths.

16.4.1 Path Ensemble

The starting point is the time correlation function $C(t)$ related to the transitions between states A and B:

$$C(t) = \frac{\langle h_A(x_0) h_B(x_t) \rangle}{\langle h_A \rangle} \approx \langle h_B \rangle \left[1 - \exp(-t/\tau_R)\right], \tag{16.4.1}$$

where x_t are the coordinates and impulses of all particles at time t, and h_A and h_B are characteristic functions that indicate whether the system is in A or B, respectively:

$$h_{A,B}(x) = \begin{cases} 1 & \text{if } x \text{ in A,B} \\ 0 & \text{otherwise.} \end{cases} \tag{16.4.2}$$

The correlation function in equation (16.4.1) gives the conditional probability of finding the system in state B at time t provided it started at $t = 0$ in state A. Also here we assume that these transitions are rare events. As a result the rate constant follows from the plateau of the time derivative:

$$k(t) = \dot{C}(t). \tag{16.4.3}$$

The definitions in this section differ slightly from those introduced in section 16.1. For example, the correlation functions in equations (16.4.1) and (16.1.11) differ by a constant. Since this constant does not affect the time derivative; therefore $k(t)$ in equations (16.4.3) and (16.2.1) are equivalent. Since the function x_t is fully determined by the initial condition x_0, the ensemble averages in equation (16.4.1) can be written as an integration over the initial conditions weighted with the equilibrium distribution $\mathcal{N}(x_0)$:

$$C(t) = \frac{\int dx_0 \mathcal{N}(x_0) h_A(x_0) h_B(x_t)}{\int dx_0 \mathcal{N}(x_0) h_A(x_0)}. \tag{16.4.4}$$

We can also look at equation (16.4.1) as an ensemble average of $h_B(x_t)$:

$$\begin{aligned} \langle C(t) \rangle_{A,t} &= \langle h_B(t) \rangle_{A,t} \\ &= \frac{\int dx_0 \mathcal{N}(x_0) h_A(x_0) h_B(x_t)}{\int dx_0 \mathcal{N}(x_0) h_A(x_0)} \\ &= \frac{\int dx_0 \mathcal{N}_{A,t}(x_0) h_B(x_t)}{\int dx_0 \mathcal{N}_{A,t}(x_0)}, \end{aligned} \tag{16.4.5}$$

where the probability of finding a particular "configuration" is given by $\mathcal{N}_{A,t}$.

Normally we associate a configuration with the position of the particles in the system. In this ensemble, however, a configuration has a completely different meaning. The characteristics of this ensemble are that we compute the average of $C(t)$ over an ensemble of paths that are trajectories that all have a fixed length t and all start at A at $t = 0$. Since we are sampling over paths this ensemble is called the path ensemble. Since these averages depend on the total length of the path (as measured by the time t) the ensemble average of h_B depends on t. A procedure for sampling this ensemble would be to perform a Molecular Dynamics simulation to generate a new path of length t and subsequently use a Monte Carlo procedure to decide whether to accept or reject this new path. In this way, we generate an ensemble of *paths* that we can use to compute ensemble averages. In the following section we will discuss the details on how to perform such a simulation.

In principle we could compute $C(t)$ from an "ordinary" path ensemble simulation. This would imply that we generate an ensemble of paths of length t that start at A and we would count all the paths that are at time t in B. However, since the transition from A to B is a rare event, the number of paths that end in B is so small that such an approach would require very long simulations. Therefore, we need to help the system explore the regions of interest.

Let us define an order parameter $\lambda = \lambda(x)$ to define the state B. The region in which the system is in state B is given by

$$x \in B \quad \text{if} \quad \lambda_{\min} < \lambda(x) < \lambda_{\max}. \tag{16.4.6}$$

If we would, for example, consider a particle moving over a potential barrier, the position of a particle on this barrier could be a choice for $\lambda(x)$. In Chapter 7 we have seen that we can explore a region of high free energies, by dividing the order parameters in small windows and imposing that the system cannot leave such a window. If the window is sufficiently small, the system will explore all values of λ and we can compute the probability distribution of the order parameter. Combining these probability functions gives the desired distribution function (this method is also described in Example 3 in section 7.3).

To compute $C(t)$ we can use a similar trick. Let us define $P(\lambda, t)$ as the probability of finding the system with order parameter $\lambda = \lambda(x_t)$ at time t starting from A at $t = 0$:

$$P(\lambda, t) \equiv \frac{\int dx_0\, \mathcal{N}(x_0) h_A(x_0) \delta[\lambda - \lambda(x_t)]}{\int dx_0\, \mathcal{N}(x_0) h_A(x_0)}. \tag{16.4.7}$$

From this distribution we can compute $C(t)$ by integrating over those values of λ for which the system is in B, cf., equation (16.4.6):

$$C(t) = \int_{\lambda_{\min}}^{\lambda_{\max}} d\lambda\, P(\lambda, t). \tag{16.4.8}$$

To compute $P(\lambda, t)$, we divide the order parameter λ into windows, and a given window is defined by $\lambda \in \left[\lambda^{\min}[i], \lambda^{\max}[i]\right]$. The window potential,

$\mathcal{W}_i(x_t)$, corresponding to window i is

$$\mathcal{W}_i(x_t) = \begin{cases} \infty & \lambda(x_t) < \lambda^{min}[i] \\ 0 & \lambda^{min}[i] < \lambda(x_t) < \lambda^{max}[i], \\ \infty & \lambda(x_t) > \lambda^{max}[i] \end{cases} \tag{16.4.9}$$

where we allow neighboring windows to overlap. The probability of finding the order parameter in window i is given by

$$P(\lambda, t : i) \equiv \frac{\int dx_0 \, \mathcal{N}(x_0) h_A(x_0) \exp[\mathcal{W}(x_t)] \delta[\lambda - \lambda(x_t)]}{\int dx_0 \, \mathcal{N}(x_0) h_A(x_0) \exp[\mathcal{W}(x_t)]}. \tag{16.4.10}$$

This probability distribution can also be written as an ensemble average:

$$\begin{aligned} \langle \delta[\lambda - \lambda(x_t)] \rangle_{A,\mathcal{W},t} &= P(\lambda, t : i) \\ &= \frac{\int dx_0 \, \mathcal{N}_{A,\mathcal{W},t}(x_0) \delta[\lambda - \lambda(x_t)]}{\int dx_0 \, \mathcal{N}_{A,\mathcal{W},t}(x_0)}. \end{aligned} \tag{16.4.11}$$

The function $P(\lambda, t : i)$ gives the probability that a trajectory of length t that starts in A has at time t a value of λ in the particular interval. This function can also be computed from simulations in the path ensemble. The additional feature is that the window potential (16.4.9) gives an additional constraint that paths leaving the window will be rejected in the Monte Carlo procedure.

In principle we can use the window approach to compute the correlation function $C(t)$ for various values of t. The rate constant follows from differentiating this function using equation (16.4.3). However, this is a relatively time-consuming procedure. Dellago *et al.* [468] have developed a more efficient method to compute this function. The function $C(t)$ can be written as the product of the probability of finding a particle in B at time t' times a correction factor:

$$\begin{aligned} C(t) &= \frac{\langle h_A(x_0) h_B(x_t) \rangle}{\langle h_A(x_0) \rangle} \\ &= \frac{\langle h_A(x_0) h_B(x_t) \rangle}{\langle h_A(x_0) h_B(x'_t) \rangle} \times \frac{\langle h_A(x_0) h_B(x'_t) \rangle}{\langle h_A(x_0) \rangle} \\ &= \frac{\langle h_A(x_0) h_B(x_t) \rangle}{\langle h_A(x_0) h_B(x'_t) \rangle} \times C(t'). \end{aligned} \tag{16.4.12}$$

The calculation of this correction, which is the ratio of paths that end in B at time t and t', respectively, is as complex as the calculation of $C(t)$. However, we can cast this term in a more convenient form. For this, let us define a slightly different path ensemble, namely the ensemble of paths that have visited B at least once in the time interval $t \in [0; T]$. An ensemble average in this ensemble can be written as

$$\langle h_B(x_t) \rangle_{A,H_B(T)} = \frac{\int dx_0 \mathcal{N}(x_0) h_A(x_0) H_B(x_T) h_B(x_t)}{\int dx_0 \mathcal{N}(x_0) h_A(x_0) H_B(x_T)}, \tag{16.4.13}$$

where $H_B(T)$ is a characteristic function that has the value 1 if in the interval $[0,T]$ the path has visited B; otherwise this function is 0. The difference with the previous ensemble is that a path does not have to end in B. We can use the fact that for all trajectories with length $t \in [0;T]$ we have

$$h_B(x_t) = h_B(x_t)H_B(x_0, x_T).$$

If at time t the path is in B both sides of the equation are unity and if at time t the path is not in B both sides are 0. If we substitute this relation into equation (16.4.12), we obtain

$$\begin{aligned} C(t) &= \frac{\langle h_A(x_0)h_B(x_t)H_B(x_T)\rangle}{\langle h_A(x_0)H_B(x_T)\rangle} \times \frac{\langle h_A(x_0)H_B(x_t)\rangle}{\langle h_A(x_0)h_B(x_t')H_B(x_T)\rangle} \times C(t') \\ &= \frac{\langle h_B(x_t)\rangle_{A,H_B(T)}}{\langle h_B(x_t')\rangle_{A,H_B(T)}} \times C(t'). \end{aligned} \tag{16.4.14}$$

In this equation we have rewritten the correction factor in terms of two ensemble averages. The nice feature of this ensemble average is that both averages can be obtained from a single simulation of paths with length T.

For the rate constant $k(t)$ we have

$$\begin{aligned} k(t) &= \frac{dC(t)}{dt} \\ &= C(t')\frac{1}{\langle h_B(x_t')\rangle_{A,H_B(T)}} \frac{d\langle h_B(x_t)\rangle_{A,H_B(T)}}{dt} \\ &= \eta(t,t')C(t'). \end{aligned} \tag{16.4.15}$$

With this result the calculation of the rate constant is done in two steps. The first step is the calculation of $\eta(t,t')$ using the path ensemble as defined by equation (16.4.13). The second step is the calculation of $C(t')$ using the window sampling approach in the path ensemble defined by equation (16.4.11). It is important to note that these two calculations are done in slightly different path ensembles.

At this point it is important to note that transition path sampling only requires information on the reactant and product state, and the reaction rate is a result of the calculation. By definition, the reaction path is also a result of the simulation. This can be very important for systems in which one does not have a good idea what to use as an order parameter. Analyzing the paths that have been generated in a path ensemble simulation might give indications what could be an appropiate order parameter.

16.4.2 Monte Carlo Simulations

In the previous section we have shown that the calculation of a rate constant requires the two types of path ensemble simulations. The first ensemble, as defined by equation (16.4.11), uses paths of length t that start in A. This ensemble is used to compute the correlation function $C(t)$. The window potential (16.4.9) gives as additional constraint that at time t the order parameter $\lambda(x_t)$ should have a value in the particular window. The probability distribution in this ensemble is

$$\mathcal{N}(A, W) \propto \mathcal{N}(x_0)h_A(x_o)\exp[-W(x_t)]. \tag{16.4.16}$$

The second ensemble, as defined by equation (16.4.13), uses paths with a length T that also start at A. The additional constraint in this ensemble is that in the interval the path should have been in B (but can leave as well). This ensemble is used to compute the correction factor $\eta(t, t')$. The probability distribution in this ensemble is

$$\mathcal{N}(A, H_B) \propto \mathcal{N}(x_0)h_A(x_o)H_b(x_T). \tag{16.4.17}$$

Acceptance Rules

Let us start with the derivation of the acceptance rules for a Monte Carlo simulation in these two ensembles. The old path is denoted by o and the new path by n. The ratio of the acceptance rules follows from the condition of detailed balance (see section 5.1):

$$\frac{\text{acc}(o \to n)}{\text{acc}(n \to o)} = \frac{\mathcal{N}(n)\alpha(n \to o)}{\mathcal{N}(o)\alpha(o \to n)}, \tag{16.4.18}$$

where $\alpha(o \to n)$ is the *a priori* probability of generating configuration n and $\mathcal{N}(n)$ the desired probability distribution, i.e., equations (16.4.16) or (16.4.17), depending on which path ensemble is used. In the Monte Carlo moves that are discussed below the *a priori* probability of generating configuration n is equal to the generation of o and therefore the acceptance rules reduce to

$$\frac{\text{acc}(o \to n)}{\text{acc}(n \to o)} = \frac{\mathcal{N}(n)}{\mathcal{N}(o)}. \tag{16.4.19}$$

Shooting Moves

We now have to discuss the way we generate a new path. At first sight it may be logical to make a small change in the initial condition x_0 and perform a Molecular Dynamics simulation of length T or t, and to use the acceptance rule to accept or reject this new path. However, this method will work only for relatively short paths. In particular for paths that visit B, a small change in the initial conditions may result in a very large change of the end position; and a path may not end in B. A better strategy is to make a change in the position somewhere in the middle of the path and integrate forwards to compute x_t and backwards in time to compute x_0. Since both A and B are stable states, these points will "attract" paths and therefore such a shooting move in the middle of a path will not deviate enormously from the initial path.

Shifting Moves

We have seen that it is very unlikely to obtain an acceptable path by a small change in the initial positions or velocities. However, we can also change the initial time by shifting initial time $t_0 \to t_0 + \Delta t$ and the final time $t \to t + \Delta t$, where Δt can be positive or negative. Since for this shifting move we do not have to compute the complete path, but only the increment; i.e., do a Molecular Dynamics simulation from t to $t + \Delta t$ if Δt is positive, or backwards in time from t_0 to $t_0 + \Delta t$ if Δt is negative. One has to be careful: since these shifting moves are not ergodic one has to combine such moves with shooting moves. Shifting moves are mostly used to impove the statistics.

Case Study 24 (Ideal Gas Particle in a Two-Dimensional Potential)
To illustrate the path sampling method, consider a system containing a single particle in the following simple two-dimensional potential [467]:

$$\begin{aligned} V(x,y) &= \left(4\left(1-x^2-y^2\right)^2 + 2\left(x^2-2\right)^2 + \left(\left(x+y\right)^2-1\right)^2 \right.\\ &\quad \left. + \left(\left(x-y\right)^2-1\right)^2 - 2\right)/6. \end{aligned} \tag{16.4.20}$$

Note that $V(x,y) = V(-x,y) = V(x,-y)$. Figure 16.5 shows that this potential consists of two stable regions around the points $(-1,0)$, which we call A, and $(1,0)$, which we call region B. To be more specific, all points within a distance of 0.7 from $(-1,0)$ or $(1,0)$ are defined to be in region A or B, respectively. At a temperature of $T = 0.1$ transitions from A to B are rare events.

To compute the rate of transitions from A to B we used path ensemble simulations. The initial distribution $\mathcal{N}(x_0)$ was chosen to be canonical, i.e.,

$$\mathcal{N}(x_0) \propto \exp\left[-\beta\mathcal{H}(x_0)\right].$$

A trajectory was generated using standard Molecular Dynamics simulations (see Chapter 4). The equations of motion were integrated using the velocity-Verlet algorithm with a time step of 0.002.

The first step was the calculation of the coefficient $\eta(t,t')$. This involves the computation of the path ensemble averages $\langle h_B(x_t)\rangle_{A,H_B(T)}$ for various times t. The result of such a simulation is shown in Figure 16.6 for $T = 4.0$ and $T = 3.6$. An important question is whether the time T is long enough.

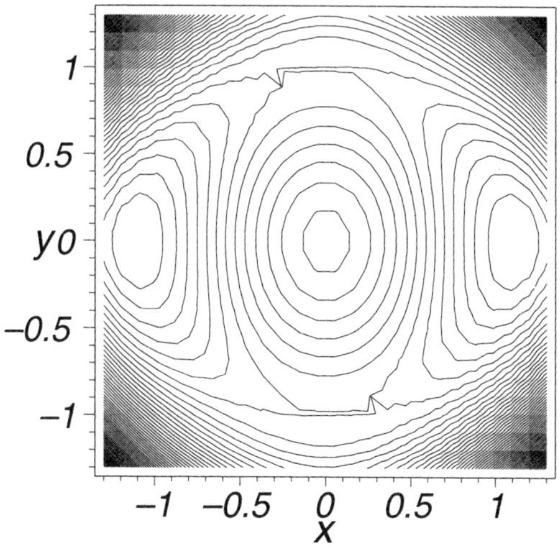

Figure 16.5: Contour plot of the function $V(x, y)$ defined by equation (16.4.20). The two minima are at $(-1, 0)$, A, and $(0, 1)$, B. These minima are separated by a potential energy barrier.

Since we are interested in the plateau of $k(t)$, the function $\langle h_B(x_t)\rangle_{A,H_B(T)}$ must have become a straight line for large values of t. If this function does not show a straight line, the value of T was probably too short, the process is not a rare event, or the process cannot be described by a single hopping rate. The consistency of the simulations can be tested by comparing the results with a simulation using a shorter (or longer, but this is more expensive) T. Figure 16.6 shows that the results of the two simulations are consistent.

The next step is the calculation of the correlation function $C(t)$. For the calculation of $P(\lambda, t)$, we have defined the order parameter λ as the distance from point B:

$$\lambda = 1 - \frac{|\mathbf{r} - \mathbf{r_B}|}{|\mathbf{r_A} - \mathbf{r_B}|}, \tag{16.4.21}$$

in which $\mathbf{r_B} = (1, 0)$. In this way, the region B is defined by $0.65 < \lambda \leq 1$ and the whole phase space is represented by $\langle -\infty, 1]$. In Figure 16.7 (left), we have plotted $P(\lambda, i, t = 3.0)$ as a function of λ for different slices i. Recombining the slices leads to Figure 16.7 (right). The value of $C(t = 3.0)$ can be obtained by integrating over region B:

$$C(t) = \int_B d\lambda P(\lambda, t). \tag{16.4.22}$$

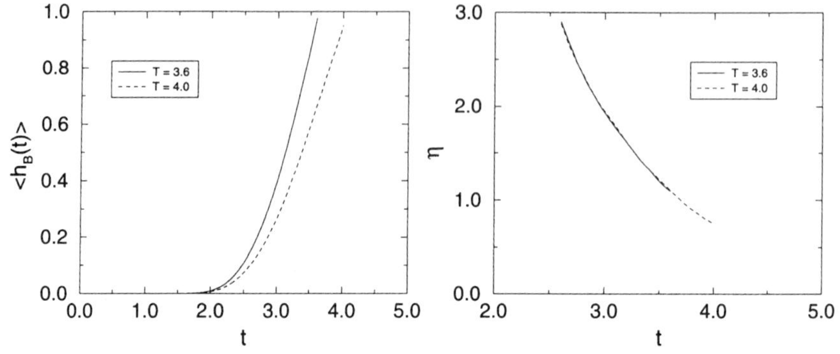

Figure 16.6: $h_B(t)$ (left) and $\eta(t)$ (right) as a function of time for various values of the total path length T.

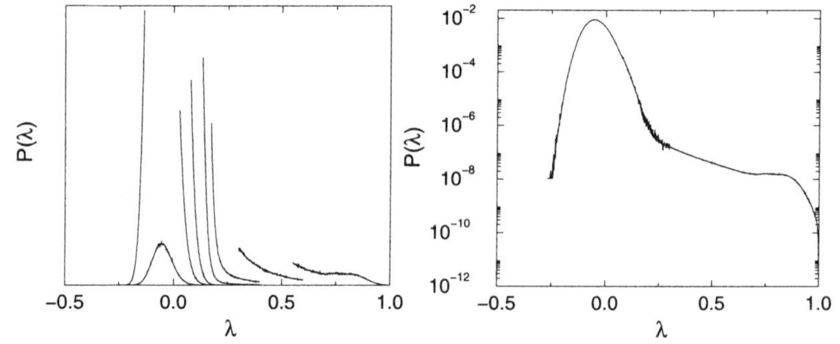

Figure 16.7: (left) $P(\lambda, i, t = 3.0)$ for all slices i. (right) $P(\lambda, t = 3.0)$ when all slices i are combined. The units on the y axis are such that $\int_{-\infty}^{1} d\lambda\, P(\lambda, t) = 1$.

Combining the results gives for the total crossing rate

$$k = \eta(t)\, C(t). \tag{16.4.23}$$

Using $t = 3.0$ leads to $\eta(2.0) = 1.94$, $C(3.0) = 4.0\times10^{-6}$, and $k = 8.0\times10^{-6}$.

Example 25 (Transition Path Sampling with Parallel Tempering)
Transition path sampling (TPS) is a technique that allows us to compute the rate of a barrier-crossing process without *a priori* knowledge of the reaction coordinate or the transition state. However, when there are many distinct pathways that lead from one stable state to another, then it can be difficult to sample all possible pathways within the time scale of a single simulation. Vlugt and Smit [470] have shown that parallel tempering (see section 14.1)

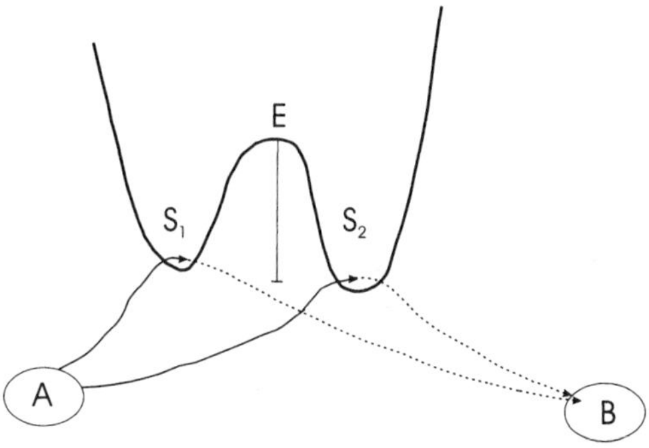

Figure 16.8: Schematic representation of two different transition paths from state A to state B. A and B are two stable states, separated by a free energy barrier. There are two dynamical pathways for the system to go from A to B, one path crosses the barrier via saddle point S_1 while the other path crosses via saddle point S_2. If the (free) energy barrier E between the two paths is much larger than $k_B T$, it is unlikely that path 1 will evolve to path 2 in a single transition path simulation. Note that the energy barrier between two paths (E) is not the same as the energy barrier along the path.

can be used to speed up the sampling of transition pathways that are separated by high free energy barriers.

The objective of TPS is to sample all relevant transition paths within a single simulation. This becomes difficult when different transition paths lead to distinct saddle points in the free energy surface. To be more precise, problems arise when the (free) energy barrier between two saddle points is much higher than $k_B T$ (see Figure 16.8). Of course, the sampling problem would be much less serious if one could work at much higher temperatures where the transition path can cross the barriers separating the saddle points.

Parallel tempering exploits the possibility of generating "transitions" between different saddle points at high temperatures, to improve the sampling efficiency at low temperatures. As was already discussed in section 14.1, parallel tempering can be used to switch between systems at various temperatures. As an illustration of this combined parallel tempering and transition path sampling approach, we consider the example discussed in ref. [470]: a two-dimensional system containing a linear chain of 15 repulsive Lennard-Jones particles. The nonbonded interactions are given by

$$u_{rep}(r) = \begin{cases} 1 + 4\left[r^{-12} - r^{-6}\right] & r \leq r_{rep} \\ 0 & r > r_{rep} \end{cases}, \tag{16.4.24}$$

in which r is the distance between two particles and $r_{rep} \equiv 2^{1/6}$ (we have used the Lennard-Jones σ as our unit of length). Neighboring particles i and j in the chain interact through a double-well potential:

$$u_{dw}(r_{ij}) = h\left[1 - \frac{(r_{ij} - w - r_{rep})^2}{w^2}\right]^2, \qquad (16.4.25)$$

where r_{ij} denotes the distance between the neighboring particles i and j. This potential has two equivalent minima, one for $r_{ij} = r_{rep}$ and the other for $r_{ij} = r_{rep} + 2w$ ($w > 0$).

The chain can have a compact state, A, if all bonds are in the first minimum, or an extended state, B, if all bonds are in the second minimum. We wish to express the transition rate from A to B as the sum of the rates of the contributions due to all distinct transition paths. Clearly, there are many (in this case 14!) distinct pathways that lead from the compact state to the fully extended state (e.g., first stretch bond $2-3$, then $1-2$, then $6-7$, etc.). Without parallel tempering it would take a prohibitively long time to obtain a representative sampling of all transition paths at low temperatures. In the present case, the number of distinct reaction paths is too large to be sampled even *with* parallel tempering. However, for many problems the number of relevant paths is small and the present approach can be used to compute the rate constant. For more details the reader is referred to ref. [470].

Example 26 (Ion Pair Dissociation)
The dissociation of an Na^+Cl^- pair in water is an example of an activated process. It is of particular interest to understand the effect of the water molecules on the dynamics of this process.

As a first guess, one can use the ionic separation as a reaction coordinate:

$$r_{ion} \equiv |\mathbf{r}_{Na^+} - \mathbf{r}_{Cl^-}|.$$

The free energy as a function of this reaction coordinate is shown schematically in Figure 16.9. Once we have computed the free energy barrier, we could, in principle, use the Bennett-Chandler approach to compute the reaction rate (see section 16.2). However, for this system one would observe a very small transmission coefficient, which suggests that the chosen reaction coordinate does not provide an adequate description of the dynamics of this reaction.

Figure 16.10 explains how an unfortunate choice of the reaction coordinate may result in a low transmission coefficient in the Bennett-Chandler expression for the rate constant. But even if the reaction coordinate is well chosen, we may still get a low transmission coefficient. If the free energy landscape of the dissociation reaction looks like Figure 16.10(a), the progress of the reaction would correlate directly with the ionic separation. However, it could still be that the system exhibits diffusive behavior near the transition

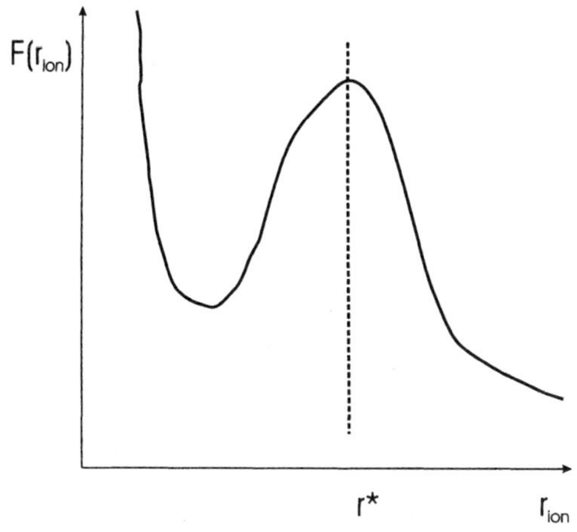

Figure 16.9: Free energy as a function of the ionic separation r_{ion}.

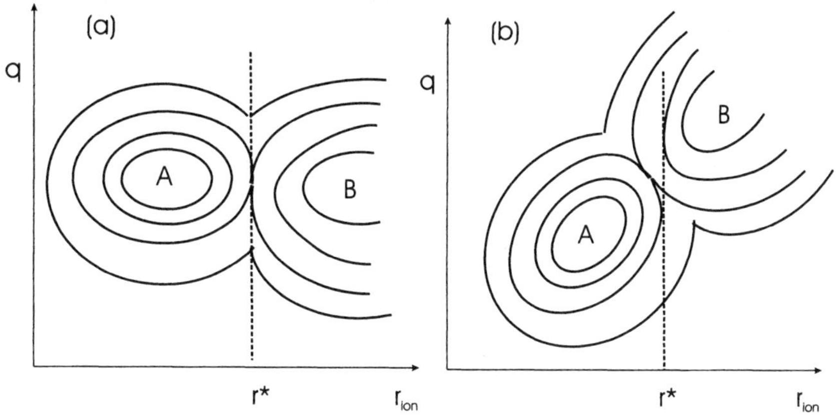

Figure 16.10: Two possible scenarios for the ion dissociation; the two figures show the two-dimensional free energy landscape in a contour plot. A is the stable associated state and state B the dissociated state, the dotted line corresponds to the dividing surface as defined by the maximum of the free energy profile (see Figure 16.9), r_{ion} is the reaction coordinate while q represents all other degrees of freedom. In (a) one sees that the dividing surface nicely separates the two stable basins, while in (b) a point of r^* "belongs" either to the A basin or to the "B" basin.

state. If this is the case, one would obtain better statistics using the diffusive barrier crossing method described in section 16.3.

Another possible scenario is shown in Figure 16.10(b). Here, we have a situation in which the ionic separation is not a good reaction coordinate. Unlike the situtation in Figure 16.10(a), the dividing surface does not discriminate the two stable states. Apparently there is another relevant coordinate (denoted by q) that is an (as yet unknown) function of positions of the solvent particles.

Since we do not know what this additional order parameter is, this is an ideal case to use transition path sampling (TPS), as in TPS we do not have to make an *a priori* choice of reaction coordinate.

Geissler *et al.* [471] used transition path sampling to generate some 10^3 reaction paths for this process. These paths were subsequently analyzed to obtain the *transition state ensemble.* This is the ensemble of configurations on the reaction paths that have the following "transition-state" property: half of the trajectories that are initiated at configurations that belong to this ensemble end up on the product side, and the other half on the reactant side. Although all trajectories originate from the same configuration, they have different initial velocities (drawn from the appropriate Maxwellian distribution). In general, P_B, the probability that a trajectory starting from an arbitrary configuration will end up in state B is different from 0.5.

Geissler *et al.* showed that $P_B \approx 0$ for most configurations that contained fivefold-coordinated sodium ions. Conversely, $P_B \approx 1$, for configurations with sixfold-coordinated sodium ions. For the Cl^- ion, no such effect was found. This indicates that, in order to reach the transition state from the associated state, water molecules have to enter into the first solvation shell of the sodium ions. The water coordination of the Na^+ ion was the order parameter that was missing in the simple analysis. This example illustrates how TPS can be used to elucidate unknown "reaction" mechanisms.

16.5 Searching for the Saddle Point

In section 16.4, we described a general procedure for finding barrier-crossing rates. In principle, this procedure should work even if we have no knowledge about the reaction coordinate. However, in practice, the calculations may become very time consuming. For this reason, a variety of techniques that aim to identify the relevant reaction coordinate have been proposed. Below, we briefly describe some of the methods that have been proposed. More details (and more references) can be found in ref. [459]. The methods for searching for a reaction path, whihc we discuss, have been designed for situations where the free energy barrier separating "reactant" and "product" is energy-dominated. This is often the case, but certainly not always (see, e.g., [427]).

One such energy-based scheme is the so-called "nudged elastic band" (NEB) method proposed by Jonsson and collaborators [472]. This method aims to find the lowest-energy path to the saddle point that separates the reactant basin from the product basin. The NEB method assumes that both the reactant and product states are known. A number of replicas of the original system are now prepared. These replicas are initially located equidistantly along the "linear" path from reactant to product. The position of this originally linear path is now relaxed to find the reaction path. This is achieved as follows: the replicas are connected by harmonic springs that tend to keep them equally spaced—this is the "elastic band." In addition all replicas experience the gradient of the intermolecular potential that tends to drive them to a minimum of the potential energy. However, these gradient forces are only allowed to act perpendicular to the local tangent. Conversely, the elastic band forces are only allowed to act along the local tangent. As a consequence, the intermolecular forces move the elastic band laterally until the tranverse forces vanish (i.e., when it is a minimum energy path) while the longitudinal forces prevent all replicas from collapsing into the reactant or product state. For more details and further refinements, see [473–475].

A technique that is similar in spirit, but very different in execution, to the NEB method is the activation-relaxation technique developed by Barkema and Mousseau [476]. This method also aims to find the lowest energy path to the saddle point. Unlike the NEB method, this scheme does not make use of any *a priori* knowledge of the product basin. To find the saddle point, the system is forced to move "uphill" against the potential energy gradient. However, if we would simply let the system move in a direction opposite to the force that acts on it, we would reach a potential energy *maximum*, rather than a saddle point. Hence, in the method of [476], the force that acts on the system is only inverted (and then only fractionally) along the vector in configuration space that connects the position of the initial energy minimum (i.e., the lowest energy initial state) with the present position of the system. In all other directions, the original forces keep acting on the system. The aim of this procedure is to force the system to stay close to the lowest energy trajectory towards the saddle point. Often there is more than one saddle point. In that case, the initial displacement of the system from the bottom of the reactant basin will determine which saddle point will be reached. Note that we cannot tell *a priori* whether the saddle point that is found will indeed be the relevant one. The true transition state can only be found by attempting many different initial displacements, and by computing the energy and (in the case of a quasi-harmonic energy landscape) the entropy of the saddle point.

The only reason special techniques are needed to simulate activated processes is simply that rare events are ... rare. If, somehow, one could artificially increase the frequency of rare events in a controlled way, this would allow us to use standard simulation techniques to study activated processes. Voter and collaborators [477,478] have explored this route. The idea behind the approach of Voter is that the rate of activated processes can be increased either by artificially lowering the energy difference between the top of the barrier and the reactant basin ("hyperdynamics" [477]) or by increasing the temperature ("temperature-accelerated dynamics" [478]). The trick is to apply these modifications in such a way that it is possible to correct for the effect that they have on the crossing rate. In both schemes, the essence of the correction is that the rate k_i^B, at which the system crosses a point i in the biased system, is higher than the corresponding rate k_i^u in the unbiased system. To recover the unbiased rate, the biased rate should be multiplied by a factor $P^u_{\text{Boltzmann}}(i)/P^B_{\text{Boltzmann}}(i)$, where $P^u_{\text{Boltzmann}}$ ($P^B_{\text{Boltzmann}}$) is the unbiased (biased) Boltzmann weight of configuration i. For more details, the reader is referred to refs. [477, 478]. To recover the unbiased rate, the biased rate should be multiplied by a factor $P^u_{\text{Boltzmann}}(i)/P^B_{\text{Boltzmann}}(i)$, where $P^u_{\text{Boltzmann}}$ ($P^B_{\text{Boltzmann}}$) is the unbiased (biased) Boltzmann weight of configuration i.

An additional "linear" speed-up of the rate calculations can be achieved by performing n barrier-crossing calculations in parallel [479]. Although this approach does not reduce the total amount of CPU time required, it does reduce the wall-clock time of the simulation. For more details, the reader is referred to refs. [477–479].

Chapter 17

Dissipative Particle Dynamics

This book focuses on molecular simulation techniques. Thus far, we focused on those simulation techniques that, if we had infinite computing power at our disposal, would yield the exact equilibrium properties of the molecular model under study. The Dissipative Particle Dynamics (DPD) technique that we discuss in the present chapter is different in spirit: it is, by construction, an approximate, coarse-grained scheme. Such models are used when we need to study the behavior of a system containing very many molecules for a very long time. For example, colloidal suspensions are dispersions of mesoscopic (10 nm – 1 μm) solid particles. These particles themselves consist of millions, or even billions, of atoms. Furthermore, the number of solvent molecules per colloid is comparable or even larger. Clearly, a Molecular Dynamics simulation that follows the behavior of several thousand colloids over an experimentally relevant time interval (milliseconds to seconds) would be prohibitively expensive. This is why colloidal suspensions are always modeled using a coarse-grained model. The simplest coarse-grained model for colloidal suspensions is the hard-sphere fluid. This model can be used to approximate the static properties of dispersions of uncharged, spherical colloids with negligible dispersion interactions.[1] However, if one is interested in the colloidal dynamics, the solvent cannot be ignored. Colloidal particles undergo many collisions with the solvent and these collisions are responsible for the Brownian motion of the colloid. In contrast, in a conventional MD simulation of hard-sphere dynamics, the motion of the spheres in between collisions is purely ballistic. This shows that, in order to model

[1] In fact, several predictions of the early simulations on hard-sphere fluids have been confirmed with experiments on suspensions of such colloids.

colloidal dynamics, we need to account for the effect of the solvent on the colloidal motion.

However, we know that Brownian motion of uncharged colloids does not depend on the atomic details of the solvent. The only properties of the medium that matter are its temperature, density, and viscosity. This suggests that a fully atomic model is not needed to study colloidal motion. What we need is a "cheap" model of a solvent that has the following features:

- It exhibits hydrodynamic behavior.
- It has thermal fluctuations that can drive Brownian motion.
- It is cheap to simulate.

The Dissipative Particle Dynamics method was introduced by Hoogerbrugge and Koelman [480, 481] with these requirements in mind. It should be stressed, however, that the DPD method is not unique. Alternative schemes for achieving the same objective, such as the fluctuating lattice-Boltzmann method of Ladd [482] and, more recently, a hybrid MD-cellular-automaton model of Malevanets and Kapral [483], have been proposed in the literature. Both the DPD approach and the method of ref. [483] bear strong similarities to Molecular Dynamics. In the DPD approach, the forces due to individual solvent molecules are lumped together to yield effective friction and a fluctuating force between moving fluid elements. While this approach does not provide a correct atomistic description of the molecular motion, it has the advantage that it does reproduce the correct hydrodynamic behavior on long length and time scales.

17.1 Description of the Technique

The basic DPD algorithm is very similar to MD: the difference is that, in addition to the conservative force acting between particles, the total force on a particle i now also contains a dissipative force and a random force:

$$\mathbf{F}_i = \sum_{j \neq i} \left[\mathbf{f}^C(\mathbf{r}_{ij}) + \mathbf{f}^D(\mathbf{r}_{ij}, \mathbf{v}_{ij}) + \mathbf{f}^R(\mathbf{r}_{ij}) \right]. \tag{17.1.1}$$

The conservative force $\mathbf{f}_{ij}^C$ can, for instance, be derived from a pair potential[2] that acts between particles i and j. The dissipative force $\mathbf{f}_{ij}^D$ corresponds to a frictional force that depends both on the positions and the relative velocities of the particles:

$$\mathbf{f}^D(\mathbf{r}_{ij}, \mathbf{v}_{ij}) = -\gamma \omega^D(r_{ij})(\mathbf{v}_{ij} \cdot \hat{\mathbf{r}}_{ij})\hat{\mathbf{r}}_{ij}, \tag{17.1.2}$$

[2]More recent versions of DPD make use of many-body forces. However, somewhat surprisingly, this does not change the structure of the algorithm (see [484]).

where $\mathbf{r}_{ij} = \mathbf{r}_i - \mathbf{r}_j$ and $\mathbf{v}_{ij} = \mathbf{v}_i - \mathbf{v}_j$, and $\hat{\mathbf{r}}_{ij}$ is the unit vector in the direction of $\mathbf{r}_{ij}$. γ is a coefficient controlling the strength of the frictional force between the DPD particles. $\omega^D(r_{ij})$ describes the variation of the friction coefficient with distance. Possible choices for the r-dependence of the friction coefficient will be discussed later. The random force $\mathbf{f}^R(\mathbf{r}_{ij})$ is of the form:

$$\mathbf{f}^R(\mathbf{r}_{ij}) = \sigma\omega^R(r_{ij})\xi_{ij}\hat{\mathbf{r}}_{ij}. \tag{17.1.3}$$

σ determines the magnitude of the random pair force between the DPD particles. ξ_{ij} is a random variable with Gaussian distribution[3] and unit variance and $\xi_{ij} = \xi_{ji}$, while $\omega^R(r_{ij})$ describes the variation of the random force with distance. The functions $\omega^R(r)$ and $\omega^D(r)$ cannot be chosen independently. In order for the configurations of the systems to occur with the proper Boltzmann weight, the following relation must be satisfied:

$$\omega^D(r_{ij}) = \left[\omega^R(r_{ij})\right]^2. \tag{17.1.4}$$

γ and σ are related to the temperature according to

$$\sigma^2 = 2k_BT\gamma. \tag{17.1.5}$$

At first sight, the DPD method bears a strong resemblance to the so-called Brownian dynamics method (see, e.g., [19]); both schemes employ a combination of random and dissipative forces. However, in Brownian dynamics the frictional and random forces do not conserve momentum. In fact, the only property that is conserved in Brownian dynamics is the total number of particles. In DPD, however, the particular functional forms of the frictional and random forces ensure that all forces obey action-equals-reaction, and hence the model conserves momentum. This is essential for recovering the correct "hydrodynamic" (Navier-Stokes) behavior on sufficiently large length and time scales.[4]

17.1.1 Justification of the Method

The first step in the justification of the method is to investigate the relation between DPD and thermodynamics [487]. The original method was proposed using heuristic arguments. Español and Warren [487] provided a proper statistical-mechanical basis. Español and Warren have shown that the DPD can be written in the form of a Fokker-Planck equation [47]:

$$\partial_t\mathcal{N}(r,p;t) = \mathcal{L}_C\mathcal{N}(r,p;t) + \mathcal{L}_D\mathcal{N}(r,p;t), \tag{17.1.6}$$

[3]Groot and Warren [485] found that a uniform distribution with unit variance gave similar results.

[4]An alternative approach for imposing both momentum conservation and proper Boltzmann sampling has been proposed by Lowe [486].

where $\mathcal{L}_C$ is the usual Liouville operator of a Hamiltonian system interacting with conservative forces $\mathbf{F}^C$,

$$\mathcal{L}_C \equiv -\left[\sum_i \frac{\mathbf{p}_i}{m}\frac{\partial}{\partial \mathbf{r}_i} + \sum_{i,j} \mathbf{f}_{ij}^C \frac{\partial}{\partial \mathbf{p}_i}\right], \tag{17.1.7}$$

and the operator $\mathcal{L}_D$ takes into account the effects of the dissipative and the random forces:

$$\mathcal{L}_D \equiv \sum_{i,j} \hat{\mathbf{r}}_{ij}\frac{\partial}{\partial \mathbf{p}_i}\left[\gamma\omega_D(r_{ij})(\hat{\mathbf{r}}_{ij}\cdot\mathbf{v}_{ij}) + \sigma^2\omega_R^2(r_{ij})\hat{\mathbf{r}}_{ij}\left(\frac{\partial}{\partial \mathbf{p}_i} - \frac{\partial}{\partial \mathbf{p}_j}\right)\right]. \tag{17.1.8}$$

The derivation of these equations uses techniques developed for stochastic differential equations. The importance of casting the DPD equations in a Fokker-Planck form is that we can use the theory of Markov processes to prove that the system evolves to an equilibrium distribution; i.e., the steady-state solution of equation (17.1.6) corresponds to

$$\partial_t \mathcal{N}_{eq}(r, p; t) = 0.$$

To make the connection with statistical mechanics, the steady-state solution should correspond to the canonical distribution:

$$\begin{aligned}\mathcal{N}_{eq}(r, p; t) &= \frac{1}{Q_{NVT}}\exp[-\beta\mathcal{H}(r, p)]\\ &= \frac{1}{Q_{NVT}}\exp\left[-\beta\left(\sum_i p_i/2m_i + V(r)\right)\right],\end{aligned}$$

where $V(r)$ is the potential that gives rise to the conservative forces, $\mathcal{H}(r, p)$ is the Hamiltonian, and Q_{NVT} is the partition function of the NVT ensemble. By definition, this equilibrium distribution satisfies

$$\mathcal{L}_C\mathcal{N}_{eq}(r, p; t) = 0.$$

We therefore need to ensure that

$$\mathcal{L}_D\mathcal{N}_{eq}(r, p; t) = 0.$$

This is achieved by imposing that

$$\omega_R^2(r) = \omega_D(r) \qquad \text{and} \qquad \sigma^2 = 2k_BT\gamma.$$

This is precisely the choice made in the previous section.

We still have to show that, on sufficiently large lengths and time scales, the DPD fluid obeys the Navier-Stokes equation of hydrodynamics. At present, there exists no rigorous demonstration that this is true for an arbitrary

DPD fluid. However, all existing numerical studies suggest that, in the limit where the integration time step $\delta t \to 0$, the large-scale behavior of the DPD fluid is described by the Navier-Stokes equation. The kinetic theory for the transport properties of DPD fluids [488–491] supports this conclusion. One interesting limit of the DPD model is the "dissipative ideal gas," i.e., a DPD fluid without the conservative forces. The static properties of this fluid are those of an ideal gas. However, its transport behavior is that of a viscous fluid.

The advantage of DPD over conventional (atomistic) MD is that it involves a coarse-grained model. This makes the technique useful when studying the mesoscopic structure of complex liquids. However, if we are only interested in static properties, we could have used standard MC or MD on a model with the same conservative forces, but without dissipation. The real advantage of DPD shows up when we try to model the dynamics of complex liquids.

17.1.2 Implementation of the Method

A DPD simulation can be implemented in any working Molecular Dynamics program. The only subtlety is in the integration of the equations of motion. As the forces between the particles depend on their relative velocities, the standard velocity-Verlet scheme cannot be used.

In their original publication Hoogerbrugge and Koelman [480] used an Euler-type algorithm to integrate the equations of motion. However, Marsh and Yeomans found that, with such an algorithm, the effective equilibrium temperature depends on the time step that is used in a DPD simulation [492]. Only in the limit of the time step approaching zero was the correct equilibrium temperature recovered. A similar result was obtained by Groot and Warren [485] using a modified velocity-Verlet algorithm. There is, however, an important feature missing in these algorithms. If we compare the DPD integration schemes with those used in a Molecular Dynamics simulation, all "good" MD schemes are intrinsically time-reversible while the above DPD schemes are not. Pagonabarraga *et al.* [493] argued that time reversibility is also important in a DPD simulation, since only with a time-reversible integration scheme can detailed balance be obeyed. In the Leap-Frog scheme (see section 4.3.1) the velocities are updated using

$$v(t+\Delta t/2) = v(t-\Delta t/2) + \Delta t\frac{f(t)}{m}, \tag{17.1.9}$$

and the positions using

$$r(t+\Delta t) = r(t) + \Delta t v(t+\Delta t/2). \tag{17.1.10}$$

In DPD, the force at time t depends on the velocities at time t. The velocity at time t is approximated by

$$v(t) = \frac{v(t + \Delta t/2) + v(t - \Delta t/2)}{2}.$$

This implies that the term $v(t + \Delta t/2)$ in equation (17.1.9) appears on both sides of the equations. In the scheme of Pagonabarraga *et al.*, these equations were solved self-consistently; i.e., the value for $v(t + \Delta t/2)$ calculated from equation (17.1.9) has to be the same as the value for $v(t + \Delta t/2)$ used to calculate the force at time t. This implies that we have to perform several iterations before the equations of motion can be solved. This self-consistent scheme implies that the equations are solved in such a way that time reversibility is preserved.

To see how this time reversibility is related to detailed balance, we consider a single DPD step as a step in a Monte Carlo simulation. If we have only conservative forces, DPD is identical to standard Monte Carlo. In fact we can use the hybrid Monte Carlo scheme (see section 14.2) for the DPD particles. For hybrid Monte Carlo it is essential that a time-reversible algorithm is used to integrate the equations of motion. In the case of hybrid Monte Carlo detailed balance implies that if we reverse the velocities, the particles should return to their original positions. If this is not the case detailed balance is not obeyed. If we use a noniterative scheme to solve the equations of motion in our DPD scheme, the velocity that we calculate at time t is not consistent with the velocity that is used to compute the force at this time. Hence, if we reverse the velocities the particles do not return to their original positions and detailed balance is not obeyed. In Case Study 25 the DPD method is illustrated with a few examples.

Case Study 25 (Dissipative Particle Dynamics)

To illustrate the DPD technique, we have simulated a system of two components (1 and 2). The conservative force is a soft repulsive force given by

$$\mathbf{f}_{ij}^{C} = \begin{cases} a_{ij}(1 - r_{ij})\hat{\mathbf{r}}_{ij} & r_{ij} < r_c \\ 0 & r_{ij} \geq r_c \end{cases}, \tag{17.1.11}$$

in which $r_{ij} = \|\mathbf{r}_{ij}\|$ and r_c is the cutoff radius of the potential. The random forces are given by equation (17.1.3) and the dissipative forces by equation (17.1.2). The total force on a particle equals the sum of the individual forces:

$$\mathbf{f}_i = \sum_{i \neq j} \left(\mathbf{f}_{ij}^{C} + \mathbf{f}_{ij}^{S} + \mathbf{f}_{ij}^{R} + \mathbf{f}_{ij}^{D} \right). \tag{17.1.12}$$

To obtain a canonical distribution we use

$$\sigma^2 = 2\gamma k_B T$$

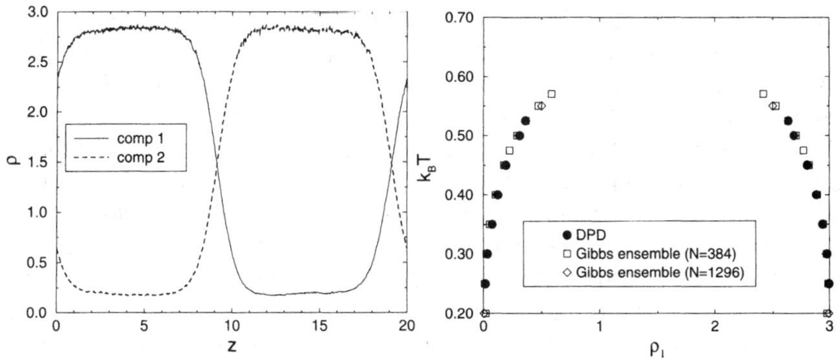

Figure 17.1: (left) Density profile for $k_B T = 0.45$. (right) Phase diagram as calculated using DPD and Gibbs ensemble simulations. Both techniques result in the same phase diagram, but the Gibbs ensemble technique needs less particles due to the absence of a surface. In the DPD simulations, we have used a box of $10 \times 10 \times 20$ (in units of r_c^3). The time step of the integration was $\Delta t = 0.03$.

$$w^D(r_{ij}) = \left[w^R(r_{ij})\right]^2.$$

A simple but useful choice is [485]

$$w^D(r_{ij}) = \begin{cases} (1 - r_{ij}/r_c)^2 & r_{ij} < r_c \\ 0 & r_{ij} \geq r_c \end{cases}$$

with $r_c = 1$. The simulations were performed with $\rho = 3.0$ and $\sigma = 1.5$. We have chosen $a_{ii} = a_{jj} = 25$ and $a_{ij,i \neq j} = 30$. This system will separate into two phases. In the example shown in Figure 17.1, we have chosen the z-direction perpendicular to the interface. The left part of Figure 17.1 shows typical density profiles of the two components. In Figure 17.1 (right), we have plotted the concentration of one of the components in the coexisting phases.

Since we can write down a Hamiltonian for a DPD system, we can also perform standard Monte Carlo simulations [494]. For example, we can also use a Gibbs ensemble simulation (see Chapter 8) to compute the phase diagram. As expected, Figure 17.1 shows that both techniques give identical results. Of course, due to the presence of an interface one needs many more particles in such a DPD simulation.

Thermodynamic quantities are calculated using only the conservative force. The pressure of the system is calculated using

$$p = \rho k_B T + \frac{1}{3V} \sum_{i>j} \left\langle \mathbf{r}_{ij} \cdot \mathbf{f}_{ij}^C \right\rangle.$$

If we have an interface in our system we can compute the interfacial tension from the pressure tensor. In a homogeneous system at equilibrium the thermodynamic pressure is constant and equal in all directions. For an inhomogeneous system hydrodynamic equilibrium requires that the component of the pressure tensor normal to the interface is constant throughout the system. The components tangential to the interface can vary in the interfacial region, but must be equal to the normal component in the bulk liquids.

For an inhomogeneous fluid there is no unambiguous way to compute the normal (p_n) and tangential (p_t) components of the pressure tensor [495–497]. Here, we have used the Kirkwood-Buff convention [498]. The system is divided into N_{sl} equal slabs parallel to the x, y plane. The local normal ($p_n(k)$) and tangential ($p_t(k)$) components of the pressure tensor are given by [207]

$$p_n(k) = k_B T \langle \rho(k) \rangle - \frac{1}{V_{sl}} \left\langle \sum_{(i,j)}^{(k)} \frac{z_{ij}^2}{r_{ij}} \frac{dU(r_{ij})}{dr} \right\rangle, \tag{17.1.13}$$

and

$$p_t(k) = k_B T \langle \rho(k) \rangle - \frac{1}{2V_{sl}} \left\langle \sum_{(i,j)}^{(k)} \frac{x_{ij}^2 + y_{ij}^2}{r_{ij}} \frac{dU(r_{ij})}{dr} \right\rangle, \tag{17.1.14}$$

where $\langle \rho(k) \rangle$ is the average density in slab k, $V_{sl} = L_x L_y L_z / N_{sl}$ is the volume of a slab, $U(r)$ is the intermolecular potential from which the conservative forces can be derived. $\sum_{(i,j)}^{(k)}$ means that the summation runs over all pairs of particles i, j for which the slab k (partially) contains the line that connects the particles i and j. Slab k gets a contribution $1/N_o$ from a given pair (i, j), where N_o is the total number of slabs which intersect this line.

It can be shown that the definition of the interfacial tension (γ) is free from ambiguities [497]. The interfacial tension can be calculated by integrating the difference of the normal and tangential components of the pressure tensor across the interface. In the case of our system with two interfaces, γ reads

$$\gamma = \frac{1}{2} \int_0^{L_z} dz \, [p_n(z) - p_t(z)] \, . \tag{17.1.15}$$

The factor $\frac{1}{2}$ corrects the two interfaces we have because of the periodic boundary conditions.

In Figure 17.2, we have plotted the surface tension of the system as a function of temperature. Clearly, the surface tension decreases with increasing temperature. Close to the critical point, the driving force for the formation of a surface (surface tension) is very low and therefore it is not possible to form an interface close to the critical point.

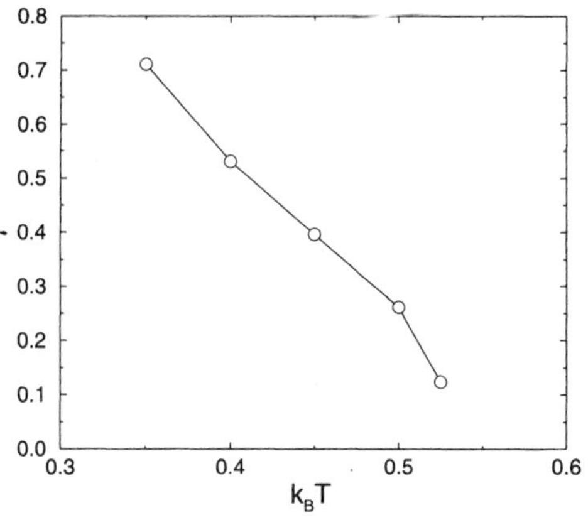

Figure 17.2: Surface tension as a function of temperature.

17.1.3 DPD and Energy Conservation

The name dissipative particle dynamics implies that during a simulation energy is dissipated. The fact that energy is not conserved in a DPD simulation can be a disadvantage for some applications. For example, if we have a system in which there is a temperature gradient, such a temperature gradient can only be sustained artificially in a DPD simulation [489]. A solution to this is to introduce an additional variable characterizing the internal energy in a DPD simulation [499,500].

To this end, we associate with every DPD particle an internal-energy "reservoir." This reservoir absorbs or releases the energy that would normally go into the internal degrees of freedom of the group of molecules that are represented by a single DPD particle. During a collision the energy of this reservoir can increase or decrease.

The internal energy of a DPD particle is denoted by ϵ_i. With this variable we can associate an entropy variable $s_i = s(\epsilon_i)$ and temperature $T_i = [\partial s_i/\partial \epsilon_i]^{-1}$. The next step is to introduce an equation of motion for this internal energy. There are two processes that can change the internal energy. The first is "heat conduction" caused by temperature differences between two particles. The second is "viscous heating" due to the frictional forces.

If the DPD particles are at rest, the time derivative of the internal energy is given by

$$d\epsilon_i = \sum_j \kappa_{ij} \left(\frac{1}{T_i} - \frac{1}{T_j} \right) \omega(r_{ij}) dt + \sum_j \xi_{ij}^T, \qquad (17.1.16)$$

where $\omega(r)$ is a function determining the range of influence between particles. ξ_{ij}^T is a random heat flux that occurs spontaneously because of thermal fluctuations.

We can associate a total mechanical energy to a system of DPD particles:

$$E_{mec} = \sum_{i<j} U(r_{ij}) + \sum_i \frac{1}{2} m_i v_i^2.$$

If we consider the change in mechanical energy, an increase in the potential energy would be compensated with a decrease in the kinetic energy if there were no dissipative and random forces. We now impose that the sum of the internal energy and the mechanical energy is constant:

$$dE = dE_{mec} + d\sum_i \epsilon_i = 0.$$

This implies for the time derivative of the internal energy:

$$\begin{aligned} d\epsilon_i &= \frac{m_i}{2}\left[\sum_j \left[\gamma\omega^D(r_{ij})(\mathbf{v}_{ij}\cdot\hat{\mathbf{r}}_{ij})^2 - \sigma^2(\omega^R(r_{ij}))^2\right]dt \right. \\ &\quad \left. - \sum_j \sigma\omega^R(r_{ij})(\hat{\mathbf{r}}_{ij}\cdot\mathbf{v}_{ij})\xi_{ij}\right]. \end{aligned} \tag{17.1.17}$$

The total energy change is a combination of convection (17.1.16) and viscous heating (17.1.17). For details on the implementation of these methods, the reader is referred to refs. [499–504].

Example 27 (Hydrodynamics and Phase Separation)

Block copolymers resemble surfactants in the sense that two chemically different units, say A and B, are linked together by a chemical bond. If the interactions between these units are very different, the system would phase separate were it not for the fact that adjacent units are connected by chemical bonds. However, one often observes microphase separation in such systems that results in a microscopic texture of alternating A-rich and B-rich domains. These domains can then arrange into larger-scale structures such as spheres, rods, sheets, and perforated sheets, or other, more complicated, sponge-like structures.

The dynamics of the formation of these microstructures has been studied experimentally and theoretically. However, the effect of the hydrodynamics on the kinetics of formation of these structures is not fully understood. Groot *et al.* [505] compared the formation of these microstructures using dissipative particle dynamics (DPD) and Brownian dynamics. Brownian dynamics

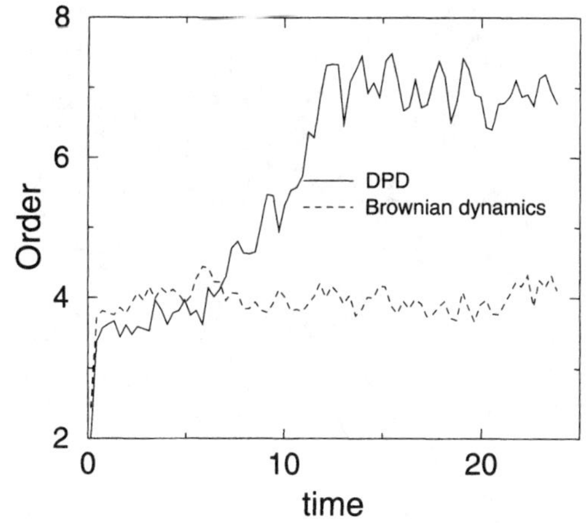

Figure 17.3: Order parameter expressing the configurational entropy of the structure (see ref. [505] for details) as a function of time for a DPD simulation and a Brownian dynamic simulation. The DPD simulation forms the hexagonal structure, while the Brownian dynamic simulation remains trapped in a phase with interconnecting tubes (see the structures shown in Figure 17.4).

(BD) is a technique that resembles DPD but in which each particle feels a random force and a drag force relative to a fixed background. The important difference between BD and DPD is that BD does not satisfy Newton's third law and hence it does not conserve momentum. As a consequence, BD can never reproduce hydrodynamic behavior. In contrast, DPD is similar to MD in that it does conserve momentum. It can therefore be used to model hydrodynamic interactions. A comparison of the predictions of a DPD and a BD simulation for the time evolution of the same model tells us something about the importance of the hydrodynamics for the process under consideration. Groot applied the two techniques to study the formation of microstructures in block copolymers. For such systems, Molecular Dynamics would require too much CPU time.

Groot *et al.* [505] studied the formation of the hexagonal phase from an initially random A_3B_7 block-copolymer liquid. In Figure 17.3 the evolution of the order parameter characterizing the various phases is plotted. This figure shows that in the DPD simulation a hexagonal phase is formed, while the system modeled by BD remains trapped in a metastable state. Apparently, hydrodynamic interactions are needed to facilitate the crossing of kinetic barriers.

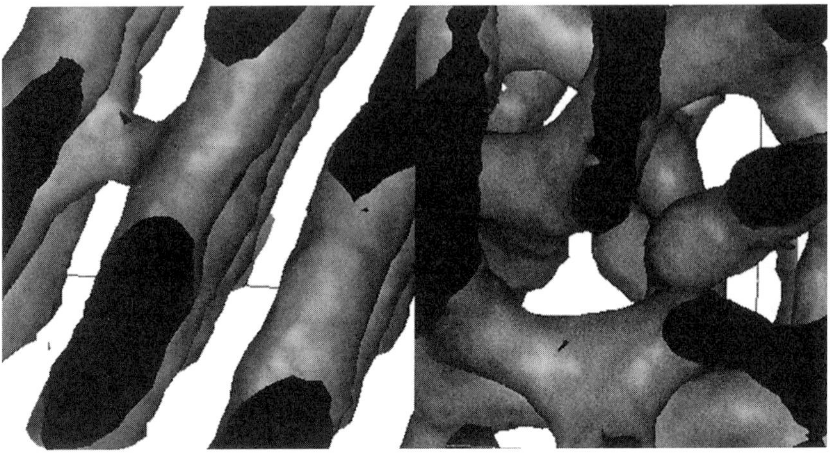

Figure 17.4: Structures obtained from DPD simulation (left) and a Brownian dynamic simulation (right) of a system of A_3B_7 block copolymers, after the same number of times steps. In the DPD simulation a hexagonal phase is observed while the Brownian dynamics simulation is trapped in a disordered phase of interconnected tubes. The figure shows the dividing surface and is based on data provided by Dr. R.D. Groot.

17.2 Other Coarse-Grained Techniques

In addition to the dissipative particle dynamics method described in the previous section, there exists a host of coarse-grained methods to model the mesoscopic dynamics of fluids. Most of these are based on a continuum description of the fluid under consideration. Such techniques fall outside the scope of this book. Here, we briefly touch upon two mesoscopic simulation schemes that are, like DPD, inherently particle based. The first is the so-called Lattice Boltzmann (LB) method [506–508]. As the name suggests, the LB method is a lattice scheme. It is based on a lattice gas cellular automaton model of a fluid [509]. In such a model, the fluid is represented by a regular lattice. Every lattice point has z nearest neighbors (in two dimensions, z is usually equal to 6, and in 3D, the most frequently used models have $z = 18$). Not only space is discretized, but time too. On every lattice point, there can be at most one particle moving in the direction linking the lattice point to any given nearest neighbor. In a single time step, a particle moves along its link from its original lattice site to the corresponding link on the nearest-neighbor lattice site. This propagation step is carried out for all particles simultaneously. The next step is the collision step. During collisions, the total number of particles and the total momentum (and, in certain models, the total energy) on a given lattice site is maintained, but apart from this constraint, particles can change their velocity. There is considerable freedom in selecting the collision rules, as long as they maintain the full symmetry of the lattice. With this very simple model, it is possible to mimic hydrodynamic behavior. However, the cellular automaton method is very "noisy" and suffers from a number of other practical drawbacks. Moreover, by construction, the lattice model lacks Galilean invariance. The lattice Boltzmann method was devised to overcome some (but not all) of the problems of lattice gas cellular automata.

In the most naive version, one can think of the lattice Boltzmann model as a preaveraged version of a lattice gas cellular automaton [506]. In this preaveraging, the number of particles on a given link is replaced by the particle density on that link. Note that the particle number is either zero or one, but the density is a real number. In addition, the resulting equations are greatly simplified if the collision operator (i.e., the function that described how the post-collision state of a lattice point depends on the precollision state) is linearized in the deviation of the particle densities from their (local) equilibrium value. Finally, there is no need to restrict the LB collision operators to forms that can be derived from an underlying cellular automaton [507,508]. It is, however, essential that the collision operator satisfies the conservation laws and the symmetries of the original model.

For the simulation of complex flows, the LB method is much more efficient than the original cellular automaton model. However, one aspect is missing in the LB approach, namely the intrinsic fluctuations that result from the discreteness in the number of particles. As a consequence, a normal LB model does not exhibit "thermal fluctuation." Often this is an advantage. But if one is interested in Brownian motion of suspended particles [510], or in the decay of spontaneous stress fluctuations [511], then it is essential to re-introduce stochastic fluctuations. Such an approach has been proposed by Ladd [482]. The LB method is very useful for studying flows in complex geometries (e.g., flow in porous media or the dynamics of colloidal suspensions). The method has also been extended to describe mixtures (see, e.g., [512]).

The aim of the lattice Boltzmann method and related techniques is to provide a computationally cheap, particle-based representation of mesoscopic fluid flows. The LB method that we sketched above achieves this aim by averaging out essentially all information about individual fluid particles. There exist other computational techniques that, like DPD, are truly particle-based. In these models, it becomes essential to make the dynamics of the fluid particles as cheap as possible. We briefly sketch two closely related schemes. The first is the so-called Direct Monte Carlo Simulation Method (DSMC) of Bird [513]. This method was developed to simulate flow of relatively dilute gases. In the DSMC method, particles move ballistically between collisions. Collisions are carried out stochastically. That is, the number of collisions per time step is fixed by the known collision frequency at the specified density. The selection of collision partners is stochastic. To this end, space is divided into cells. Collision partners are then randomly selected pairs within one cell. The precise collision dynamics depends on the molecular model that is used. However, in all cases, collisions conserve linear momentum and energy. An extension of the DSMC method to dense gases has been proposed by Alexander *et al.* [514]. For more details on the DSMC method in general, see the book and review by Bird [513,515], and a tutorial article by Alexander and Garcia [516]. Recently, Malevanets and Kapral proposed a particle-based method that is very similar in spirit to DSMC [483]. This method was designed to model the dynamics of macromolecules in an unstructured solvent. To this end, the forces between the macromolecules are taken into account explicitly, as is the interaction between solute and solvent. However, the solvent-solvent interactions are accounted for stochastically. This means that normal Molecular Dynamics is used to update the solvent and solute positions and momenta, while ignoring the solvent-solvent interaction. After this normal MD step, the solvent-solvent interaction is accounted for in a way that is reminiscent of the DMSC scheme: space is divided into cells and the solvent-solvent collisions take place inside these cells. Unlike the DSMC scheme, the Malevanets-Kapral method uses many-body collisions between the solvent particles. The motion of the center of mass of all particles in a cell is left unchanged, but a uniform, random rotation is applied to all relative velocities of the particles in the cell. As the relative velocities do not change in size, the total energy is conserved, while the conservation of the center-of-mass velocity takes care of the momentum conservation.

Part V

Appendices

Appendix A

Equations of Motion from the Lagrangian or Hamiltonian

Knowledge of Newton's equations of motion is sufficient to understand the basis of the Molecular Dynamics method. However, many of the more advanced simulation techniques make use of the Lagrangian and Hamiltonian formulations of classical mechanics. Here we briefly sketch the relation between these different approaches (see also [517]). For a more detailed, and more rigorous description of classical mechanics, the reader is referred to the book by Goldstein [45].

The Lagrangian formulation of classical mechanics is based on a variational principle. The actual trajectory followed by a classical system in a time interval $\{t_b, t_e\}$, between an initial position x_b and a final position x_e, is the one for which the action, S, is an extremum (usually, a minimum). The classical action S for an arbitrary trajectory is defined as the time integral of the difference between the kinetic energy $\mathcal{K}$ and the potential energy $\mathcal{U}_K$ of the system, computed along that trajectory:

$$S = \int_{t_b}^{t_e} dt\,[\mathcal{K} - \mathcal{U}]\,. \tag{A.0.1}$$

Before considering the general Lagrangian equations of motion that follow from this extremum principle, let us first consider a few simple examples.

The first case is that of a single particle that moves in the absence of an external potential, i.e., $\mathcal{U} = 0$. As the particle has to move from x_b to x_e in a time interval $t_e - t_b$, we already know its average velocity: v_{av}. If the particle would always move with this average velocity, it would follow a straight trajectory that we denote by $\bar{x}(t)$. Let us denote the true trajectory of the particle by $x(t) = \bar{x}(t) + \eta(t)$, where $\eta(t)$ is, as yet, unknown. Then the velocity of the particle is the sum of the average velocity v_{av} and the deviation from it, $\dot{\eta}(t)$:

$$v(t) = v_{av} + \dot{\eta}(t).$$

By construction,

$$\int dt\ \dot{\eta}(t) = 0.$$

In the present example, the potential energy is always zero and hence the action S is determined by the time integral of the kinetic energy:

$$S = \frac{1}{2}m \int dt\ [v_{av} + \dot{\eta}(t)]^2 = S_{av} + \frac{1}{2}m \int dt\ \dot{\eta}^2(t).$$

Since the last term is always greater than 0, the action has its minimum if $\dot{\eta}(t) = 0$ for all t. In other words, we recover the well-known result that, in the absence of external forces, the particle moves with constant velocity. This is Newton's first law.

Next, consider a particle moving in a one-dimensional potential $U(x)$. In this case, the action is

$$S = \int_{t_b}^{t_e} dt \left[\frac{1}{2}m \left(\frac{dx(t)}{dt} \right)^2 - U(x) \right].$$

An arbitrary path, $x(t)$, can be written as the sum of the actual path that a classical particle will follow, $\bar{x}(t)$, plus the deviation from this path $\eta(t)$:

$$x(t) = \bar{x}(t) + \eta(t).$$

As before, we impose the initial and final positions of the particle and hence $\eta(t_b) = \eta(t_e) = 0$. For paths $x(t)$ that are close to the actual path, we can expand the action in powers of the (small) quantity $\eta(t)$. Actually, as $\eta(t)$ is itself a function of t, such an expansion is called a *functional expansion*. The action is extremal if the leading (linear) terms in this functional expansion vanish. Let us now consider the functional expansion of the action around the action of the true path, to linear order in $\eta(t)$:

$$\begin{aligned} S &= \int_{t_b}^{t_e} dt \, \frac{1}{2} m \left(\frac{d\bar{x}(t)}{dt} + \frac{d\eta(t)}{dt} \right)^2 - U\left[\bar{x}(t) + \eta(t)\right] \\ &\approx \int_{t_b}^{t_e} dt \, \frac{1}{2} m \left[\left(\frac{d\bar{x}(t)}{dt} \right)^2 + 2 \frac{d\bar{x}(t)}{dt} \frac{d\eta(t)}{dt} \right] - \left[U(\bar{x}(t)) + \frac{\partial U(\bar{x})}{\partial x} \eta(t) \right] \\ &= \bar{S} + \int_{t_b}^{t_e} dt \left[m \frac{d\bar{x}(t)}{dt} \frac{d\eta(t)}{dt} - \frac{\partial U(\bar{x})}{\partial x} \eta(t) \right] \\ &= \bar{S} + m \frac{d\bar{x}(t)}{dt} \eta(t) \Big|_{t_b}^{t_e} - \int_{t_b}^{t_e} dt \left[m \frac{d^2\bar{x}(t)}{dt^2} + \frac{\partial U(\bar{x})}{\partial x} \right] \eta(t), \end{aligned}$$

where the last step has been obtained via partial integration. Since by definition $\eta(t) = 0$ at the boundaries, the second term on the right-hand side vanishes. The action has its extremum if the integrand in the last line of the above equation vanishes for arbitrary $\eta(t)$. This condition can be satisfied if and only if

$$m \frac{d^2\bar{x}(t)}{dt^2} = -\frac{\partial U(\bar{x})}{\partial x}, \tag{A.0.2}$$

which is Newton's second law. In other words, Newton's equations of motion can be derived from the statement that a particle follows a path for which the action is an extremum.

A.1 Lagrangian

There would be little point in introducing this alternative expression of the laws of classical mechanics, if it did not allow us to do more than simply rederive $F = ma$. In fact, the Lagrangian formulation of classical mechanics turns out to be very powerful. For one thing, the Lagrangian approach makes it easy to derive equations of motion in non-Cartesian coordinate frames. Suppose that we wish to use some generalized coordinates q instead of the Cartesian coordinate x. For example, consider a pendulum of length l in a uniform gravitational field. The angle that the pendulum makes with the vertical (i.e., with the direction of the gravitational field) can be used to specify its orientation. Since the path that the pendulum follows is clearly independent of the coordinates that we happen to use to specify its state, the action should be the same:

$$S = \int dt \, \mathcal{L}(x, \dot{x}) = \int dt \, \mathcal{L}(q, \dot{q}), \tag{A.1.1}$$

where the quantity $\mathcal{L}$ is called the Lagrangian. The Lagrangian is defined as the kinetic energy minus the potential energy[1]:

$$\mathcal{L} \equiv \mathcal{U}_K(\dot{q}) - \mathcal{U}_P(q). \tag{A.1.2}$$

We again introduce our actual path $\bar{q}(t)$ and the deviation $\eta(t)$ from it:

$$\begin{aligned} q(t) &= \bar{q}(t) + \eta(t) \\ \dot{q}(t) &= \dot{\bar{q}}(t) + \dot{\eta}(t). \end{aligned}$$

We can write for the Lagrangian

$$\mathcal{L}(q, \dot{q}) = \mathcal{L}(\bar{q}, \dot{\bar{q}}) + \frac{\partial \mathcal{L}(\bar{q}, \dot{\bar{q}})}{\partial \dot{q}} \dot{\eta}(t) + \frac{\partial \mathcal{L}(\bar{q}, \dot{\bar{q}})}{\partial q} \eta(t).$$

As in the previous section, we use the functional expansion of S in powers of $\eta(t)$ to derive an expression for the classical path. To this end, we substitute the Lagrangian in the expression for the action (A.1.1). Next we write a possible path of the particle as the sum of the actual path and a correction $\eta(t)$. As before, we use partial integration, and use the fact that $\eta(t)$ vanishes at the boundaries of the integration. It then follows that the action has an extremum if

$$\int dt \left[-\frac{d}{dt} \left(\frac{\partial \mathcal{L}(\bar{q}, \dot{\bar{q}})}{\partial \dot{q}} \right) + \frac{\partial \mathcal{L}(\bar{q}, \dot{\bar{q}})}{\partial q} \right] \eta(t) = 0, \tag{A.1.3}$$

which is satisfied for arbitrary $\eta(t)$ if and only if

$$\left[-\frac{d}{dt} \left(\frac{\partial \mathcal{L}(\bar{q}, \dot{\bar{q}})}{\partial \dot{q}} \right) + \frac{\partial \mathcal{L}(\bar{q}, \dot{\bar{q}})}{\partial q} \right] = 0. \tag{A.1.4}$$

This is the Lagrangian equation of motion. To cast this equation of motion in a more familiar form, we introduce the generalized momentum p associated with the generalized coordinate q:

$$p \equiv \frac{\partial \mathcal{L}(q, \dot{q})}{\partial \dot{q}}. \tag{A.1.5}$$

Substitution of this expression into equation (A.1.4) yields

$$\dot{p} = \frac{\partial \mathcal{L}(q, \dot{q})}{\partial q}. \tag{A.1.6}$$

As the above formulation is valid for any coordinate system, it should certainly hold for Cartesian coordinates. In these coordinates the Lagrangian reads

$$\mathcal{L}(x, \dot{x}) = \frac{1}{2} m \dot{x}^2 - U(x).$$

[1]The correct definition is more restrictive; see [45] for more details.

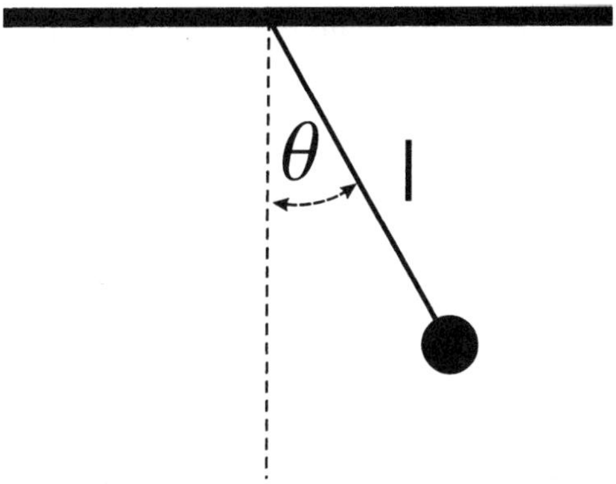

Figure A.1: A simple pendulum of length l with mass m.

The momentum associated with x is

$$p_x = \frac{\partial \mathcal{L}(x, \dot{x})}{\partial \dot{x}} = m\dot{x}$$

and the equation of motion is

$$m\ddot{x} = -\frac{\partial U(x)}{\partial x},$$

which is indeed the result we would obtain from Newton's equation of motion.

Example 28 (A Pendulum in a Gravitational Field)
Consider a simple pendulum of length l with mass m (see Figure A.1). A uniform gravitational field is acting on the pendulum and the potential energy is a simple function of the angle θ that the pendulum makes with the vertical:

$$U(\theta) = mgl\,[1 - \cos(\theta)]\,.$$

We wish to express the equations of motion in terms of the generalized coordinate θ. The Lagrangian is

$$\begin{aligned}\mathcal{L} &= \mathcal{U}_K - \mathcal{U}_P = \frac{1}{2}m\left[\dot{x}^2(t) + \dot{y}^2(t)\right] - U(\theta) \\ &= \frac{ml^2}{2}\dot{\theta}^2 - U(\theta).\end{aligned}$$

The generalized momentum is defined as

$$p_\theta = \frac{\partial \mathcal{L}}{\partial \dot{q}} = ml^2\dot{\theta}$$

and the equation of motion follows from equation (A.1.6)

$$\dot{p}_\theta = -\frac{\partial \mathcal{U}(\theta)}{\partial \theta}$$

or

$$\ddot{\theta} = -\frac{1}{ml^2}\frac{\partial \mathcal{U}(\theta)}{\partial \theta}.$$

A.2 Hamiltonian

Using the Lagrangian we have derived the equations of motion in terms of q and $\dot{q}$. Often, it is convenient to express the equations of motion in terms of q and its conjugate momentum p. To do this we can perform a Legendre transformation[2]:

$$\mathcal{H}(q, p) \equiv p\dot{q} - \mathcal{L}(q, \dot{q}, t). \tag{A.2.1}$$

This equation defines the *Hamiltonian* $\mathcal{H}$ of the system. As $\mathcal{H}$ is a function of q, p and, in general, also of t, it is clear that we can write an infinitesimal variation of $\mathcal{H}$ as

$$d\mathcal{H}(q, p) = \frac{\partial \mathcal{H}}{\partial p}dp + \frac{\partial \mathcal{H}}{\partial q}dq + \frac{\partial \mathcal{H}}{\partial t}dt. \tag{A.2.2}$$

But, using the definition of $\mathcal{H}$, we can also write

$$\begin{aligned} d\mathcal{H}(q,p) &= d(p\dot{q}) - d\mathcal{L}(q, \dot{q}) \\ &= pd\dot{q} + \dot{q}dp - \left[\frac{\partial \mathcal{L}}{\partial q}dq + \frac{\partial \mathcal{L}}{\partial \dot{q}}d\dot{q} + \frac{\partial \mathcal{L}}{\partial t}dt\right] \\ &= pd\dot{q} + \dot{q}dp - \dot{p}dq - pd\dot{q} - \frac{\partial \mathcal{L}}{\partial t}dt \\ &= \dot{q}dp - \dot{p}dq - \frac{\partial \mathcal{L}}{\partial t}dt, \end{aligned}$$

[2]In thermodynamics, Legendre transforms are used to derive various thermodynamic potentials. For example, the energy E is a natural function of the entropy S and volume V: $E = E(S, V)$, i.e., in these variables E is a thermodynamic potential. In most practical application it is more convenient to have the temperature T rather than the entropy S as independent variable. Since the temperature is the variable conjugate to the entropy ($\partial E/\partial S = T$), we can perform a Legendre transform to remove the S dependence:

$$A \equiv E - TS,$$

yielding

$$dA = dE - d(TS) = -SdT - pdV.$$

For historical reasons the Legendre transform linking the Lagrangian to the Hamiltonian has the opposite sign.

where we have used the definitions of p and $\dot{p}$, equations (A.1.5) and (A.1.6), respectively. It then follows directly that

$$\frac{\partial \mathcal{H}}{\partial p} = \dot{q} \tag{A.2.3}$$

$$\frac{\partial \mathcal{H}}{\partial q} = -\dot{p}. \tag{A.2.4}$$

These are the desired equations of motion in terms of q, p. For most systems that we consider in this book, the Lagrangian does not explicitly depend on time. In those circumstances, the Hamiltonian is conserved. This follows directly from the equations of motion:

$$\begin{aligned} \frac{d\mathcal{H}(q,p)}{dt} &= \frac{\partial \mathcal{H}}{\partial p}\dot{p} + \frac{\partial \mathcal{H}}{\partial q}\dot{q} \\ &= -\frac{\partial \mathcal{H}}{\partial p}\frac{\partial \mathcal{H}}{\partial q} + \frac{\partial \mathcal{H}}{\partial q}\frac{\partial \mathcal{H}}{\partial p} \\ &= 0. \end{aligned}$$

This conservation law expresses the fact that, in a closed system, the total energy is conserved. In Cartesian coordinates, the Hamiltonian can be written as

$$\begin{aligned} \mathcal{H}(x, p_x) &= \dot{x}p_x - \mathcal{L}(x, \dot{x}) \\ &= m\dot{x}^2 - \frac{1}{2}m\dot{x}^2 + U(x) \\ &= \frac{1}{2m}p_x^2 + U(x), \end{aligned}$$

and the Hamiltonian equations of motion reduce to Newton's equations

$$\begin{aligned} \dot{x} &= \frac{\partial \mathcal{H}}{\partial p_x} = \frac{p_x}{m} \\ \dot{p}_x &= -\frac{\partial \mathcal{H}}{\partial x} = -\frac{\partial U(x)}{\partial x}. \end{aligned}$$

The Hamiltonian equations of motion are two first-order differential equations—one for p and one for q. In contrast, the Lagrangian formalism yields a single second-order equation. However, both formalisms yield identical results. The choice between the two is dictated by considerations of mathematical convenience.

Example 29 (A Pendulum in a Gravitational Field: Part II)
We consider again the simple pendulum in a uniform gravitational field, introduced in Example 28:

$$U(\theta) = mgl\,[1 - \cos(\theta)]\,,$$

where θ is the angle that the pendulum makes with the vertical.

In Example 28 we have derived the equations of motion from the Lagrangian in terms of a second-order differential equation in θ. Now we will use Hamilton's formulation.

The Lagrangian is

$$\mathcal{L}(\theta,\dot{\theta}) = \mathcal{U}_K - \mathcal{U}_P = \frac{ml^2}{2}\dot{\theta}^2 - u(\theta).$$

The Lagrangian depends on the variables θ and $\dot{\theta}$ and in the Hamiltonian language we want to express the equations of motion in terms of θ and its conjugate momentum p_θ. This conjugate momentum is defined by equation (A.1.5)

$$p_\theta \equiv \frac{\partial \mathcal{L}(\theta,\dot{\theta})}{\partial \dot{\theta}} = ml^2\dot{\theta}.$$

The Hamiltonian follows from the Legendre transformation (A.2.1)

$$\begin{aligned} \mathcal{H} &= p_\theta\dot{\theta} - \mathcal{L}(\theta,\dot{\theta}) \\ &= \frac{p_\theta}{2ml^2} + \mathcal{U}(\theta) \\ &= \frac{1}{2}ml^2\dot{\theta}^2 + u(\theta), \end{aligned}$$

which is, of course, equal to the total energy of the pendulum.

The equations of motion follow from equations (A.2.3) and (A.2.4):

$$\begin{aligned} \dot{\theta} &= \frac{\partial \mathcal{H}}{\partial p_\theta} = \frac{p_\theta}{ml^2} \\ \dot{p_\theta} &= -\frac{\partial \mathcal{H}}{\partial \theta} = -\frac{du(\theta)}{d\theta}, \end{aligned}$$

which are the desired equations of motion in terms of two first-order differential equations.

A.3 Hamilton Dynamics and Statistical Mechanics

The Hamiltonian and Lagrangian formulations of classical mechanics yield identical results. This is not surprising as the Hamiltonian formulation was derived from the Lagrangian equations (see section A.2). Yet, the forms of the Lagrangian and Hamiltonian equations of motion are quite different: the Hamiltonian equations of motion are *first-order* differential equations for the momenta and coordinates of all particles in the system. The Lagrangian equations of motion are *second-order* equations of motion for the coordinates only. The choice of formalism is dictated by considerations of convenience. For instance, in order to derive the equations of motion of a system with constraints, the Lagrangian formalism is more convenient (see 15.1). On the other hand, the Hamiltonian expressions are to be used when establishing the connection with statistical mechanics (see Chapter 2).

A.3.1 Canonical Transformation

In the Hamiltonian formulation the generalized coordinates and momenta are independent variables. One can therefore introduce a transformation of both variables simultaneously. For example, the transformation of the coordinates q, p to Q, P is denoted by

$$\begin{aligned} Q &= Q(q,p) \\ P &= P(q,p) \end{aligned} \tag{A.3.1}$$

and the inverse transformation, Q, P into q, p, by

$$\begin{aligned} q &= q(Q,P) \\ p &= p(Q,P). \end{aligned} \tag{A.3.2}$$

Obviously, the value of any function of the phase-space coordinates is unaffected by the coordinate transformation. In the case of the Hamiltonian, this implies that

$$\mathcal{H}(q,p) \equiv \mathcal{H}\left[Q(p,q), P(q,p)\right] \equiv \mathcal{H}'(Q,P). \tag{A.3.3}$$

In general, the equations of motion in the new coordinates are not of the canonical form, unless the coordinate transformation is *canonical*.[3] If the coordinate transformation is canonical, the equations of motion for the new phase-space coordinates Q, P are

$$\dot{Q} = \left(\frac{\partial \mathcal{H}'(Q,P)}{\partial P}\right) \tag{A.3.4}$$

$$\dot{P} = -\left(\frac{\partial \mathcal{H}'(Q,P)}{\partial Q}\right). \tag{A.3.5}$$

From equation (A.3.1) and the Hamilton equations of motion for the coordinates q, p, it follows that

$$\begin{aligned} \dot{Q} &= \left(\frac{\partial Q(q,p)}{\partial q}\right)\dot{q} + \left(\frac{\partial Q(q,p)}{\partial p}\right)\dot{p} \\ &= \left(\frac{\partial Q(q,p)}{\partial q}\right)\left(\frac{\partial \mathcal{H}(q,p)}{\partial p}\right) - \left(\frac{\partial Q(q,p)}{\partial p}\right)\left(\frac{\partial \mathcal{H}(q,p)}{\partial q}\right), \end{aligned}$$

[3]Since time does not appear explicitly in these equations, we are defining a so-called restricted canonical transformation.

Using equation (A.3.3), we can write

$$\left(\frac{\partial \mathcal{H}'(Q,P)}{\partial P}\right) = \left(\frac{\partial \mathcal{H}(q,p)}{\partial p}\right)\left(\frac{\partial p(P,Q)}{\partial P}\right) + \left(\frac{\partial \mathcal{H}(q,p)}{\partial q}\right)\left(\frac{\partial q(P,Q)}{\partial P}\right).$$

This equation can only be equal to expression (A.3.4) for $\dot{Q}$ if

$$\left(\frac{\partial Q(q,p)}{\partial q}\right) = \left(\frac{\partial p(Q,P)}{\partial P}\right) \quad \text{and} \quad \left(\frac{\partial Q(q,p)}{\partial p}\right) = -\left(\frac{\partial q(Q,P)}{\partial P}\right). \tag{A.3.6}$$

Similarly, we can start with $\dot{P}$, and derive two other conditions:

$$\left(\frac{\partial P(q,p)}{\partial q}\right) = -\left(\frac{\partial p(Q,P)}{\partial Q}\right) \quad \text{and} \quad \left(\frac{\partial P(q,p)}{\partial p}\right) = \left(\frac{\partial q(Q,P)}{\partial Q}\right). \tag{A.3.7}$$

These two equations define the condition for a canonical transformation.

Maxwell's relation

A.3.2 Symplectic Condition

We can express the above conditions for a canonical transformation in a single equation, by using a matrix notation. Let η be a 2N-dimensional vector containing the generalized coordinates q_i and momenta p_i of the N particles (for the sake of simplicity, we consider a one-dimensional system). Hamilton's equations of motion (A.2.3) and (A.2.4) can be written as

$$\dot{\eta} = \omega \frac{\partial \mathcal{H}}{\partial \eta}, \tag{A.3.8}$$

where ω is an antisymmetric matrix defined as

$$\omega = \begin{pmatrix} 0 & 1 \\ -1 & 0 \end{pmatrix}.$$

In a similar way we can define ξ to be the 2N-dimensional vector containing the generalized coordinates Q_i and P_i. Using the matrix notation the transformation (A.3.1) from Q, P to q, p is written as

$$\xi = \xi(\eta).$$

For the time derivatives of ξ, we can write

$$\dot{\xi} = \mathsf{M}\dot{\eta},$$

where M is the Jacobian matrix of the transformation. The elements of this matrix are

$$M_{ij} = \frac{\partial \xi_i}{\partial \eta_j}. \tag{A.3.9}$$

We can write, using equation (A.3.8), for the time derivatives of ξ

$$\dot{\xi} = \mathsf{M}\omega \frac{\partial \mathcal{H}}{\partial \eta}. \tag{A.3.10}$$

In a similar way, we can define the inverse transformation (A.3.2)

$$\eta = \eta(\xi).$$

Since

$$\mathcal{H}(p, q) = \mathcal{H}(P, Q),$$

we can write

$$\frac{\partial \mathcal{H}(\eta)}{\partial \eta_i} = \sum_j \frac{\partial \mathcal{H}(\xi)}{\partial \xi_j} \frac{\partial \xi_j}{\partial \eta_i}. \tag{A.3.11}$$

If we define the transposed matrix[4] of M as defined in equation (A.3.9),

$$\tilde{\mathsf{M}}_{ij} = \frac{\partial \xi_j}{\partial \eta_i}.$$

This allows us to rewrite equation (A.3.11) in matrix notation as

$$\frac{\partial \mathcal{H}(\eta)}{\partial \eta} = \tilde{\mathsf{M}} \frac{\partial \mathcal{H}(\xi)}{\partial \xi}. \tag{A.3.12}$$

If we combine equations (A.3.10) and (A.3.12), we have

$$\dot{\xi} = \mathsf{M}\omega\tilde{\mathsf{M}} \frac{\partial \mathcal{H}}{\partial \xi}.$$

This expression for the equations of motion is valid for any set of variables ξ that are being transformed (independently of time) from the set η. Such a transformation is canonical if the equations of motion in the new coordinates have the canonical form:

$$\dot{\xi} = \omega \frac{\partial \mathcal{H}}{\partial \xi}.$$

This can only be the case if M satisfies the condition

$$\mathsf{M}\omega\tilde{\mathsf{M}} = \omega. \tag{A.3.13}$$

This condition is often called the *symplectic condition*. A matrix M that satisfies this condition is called a *symplectic matrix*.[5]

[4]One can obtain the transposed matrix of a given matrix A by interchanging rows and columns, i.e., $\tilde{a}_{ij} = a_{ji}$.

[5]To see that this condition is identical to equations (A.3.6) and (A.3.7), we have to multiply this equation from the right with the inverse matrix of $\tilde{\mathsf{M}}$:

$$\mathsf{M}\omega = \omega\tilde{\mathsf{M}}^{-1}.$$

A.3.3 Statistical Mechanics

Using the symplectic notation for a canonical transformation, we consider the implications for statistical mechanics. In the microcanonical ensemble, the partition function in a three-dimensional system is defined as

$$\Omega_{N,V,E} = \frac{1}{h^{3N}N!}\int \mathbf{dp}^N\mathbf{dq}^N\delta\left(\mathcal{H}(p,q)-E\right), \tag{A.3.14}$$

where h is Planck's constant and the delta-function restricts the integration to the hypersurface in phase space defined by $\mathcal{H}(p,q) = E$. We can re-express this integral in terms of other phase-space coordinates, but then we have to take into account that a volume element in the two coordinate sets needs not be the same. The volume element associated with η is

$$d\eta = dq_1 \ldots dq_N dp_1 \ldots dp_N$$

and to ξ

$$d\xi = dQ_1 \ldots dQ_N dP_1 \ldots dP_N.$$

These two volume elements are related via the Jacobian matrix of the transformation matrix

$$d\eta = |\mathrm{Det}(M)|\, d\xi. \tag{A.3.15}$$

This equation shows that, in general, a coordinate transformation will result in the appearance of a Jacobian in the partition function:

$$\Omega_{N,V,E} = \frac{1}{h^{3N}N!}\int \mathbf{dP}^N\mathbf{dQ}^N\,|\mathrm{Det}(M)|\,\delta\left(\mathcal{H}'(P,Q)-E\right). \tag{A.3.16}$$

When computing ensemble averages in coordinate systems other than the original Cartesian one, the Jacobian of the transformation may be different from one, and should be taken into account. In what follows, we denote the Jacobian $|\mathrm{Det}(M)|$ by the symbol ω.

For a transformation that is canonical, i.e., obeys condition (A.3.13), the absolute value of the Jacobian is one. To derive this result we take the determinant on both sides of the symplectic condition (A.3.13)

$$\begin{aligned}\mathrm{Det}(M\omega\tilde{M}) &= \mathrm{Det}(\omega)\\ \mathrm{Det}^2(M)\mathrm{Det}(\omega) &= \mathrm{Det}(\omega).\end{aligned}$$

This equation can only be true if the determinant of M is ± 1, which implies that for a canonical transformation the absolute value of the Jacobian associated with this transformation must be one.

The natural time evolution in phase space of a classical system may be considered as a coordinate transformation:

$$\eta(t_0) \to \eta(t).$$

One important property of a Hamiltonian system is that the natural time evolution corresponds to a symplectic coordinate transformation. We can consider the transformation from $\eta(t_0)$ to $\eta(t)$ as a sequence of infinitesimal transformations with time step δt. Suppose that we define the evolution of the coordinates during the time interval δt as transformation of coordinates from η to ξ:

$$\begin{aligned}\xi &= \xi(\eta) \\ &= \eta(t+\delta t) \\ &= \eta(t) + \dot{\eta}(t)\delta t.\end{aligned}$$

The Jacobian of this transformation is

$$\begin{aligned}\mathsf{M} &\equiv \frac{\partial \xi}{\partial \eta} \\ &= 1 + \delta t \frac{\partial}{\partial \eta}\left(\omega \frac{\partial \mathcal{H}}{\partial \eta}\right) \\ &= 1 + \delta t \omega \frac{\partial^2 \mathcal{H}}{\partial \eta \partial \eta},\end{aligned}$$

where

$$\left(\frac{\partial^2 \mathcal{H}}{\partial \eta \partial \eta}\right)_{ij} = \frac{\partial^2 \mathcal{H}}{\partial \eta_i \partial \eta_j}.$$

Taking into account that ω is an antisymmetric matrix, we can write for the transpose of the matrix M:

$$\tilde{\mathsf{M}} = 1 - \frac{\partial^2 \mathcal{H}}{\partial \eta \partial \eta}\omega.$$

Substitution of this expression for the Jacobian into the symplectic condition (A.3.13) yields (to first order in δt)

$$\begin{aligned}\mathsf{M}\omega\tilde{\mathsf{M}} &= \left(1 + \delta t \omega \frac{\partial^2 \mathcal{H}}{\partial \eta \partial \eta}\right)\omega\left(1 - \delta t \frac{\partial^2 \mathcal{H}}{\partial \eta \partial \eta}\omega\right) \\ &\approx \omega + \delta t \omega \frac{\partial^2 \mathcal{H}}{\partial \eta \partial \eta}\omega - \omega \delta t \frac{\partial^2 \mathcal{H}}{\partial \eta \partial \eta}\omega \\ &= \omega.\end{aligned}$$

Hence the symplectic condition holds for the evolution of η during an infinitesimal time interval. As we can consider the evolution of η during a finite interval, as a sequence of canonical transformations of infinitesimal steps, the total time evolution also satisfies the symplectic condition.

One may view the Hamiltonian as the generator of a canonical transformation acting on all points in phase space. As the Jacobian of a canonical transformation is equal to 1, the size of a volume element in phase space does not change during the natural time evolution of a Hamiltonian system. Moreover, the density $f(q(t), p(t))$ around any point in phase space also remains constant during the time evolution. To see this, consider a volume V in phase space bounded by a surface S. During time evolution, the surface moves and so do all points inside the surface. However, a point cannot *cross* the surface. The reason is simple: if two trajectories in phase space would cross, it would imply that there are two trajectories that start from the same phase-space point. But this is impossible, as it would mean that a trajectory starting from this point is not uniquely specified by its initial conditions. Hence, the number of phase-space points inside any volume does not change in time. As the volume itself is also constant, this implies that the phase-space density (i.e., the number of points per unit volume) is constant. In other words: the phase-space density of a Hamiltonian system behaves like an incompressible fluid:

$$\frac{df}{dt} = 0. \tag{A.3.17}$$

While the exact solution of Hamilton's equations of motion will satisfy the incompressibility condition, discrete, numerical schemes will—in general—violate it. As before, we can consider any numerical MD algorithm (e.g., Verlet, velocity Verlet, ...) as a transformation from $(q(t), p(t))$ to $(q(t + \Delta t), q(t + \Delta t))$. We can then compute the Jacobian of this transformation, and check whether it is equal to 1 (see sections 4.3 and 4.3.3). For all "good" algorithms to solve Newton's equations of motion, the Jacobian of the transformation from $(q(t), p(t))$ to $(q(t + \Delta t), q(t + \Delta t))$ is equal to 1—such algorithms are said to be "area preserving." It should be noted that the symplectic condition implies more than just the area-preserving properties. Unfortunately, these other consequences do not have such a simple intuitive interpretation. When we say that it is desirable that an algorithm be symplectic, we mean more than that it should be area preserving—it should really satisfy the symplectic condition. Fortunately, in many cases the symplectic nature of an algorithm is easy to demonstrate by making use of the fact that *any* set of classical Hamiltonian equations of motion satisfies the symplectic condition. An algorithm that can be written as a sequence of exact time evolutions generated by simple Hamiltonians is therefore necessarily simplectic. An example is the Verlet algorithm. As discussed in section 4.3.3, this algorithm can be viewed as a sequence of exact propagations using either the kinetic part of the Hamiltonian or the potential part. Either propagation satisfies the symplectic condition. Hence the Verlet algorithm as a whole is symplectic. For an accessible discussion of symplectic dynamics in general, see ref. [518]. A discussion of symplectic integrators for Molecular Dynamics simulations can be found in ref. [519].

Appendix B

Non-Hamiltonian Dynamics

The classical Newtonian equations of motion describe the time evolution of a system of N particles in a volume V, at a total energy E. Often it is more convenient to keep other thermodynamic variables constant, e.g., the temperature T or the pressure P. This is easily achieved in a Monte Carlo simulation, but it is more subtle in the case of Molecular Dynamics simulations. One way to impose the condition of constant temperature (say) is to make use of an extended Lagrangian, from which the equations of motion are then derived (see Chapter 6). While the mechanical consequences of extending the Lagrangian are straightforward, the effects on the statistical mechanics of the system are less obvious. The reason is that, in general, these extended Lagrangians cannot be transformed into a Hamiltonian form. This implies that we cannot make a connection to statistical mechanics using the methods introduced in Appendix A.3.3. A systematic procedure for extending the techniques of classical statistical mechanics to non-Hamiltonian systems was proposed by Tuckerman *et al.* [137, 520]. In the present Appendix we sketch the general approach for analyzing extended Lagrangian systems. We will, however, skip most of the derivations. For a more complete and more rigorous derivation, using the mathematical techniques of differential geometry, the reader is referred to [135].

B.1 Theoretical Background

In general, the dynamics that results from solving non-Hamiltonian equations of motion is not area preserving. As we have seen in Appendix A.3.3, solving the equations of motion can be considered as a coordinate transformation. If the system is Hamiltonian, any volume element in phase space that is thus transformed may change its shape, but not its volume. In contrast, for a non-Hamiltonian system, we have to take into account the Jacobian of the transformation associated with the evolution of η, $\eta(t_0) \to \eta_t$, viz. equation (A.3.15),

$$d\eta_t = J(\eta_t; \eta_0) d\eta_0,$$

where the subscript 0 indicates the phase-space volume at $t = 0$ and J is the determinant of the Jacobian matrix M of the transformation.

The motion in phase space of a Hamiltonian system is similar to that of an incompressible liquid: in time the volume of this "liquid" does not change. In contrast, a non-Hamiltonian system is compressible. This compressibility must be taken into account when considering the generalization of the Liouville theorem to non-Hamiltonian systems.

The compressibility can be derived from the time dependence of the Jacobian

$$\frac{dJ(\eta_t; \eta_0)}{dt} = \kappa(\eta_t, t) J(\eta_t; \eta_0) \tag{B.1.1}$$

in which $\kappa(\eta_t, t)$, the *phase space compressibility* of the dynamical system, is defined:

$$\kappa(\eta_t, t) \equiv \nabla_\eta \cdot \dot{\eta}. \tag{B.1.2}$$

Equation (B.1.1) has as formal solution

$$J(\eta_t; \eta_0) = \exp\left(\int_0^t \kappa(\eta_s, s) ds\right).$$

If we define $w(\eta_t, t)$ as the primitive function of $\kappa(\eta_t, t)$, then we can write

$$\begin{aligned} J(\eta_t; \eta_0) &= \exp\left[w(\eta_t, t) - w(\eta_0, 0)\right] \\ &\equiv \frac{\sqrt{g(\eta_0, 0)}}{\sqrt{g(\eta_t, t)}}, \end{aligned}$$

where the last line defines the quantity $\sqrt{g}$. Recall that

$$\begin{aligned} d\eta_t &= J(\eta_t; \eta_0) d\eta_0 \\ &= \frac{\sqrt{g(\eta_0, 0)}}{\sqrt{g(\eta_t, t)}} d\eta_0. \end{aligned}$$

Hence,

$$\sqrt{g(\eta_t, t)} d\eta_t = \sqrt{g(\eta_0, 0)} d\eta_0,$$

which defines an invariant measure in phase space. This result can be used to derive the new form of the Liouville equation for non-Hamiltonian systems. The important point here is that the phase-space distribution, $f(\eta)$, the function in which we are interested in, which gives the probability density in phase space, should be kept separate from the phase-space metric, $\sqrt{g}$, which ensures that the volume of phase space of a non-Hamiltonian system is invariant under time evolution,

$$\frac{\partial f\sqrt{g}}{\partial t} + \nabla \cdot (f\sqrt{g}\dot{\eta}) = 0. \tag{B.1.3}$$

The expression corresponding to an ensemble average is

$$\langle A \rangle = \frac{\int d\eta \sqrt{g(\eta)} A(\eta) f(\eta)}{\int d\eta \sqrt{g(\eta)} f(\eta)}. \tag{B.1.4}$$

Assuming that there are n_c conservation laws, $\Lambda_k(\eta') = C_k$ for $k = 1, \ldots, n_c$, the partition function of the non-Hamiltonian system is given by

$$\Omega(C_1, \ldots, C_n) = \int d\eta' \sqrt{g(\eta')} \prod_{k=1}^{n_c} \delta\left(\Lambda_k(\eta') - C_k\right). \tag{B.1.5}$$

In many applications, one obtains the correct (NVT or NPT) partition function from the above "microcanonical" partition function, by carrying out the integration over the unphysical variables that have been introduced to represent the effect of a thermostat or barostat. In order to do this properly, it is essential to identify *all* conservation laws. Moreover, it is useful to eliminate from the analysis all those coordinates that are linearly dependent on other variables and variables that are "driven." Variables are called "driven" when they do not influence the time evolution of (and are not coupled through a conservation law) the physical variables of interest in the system, even though their own time evolution may depend on these last variables.

B.2 Non-Hamiltonian Simulation of the N,V,T Ensemble

We now apply the methods of non-Hamiltonian dynamics to analyze the Nosé-Hoover algorithm and the Nosé-Hoover chains that have been discussed in sections 6.1.2 and 6.1.3. Our discussion of these algorithms is only intended as an illustration that there exist systematic techniques for predicting the phase-space density generated by a particular non-Hamiltonian dynamics scheme. Such an analysis is essential when one is considering the use of thermostats or barostats in MD simulations. While we show a few simple examples, we refer the reader to ref. [135] for a more comprehensive discussion.

B.2.1 The Nosé-Hoover Algorithm

In section 6.1.2 we showed that the Nosé-Hoover algorithm generates non-Hamiltonian dynamics. The Nosé-Hoover equations can be written as

$$\begin{aligned}\dot{\mathbf{r}}_i &= \mathbf{p}_i/m_i\\ \dot{\mathbf{p}}_i &= \mathbf{F}_i - \frac{p_\xi}{Q}\mathbf{p}_i\\ \dot{\xi} &= p_\xi/Q\\ \dot{p}_\xi &= \sum_i p_i^2/m_i - k_B TL,\end{aligned}$$

where L is a parameter that has to be determined to generate the canonical distribution.

To analyze the dynamics of this system, we have to determine the conservation laws and the nondriven variables. Let us consider first the case in which we assume that we only have conservation of energy, viz. equation (6.1.6)

$$\begin{aligned}H_{\text{Nose}} &= \sum_{i=1}^{N}\frac{\mathbf{p}_i^2}{2m_i} + \mathcal{U}(\mathbf{r}^N) + \frac{p_\xi^2}{2Q} + Lk_BT\xi\\ &= \mathcal{H}(\mathbf{r},\mathbf{p}) + \frac{p_\xi^2}{2Q} + Lk_BT\xi = C_1,\end{aligned}$$

where $\mathcal{H}(\mathbf{r},\mathbf{p})$ is the physical Hamiltonian. If we use $\eta = (\mathbf{r}^N, \mathbf{p}^N, \xi, p_\xi)$, the phase-space compressibility of this system can be written as

$$\begin{aligned}\kappa(\eta) &= \nabla_\eta \cdot \dot{\eta}\\ &= \sum_i \nabla_{\mathbf{r}_i}\cdot\dot{\mathbf{r}}_i + \sum_i \nabla_{\mathbf{p}_i}\cdot\dot{\mathbf{p}}_i + \nabla_\xi\cdot\dot{\xi} + \nabla_{p_\xi}\cdot\dot{p}_\xi\\ &= \sum_i \nabla_{\mathbf{p}_i}\cdot\dot{\mathbf{p}}_i\\ &= -dNp_\xi/Q = -dN\dot{\xi}.\end{aligned}$$

Hence, it follows that the metric $\sqrt{g}$ is given by

$$\sqrt{g} = \exp\left[-\int \kappa dt\right] = \exp(dN\xi).$$

Substitution of this metric in the expression for the partition function gives:

$$\begin{aligned}\Omega_T(N,V,C_1) &= \int d^N\mathbf{p}\int d^N\mathbf{r}\int dp_\xi\int d\xi\\ &\quad\times\exp(dN\xi)\delta\left(\mathcal{H}(\mathbf{r},\mathbf{p}) + \frac{p_\xi^2}{2Q} + Lk_BT\xi - C_1\right).\end{aligned}$$

In this expression the integration over ξ and p_ξ can be performed analytically. Because of the δ-function, integration over ξ gives as condition

$$\xi = \frac{1}{Lk_BT}\left(C_1 - \mathcal{H}(\mathbf{r},\mathbf{p}) - \frac{p_\xi^2}{2Q}\right).$$

Substitution of this condition in the partition function gives

$$\begin{aligned}
\Omega_T(N,V,C_1) &= \frac{1}{Lk_BT}\int d^N\mathbf{p}\int d^N\mathbf{r}\int dp_\xi \\
&\quad \times \exp\left[\frac{dN}{Lk_BT}\left(C_1 - \mathcal{H}(\mathbf{r},\mathbf{p}) - \frac{p_\xi^2}{2Q}\right)\right] \\
&= \frac{\exp\left(dNC_1/Lk_BT\right)}{Lk_BT}\int dp_\xi \exp\left(-dNp_\xi^2/2QL\right) \\
&\quad \times \int d^N\mathbf{p}\int d^N\mathbf{r}\exp\left(-\beta dN\mathcal{H}(\mathbf{r},\mathbf{p})/L\right) \\
&\propto Q(N,V,T),
\end{aligned}$$

where the last equality only holds provided that we choose L equal to dN. The integration over p_ξ yields a constant prefactor that has no physical importance. In section 6.1.2, we derived a similar result for the Nosé equations in terms of its real variables. The present demonstration that the Nosé-Hoover equations lead to a canonical distribution is completely different from Nosé's original argument; yet the end result is the same. Note, however, that we have assumed that there is only a single conservation law, viz. conservation of H_{Nose}. In general, there will be more conserved quantities. For instance, if we consider a system in the absence of external forces, then the total linear momentum is also conserved. This will affect the phase-space distribution. In the Nosé-Hoover dynamics, conservation of total momentum reads

$$\begin{aligned}
\frac{d\mathbf{P}e^\xi}{dt} &= e^\xi\left(\dot{\mathbf{P}} + \mathbf{P}\dot{\xi}\right) = e^\xi\left(\dot{\mathbf{P}} + \mathbf{P}\frac{p_\xi}{Q}\right) \\
&= e^\xi\left(\sum_i\left(\mathbf{F}_i - \frac{p_\xi}{Q}\mathbf{p}_i\right) + \mathbf{P}\frac{p_\xi}{Q}\right) \\
&= 0,
\end{aligned}$$

and hence

$$\mathbf{P}e^\xi = \mathbf{K}, \tag{B.2.1}$$

where $\mathbf{P} = \sum_i \mathbf{p}_i$ is the center-of-mass momentum of the system and $\mathbf{K}$ is an arbitrary constant vector.

To continue the analysis, we should eliminate the driven variables from our system. The center-of-mass position, **R**, and momentum, **P**, have no influence on the other variables. We can eliminate these by considering the positions and momenta relative to the center of mass of the system, $\mathbf{r}'$ and $\mathbf{p}'$, respectively. Note, however, that the magnitude of the center-of-mass momentum is coupled to the other variables through a conservation law and cannot be eliminated from the analysis. The components of the center-of-mass momentum **P** are linearly dependent.[1] Therefore of the d components only one component can be chosen independently, or we can take as independent variable $P = \left(\sum_\alpha P_\alpha^2\right)^{1/2}$.

We now have to perform a transformation of our systems to the variables $\{\mathbf{p}', \mathbf{P}, \mathbf{r}', \mathbf{R}\}$; the resulting equations of motion are

$$
\begin{aligned}
\dot{\mathbf{r}}_i' &= \mathbf{p}'_i/m_i' \\
\dot{\mathbf{p}}_i' &= \mathbf{F}'_i - \frac{p_\xi}{Q}\mathbf{p}'_i \\
\dot{P}_i &= -\frac{p_\xi}{Q} P_i \\
\dot{\xi} &= \frac{p_\xi}{Q} \\
\dot{p}_\xi &= \sum_i^{N-1} \frac{p'^2_i}{m_i'} + \frac{P^2}{M} - k_B TL.
\end{aligned}
$$

The equations of motion have two[2] conservation laws:

$$
\begin{aligned}
\mathcal{H}(\mathbf{p}', \mathbf{r}', P) + \frac{p_\xi^2}{2Q} + LK_B T\xi &= C_1 \\
P \exp(\xi) &= C_2.
\end{aligned}
$$

In the first conservation law we have used

$$
\mathcal{H}(\mathbf{r}', \mathbf{p}', P) = \sum_{i=1}^{N-1} \frac{\mathbf{p}'^2_i}{2m_i'} + \frac{P^2}{2M} + \mathcal{U}\left(\mathbf{r}'^N\right) = \mathcal{H}(\mathbf{r}, \mathbf{p}).
$$

To compute the partition function, we have to determine the metric from the

[1] To see this, consider the components of equation (B.2.1):

$$
\frac{P_x}{K_x} = \frac{P_y}{K_y} = \frac{P_z}{K_z} = e^{\xi},
$$

which shows that only one of the components is independent.

[2] Because we have replaced the d center-of-mass momenta components by a single variable P only one conservation law for the momenta is left.

compressibility:

$$\begin{aligned}\kappa &= \sum_i^{N-1} \nabla_{\mathbf{r}'_i} \cdot \dot{\mathbf{r}}'_i + \sum_i^{N-1} \nabla_{\mathbf{p}'_i} \cdot \dot{\mathbf{p}}'_i + \nabla_P \cdot \dot{P} + \nabla_\xi \cdot \dot{\xi} + \nabla_{p_\xi} \cdot \dot{p}_\xi \\ &= -\left[d(N-1)+1\right] \dot{\xi}.\end{aligned}$$

From which the metric follows directly:

$$\sqrt{g} = \exp\{[d(N-1)+1]\,\xi\}.$$

The partition function Ω contains two δ-functions that express the two conservation laws:

$$\Omega_T(N, V, C_1, C_2) = \int d^{N-1}\mathbf{p}' \int d^{N-1}\mathbf{r}' \int dP \int dp_\xi \int d\xi$$
$$\times \exp\{[d(N-1)+1]\,\xi\}\,\delta\left(\mathcal{H}(\mathbf{r}', \mathbf{p}', P) + \frac{p_\xi^2}{2Q} + Lk_BT\xi - C_1\right)\delta\left(e^\xi P - C_2\right).$$

The second δ-function imposes that

$$\xi = \ln(C_2/P).$$

Hence, integrating over ξ yields

$$\Omega_T(N, V, C_1, C_2) = \frac{1}{C_2}\int d^{N-1}\mathbf{p}' \int d^{N-1}\mathbf{r}' \int dP \int dp_\xi$$
$$\times \quad \left(\frac{C_2}{P}\right)^{d(N-1)+1} \delta\left(\mathcal{H}(\mathbf{r}', \mathbf{p}', P) + \frac{p_\xi^2}{2Q} + Lk_BT\ln(C_2/P) - C_1\right).$$

The remaining δ-function fixes p_ξ:

$$p_\xi = \{2Q\,[C_1 - \mathcal{H}(\mathbf{r}', \mathbf{p}', P) - Lk_BT\ln(C_2/P)]\}^{-\frac{1}{2}}.$$

Integration over p_ξ then results in the following expression for Ω:

$$\Omega_T(N, V, C_1, C_2) = \frac{\sqrt{2Q}}{C_2}\int d^{N-1}\mathbf{p}' \int d^{N-1}\mathbf{r}'$$
$$\times \int dP \left(\frac{C_2}{P}\right)^{d(N-1)+1} [C_1 - \mathcal{H}(\mathbf{r}', \mathbf{p}', P) - Lk_BT\ln(C_2/P)]^{-\frac{1}{2}}.$$

Note that this is *not* the partition function for a canonical ensemble. This problem was first pointed out by Cho *et al.* [521]. Only in the case that $C_2 =$

0 can the conventional Nosé-Hoover generate a canonical distribution. If $C_2 = 0$ the partition function reads

$$\Omega_T(N, V, C_1, 0) = \int d^{N-1}\mathbf{p}' \int d^{N-1}\mathbf{r}' \int dP \int dp_\xi \int d\xi$$
$$\times \exp\{[d(N-1)+1]\,\xi\}\,\delta\left(\mathcal{H}(\mathbf{r}', \mathbf{p}', P) + \frac{p_\xi^2}{2Q} + Lk_BT\xi - C_1\right)\delta\left(e^{\xi}P - 0\right).$$

The δ-function imposes that $P = 0$. Integration over P then yields

$$\begin{aligned}\Omega_T(N, V, C_1, 0) &= \int d^{N-1}\mathbf{p}' \int d^{N-1}\mathbf{r}' \int dp_\xi \int d\xi \\ &\quad \times \exp\{[d(N-1)+1]\,\xi\}\exp(-\xi) \\ &\quad \times \delta\left(\mathcal{H}(\mathbf{r}', \mathbf{p}') + \frac{p_\xi^2}{2Q} + Lk_BT\xi - C_1\right).\end{aligned}$$

The other δ-function fixes ξ

$$\xi = -\frac{\beta}{L}\left[\mathcal{H}(\mathbf{r}', \mathbf{p}') + \frac{p_\xi^2}{2Q} - C_1\right],$$

and we finally obtain

$$\begin{aligned}\Omega_T(N, V, C_1, 0) &= \frac{\exp\left[\beta d(N-1)C_1/L\right]}{Lk_BT} \\ &\quad \times \int dp_\xi \exp\left[-\beta d(N-1)p_\xi^2/(2QL)\right] \\ &\quad \times \int d^{N-1}\mathbf{p}' \int d^{N-1}\mathbf{r}' \exp\left[-\beta\frac{d(N-1)}{L}\mathcal{H}(\mathbf{r}', \mathbf{p}')\right] \\ &\propto \int d^{N-1}\mathbf{p}' \int d^{N-1}\mathbf{r}' \exp\left[-\beta\frac{d(N-1)}{L}\mathcal{H}(\mathbf{r}', \mathbf{p}')\right].\end{aligned}$$

Clearly, if we choose $L = d(N-1)$, then the correct canonical partition function is recovered. In practice, most conventional Nosé-Hoover simulations are performed with a fixed center of mass and therefore obey the condition $P = 0$.

B.2.2 Nosé-Hoover Chains

The calculation in the previous section shows that in the case of more than one conservation law, the Nosé-Hoover algorithm does not give us enough "flexibility" to deal with more than one δ-function in the partition function.

To recover the canonical partition function one needs more that one thermostat. This is exactly what the Nosé-Hoover chains algorithm of Martyna *et al.* [136] is able to do. For M chains the Nosé-Hoover chains equations of motion are given by (see also section 6.1.3)

$$
\begin{aligned}
\dot{\mathbf{r}}_i &= \frac{\mathbf{p}_i}{m_i} \\
\dot{\mathbf{p}}_i &= \mathbf{F}_i - \frac{p_{\xi_1}}{Q_1}\mathbf{p}_i \\
\dot{\xi}_k &= \frac{p_{\xi_k}}{Q_k} \qquad k = 1, \ldots, M \\
\dot{p}_{\xi_1} &= \left(\sum_i \frac{p_i^2}{m_i} - Lk_BT\right) - \frac{p_{\xi_2}}{Q_2}p_{\xi_1} \\
\dot{p}_{\xi_k} &= \left[\frac{p_{\xi_{k-1}}^2}{Q_{k-1}} - k_BT\right] - \frac{p_{\xi_{k+1}}}{Q_{k+1}}p_{\xi_k} \\
\dot{p}_{\xi_M} &= \left[\frac{p_{\xi_{M-1}}^2}{Q_{M-1}} - k_BT\right].
\end{aligned}
$$

For these equations of motion the conserved energy is

$$
H_{NHC} = \mathcal{H}(\mathbf{r}, \mathbf{p}) + \sum_{k=1}^{M} \frac{p_{\xi_k}^2}{2Q_k} + Lk_BT\xi_1 + \sum_{k=2}^{M} k_BT\xi_k. \tag{B.2.2}
$$

Of these M chains only ξ_1 and the thermostat center, $\xi_c = \sum_{k=2}^{M} \xi_k$, are independently coupled to the dynamics. The remaining chain variables are driven. Therefore the M chains add only two additional degrees of freedom to the system.

Let us now analyze this method with a chain of two thermostats, $M = 2$ and ($\xi_c = \xi_2$), for a system without external forces, $\sum_i \mathbf{F}_i = \mathbf{0}$. In the previous section we have shown that the conservation laws are the total energy and one variable of the total momentum:

$$
\begin{aligned}
\mathcal{H}(\mathbf{r}', \mathbf{p}', P) + \frac{p_{\xi_1}^2}{2Q_1} + \frac{p_{\xi_c}^2}{2Q_c} + Lk_BT\xi_1 + k_BT\xi_c &= C_1 \\
P\exp(\xi_1) &= C_2,
\end{aligned}
$$

in which we have again introduced a coordinate system with respect to the position and momentum of the center of mass $\{\mathbf{r}', \mathbf{p}'\}$ and the independent variable of the total center-of-mass momentum P. For the Hamiltonian $\mathcal{H}(\mathbf{r}', \mathbf{p}', P)$, we have

$$
\mathcal{H}(\mathbf{r}', \mathbf{p}', P) = \sum_{i=1}^{N-1} \frac{\mathbf{p}'^2_i}{2m'_i} + \frac{P^2}{2M} + \mathcal{U}\left(\mathbf{r}'^N\right) = \mathcal{H}(\mathbf{r}, \mathbf{p}).
$$

The next step is to determine the compressibility

$$\begin{aligned}\kappa &= \sum_i^{N-1} \nabla_{\mathbf{r}'_i} \cdot \dot{\mathbf{r}}'_i + \sum_i^{N-1} \nabla_{\mathbf{p}'_i} \cdot \dot{\mathbf{p}}'_i + \nabla_P \cdot \dot{P} + \nabla_{\xi_1} \cdot \dot{\xi}_1 + \nabla_{p_{\xi_c}} \cdot \dot{p}_{\xi_c} \\ &= 0 - d(N-1)\frac{p_{\xi_1}}{Q_1} - \frac{p_{\xi_1}}{Q_1} + 0 - \frac{p_{\xi_c}}{Q_c} = -(dN - d + 1)\dot{\xi}_1 - \dot{\xi}_c,\end{aligned}$$

from which the following metric can be derived:

$$\sqrt{g} = \exp\left[(dN - d + 1)\xi_1 + \xi_c\right].$$

To derive the corresponding partition function this metric and the conservation laws have to be substituted in the general expression for the partition function (B.1.5):

$$\begin{aligned}\Omega_T(N, V, C_1, C_2) &= \int d^{N-1}\mathbf{p}' \int d^{N-1}\mathbf{r}' \int dP \int dp_{\xi_1} \int d\xi_1 \int dp_{\xi_c} \int d\xi_c \\ &\times \exp\left[(dN - d + 1)\xi_1 + \xi_c\right] \delta\left(e^{\xi_1} P - C_2\right) \\ &\times \delta\left(\mathcal{H}(\mathbf{r}', \mathbf{p}', P) + \frac{p_{\xi_1}^2}{2Q_1} + \frac{p_{\xi_c}^2}{2Q_c} + Lk_B T \xi_1 + k_B T \xi_c - C_1\right).\end{aligned}$$

To obtain the physical important part, we have to integrate over ξ_1 and ξ_c. The δ-function in the integration over ξ_1 gives as condition

$$\xi_1 = \ln(C_2/P).$$

Substitution of this expression into the partition function gives

$$\begin{aligned}\Omega_T(N, V, C_1, C_2) &= \frac{1}{C_2} \int d^{N-1}\mathbf{p}' \int d^{N-1}\mathbf{r}' \int dP \int dp_{\xi_1} \int dp_{\xi_c} \int d\xi_c \\ &\times \left(\frac{C_2}{P}\right)^{dN-d+1} \exp(\xi_c) \\ &\times \delta\left(\mathcal{H}(\mathbf{r}', \mathbf{p}', P) + \frac{p_{\xi_1}^2}{2Q_1} + \frac{p_{\xi_c}^2}{2Q_c} + Lk_B T \ln(C_2/P) + k_B T \xi_c - C_1\right).\end{aligned}$$

The second δ-function gives as condition for ξ_c

$$\xi_c = -\beta\left[\mathcal{H}(\mathbf{r}', \mathbf{p}', P) + \frac{p_{\xi_1}^2}{2Q_1} + \frac{p_{\xi_c}^2}{2Q_c} + Lk_B T \ln(C_2/P) - C_1\right],$$

which gives for the partition function

$$\begin{aligned}
\Omega_T(N,V,C_1,C_2) &= \frac{\exp(C_1/K_BT)}{k_BTC_2}\int d^{N-1}\mathbf{p}'\int d^{N-1}\mathbf{r}'\int dP\int dp_{\xi_1}\int dp_{\xi_c} \\
&\times\left(\frac{C_2}{P}\right)^{dN-d+1}\exp\left[-\beta\mathcal{H}(\mathbf{r}',\mathbf{p}',P)\right] \\
&\times\exp\left\{-\beta\left[\frac{p_{\xi_1}^2}{2Q_1}+\frac{p_{\xi_c}^2}{2Q_c}\right]\right\}\exp\left[-\beta Lk_BT\ln(C_2/P)\right] \\
&= \int dp_{\xi_1}\int dp_{\xi_c}\exp\left\{-\beta\left[\frac{p_{\xi_1}^2}{2Q_1}+\frac{p_{\xi_c}^2}{2Q_c}\right]\right\}\int d^{N-1}\mathbf{p}'\int dP \\
&\times\int d^{N-1}\mathbf{r}'\exp\left[-\beta\mathcal{H}(\mathbf{r}',\mathbf{p}',P)\right]\left(\frac{P}{C_2}\right)^{L-d(N-1)-1}.
\end{aligned}$$

The integration over p_{ξ_1} and p_{ξ_c} only changes the prefactor. Since P is the modulus of the center-of-mass momentum vector, it is a polar coordinate. The integration over polar coordinates requires P^{d-1}. With the choice $L = dN$, the correct partition function is recovered.

B.3 The N,P,T Ensemble

For the N,P,T ensemble the equations of motion are (see also section 6.2)

$$\begin{aligned}
\dot{\mathbf{r}}_i &= \frac{\mathbf{p}_i}{m_i}+\frac{p_\epsilon}{W}\mathbf{r}_i \\
\dot{\mathbf{p}}_i &= \mathbf{F}_i-\left(1+\frac{1}{N}\right)\frac{p_\epsilon}{W}\mathbf{p}_i-\frac{p_{\xi_1}}{Q_1}\mathbf{p}_i \\
\dot{V} &= dV\frac{p_\epsilon}{W} \\
\dot{p}_\epsilon &= dV\left(P_{\text{int}}-P_{\text{ext}}\right)+\frac{1}{N}\sum_{i=1}^{N}\frac{\mathbf{p}_i^2}{m_i}-\frac{p_{\xi_1}}{Q_1}p_\epsilon \\
\dot{\xi}_k &= \frac{p_{\xi_k}}{Q_k} \qquad \text{for } k=1,\ldots,M \\
\dot{p}_{\xi_1} &= \sum_{i=1}^{N}\frac{\mathbf{p}_i^2}{m_i}+\frac{p_\epsilon^2}{W}-(dN+1)k_BT-\frac{p_{\xi_2}}{Q_2}p_{\xi_1} \\
\dot{p}_{\xi_k} &= \frac{p_{\xi_{k-1}}^2}{Q_{k-1}}-k_BT-\frac{p_{\xi_{k+1}}}{Q_{k+1}}p_{\xi_k} \qquad \text{for } k=2,\ldots,M-1 \\
\dot{p}_{\xi_M} &= \frac{p_{\xi_{M-1}}^2}{Q_{M-1}}-k_BT.
\end{aligned}$$

To analyze the dynamics of this system we have to consider two cases. The case in which the sum of the forces is zero, $\sum_i \mathbf{F}_i = \mathbf{0}$, implies that we have additional conservation forces. The second case, $\sum_i \mathbf{F}_i \neq \mathbf{0}$, has only one conserved quantity; conservation of "energy":

$$H' - C_1 = \mathcal{H}(\mathbf{p}, \mathbf{r}) + \frac{p_\epsilon^2}{W} + \sum_{k=1}^{M} \frac{p_{\xi_k}^2}{Q_k} + (dN+1)k_B T\xi_1 + k_B T\xi_c + P_{ext}V - C_1 = 0,$$

where ξ_c is defined as the center of the thermostat

$$\xi_c = \sum_{k=2}^{M} \xi_k.$$

We first consider the case $\sum_i \mathbf{F}_i \neq \mathbf{0}$. To analyze its dynamics we have to compute the compressibility. The independent variables are[3] $\eta = \mathbf{p}^N, \mathbf{r}^N, \xi_1, \xi_c, p_{\xi_1}, p_{\xi_c}, V, p_\epsilon$:

$$\kappa = \nabla \cdot \eta = -(dN+1)\frac{p_{\xi_1}}{Q_1} - \frac{p_{\xi_c}}{Q_c},$$

which gives as phase-space metric

$$\sqrt{g} = \exp[(dN+1)\xi_1 + \xi_c].$$

We can now write for the partition function

$$\Omega_{T,P_{ext}}(N, C_1) = \int dV \int d^N\mathbf{p} \int d^N\mathbf{r} \int dp_{\xi_1} \int d\xi_1 \times \int dp_{\xi_c} \int d\xi_c \int dp_\epsilon \exp\left[(dN+1)\xi_1 + \xi_c\right] \delta\left(H' - C_1\right).$$

The delta function gives as condition for ξ_1

$$\xi_1 = \frac{1}{(dN+1)K_B T}\left(C_1 - \mathcal{H}(\mathbf{p}, \mathbf{r}) - \frac{p_\epsilon^2}{W} - \sum_{k=1}^{M} \frac{p_{\xi_k}^2}{Q_k} - k_B T\xi_c - P_{ext}V\right).$$

Substitution of this expression into the partition function gives

$$\begin{aligned}\Omega_{T,P_{ext}}(N, C_1) &= \frac{\exp(\beta C_1)}{(dN+1)K_B T} \int dV \int d^N\mathbf{p} \int d^N\mathbf{r} \int dp_{\xi_1} \int d\xi_1 \\ &\quad \times \int dp_{\xi_c} \int d\xi_c \int dp_\epsilon \\ &\quad \times \exp\left[-\beta\left(\mathcal{H}(\mathbf{p}, \mathbf{r}) + \frac{p_\epsilon^2}{W} + \sum_{k=1}^{M} \frac{p_{\xi_k}^2}{Q_k} + P_{ext}V\right)\right] \\ &\propto \int dV \exp(-\beta P_{ext}V) \int d^N\mathbf{p} \int d^N\mathbf{r} \exp\left[-\beta\mathcal{H}(\mathbf{p}, \mathbf{r})\right].\end{aligned}$$

[3]A Nosé-Hoover chain of length M has two independent variables, we use ξ_1 and ξ_c.

The integration over ξ_c gives a constant, which can be infinite but has no physical importance. This demonstrates that the desired distribution is generated.

At this point we would like to emphasize that the original Nosé-Hoover algorithm does not generate this distribution. The reason is that the metric for this algorithm generates an additional $1/V$ term in the partition function. With the algorithm of Martyna *et al.* this term is removed. This point is explained in detail in ref. [135].

For the case $\sum_i \mathbf{F}_i = \mathbf{0}$, we have as additional conservation laws for the total momentum $\mathbf{P}$

$$\mathbf{P} \exp\left[(1 + 1/N)\,\epsilon + \xi_1\right] = \mathbf{K}.$$

Similar to the N,V,T ensemble the components of P are linearly dependent and the center-of-mass coordinates have to be eliminated from the analysis. This results in a set of equations of motion in coordinates relative to the center of mass. The details of this proof can be found in ref. [135]. Similar to the N,V,T ensemble, if we use $\mathbf{K} = \mathbf{0}$, we generate an $(N-1)PT$ ensemble.

Appendix C

Linear Response Theory

The Green-Kubo relations presented in section 4.4 are but an example of the relation between transport phenomena and time-correlation functions. In fact, Onsager was the first to suggest that a disturbance created in a system by a weak external perturbation decays in the same way as a spontaneous fluctuation in equilibrium. The theory that provides this link between correlation functions and response to weak perturbations is called *linear response theory*. In this Appendix, we shall give a very simple introduction to linear response theory, mainly to illustrate the "mechanical" basis of Onsager's regression hypothesis. For a more detailed discussion, the reader is referred to any modern textbook on statistical mechanics, such as [44]. A simple introduction (very similar to the one presented here) is given in the book by Chandler [187], while an extensive discussion of linear response theory in the context of the theory of liquids is given in [79].

C.1 Static Response

First, we consider the static response of a system to a weak applied field. The field could be an electric field, for instance, and the response might be the electric current or, for a nonconducting material, the electric polarization. Suppose that we are interested in the response of a property that can be expressed as the ensemble average of a dynamical variable A. In the presence of an external perturbation, the average of A changes from its equilibrium value $\langle A \rangle_0$ to $\langle A \rangle_0 + \langle \Delta A \rangle$. Next, we must specify the perturbation. We assume that the perturbation also can be written as an explicit function of the coordinates (and, possibly, momenta) of the particles in the system. The effect of the perturbation is to change the Hamiltonian $\mathcal{H}_0$ of the system, to $\mathcal{H}_0 - \lambda B(\mathbf{p}^N, \mathbf{q}^N)$. For instance, in the case of an electric field along the x direction, the change in $\mathcal{H}$ would be $\Delta\mathcal{H} = -E_x M_x(\mathbf{q}^N)$, where M_x is the x component of the total dipole moment of the system. The electric field E_x corresponds to the parameter λ. We can immediately write down the general expression for $\langle \Delta A \rangle$:

$$\langle A \rangle_0 + \langle \Delta A \rangle = \frac{\int d\Gamma \, \exp[-\beta(\mathcal{H}_0 - \lambda B)]A}{\int d\Gamma \, \exp[-\beta(\mathcal{H}_0 - \lambda B)]},$$

where we have used the symbol Γ to denote $\{\mathbf{p}^N, \mathbf{q}^N\}$, the phase-space coordinates of the system. Let us now compute the part of $\langle \Delta A \rangle$ that varies linearly with λ. To this end, we compute

$$\left(\frac{\partial \langle \Delta A \rangle}{\partial \lambda} \right)_{\lambda=0}.$$

Straightforward differentiation shows that

$$\left(\frac{\partial \langle \Delta A \rangle}{\partial \lambda} \right)_{\lambda=0} = \beta \{ \langle AB \rangle_0 - \langle A \rangle_0 \langle B \rangle_0 \}. \tag{C.1.1}$$

To take again the example of the electric polarization, let us compute the change in dipole moment of a system due to an applied field E_x:

$$\langle \Delta M_x \rangle = E_x \left(\frac{\partial M_x}{\partial E_x} \right)_{E_x=0} = \beta E_x \left\{ \langle M_x^2 \rangle - \langle M_x \rangle^2 \right\}.$$

Suppose that we wish to compute the electric susceptibility of an ideal gas of nonpolarizable dipolar molecules with dipole moment μ. In that case,

$$\begin{aligned} \left\{ \langle M_x^2 \rangle - \langle M_x \rangle^2 \right\} &= \sum_{i,j=1}^{N} \langle \mu_x^i \mu_x^j \rangle \\ &= N \left\langle (\mu_x^i)^2 \right\rangle \\ &= \frac{N\mu^2}{3}, \end{aligned}$$

and hence,

$$P_x \equiv \frac{M_x}{V} = \frac{\mu^2 \rho}{3k_B T} E_x.$$

Of course, this example is special because it can be evaluated exactly. But, in general, we can compute the expression (C.1.1) only for the susceptibility, numerically. It should also be noted that, actually, the computation of the dielectric susceptibility is quite a bit more subtle than suggested in the preceding example (see, e.g., the discussion in the book of Allen and Tildesley [19] and the article by McDonald in [39]).

C.2 Dynamic Response

Thus far, we considered only the static response to a constant perturbation. Let us now consider a very simple time-dependent perturbation. We begin by preparing the system in the presence of a very weak, constant perturbation (λB). The static response of A to this perturbation is given by equation (C.1.1). At time $t = 0$, we discontinuously switch off the external perturbation. The response ΔA will now decay to 0. We can write an expression for the average of ΔA at time t:

$$\langle \Delta A(t) \rangle = \frac{\int d\Gamma \, \exp[-\beta(\mathcal{H}_0 - \lambda B)] A(t)}{\int d\Gamma \, \exp[-\beta(\mathcal{H}_0 - \lambda B)]},$$

where $A(t)$ is the value of A at time t if the system started at point Γ in phase space and then evolved according to the natural time evolution of the *unperturbed* system. For convenience, we have assumed that the average of A in the unperturbed system vanishes. In the limit $\lambda \to 0$, we can write

$$\begin{aligned} \langle \Delta A(t) \rangle &= \beta\lambda \frac{\int d\Gamma \, \exp[-\beta \mathcal{H}_0] B A(t)}{\int d\Gamma \, \exp[-\beta \mathcal{H}_0]} \\ &= \beta\lambda \langle B(0) A(t) \rangle . \end{aligned} \quad \text{(C.2.1)}$$

To give a specific example, consider once again a gas of dipolar molecules in the presence of a weak electric field E_x. The perturbation is equal to $-E_x M_x$. At time $t = 0$, we switch off the electric field. When the field was still on, the system had a net dipole moment. When the field is switched off, this dipole moment decays:

$$\langle M_x(t) \rangle = E_x \beta \langle M_x(0) M_x(t) \rangle .$$

In words, the decay of the macroscopic dipole moment of the system is determined by the dipole correlation function, which describes the decay of spontaneous fluctuations of the dipole moment in equilibrium. This relation between the decay of the response to an external perturbation and the decay of fluctuations in equilibrium is an example of Onsager's regression hypothesis.

It might seem that the preceding example of a constant perturbation that is suddenly switched off is of little practical use, because we are interested in the effect of an arbitrary time-dependent perturbation. Fortunately, in the linear regime that we are considering, the relation given by equation (C.2.1) is enough to derive the general response.

To see this, let us consider a time-dependent external field $f(t)$ that couples to a mechanical property B; that is,

$$\mathcal{H}(t) = \mathcal{H}_0 - f(t) B.$$

To linear order in $f(t)$, the most general form of the response of a mechanical property A to this perturbation is

$$\langle \Delta A(t) \rangle = \int_{-\infty}^{\infty} dt' \chi_{AB}(t,t') f(t'), \tag{C.2.2}$$

where χ_{AB}, the "after-effect" function, describes the linear response. We know several things about the response of the system that allow us to simplify equation (C.2.2). First of all, the response must be *causal*; that is, there can be no response *before* the perturbation is applied. As a consequence,

$$\chi_{AB}(t,t') = 0 \text{ for } t < t'.$$

Second, the response at time t to a perturbation at time t' depends only on the time *difference* $t - t'$. Hence,

$$\langle \Delta A(t) \rangle = \int_{-\infty}^{t} dt' \chi_{AB}(t - t') f(t'). \tag{C.2.3}$$

Note that, once we know χ, we can compute the linear response of the system to an arbitrary time-dependent perturbing field $f(t')$. To find an expression for χ_{AB}, let us consider the situation described in equation (C.2.1), namely, an external perturbation that has a constant value λ until $t = 0$ and 0 from then on. From equation (C.2.3), it follows that the response to such a perturbation is

$$\begin{aligned} \langle \Delta A(t) \rangle &= \lambda \int_{-\infty}^{0} dt' \chi_{AB}(t - t') \\ &= \lambda \int_{t}^{\infty} d\tau\, \chi_{AB}(\tau). \end{aligned} \tag{C.2.4}$$

If we compare this expression with the result of equation (C.2.1), we see immediately that

$$\int_{t}^{\infty} d\tau\, \chi_{AB}(\tau) = \beta\lambda \langle B(0) A(t) \rangle$$

or

$$\chi_{AB}(t) = \begin{cases} -\beta \langle B(0) \dot{A}(t) \rangle & \text{for } t > 0 \\ 0 & \text{for } t \leq 0. \end{cases} \tag{C.2.5}$$

To give a specific example, consider the mobility of a molecule in an external field F_x. The Hamiltonian in the presence of this field is

$$\mathcal{H} = \mathcal{H}_0 - F_x x.$$

The phenomenological expression for the steady-state velocity of a molecule in an external field is

$$\langle v_x(t)\rangle = \mu F_x. \tag{C.2.6}$$

If we derive this relation in terms of correlation functions, we shall find a microscopic expression for the mobility μ. From equations (C.2.3) through (C.2.5), we have

$$\begin{aligned}
\langle v_x(t)\rangle &= F_x \int_{-\infty}^{t} dt' \chi_{v_x x}(t-t') \\
&= F_x \int_0^{\infty} d\tau \chi_{v_x x}(\tau) \\
&= -\beta F_x \int_0^{\infty} d\tau \, \langle x(0)\dot{v}_x(\tau)\rangle \\
&= +\beta F_x \int_0^{\infty} d\tau \, \langle v_x(0) v_x(\tau)\rangle .
\end{aligned} \tag{C.2.7}$$

In the last line of equation (C.2.7), we used the stationarity property of time-correlation functions:

$$\frac{d}{dt}\langle A(t)B(t+t')\rangle = 0.$$

Carrying out the differentiation, we find that

$$\left\langle \dot{A}(t)B(t+t')\right\rangle = -\left\langle A(t)\dot{B}(t+t')\right\rangle .$$

Combining equations (C.2.6) and (C.2.7), we find that

$$\mu = \beta \int_0^{\infty} dt \, \langle v_x(0) v_x(t)\rangle . \tag{C.2.8}$$

If we compare this result with the Green-Kubo relation for the self-diffusion coefficient (4.4.11), we recover the Einstein relation, $\mu = \beta D$.

C.3 Dissipation

Many experimental techniques probe the dynamics of a many-body system by measuring the absorption of some externally applied field (e.g., visible light, infrared radiation, microwave radiation). Linear response theory allows us to establish a simple relation between the absorption spectrum and the Fourier transform of a time-correlation function. To see this, let us again consider an external field that is coupled to a dynamical variable $A(\mathbf{p}^N, \mathbf{q}^N)$. The time-dependent Hamiltonian of the system is

$$\mathcal{H}(t) = \mathcal{H}_0 - f(t)A(\mathbf{p}^N, \mathbf{q}^N).$$

Note that the only quantity explicitly time dependent is $f(t)$. As the Hamiltonian depends on time, the total energy E of the system also changes with time:

$$E(t) = \langle \mathcal{H}(t) \rangle .$$

Let us compute the average rate of change of the energy of the system. This is the amount of energy absorbed (or emitted) by the system, per unit of time:

$$\begin{aligned} \frac{\partial E}{\partial t} &= \left\langle \frac{d\mathcal{H}}{dt} \right\rangle \qquad \text{(C.3.1)} \\ &= \left\langle \sum_i \left(\dot{\mathbf{q}}_i \frac{\partial \mathcal{H}}{\partial \mathbf{q}_i} + \dot{\mathbf{p}}_i \frac{\partial \mathcal{H}}{\partial \mathbf{p}_i} \right) + \frac{\partial \mathcal{H}}{\partial t} \right\rangle . \end{aligned}$$

But, from Hamilton's equations of motion, we have

$$\dot{\mathbf{q}}_i = \frac{\partial \mathcal{H}}{\partial \mathbf{p}_i}$$

and

$$\dot{\mathbf{p}}_i = -\frac{\partial \mathcal{H}}{\partial \mathbf{q}_i} .$$

As a consequence, equation (C.3.1) simplifies to

$$\begin{aligned} \frac{\partial E}{\partial t} &= \left\langle \frac{\partial \mathcal{H}}{\partial t} \right\rangle \\ &= -\left\langle \dot{f}(t) A(\mathbf{p}^N, \mathbf{q}^N) \right\rangle \\ &= -\dot{f}(t) \langle A(t) \rangle . \qquad \text{(C.3.2)} \end{aligned}$$

Note, however, that $\langle A(t) \rangle$ itself is the response to the applied field f:

$$\langle A(t) \rangle = \int_{-\infty}^{\infty} dt' \, \chi_{AA}(t - t') f(t').$$

Let us now consider the situation where $f(t)$ is a periodic field with frequency ω (e.g., monochromatic light). In that case, we can write $f(t)$ as

$$f(t) = \text{Re}\, f_\omega e^{i\omega t}$$

and

$$\dot{f}(t) = \frac{i\omega}{2} \left(f_\omega e^{i\omega t} - f_\omega^* e^{-i\omega t} \right) .$$

The average rate of energy dissipation is

$$\begin{aligned} \frac{\partial E}{\partial t} &= -\dot{f}(t) \langle A(t) \rangle \qquad \text{(C.3.3)} \\ &= -\dot{f}(t) \int_{-\infty}^{\infty} dt' \, \chi_{AA}(t - t') f(t'). \end{aligned}$$

For a periodic field, we have

$$\begin{aligned}
\int_{-\infty}^{\infty} dt' \chi_{AA}(t-t')f(t') &= \frac{f_\omega e^{i\omega t}}{2}\int_{-\infty}^{\infty} dt' \chi_{AA}(t-t')e^{i\omega(t'-t)} \\
&\quad + \frac{f_\omega^* e^{-i\omega t}}{2}\int_{-\infty}^{\infty} dt' \chi_{AA}(t-t')e^{-i\omega(t'-t)} \\
&= \pi\left[f_\omega e^{i\omega t}\chi_{AA}(\omega) + f_\omega^* e^{-i\omega t}\chi_{AA}(-\omega)\right],
\end{aligned} \quad (C.3.4)$$

where

$$\chi_{AA}(\omega) \equiv \frac{1}{2\pi}\int_0^{\infty} dt\, \chi_{AA}(t)e^{-i\omega t}.$$

To compute $\dot{E}$, the rate of change of the energy, we must average $\langle \partial\mathcal{H}/\partial t\rangle$ over one period, $T\ (= 2\pi/\omega)$, of the field:

$$\begin{aligned}
\dot{E} &= \frac{-\pi}{2T}\int_0^T dt\ \{i\omega(f_\omega e^{i\omega t} - f_\omega^* e^{-i\omega t}) \\
&\quad \times \left[f_\omega e^{i\omega t}\chi_{AA}(\omega) + f_\omega^* e^{-i\omega t}\chi_{AA}(-\omega)\right]\} \\
&= -\pi\omega\, |f_\omega|^2\, \frac{\chi_{AA}(\omega) - \chi_{AA}(-\omega)}{2i} \\
&= -\pi\omega\, |f_\omega|^2\, \mathrm{Im}\left[\chi_{AA}(\omega)\right].
\end{aligned} \quad (C.3.5)$$

We use the relation between $\chi_{AA}(t)$ and the autocorrelation function (C.2.5) of A:

$$\chi_{AA}(\omega) = \frac{1}{2\pi}\int_0^{\infty} dt\, e^{-i\omega t}\left[-\beta\left\langle A(0)\dot{A}(t)\right\rangle\right].$$

The imaginary part of $\chi_{AA}(\omega)$ is given by

$$\begin{aligned}
\mathrm{Im}\left[\chi_{AA}(\omega)\right] &= \frac{\beta}{2\pi}\int_0^{\infty} dt\ \sin(\omega t)\left\langle A(0)\dot{A}(t)\right\rangle \\
&= -\frac{\beta}{4\pi}\int_{-\infty}^{\infty} dt\ \omega\cos(\omega t)\left\langle A(0)A(t)\right\rangle.
\end{aligned} \quad (C.3.6)$$

Finally, we obtain

$$\dot{E} = |f_\omega|^2\frac{\beta\omega^2}{4}\int_{-\infty}^{\infty} dt\ \cos(\omega t)\left\langle A(0)A(t)\right\rangle. \quad (C.3.7)$$

Hence, from knowledge of the autocorrelation function of the quantity that couples with the applied field, we can compute the shape of the absorption spectrum. This relation was derived assuming classical dynamics and therefore is valid only if $\hbar\omega \ll k_B T$. However, it is also possible to derive a quantum-mechanical version of linear response theory that is valid for arbitrary frequencies (see, e.g., [44]).

To give a specific example, let us compute the shape of the absorption spectrum of a dilute gas of polar molecules. In that case, the relevant correlation function is the dipole autocorrelation function:

$$\langle M_x(0)M_x(t)\rangle = \frac{N}{3}\langle \mu(0)\cdot\mu(t)\rangle .$$

For molecules that rotate almost freely (*almost*, otherwise there would be no dissipation), $\mu(0)\cdot\mu(t)$ depends on time, because each molecule rotates. For a molecule with a rotation frequency ω, we have

$$\mu(0)\cdot\mu(t) = \mu^2\cos(\omega t),$$

and for an assembly of molecules with a thermal distribution of rotational velocities $P(\omega)$, we have

$$\langle \mu(0)\cdot\mu(t)\rangle = \mu^2\int d\omega P(\omega)\cos(\omega t).$$

The rate of absorption of radiation is then given by

$$\dot{E} = \frac{\pi\beta\omega^2 N\mu^2}{12}P(\omega)\left|f_\omega\right|^2 . \tag{C.3.8}$$

For more details about the relation between spectroscopic properties and time correlation functions, the reader is referred to the article by Madden in [39].

In the preceding derivation of linear response theory, we assumed that we prepare the system in an equilibrium state with the perturbation *on* and then allow the system to relax to a new equilibrium state with the perturbation *off*. However, this will not always work. Consider, for instance, electrical conductivity. In that case, the perturbation is an electrical field that will cause a current to flow in the system. Hence, the state in which we prepared the system with the field on is *not* an equilibrium state but a steady *nonequilibrium* state. The same holds, for instance, for a system under steady shear. It would seem that, in such circumstances, one cannot use the framework of linear response theory in its simplest form to derive transport coefficients such as the electrical conductivity σ or the viscosity η.

C.3.1 Electrical Conductivity

Fortunately, things are not quite as bad as that. Consider, for example, electrical conductivity. Indeed, if we put a conducting system in an external field, we will generate a nonequilibrium steady state. However, what we *can* do is to perturb the system by switching on a weak, uniform *vector potential* **A**. The Hamiltonian of the system with the vector potential switched on is

$$\mathcal{H}' = \sum_{i=1}^{N} \frac{1}{2m_i}\left(\mathbf{p}_i - \frac{e_i}{c}\mathbf{A}\right)^2 + \mathcal{U}_{\text{pot}}. \tag{C.3.9}$$

The system described by this Hamiltonian satisfies the same equations of motion as the unperturbed system (**A** is a gauge field) and the system will be in an equilibrium state at $t = 0$. We then abruptly switch off the vector potential. From electrodynamics, we know that a time-dependent vector potential generates an electric field:

$$\mathbf{E} = -\frac{1}{c}\dot{\mathbf{A}}. \tag{C.3.10}$$

In the present case, the electrical field will be an infinitesimal δ spike at $t = 0$:

$$\mathbf{E}(t) = \frac{1}{c}\mathbf{A}\delta(t). \tag{C.3.11}$$

We can compute the current that results in the standard way. We note that we can write $\mathcal{H}'$ in equation (C.3.9) as

$$\begin{aligned}
\mathcal{H}' &= \mathcal{H}_0 - \sum_{i=1}^{N} \frac{e_i}{cm_i}\mathbf{p}_i \cdot \mathbf{A} + \mathcal{O}(A^2) \\
&= \mathcal{H}_0 - \frac{\mathbf{A}}{c}\int d\mathbf{r} \sum_{i=1}^{N} \frac{e_i}{m_i}\mathbf{p}_i\delta(\mathbf{r}_i - \mathbf{r}) \\
&= \mathcal{H}_0 - \frac{\mathbf{A}}{c}\int d\mathbf{r}\, \mathbf{j}(\mathbf{r}),
\end{aligned} \tag{C.3.12}$$

where $\mathbf{j}(\mathbf{r})$ denotes the current density at point $\mathbf{r}$. The average current density at time t due to the perturbation is given by

$$\langle \mathbf{j}(t)\rangle = \frac{\mathbf{A}}{cVk_BT}\int d\mathbf{r}d\mathbf{r}' \left\langle \mathbf{j}(\mathbf{r},0)\mathbf{j}(\mathbf{r}',t)\right\rangle. \tag{C.3.13}$$

The phenomenological expression for the current response to an applied δ-function electric field spike is (see equation (C.2.3))

$$\begin{aligned}
\langle \mathbf{j}(t)\rangle &= \int_{-\infty}^{t} dt'\sigma(t-t')E(t') \\
&= \sigma(t)\frac{A}{c}.
\end{aligned} \tag{C.3.14}$$

From this it immediately follows that

$$\sigma(t) = \frac{1}{V k_B T} \int d\mathbf{r} d\mathbf{r}' \, \langle \mathbf{j}(\mathbf{r}, 0) \mathbf{j}(\mathbf{r}', t) \rangle . \tag{C.3.15}$$

The dc conductivity is then given by

$$\sigma(\omega = 0) = \frac{1}{V k_B T} \int_0^\infty dt \int d\mathbf{r} d\mathbf{r}' \, \langle \mathbf{j}(\mathbf{r}, 0) \mathbf{j}(\mathbf{r}', t) \rangle . \tag{C.3.16}$$

C.3.2 Viscosity

The corresponding linear response expression for the viscosity seems more subtle because shear usually is not interpreted in terms of an external field acting on all molecules. Still, we can use, by analogy to the electrical conductivity case, a canonical transformation, the time derivative that corresponds to uniform shear. To achieve this, we consider a system of N particles with coordinates $\mathbf{r}^M$ and Hamiltonian

$$\mathcal{H}_0 = \sum_{i=1}^{N} p_i^2/(2m_i) + \mathcal{U}(\mathbf{r}^N). \tag{C.3.17}$$

Now consider another system described by a set of coordinates $\mathbf{r}'^N$ related to $\mathbf{r}^N$ by a linear transformation:

$$\mathbf{r}'_i = \mathsf{h}\mathbf{r}_i . \tag{C.3.18}$$

The Hamiltonian for the new system can be written as

$$\mathcal{H}_1 = \sum_{i=1}^{N} \frac{1}{2m_i} \mathbf{p}'_i \cdot \mathsf{G}^{-1} \cdot \mathbf{p}'_i + \mathcal{U}(\mathbf{r}'^N), \tag{C.3.19}$$

where G, the metric tensor, is defined as

$$\mathsf{G} \equiv \mathsf{h}^T \cdot \mathsf{h}. \tag{C.3.20}$$

We assume that h differs infinitesimally from the unit matrix I:

$$\mathsf{h} = \mathsf{I} + \epsilon. \tag{C.3.21}$$

In the case that we are interested in the effect of uniform shear, for instance, we could choose $\epsilon_{xy} = \epsilon$, while all other elements of $\epsilon_{\alpha\beta}$ are 0. Now consider the case that we equilibrate the system with Hamiltonian $\mathcal{H}_1$, and at time $t = 0$, we switch off the infinitesimal deformation ϵ. This means that, at $t = 0$, the system experiences a δ-function spike in the shear rate

$$\frac{\partial v_x}{\partial y} = -\epsilon\delta(t). \tag{C.3.22}$$

We can compute the time-dependent response of the shear stress, $\sigma_{xy}(t)$, to the sudden change from $\mathcal{H}_1$ to $\mathcal{H}_0$:

$$\langle\sigma_{xy}(t)\rangle = -\epsilon\frac{1}{Vk_BT}\langle\sigma_{xy}(0)\sigma_{xy}(t)\rangle\,. \tag{C.3.23}$$

By combining equations (C.3.22) and (C.3.23) with equation (C.2.3), we immediately see that the steady-state stress that results from a steady shear is given by

$$\sigma_{xy} = \frac{\partial v_x}{\partial y}\times\frac{1}{Vk_BT}\int_0^\infty dt\,\langle\sigma_{xy}(0)\sigma_{xy}(t)\rangle\,, \tag{C.3.24}$$

and the resulting expression for the shear viscosity η is

$$\eta = \frac{1}{Vk_BT}\int_0^\infty dt\,\langle\sigma_{xy}(0)\sigma_{xy}(t)\rangle\,. \tag{C.3.25}$$

C.4 Elastic Constants

A liquid flows under the influence of shear forces. A solid does not. Rather, any small deformation of a solid induces an elastic response (stress) that counteracts it. This elastic stress is proportional to the applied deformation (strain). The constants of proportionality between stress and strain (to be defined more precisely below) are called the elastic constants. Below we discuss how to measure these constants by computer simulation. For the sake of simplicity, we limit the discussion to crystals on isotropic (hydrostatic) pressure.

When considering the effect of the strain on the free energy of a solid, it is essential to introduce the so-called *Lagrangian strain tensor* (see, e.g., [105]). The reason is that, on a local scale, all changes in free energy are due to changes in the *distances* between the particles that make up the solid. And the quantity that measures this change is precisely the Lagrangian strain. We start with the relation between new and old coordinates due to an elastic deformation:

$$\mathbf{r}' = (1+\epsilon)\mathbf{r}, \tag{C.4.1}$$

where

$$\epsilon_{\alpha\beta} \equiv \frac{\partial u_\alpha}{\partial x_\beta} \tag{C.4.2}$$

is the (conventional) strain tensor. It measures the variation of the displacement field $\mathbf{u}$ with the original coordinate $\mathbf{r}$. Due to the strain, the distance r_{ij} separating two points i and j in the solid is changed. The new squared distance is then related to the old distance by

$$\begin{aligned} r_{ij}'^2 &= \mathbf{r}_{ij}\left(1+\epsilon^T\right)\left(1+\epsilon\right)\mathbf{r}_{ij} \\ &= \mathbf{r}_{ij}(1+\epsilon^T+\epsilon+\epsilon^T\epsilon)\mathbf{r}_{ij} \\ &\equiv \mathbf{r}_{ij}(1+2\eta)\mathbf{r}_{ij}. \end{aligned}$$

This defines the Lagrangian strain η. The new volume V' of the system is related to the original volume V_0 by

$$V' = V_0 \det(1+\epsilon) \tag{C.4.3}$$

or

$$V' = V_0\sqrt{\det(1+2\eta)}. \tag{C.4.4}$$

We now expand the Helmholtz free-energy (F) per unit of (undeformed) volume (V) in powers of the Lagrangian strain parameters η:

$$\begin{aligned} F(\eta)/V &= V^{-1}\left[F(0)+\frac{\partial F}{\partial\eta_{\alpha\beta}}\eta_{\alpha\beta}+\frac{1}{2}\frac{\partial^2 F}{\partial\eta_{\alpha\beta}\partial\eta_{\gamma\delta}}\eta_{\alpha\beta}\eta_{\gamma\delta}+\cdots\right] \\ &= V^{-1}F(0)+C^{(1)}_{\alpha\beta}\eta_{\alpha\beta}+\frac{1}{2}C^{(2)}_{\alpha\beta\gamma\delta}\eta_{\alpha\beta}\eta_{\gamma\delta}+\cdots. \end{aligned} \tag{C.4.5}$$

This equation defines the (second-order) elastic constants $C^{(2)}_{\alpha\beta\gamma\delta}$. To compute the elastic constants numerically, we need a *microscopic* expression for the η-dependence of F. To derive such a relation, we must consider in detail what a deformation of the system does to the partition function. Let us first consider the deformed system. The partition function of this system (ignoring constants, such as h^{-3N}) is equal to

$$Q(\eta) = \int d\mathbf{p}^N d\mathbf{r}^N \exp\left[-\beta\mathcal{H}\left(\mathbf{p}^N,\mathbf{r}^N\right)\right]. \tag{C.4.6}$$

This partition function depends on the deformation through the boundary conditions of the integral over the coordinates. This is not very convenient when computing derivatives with respect to the strain. Therefore, we first express the partition function of the deformed system in terms of coordinates and momenta of the original, undeformed system. We can express the coordinates ($\mathbf{r}_i$) and velocities ($\dot{\mathbf{r}}_i$) in this system in terms of the strain tensor $\mathsf{h} \equiv (1+\epsilon)$, and the original coordinates ($\mathbf{r}_{0,i}$) and velocities ($\dot{\mathbf{r}}_{0,i}$):

$$\begin{aligned} \mathbf{r}_i &= \mathsf{h}\mathbf{r}_{0,i} \\ \dot{\mathbf{r}}_i &= \mathsf{h}\dot{\mathbf{r}}_{0,i} \end{aligned} \tag{C.4.7}$$

The kinetic energy, $\mathcal{K} = \sum m_i\dot{\mathbf{r}}_i^2$, can be written as

$$\begin{aligned} \mathcal{K} &= \sum\frac{1}{2}m_i\dot{\mathbf{r}}_i^2 \\ &= \sum\frac{1}{2}m_i\dot{\mathbf{r}}_{0,i}(\mathsf{h}^T\mathsf{h})\dot{\mathbf{r}}_{0,i} \equiv \sum\frac{1}{2}m_i\dot{\mathbf{r}}_{0,i}\cdot\mathsf{G}\cdot\dot{\mathbf{r}}_{0,i}, \end{aligned} \tag{C.4.8}$$

where $h^T = (1 + \epsilon^T)$ is the transverse of h and $G = h^T h$ is the metric tensor. From the definition of h it follows that $G = (1 + 2\eta)$. We can now write down the generalized momentum $\mathbf{p}_{0,i}$ conjugate to the coordinate $\mathbf{r}_{0,i}$ (see section A):

$$p_{0,i}^{\alpha} = \left(\frac{\partial \mathcal{K}}{\partial \dot{r}_{0,i}^{\alpha}}\right) = m_i G_{\alpha\beta} \dot{r}_{0,i}^{\beta} \tag{C.4.9}$$

and hence

$$\begin{aligned}
\mathcal{K} &= \sum \frac{1}{2} m_i \dot{\mathbf{r}}_{0,i} \cdot G \cdot \dot{\mathbf{r}}_{0,i} \\
&= \sum \frac{1}{2m_i} \mathbf{p}_{0,i} \cdot G^{-1} \cdot \mathbf{p}_{0,i} \\
&= \sum \frac{1}{2m_i} \mathbf{p}_{0,i} \cdot (1 + 2\eta)^{-1} \cdot \mathbf{p}_{0,i}.
\end{aligned} \tag{C.4.10}$$

As

$$\begin{aligned}
\mathbf{p}_i &= m_i \dot{\mathbf{r}}_i = m_i h \dot{r}_{0,i} = (h^T)^{-1} \mathbf{p}_{0,i} \\
\mathbf{r}_i &= h \mathbf{r}_{0,i},
\end{aligned} \tag{C.4.11}$$

the Jacobian of the transformation between $\{\mathbf{p}^N, \mathbf{r}^N\}$ and $\{\mathbf{p}_0^N, \mathbf{r}_0^N\}$ is equal to 1. Hence, we can write

$$\begin{aligned}
Q(\eta) &= \int d\mathbf{p}^N d\mathbf{r}^N \exp\left[-\beta \mathcal{H}\left(\mathbf{p}^N, \mathbf{r}^N\right)\right] \qquad \text{(C.4.12)} \\
&= \int d\mathbf{p}_0^N d\mathbf{r}_0^N \exp\left\{-\beta \left[\sum \frac{1}{2m_i} \mathbf{p}_{0,i} \cdot (1 + 2\eta)^{-1} \cdot \mathbf{p}_{0,i} + \mathcal{U}\left(\mathbf{r}_0^N; \eta\right)\right]\right\}.
\end{aligned}$$

Now the dependence of $Q(\eta)$ on η is only contained in the Hamiltonian. We can now explicitly carry out the differentiation with respect to η. Using

$$\begin{aligned}
\left(\frac{\partial u}{\partial \eta_{\alpha\beta}}\right) &= \sum_{i<j} \left(\frac{\partial u}{\partial r_{ij}^2}\right) \left(\frac{\partial r_{ij}^2}{\partial \eta_{\alpha\beta}}\right) = \sum_{i<j} \left(\frac{\partial u}{\partial r_{ij}}\right) \frac{r_{0,ij}^{\alpha} r_{0,ij}^{\beta}}{r_{ij}} \\
&= \left(\mathbf{h}^{-1} \sum_{i<j} \left(\frac{\partial u}{\partial r_{ij}}\right) \frac{\mathbf{r}_{ij} \mathbf{r}_{ij}}{r_{ij}} (\mathbf{h}^T)^{-1}\right)_{\alpha\beta}
\end{aligned} \tag{C.4.13}$$

and

$$\begin{aligned}
\sum \frac{1}{2m_i} \mathbf{p}_{0,i} \cdot \left(\frac{\partial G^{-1}}{\partial \eta_{\alpha\beta}}\right) \cdot \mathbf{p}_{0,i} &= -\sum \frac{1}{m_i} \left(\mathbf{p}_{0,i} \cdot G^{-1}\right)_{\alpha} \left(G^{-1} \cdot \mathbf{p}_{0,i}\right)_{\beta} \\
&= -\left(h^{-1} \left(\sum \frac{1}{m_i} \mathbf{p}\mathbf{p}\right) (\mathbf{h}^T)^{-1}\right)_{\alpha\beta},
\end{aligned} \tag{C.4.14}$$

we obtain

$$\left(\frac{\partial F}{\partial \eta_{\alpha\beta}}\right) = -\left(h^{-1}\left\langle \sum \frac{1}{m_i}\mathbf{pp} + \sum_{j<i} \mathbf{r}_{ij}\mathbf{f}_{ij}\right\rangle (h^T)^{-1}\right)_{\alpha\beta} . \qquad (C.4.15)$$

From this it follows immediately that

$$\begin{aligned} C^{(1)}_{\alpha\beta} &\equiv \left(\frac{\partial F}{\partial \eta_{\alpha\beta}}\right) \\ &= \frac{V'}{V}\left[(1+\epsilon)^{-1}\,\sigma\,\left(1+\epsilon^T\right)^{-1}\right]_{\alpha\beta} \\ &= \sqrt{\det(1+2\eta)}\left[(1+\epsilon)^{-1}\sigma(1+\epsilon^T)^{-1}\right]_{\alpha\beta}, \end{aligned} \qquad (C.4.16)$$

where

$$\sigma_{\gamma\delta} = -\frac{1}{V'}\sum_i \left[\frac{p_{i\gamma}p_{i\delta}}{m_i} + \sum_{j<i} r_{ij\gamma}f_{ij\delta}\right] \qquad (C.4.17)$$

denotes the microscopic stress in the deformed system. Note that σ can be measured in a simulation, while ϵ is fixed by the applied strain. For an undeformed system $C^{(1)}$ is simply equal to $-P$, where P is the hydrostatic pressure. From equation (C.4.16), it also follows that the constant of proportionality between the stress σ_{ij} and the linear strain ϵ_{rs} is given by

$$\left(\frac{\partial \sigma_{\alpha\beta}}{\partial \epsilon_{\gamma\delta}}\right) = (\sigma_{\alpha\delta}\delta_{\beta\gamma} + \sigma_{\beta\delta}\delta_{\alpha\gamma} - \sigma_{\alpha\beta}\delta_{\gamma\delta}) + C^{(2)}_{\alpha\beta\gamma\delta}. \qquad (C.4.18)$$

To determine the second-order elastic constants, $C^{(2)}_{\alpha\beta\gamma\delta}$, we must determine the initial linear dependence of $C^{(1)}_{\alpha\beta}$ on $\eta_{\gamma\delta}$. This technique for measuring the elastic constants is simple and also quite accurate (see, e.g., [522]). However, several computations are needed to measure the different elastic constants. The lower the symmetry of the crystal, the more calculations are needed. This can be avoided by directly considering the microscopic expression for the $C^{(2)}_{\alpha\beta\gamma\delta}$. Such an expression was derived by Squire *et al.* [523]:

$$\begin{aligned} C^{(2)}_{\alpha\beta\gamma\delta} &= -\frac{1}{Vk_BT}\langle \Delta\sigma_{\alpha\beta}\,\Delta\sigma_{\gamma\delta}\rangle + 2\rho k_B T(\delta_{\alpha\gamma}\delta_{\beta\delta} + \delta_{\alpha\delta}\delta_{\beta\gamma}) \\ &\quad +4\sum_{i<j,k<l}\left\langle \frac{\partial^2 U}{\partial r_{ij}^2 \partial r_{kl}^2} r_{ij\alpha}r_{ij\beta}r_{kl\gamma}r_{kl\delta}\right\rangle . \end{aligned} \qquad (C.4.19)$$

Using equation (C.4.19), it takes only a single simulation to measure *all* elastic constants. Unfortunately, the statistical errors in the evaluation of this fluctuation expression are usually larger than those obtained when computing equation (C.4.16). The problems with statistics appear even worse in constant-stress MD simulations on the same system, where the elastic compliances (rather than the moduli) are determined from fluctuations in the box shape [524]. Equation (C.4.19) can only be used if the intermolecular potentials are everywhere continuous. However, a fluctuation expression that works for hard-core systems has been developed by Farago and Kantor [525]. More details about the numerical evaluation of elastic constants can be found in [523], [286], [525], and [526], while the general framework of elastic properties at finite temperature is discussed in [105].

Appendix D

Statistical Errors

It is often stated that a computer simulation generates "exact" data for a given model. However, this is true only if we can perform an infinitely long simulation. In practice we have neither the budget nor the patience to approximate such a simulation. Therefore, the results of a simulation will always be subjected to statistical errors, which have to be estimated.

D.1 Static Properties: System Size

Let us consider the statistical accuracy of the measurement of a dynamical quantity A in a Molecular Dynamics simulation (the present discussion applies, with minor modifications, to Monte Carlo simulations). During a simulation of total length T, we obtain the following estimate for the equilibrium-average of A:

$$A_\tau = \frac{1}{\tau}\int_0^\tau dt\, A(t), \tag{D.1.1}$$

where the subscript on A_τ refers to averaging over a finite "time" τ. If the ergodic hypothesis is justified then $A_\tau \to \langle A \rangle$, as $\tau \to \infty$, where $\langle A \rangle$ denotes the ensemble average of A. Let us now estimate the variance in A_τ, $\sigma^2(A)$:

$$\begin{aligned}\sigma^2(A) &= \langle A_\tau^2 \rangle - \langle A_\tau \rangle^2 \\ &= \frac{1}{\tau^2}\int_0^\tau\int_0^\tau dtdt'\ \langle [A(t) - \langle A \rangle]\,[A(t') - \langle A \rangle] \rangle\,. \end{aligned} \tag{D.1.2}$$

Note that $\langle [A(t) - \langle A \rangle][A(t') - \langle A \rangle] \rangle$ in equation (D.1.2) is simply the time correlation function of fluctuations in the variable A. Let us denote this correlation function by $C_A(t - t')$. If the duration of the sampling τ is much larger than the characteristic decay time t_A^c of C_A, then we may rewrite equation (D.1.2) as

$$\begin{aligned} \sigma^2(A) &\approx \frac{1}{\tau}\int_{-\infty}^{\infty} dt\, C_A(t) \\ &\approx \frac{2t_A^c}{\tau} C_A(0). \end{aligned} \tag{D.1.3}$$

In the last equation we have used the definition of t_A^c as half the integral from $-\infty$ to ∞ of the normalized correlation function $C_A(t)/C_A(0)$. The relative variance in A_τ therefore is given by

$$\frac{\sigma^2(A)}{\langle A\rangle^2} \approx (2t_A^c/\tau)\frac{\langle A^2\rangle - \langle A\rangle^2}{\langle A\rangle^2}. \tag{D.1.4}$$

Equation (D.1.4) clearly shows that the root-mean-square error in A_τ is proportional to $\sqrt{t_A^c/\tau}$. This result is hardly surprising. It simply states the well-known fact the variance in a measured quantity is inversely proportional to the number of uncorrelated measurements. In the present case, this number is clearly proportional to τ/t_A^c. This result may appear to be trivial, but it is nevertheless very important, because it shows directly how the lifetime and amplitude of fluctuations in an observable A affect the statistical accuracy. This is of particular importance in the study of fluctuations associated with hydrodynamical modes or pretransitional fluctuations near a symmetry-breaking phase transition. Such modes usually have a characteristic lifetime proportional to the square of their wavelengths. To minimize the effects of the finite system size on such phase transitions, it is preferable to study systems with a box size L large compared with all relevant correlation lengths in the system. However, due to the slow decay of long-wavelength fluctuations, the length of the simulation needed to keep the relative error fixed should be proportional to L^2. As the CPU time for a run of fixed length is proportional to the number of particles (at best), the CPU time needed to maintain constant accuracy increases quite rapidly with the linear dimensions of the system (e.g., as L^5 in three dimensions).

Another aspect of equation (D.1.4) is not immediately obvious; namely, it makes a difference whether the observable A can be written as a sum of uncorrelated single-particle properties. If this is the case, then it is easy to see that the ratio $(\langle A^2\rangle - \langle A\rangle^2)/\langle A\rangle^2$ is inversely proportional to the number of particles, N. To see this, consider the expressions for $\langle A\rangle$ and $\langle A^2\rangle - \langle A\rangle^2$ in this case:

$$\langle A\rangle = \sum_{i=1}^{N} \langle a_i\rangle = N\langle a\rangle \tag{D.1.5}$$

and

$$\langle A^2 \rangle - \langle A \rangle^2 = \sum_{i=1}^{N} \sum_{j=1}^{N} \langle [a_i - \langle a \rangle]\,[a_j - \langle a \rangle] \rangle . \tag{D.1.6}$$

If the fluctuations in a_i and a_j are uncorrelated, then we find that

$$\frac{\langle A^2 \rangle - \langle A \rangle^2}{\langle A \rangle^2} = \frac{1}{N} \frac{\langle a^2 \rangle - \langle a \rangle^2}{\langle a \rangle^2} . \tag{D.1.7}$$

From equation (D.1.7) it is clear that the statistical error in a single-particle property is inversely proportional to $\sqrt{N}$. Hence, for single-particle properties, the accuracy improves as we go to larger systems (at fixed length of the simulation). In contrast, no such advantage is to be gained when computing truly collective properties.

D.2 Correlation Functions

We can apply essentially the same arguments to estimate the statistical errors in time correlation functions. Suppose that we wish to measure the (auto)correlation function[1] of the dynamical quantity A. To obtain an estimate of $C_A(\tau) \equiv \langle A(0)A(\tau) \rangle$, we average the product $A(t)A(t+\tau)$ over the initial time t. Suppose that the length of our run is τ_0, then our estimate of $C_A(\tau)$ is

$$\overline{C}_A(\tau) = 1/\tau_0 \int_0^{\tau_0} dt\ A(t)A(t+\tau),$$

where the bar over C_A denotes the average over a finite time τ_0. Next, we consider the variance in $\overline{C}_A(\tau)$ [527]:

$$\begin{aligned} &\langle \overline{C}_A(\tau)^2 \rangle - \langle \overline{C}_A(\tau) \rangle^2 \\ &= (1/\tau_0{}^2) \int_0^{\tau_0} \int_0^{\tau_0} dt' dt''\ \langle A(t')A(t'+\tau)A(t'')A(t''+\tau) \rangle \\ &\quad -(1/\tau_0{}^2) \int_0^{\tau_0} \int_0^{\tau_0} dt' dt''\ \langle A(t')A(t'+\tau) \rangle \langle A(t'')A(t''+\tau) \rangle . \end{aligned} \tag{D.2.1}$$

The first term on the right-hand side of equation (D.2.1) contains a fourth-order correlation function. To simplify matters, we shall assume that the fluctuations of A follow a Gaussian distribution. This is not the simple Gaussian distribution that describes, for instance, the Maxwell distribution

[1] The extension to cross-correlation functions of the type $\langle A(t)B(0) \rangle$ is straightforward and left as an exercise to the reader.

of particle velocities in equilibrium, but a multidimensional (in fact, infinite-dimensional) distribution that describes all correlations between fluctuations of A at different times. For the simple case that we consider only real fluctuations at discrete times, this distribution would be of the following form:

$$P(A(t_1), A(t_2), \cdots, A(\tau_{0n})) = \text{const.} \times \exp\left[-\frac{1}{2}\sum_{i,j} A(t_i)\alpha(t_i - t_j)A(t_j)\right],$$

where the matrix $\alpha(t_i - t_j)$ is simply the inverse of the (discrete) time correlation function $C_A(t_i - t_j)$. For Gaussian variables, we can factorize all higher-order correlation functions. In particular,

$$\begin{aligned} &\langle A(t')A(t'+\tau)A(t'')A(t''+\tau)\rangle \\ &\quad = \langle A(t')A(t'+\tau)\rangle \langle A(t'')A(t''+\tau)\rangle \\ &\quad + \langle A(t')A(t'')\rangle \langle A(t'+\tau)A(t''+\tau)\rangle \\ &\quad + \langle A(t')A(t''+\tau)\rangle \langle A(t'+\tau)A(t')\rangle . \end{aligned} \tag{D.2.2}$$

Inserting equation (D.2.2) in equation (D.2.1), we get

$$\begin{aligned} &\langle \overline{C}_A(\tau)^2\rangle - \langle \overline{C}_A(\tau)\rangle^2 \\ &= (1/\tau_0^2)\int_0^{\tau_0}\int_0^{\tau_0} dt'dt'' \; \langle A(t')A(t'')\rangle \langle A(t'+\tau)A(t''+\tau)\rangle \\ &+ (1/\tau_0^2)\int_0^{\tau_0}\int_0^{\tau_0} dt'dt'' \; \langle A(t')A(t''+\tau)\rangle \langle A(t'+\tau)A(t'')\rangle \\ &= (1/\tau_0^2)\int_0^{\tau_0}\int_0^{\tau_0} dt'dt'' \; \langle A(t'-t'')A(0)\rangle^2 \\ &+ (1/\tau_0^2)\int_0^{\tau_0}\int_0^{\tau_0} dt'dt'' \; \langle A(t'-t''-\tau)A(0)\rangle \langle A(t'-t''+\tau)A(0)\rangle . \end{aligned} \tag{D.2.3}$$

Again, we consider the case where the length of the simulation, τ_0, is much longer than the characteristic decay time of the fluctuations of A. In that case, we can write

$$\begin{aligned} &\langle \overline{C}_A(\tau)^2\rangle - \langle \overline{C}_A(\tau)\rangle^2 \\ &= (1/\tau_0)\int_{-\infty}^{\infty} dx \left(\langle A(x)A(0)\rangle^2 + \langle A(x-\tau)A(0)\rangle \langle A(x+\tau)A(0)\rangle\right), \end{aligned} \tag{D.2.4}$$

where we have defined the variable x as $t' - t''$. Let us now consider two limiting cases, $\tau = 0$ and $\tau \to \infty$. For $\tau = 0$, we can write

$$\begin{aligned}\left\langle \overline{C}_A(\tau)^2 \right\rangle - \left\langle \overline{C}_A(\tau) \right\rangle^2 &= (2/\tau_0) \int_{-\infty}^{\infty} dx \, \langle A(x)A(0) \rangle^2 \\ &= 4 \left\langle A^2(0) \right\rangle^2 \frac{\tau^c}{\tau_0}. \end{aligned} \tag{D.2.5}$$

The last line of this equation defines the correlation time τ^c:

$$\tau^c \equiv \frac{\int_0^\infty dx \, \langle A(x)A(0) \rangle^2}{\langle A^2(0) \rangle^2}.$$

For $\tau \to \infty$, the product

$$\langle A(x-\tau)A(0) \rangle \langle A(x+\tau)A(0) \rangle$$

vanishes, and we have

$$\left\langle \overline{C}_A(\tau)^2 \right\rangle - \left\langle \overline{C}_A(\tau) \right\rangle^2 = 2 \left\langle A^2(0) \right\rangle^2 \frac{\tau^c}{\tau_0}. \tag{D.2.6}$$

Comparison of equation (D.2.5) with equation (D.2.6) shows that the *absolute* error in $C_A(\tau)$ changes only little with τ. As a consequence, the *relative* error in time correlation functions increases rapidly as $C_A(\tau)$ decays to 0. In this derivation we have assumed that the total number of samples for each τ is equal; in case we have (many) fewer samples for large τ, this approach is not valid. In fact, if we have many fewer samples for large values of τ, we may wonder whether these values are reliable.

It should be stressed that the preceding error estimate is only approximate, because it relies on the validity of the Gaussian approximation. In specific cases (e.g., for the fluctuations of particle velocities), it is known that deviations from the Gaussian approximation occur. However, the deviations (where they are known) are usually not large, and it seems likely that error estimates based on the Gaussian approximation are on the correct order of magnitude. Very little evidence, however, supports or contradicts this belief.

A more detailed discussion of statistical errors in collective and single-particle time correlation functions can be found in [527] and [528]. Systematical techniques for measuring statistical errors in a simulation are discussed in [529] and [19].

D.3 Block Averages

The previous section showed that we can estimate the statistical error in time correlation functions on the basis of our knowledge of the time correlation function itself. Hence, no extra work is needed to arrive at an error estimate. However, as discussed in section D.1, to arrive at an error estimate for a static quantity, we need to compute the time correlation function of that quantity. As the computational effort to compute a time correlation function is larger than that required for a static average, we usually estimate statistical errors in static quantities by studying the behavior of so-called block averages. A block average is simply a time average over a finite time t_B:

$$\overline{A}_B \equiv \frac{1}{t_B} \int_0^{t_B} dt\, A(t).$$

During a simulation, we can easily accumulate block averages for a given block length t_B. After the simulation has been completed, we can compute the block averages for blocks of length $n \times t_B$ by simply averaging the block averages of n adjacent blocks of length t_B. Let us now consider the variance in the block averages for a given value of t_B:

$$\sigma^2(\overline{A}_B) = \frac{1}{n_B} \sum_{b=1}^{n_B} \left(\overline{A}_B - \langle A \rangle\right)^2.$$

If t_B is much larger than the correlation time t_A^c, we know from section D.1 that

$$\sigma^2(\overline{A}_B) \approx \left(\langle A^2 \rangle - \langle A \rangle^2\right) \frac{t_A^c}{t_B}. \tag{D.3.1}$$

But, as yet, we do not know t_A^c. We therefore compute the product

$$P(t_B) \equiv t_B \times \frac{\sigma^2(\overline{A}_B)}{\langle A^2 \rangle - \langle A \rangle^2}.$$

In the limit $t_B \gg t_A^c$, we know that $P(t_B)$ must approach t_A^c. Hence, we plot $P(t_B)$ versus t_B (or, more conveniently, $1/P(t_B)$ versus $1/t_B$) and estimate the limit of $P(t_B)$ for $t_B \to \infty$. This yields our estimate of t_A^c and thereby our error estimate for $\overline{A}$. This analysis of block averages is a very powerful tool to determine whether a simulation is long enough to yield a reliable estimate of a particular quantity: if we find that $P(t_B)$ is still strongly dependent on t_B in the limit $t_B = \tau$, then we know that our run is too short.

An alternative method for estimating the statistical error in a simulation has been developed by Flyvbjerg and Petersen [84]. Let $A_1, A_2, \ldots, A_L$ be L consecutive samples of some fluctuating quantity A of which we want to calculate its ensemble average and statistical error. We assume that all L samples are taken after the system has been equilibrated. The ensemble average is estimated from

$$\langle A \rangle \approx \bar{A} \equiv \frac{1}{L} \sum_{i=1}^{L} A_i, \tag{D.3.2}$$

and we need an estimator of the variance of

$$\sigma^2(A) = \langle A^2 \rangle - \langle A \rangle^2 \approx \frac{1}{L} \sum_{i=1}^{L} \left[A_i - \bar{A}\right]^2. \qquad \text{(D.3.3)}$$

If all L samples were uncorrelated, we could use the standard formulas of statistics to calculate this variance. However, in a simulation, the samples are correlated and we have to take this correlation into account.

The idea behind the method of Flyvbjerg and Petersen is to group the simulation data into consecutive blocks and compute an average for each of these blocks. These block averages will show less and less correlation between two consecutive blocks if the block size is made larger. In the limit that there is no detectable correlation between the blocks, the standard statistical formulas are valid and the standard deviations as a function of the block size follow these formulas. This procedure leads to a reliable estimate of the standard deviation.

To see how this method works in practice, consider the following transformation of our data set $A_1, A_2, \ldots, A_L$ into a new data set $A'_1, A'_2, \ldots, A'_{L'}$, which has half the size of the original set:

$$A'_i = 0.5(A_{2i-1} + A_{2i})$$

with

$$L' = 0.5L.$$

Note that the average of the new set $\bar{A}'$ is the same as for the original one. The variance in $\bar{A}'$ is given by

$$\sigma^2(A') = \left\langle A'^2 \right\rangle - \langle A' \rangle^2 = \frac{1}{L'} \sum_{i=1}^{L'} A'^2_i - \bar{A}'^2.$$

We can continue to perform this blocking operation, and if we have performed the simulation sufficiently long, the averages A'_i will become completely uncorrelated. If this is the case the following relation should hold:

$$\frac{\sigma^2(A')}{L'-1} \approx \text{Constant}.$$

This constant value is used as an estimate of the variance in the ensemble average. Note that, in a similar way, we can also determine the statistical error in $\sigma^2(A')$. This gives as estimate of the variance in our ensemble average

$$\sigma^2(A) = \frac{\sigma^2(A')}{L'-1} \pm \sqrt{\frac{2\sigma^4(A')}{(L'-1)^3}}. \qquad \text{(D.3.4)}$$

In Case Study 4, this method was used to compute the standard deviation of the energy in a Molecular Dynamics simulation. In Figure 4.4 a typical plot of this estimate of the variance as a function of block size is shown. For small values of M, the number of blocking operations, the data are correlated and as a consequence the variance will increase if we perform the blocking operation. For very high values of M we have only a few samples, and as a result, the statistical error in our estimate of $\sigma^2(A)$ will be large. The plateau in between gives us the value of $\sigma^2(A)$ we are interested in.

Appendix E

Integration Schemes

E.1 Higher-Order Schemes

The basic idea behind the predictor-corrector algorithms is to use information about the position and its first n derivatives at time t to arrive at a prediction for the position and its first n derivatives at time $t + \Delta t$. We then compute the forces (and thereby the accelerations) at the predicted positions. And then we find that these accelerations are *not* equal to the values that we had predicted. So we adjust our predictions for the accelerations to match the facts. But we do more than that. On the basis of the observed discrepancy between the predicted and observed accelerations, we also try to improve our estimate of the positions and the remaining $n-1$ derivatives. This is the "corrector" part of the predictor-corrector algorithm. The precise "recipe" used in applying this correction is a compromise between accuracy and stability. Here, we shall simply show a specific example of a predictor-corrector algorithm, without attempting to justify the form of the corrector part.

Consider the Taylor expansion of the coordinate of a given particle at time $t + \Delta t$:

$$r(t + \Delta t) = r(t) + \Delta t \frac{\partial r}{\partial t} + \frac{\Delta t^2}{2!} \frac{\partial^2 r}{\partial t^2} + \frac{\Delta t^3}{3!} \frac{\partial^3 r}{\partial t^3} + \cdots .$$

Using the notation

$$
\begin{aligned}
x_0(t) &\equiv r(t) \\
x_1(t) &\equiv \Delta t \frac{\partial r}{\partial t} \\
x_2(t) &\equiv \frac{\Delta t^2}{2!} \frac{\partial^2 r}{\partial t^2} \\
x_3(t) &\equiv \frac{\Delta t^3}{3!} \frac{\partial^3 r}{\partial t^3},
\end{aligned}
$$

we can write the following *predictions* for $x_0(t + \Delta t)$ through $x_3(t + \Delta t)$:

$$
\begin{aligned}
x_0(t + \Delta t) &= x_0(t) + x_1(t) + x_2(t) + x_3(t) \\
x_1(t + \Delta t) &= x_1(t) + 2x_2(t) + 3x_3(t) \\
x_2(t + \Delta t) &= x_2(t) + 3x_3(t) \\
x_3(t + \Delta t) &= x_3(t).
\end{aligned}
$$

Now that we have $x_0(t + \Delta t)$, we can compute the forces at the predicted position, and thus compute the corrected value for $x_2(t+\Delta t)$. We denote the difference between $x_2^{\text{corrected}}$ and $x_2^{\text{predicted}}$ by Δx_2:

$$\Delta x_2 \equiv x_2^{\text{corrected}} - x_2^{\text{predicted}}.$$

We now estimate "corrected" values for x_0 through x_3, as follows:

$$x_n^{\text{corrected}} = x_n^{\text{predicted}} + C_n \Delta x_2, \tag{E.1.1}$$

where the C_n are constants fixed for a given order algorithm. As indicated, the values for C_n are such that they yield an optimal compromise between the accuracy and the stability of the algorithm. For instance, for a fifth-order predictor-corrector algorithm (i.e., one that uses x_0 through x_4), the values for C_n are

$$
\begin{aligned}
C_0 &= \frac{19}{120} \\
C_1 &= \frac{3}{4} \\
C_2 &= 1 \qquad \text{(of course)} \\
C_3 &= \frac{1}{2} \\
C_4 &= \frac{1}{12}.
\end{aligned}
$$

One may iterate the predictor and corrector steps to self-consistency. However, there is little point in doing so because (1) every iteration requires a force calculation. One would be better off spending the same computer time to run with a *shorter* time step and only one iteration because (2) even if we iterate the predictor-corrector algorithm to convergence, we still do not get the *exact* trajectory: the error is still of order Δt^n for an nth-order algorithm. This is why we gain more accuracy by going to a shorter time step than by iterating to convergence at a fixed value of Δt.

E.2 Nosé-Hoover Algorithms

As discussed in section 6.1.2, it is advantageous to implement the Nosé thermostat using the formulation of Hoover, equations (6.1.24)–(6.1.27). Since the velocity also appears on the right-hand side of equation (6.1.25), this scheme cannot be implemented directly into the velocity Verlet algorithm (see also section 4.3). To see this, consider a standard constant-N,V,E simulation, for which the velocity Verlet algorithm is of the form

$$\begin{aligned} r(t+\Delta t) &= r(t) + v(t)\Delta t + f(t)\Delta t^2/(2m) \\ v(t+\Delta t) &= v(t) + \frac{f(t+\Delta t)+f(t)}{2m}\Delta t. \end{aligned}$$

When we use this scheme for the Nosé-Hoover equations of motion, we obtain for the positions and velocities

$$r_i(t+\Delta t) = r_i(t) + v(t)\Delta t + \left[f_i(t)/m_i - \xi(t)v_i(t)\right]\Delta t^2/2 \quad \text{(E.2.1)}$$

$$\begin{aligned} v_i(t+\Delta t) = {} & v_i(t) + [f_i(t+\Delta t)/m_i - \xi(t+\Delta t)v_i(t+\Delta t) \\ & + f_i(t)/m_i - \xi(t)v_i(t)]\,\Delta t/2. \end{aligned} \quad \text{(E.2.2)}$$

The first step of the velocity Verlet algorithm can be carried out without difficulty. In the second step, we first update the velocity, using the old "forces" to the intermediate value $v(t+\Delta t/2) \equiv v'$. And then we must use the new "forces" to update v':

$$v_i(t+\Delta t) = v'_i + [f_i(t+\Delta t)/m_i - \xi(t+\Delta t)v_i(t+\Delta t)]\,\Delta t/2. \quad \text{(E.2.3)}$$

In these equations $v_i(t+\Delta t)$ appears on the right- and left-hand sides; therefore, these equations cannot be integrated exactly.[1] For this reason the Nosé-Hoover method is usually implemented using a predictor-corrector scheme or solved iteratively [138]. This has a disadvantage that the solution is no longer time reversible. Martyna *et al.* [85] have developed a set of explicit reversible integrators using the Liouville approach (see section 4.3.3) for this type of extended systems.

[1] For the harmonic oscillator it is possible to find an analytic solution (see Case Study 12).

E.2.1 Canonical Ensemble

For M chains, the Nosé-Hoover equations of motion are given by (see also section 6.1.3)

$$\begin{aligned}
\dot{\mathbf{r}}_i &= \mathbf{p}_i/m_i \\
\dot{\mathbf{p}}_i &= \mathbf{F}_i - \frac{p_{\xi_1}}{Q_1}\mathbf{p}_i \\
\dot{\xi}_k &= \frac{p_{\xi_k}}{Q_k} \qquad\qquad k = 1, \ldots, M \\
\dot{p}_{\xi_1} &= \left(\sum_i p_i^2/m_i - Lk_B T\right) - \frac{p_{\xi_2}}{Q_2} p_{\xi_1} \\
\dot{p}_{\xi_k} &= \left[\frac{p^2_{\xi_{k-1}}}{Q_{k-1}} - k_B T\right] - \frac{p_{\xi_{k+1}}}{Q_{k+1}} p_{\xi_k} \\
\dot{p}_{\xi_M} &= \left[\frac{p^2_{\xi_{M-1}}}{Q_{M-1}} - k_B T\right].
\end{aligned}$$

The Liouville operator for the equation of motions is defined as (see section 4.3.3)

$$iL \equiv \dot{\eta}\frac{\partial}{\partial \eta}$$

with $\eta = (\mathbf{r}^N, \mathbf{p}^N, \xi^M, p_\xi^M)$. Using the equations of motion, $\mathbf{p}_i = m_i \mathbf{v}_i$, and $p_{\xi_k} = Q_k v_{\xi_k}$, we obtain as Liouville operator for the Nosé-Hoover chains

$$\begin{aligned}
iL_{\mathrm{NHC}} &= \sum_{i=1}^{N} \mathbf{v}_i \cdot \boldsymbol{\nabla}_{\mathbf{r}_i} + \sum_{i=1}^{N} \left[\frac{\mathbf{F}_i(\mathbf{r}_i)}{m_i}\right] \cdot \boldsymbol{\nabla}_{\mathbf{v}_i} - \sum_{i=1}^{N} v_{\xi_1} \mathbf{v}_i \cdot \boldsymbol{\nabla}_{\mathbf{v}_i} + \sum_{k=1}^{M} v_{\xi_k} \frac{\partial}{\partial \xi_k} \\
&\quad + \sum_{k=1}^{M-1} \left(G_k - v_{\xi_k} v_{\xi_{k+1}}\right) \frac{\partial}{\partial v_{\xi_k}} + G_M \frac{\partial}{\partial v_{\xi_M}}
\end{aligned}$$

with

$$\begin{aligned}
G_1 &= \frac{1}{Q_1}\left(\sum_{i=1}^{N} m_i \mathbf{v}_i^2 - Lk_B T\right) \\
G_k &= \frac{1}{Q_k}\left(Q_{k-1} v^2_{\xi_{k-1}} - k_B T\right).
\end{aligned}$$

As explained in section 4.3.3, the Liouville equation combined with the Trotter formula is a powerful technique for deriving a time-reversible algorithm for solving the equations of motion numerically. Here we will use this technique to derive such a scheme for the Nosé-Hoover thermostats. We use a simplified version; a more complete description can be found in ref. [85].

We have to make an intelligent separation of the Liouville operator. The first step is to separate the part of the Liouville operator that only involves the positions (iL_r) and the velocities (iL_v) from the parts that involve the Nosé-Hoover thermostats (iL_C):

$$iL_{NHC} = iL_r + iL_v + iL_C$$

with

$$\begin{aligned} iL_r &= \sum_{i=1}^{N} \mathbf{v}_i \cdot \nabla_{\mathbf{r}_i} \\ iL_v &= \sum_{i=1}^{N} \frac{\mathbf{F}_i(\mathbf{r}_i)}{m_i} \cdot \nabla_{\mathbf{v}_i} \\ iL_C &= \sum_{k=1}^{M} v_{\xi_k} \frac{\partial}{\partial \xi_k} - \sum_{i=1}^{N} v_{\xi_1} \mathbf{v}_i \cdot \nabla_{\mathbf{v}_i} \\ &\quad + \sum_{k=1}^{M-1} \left(G_k - v_{\xi_k} v_{\xi_{k+1}}\right) \frac{\partial}{\partial v_{\xi_k}} + G_M \frac{\partial}{\partial v_{\xi_M}}. \end{aligned}$$

There are several ways to factorize iL_{NHC} using the Trotter formula; we follow the one used by Martyna *et al.* [85]:

$$e^{(iL\Delta t)} = e^{(iL_C\Delta t/2)} e^{(iL_v\Delta t/2)} e^{(iL_r\Delta t)} e^{(iL_v\Delta t/2)} e^{(iL_C\Delta t/2)} + \mathcal{O}\left(\Delta t^3\right). \quad \text{(E.2.4)}$$

The Nosé-Hoover chain part L_C has to be further factorized. Here, we will do this for a chain of length $M = 2$; the more general case is discussed in ref. [85]. The Nosé-Hoover part of the Liouville operator for this chain length can be separated into five terms:

$$iL_C = iL_\xi + iL_{C_v} + iL_{G1} + iL_{v\xi_1} + iL_{G2},$$

where the terms are defined as

$$\begin{aligned} iL_\xi &\equiv \sum_{k=1}^{2} v_{\xi_k} \frac{\partial}{\partial \xi_k} \\ iL_{C_v} &\equiv -\sum_{i=1}^{N} v_{\xi_1} \mathbf{v}_i \cdot \nabla_{\mathbf{v}_i} \\ iL_{G1} &\equiv G_1 \frac{\partial}{\partial v_{\xi_1}} \\ iL_{v\xi_1} &\equiv -\left(v_{\xi_1} v_{\xi_2}\right) \frac{\partial}{\partial v_{\xi_1}} \\ iL_{G2} &\equiv G_2 \frac{\partial}{\partial v_{\xi_2}}. \end{aligned}$$

The factorization for the Trotter equation that we use is[2]

$$\begin{aligned}
e^{(iL_C\Delta t/2)} &= e^{(iL_{G2}\Delta t/4)} e^{(iL_{v\xi_1}\Delta t/4 + iL_{G1}\Delta t/4)} \\
&\quad \times e^{(iL_\xi \Delta t/2)} e^{(iL_{Cv}\Delta t/2)} e^{(iL_{G1}\Delta t/4 + iL_{v\xi_1}\Delta t/4)} e^{(iL_{G2}\Delta t/4)} \\
&= e^{(iL_{G_2}\Delta t/4)} \left[e^{(iL_{v\xi_1}\Delta t/8)} e^{(iL_{G1}\Delta t/4)} e^{(iL_{v\xi_1}\Delta t/8)} \right] \\
&\quad \times e^{(iL_\xi \Delta t/2)} e^{(iL_{Cv}\Delta t/2)} \\
&\quad \times \left[e^{(iL_{v\xi_1}\Delta t/8)} e^{(iL_{G1}\Delta t/4)} e^{(iL_{v\xi_1}\Delta t/8)} \right] e^{(iL_{G2}\Delta t/4)}. \qquad \text{(E.2.5)}
\end{aligned}$$

Our numerical algorithm is now fully defined by equations (E.2.4) and (E.2.5). This seemingly complicated set of equations is actually relatively easy to implement in a simulation.

To see how the implementation works, we need to know how each operator works on our coordinates $\eta = (\mathbf{r}^N, \mathbf{v}^N, \xi_1, v_{\xi_1}, \xi_2, v_{\xi_2})$. If we start at $t = 0$ with initial condition η, the position at time $t = \Delta t$ follows from

$$e^{iL_{NHC}\Delta t} f[\mathbf{r}^N, \mathbf{v}^N, \xi_1, v_{\xi_1}, \xi_2, v_{\xi_2}].$$

Because of the Trotter expansion, we can apply each term in iL_{NHC}, sequentially. For example, if we let the first term of the Liouville operator, iL_{G2}, act on the initial state η,

$$\begin{aligned}
&\exp\left(\frac{\Delta t}{4} G_2 \frac{\partial}{\partial v_{\xi_2}}\right) f\left[\mathbf{r}^N, \mathbf{p}^N, \xi_1, v_{\xi_1}, \xi_2, v_{\xi_2}\right] \\
&= \sum_{n=0}^{\infty} \frac{(G_2\Delta t/4)^n}{n!} \frac{\partial^n}{\partial v_{\xi_2}^n} f\left[\mathbf{r}^N, \mathbf{p}^N, \xi_1, v_{\xi_1}, \xi_2, v_{\xi_2}\right] \\
&= f\left[\mathbf{r}^N, \mathbf{p}^N, \xi_1, v_{\xi_1}, \xi_2, v_{\xi_2} + G_2\Delta t/4\right].
\end{aligned}$$

This shows that the effect of iL_{G2} is to shift v_{ξ_2} without affecting the other coordinates. This gives as transformation rule for this operator:

$$e^{(iL_{G2}\Delta t/4)}: \qquad v_{\xi_2} \to v_{\xi_2} + G_2\Delta t/4. \qquad \text{(E.2.6)}$$

The operators ($iL_{v\xi_1}$ and iL_{Cv}) are of the form $\exp(ax\partial/\partial x)$; such operators give a scaling of the x coordinate:

$$\begin{aligned}
\exp\left(ax\frac{\partial}{\partial x}\right) f(x) &= \exp\left(a\frac{\partial}{\partial \ln(x)}\right) f\{\exp[\ln(x)]\} \\
&= f\{\exp[\ln(x) + a]\} = f[x\exp(a)]
\end{aligned}$$

[2]The second factorization, indicated by $[\cdots]$, is used to avoid a hyperbolic sine function, which has a possible singularity. See ref. [85] for details.

If we apply this result[3] to $iL_{\nu\xi_1}$, we obtain for this operator

$$\exp\left(-\frac{\Delta t}{8}\nu_{\xi_2}\nu_{\xi_1}\frac{\partial}{\partial \nu_{\xi_1}}\right) f\left[\mathbf{r}^N, \mathbf{p}^N, \xi_1, \nu_{\xi_1}, \xi_2, \nu_{\xi_2}\right]$$
$$= f\left[\mathbf{r}^N, \mathbf{p}^N, \xi_1, \exp\left(-\frac{\Delta t}{8}\nu_{\xi_2}\right)\nu_{\xi_1}, \xi_2, \nu_{\xi_2}\right],$$

giving the transformation rule

$$e^{(iL_{\nu\xi_1}\Delta t/8)}: \qquad \nu_{\xi_1} \to \exp\left[-\nu_{\xi_2}\Delta t/8\right]\nu_{\xi_1}. \tag{E.2.7}$$

In a similar way we can derive for the other terms

$$e^{(iL_{G1}\Delta t/4)}: \qquad \nu_{\xi_1} \to \nu_{\xi_1} + G_1\Delta t/4 \tag{E.2.8}$$
$$e^{(iL_{\xi}\Delta t/2)}: \qquad \xi_1 \to \xi_1 - \nu_{\xi_1}\Delta t/2 \tag{E.2.9}$$
$$\xi_2 \to \xi_2 - \nu_{\xi_2}\Delta t/2 \tag{E.2.10}$$
$$e^{(iL_{C\nu}\Delta t/2)}: \qquad \nu_i \to \exp\left[-\nu_{\xi_1}\Delta t/2\right]\nu_i. \tag{E.2.11}$$

Finally, the transformation rules that are associated to iL_ν and iL_r are similar to the velocity Verlet algorithm, i.e.,

$$e^{(iL_\nu\Delta t/2)}: \qquad \mathbf{v}_i \to \mathbf{v}_i + \mathbf{F}_i\Delta t/(2m) \tag{E.2.12}$$
$$e^{(iL_r\Delta t)}: \qquad \mathbf{r}_i \to \mathbf{r}_i + \mathbf{v}_i\Delta t. \tag{E.2.13}$$

With these transformation rules (E.2.6)–(E.2.13) we can write down our numerical algorithm by subsequently applying the transformation rules according to the order defined by equations (E.2.4) and (E.2.5). If we start with initial coordinate $\eta(0) = (\mathbf{r}^N, \mathbf{v}^N, \xi_1, \nu_{\xi_1}, \xi_2, \nu_{\xi_2})$, we have to apply first $e^{iL_C}\eta$. Since this operator is further factorized according to equation (E.2.5), the first step in our algorithm is to apply $e^{(iL_{G2}\Delta t/4)}$. According to transformation rule (E.2.6) applying this operator on η gives as new state

$$\nu_{\xi_2}(\Delta t/4) = \nu_{\xi_2} + G_2\Delta t/4.$$

The output of this rule is the new state on which we apply the next operator in equation (E.2.5), $iL_{\nu\xi_1}$, with transformation rule (E.2.9):

$$\nu_{\xi_1}(\Delta t/8) = \exp\left[-\nu_{\xi_2}(\Delta t/4)\Delta t/8\right]\nu_{\xi_1}.$$

[3]This can be generalized, giving the identity

$$\exp\left(a\frac{\partial}{\partial g(x)}\right) f(x) = \exp\left(a\frac{\partial}{\partial g(x)}\right) f\left\{g^{-1}[g(x)]\right\}$$
$$= \exp\left(a\frac{\partial}{\partial y}\right) f\left[g^{-1}(y)\right]$$
$$= f\left\{g^{-1}[y+a]\right\} = f\left\{g^{-1}[g(x)+a]\right\}.$$

Algorithm 30 (Equations of Motion: Nosé-Hoover)

```
subroutine integrate          integrate equations of motion
                              Nosé-Hoover thermostat
call chain(uk)
call pos_vel(uk)
call chain(uk)
return
end
```

Comments to this algorithm:

1. *This subroutine solves the equations of motion for a single time step* Δt *using the Trotter equations (E.2.4) and (E.2.5).*
2. *In the subroutine* `chain` *we apply* $e^{iL_C \Delta t/4}$ *to the current state (see Algorithm 31).*
3. *In the subroutine* `pos_vel` *we apply* $e^{(iL_r + iL_p)\Delta t}$ *to the current state (see Algorithm 32).*
4. uk *is the total kinetic energy.*

The next step is to apply iL_{G1}, followed by again $iL_{v_{\xi_1}}$, etc. In this way we continue to apply all operators on the output of the previous step.

Applying the Nosé-Hoover part of the Liouville operator changes ξ_k, v_{ξ_k}, and $\mathbf{v}_i$. The other two Liouville operators change $\mathbf{v}_i$ and $\mathbf{r}_i$. This makes it convenient to separate the algorithm into two parts in which the positions and velocities of the particles and the Nosé-Hoover chains are considered separately. An example of a possible implementation is shown in Algorithm 30.

E.2.2 The Isothermal-Isobaric Ensemble

Similar to the canonical ensemble we can derive a time-reversible integration scheme for simulation in the NPT ensemble. The equations of motions are given by expressions (6.2.1)–(6.2.8):

$$\begin{aligned}
\dot{\mathbf{r}}_i &= \frac{\mathbf{p}_i}{m_i} + \frac{p_\epsilon}{W}\mathbf{r}_i \\
\dot{\mathbf{p}}_i &= \mathbf{F}_i - \left(1 + \frac{d}{dN}\right)\frac{p_\epsilon}{W}\mathbf{p}_i - \frac{p_{\xi_1}}{Q_1}\mathbf{p}_i \\
\dot{V} &= dVp_\epsilon/W
\end{aligned}$$

Algorithm 31 (Propagating the chain)

```
subroutine chain(uk)              apply equation (E.2.5)
                                  to the current position

G2=(Q1*vxi1*vxi1-T)
vxi2=vxi2+G2*delt4                Update ν_ξ2 using equation (E.2.6)
vxi1=vxi1*exp(-vxi2*delt8)        Update ν_ξ1 using equation (E.2.7)
G1=(2*uk-L*T)/Q1
vxi1=vxi1+G1*delt4                Update ν_ξ1 using equation (E.2.8)
vxi1=vxi1*exp(-vxi2*delt8)        Update ν_ξ1 using equation (E.2.7)
xi1=xi1+vxi1*delt2                Update ξ1 using equation (E.2.9)
xi2=xi2+vxi2*delt2                Update ξ2 using equation (E.2.10)
s=exp(-vxi1*delt2)                Scale factor in equation (E.2.11)
do i=1,npart
  v(i)=s*v(i)                     update v_i using equation (E.2.11)
enddo
uk=uk*s*s                         update kinetic energy
vxi1=vxi1*exp(-vxi2*delt8)        Update ν_ξ1 using equation (E.2.7)
G1=(2*uk-L*T)/Q1
vxi1=vxi1+G1*delt4                Update ν_ξ1 using equation (E.2.8)
vxi1=vxi1*exp(-vxi2*delt8)        Update ν_ξ1 using equation (E.2.7)
G2=(Q1*vxi1*vxi1-T)/Q2
vxi2=vxi2+G2*delt4                Update ν_ξ2 using equation (E.2.6)
return
end
```

Comments to this algorithm:

1. *In this subroutine* `T` *is the imposed temperature,* `delt`$=\Delta t$, `delt2` $=\Delta t/2$, `delt4` $=\Delta t/4$, *and* `delt8` $=\Delta t/8$.
2. `uk` *is the total kinetic energy.*

Algorithm 32 (Propagating the Positions and Velocities)

```
subroutine pos_vel(uk)            apply equation (E.2.4)
                                  to the current position
uk=0
do i=1,npart
  x(i)=x(i)+v(i)*delt2            update x_i using equation (E.2.13)
enddo
call force                        calculate the force
do i=1,npart
  v(i)=v(i)+f(i)*delt/m           update v_i using equation (E.2.12)
  x(i)=x(i)+v(i)*delt2            update x_i using equation (E.2.13)
  uk=uk+m*v(i)*v(i)/2             update kinetic energy
enddo
return
end
```

Comments to this algorithm:

1. *In this subroutine* `delt` $= \Delta t$ *and* `delt2` $= \Delta t/2$.
2. *The subroutine* `force` *calculates the force on the particles.*

$$
\begin{aligned}
\dot{p}_\epsilon &= dV(P_{\text{int}} - P_{\text{ext}}) + \frac{1}{N}\sum_{i=1}^{N}\frac{\mathbf{p}_i^2}{m_i} - \frac{p_{\xi_1}}{Q_1}p_\epsilon \\
\dot{\xi}_k &= \frac{p_{\xi_k}}{Q_k} \qquad \text{for } k = 1, \ldots, M \\
\dot{p}_{\xi_1} &= \sum_{i=1}^{N}\frac{\mathbf{p}_i^2}{m_i} + \frac{p_\epsilon^2}{W} - (dN+1)k_BT - \frac{p_{\xi_2}}{Q_2}p_{\xi_1} \\
\dot{p}_{\xi_k} &= \frac{p_{\xi_{k-1}}^2}{Q_{k-1}} - k_BT - \frac{p_{\xi_{k+1}}}{Q_{k+1}}p_{\xi_k} \qquad \text{for } k = 2, \ldots, M-1 \\
\dot{p}_{\xi_M} &= \frac{p_{\xi_{M-1}}^2}{Q_{M-1}} - k_BT.
\end{aligned}
$$

To derive a time-reversible numerical integration scheme to solve the equations of motion we use again the Liouville approach.

A state is characterized by the variables $\eta = (\mathbf{r}^N, \mathbf{p}^N, \epsilon, p_\epsilon, \xi^M, p_\xi^M)$. The Liouville operator is defined by

$$iL_{NPT} \equiv \dot{\eta}\frac{\partial}{\partial \eta}.$$

For a chain of length $M = 2$, using $\mathbf{p}_i = m_i \mathbf{v}_i (\neq m_i \dot{\mathbf{r}}_i)$, $p_{\xi_k} = Q_k v_{\xi_k}$, $\epsilon = (\ln V)/d$, and $p_\eta = W v_\eta$, the Liouville operator for these equations of motion can be written as

$$iL_{NPT} = iL_r + iL_v + iL_{CP},$$

in which we define the operators

$$\begin{aligned}
iL_r &= \sum_{i=1}^{N} (\mathbf{v}_i + v_\epsilon \mathbf{r}_i) \cdot \nabla_{\mathbf{r}_i} + v_\epsilon \frac{\partial}{\partial \epsilon} \\
iL_v &= \sum_{i=1}^{N} \left[\frac{\mathbf{F}_i(\mathbf{r})}{m_i} \right] \cdot \nabla_{\mathbf{v}_i} \\
iL_{CP} &= -\sum_{i=1}^{N} v_{\xi_1} \mathbf{v}_i \cdot \nabla_{\mathbf{v}_i} + \sum_{k=1}^{M} v_{\xi_k} \frac{\partial}{\partial \xi_k} + \sum_{k=1}^{M-1} (G_k - v_{\xi_k} v_{\xi_{k+1}}) \frac{\partial}{\partial v_{\xi_k}} \\
&\quad + G_M \frac{\partial}{\partial v_{\xi_M}} - \left(1 + \frac{1}{N}\right) \sum_{i=1}^{N} v_\epsilon \mathbf{v}_i \cdot \nabla_{\mathbf{v}_i} + (G_\epsilon - v_\epsilon v_{\xi_1}) \frac{\partial}{\partial v_\epsilon}
\end{aligned}$$

with

$$\begin{aligned}
G_1 &= \frac{1}{Q} \left[\sum_{i=1}^{N} m_i \mathbf{v}_i^2 + W v_\epsilon^2 - (N_f + 1) k_B T \right] \\
G_k &= \frac{1}{Q_k} \left(Q_{k-1} v_{\xi_{k-1}}^2 - k_B T \right) \\
G_\epsilon &= \frac{1}{W} \left[\left(1 + \frac{1}{N}\right) \sum_{i=1}^{N} m_i \mathbf{v}_i^2 + \sum_{i=1}^{N} \mathbf{r}_i \cdot \mathbf{F}_i(\mathbf{r}) m_i - dV \frac{\partial U(\mathbf{r}, V)}{\partial V} - dP_{ext} V \right].
\end{aligned}$$

An appropriate Trotter equation for the equations of motion is [85]

$$e^{(iL_{NPT}\Delta t)} = e^{(iL_{CP}\Delta t/2)} e^{(iL_v \Delta t/2)} e^{(iL_r \Delta t)} e^{(iL_v \Delta t/2)} e^{(iL_{CP}\Delta t/2)} + \mathcal{O}\left(\Delta t^3\right). \tag{E.2.14}$$

The operator iL_{CP} has to be further factorized:

$$iL_{CP} = iL_\xi + iL_{C_v} + iL_{G\epsilon} + iL_{v\epsilon} + iL_{G1} + iL_{v\xi_1} + iL_{G2},$$

where the terms are defined as

$$
\begin{aligned}
iL_{\xi} &\equiv \sum_{k=1}^{2} \nu_{\xi_k} \frac{\partial}{\partial \xi_k} \\
iL_{C_v} &\equiv -\sum_{i=1}^{N} \left[\nu_{\xi_1} + \left(1 + \frac{d}{dN}\right) \right] \mathbf{v}_i \cdot \boldsymbol{\nabla}_{\mathbf{v}_i} \\
iL_{G\epsilon} &\equiv G_{\epsilon} \frac{\partial}{\partial \nu_{\epsilon}} \\
iL_{\nu\epsilon} &\equiv -\left(\nu_{\xi_1} \nu_{\epsilon}\right) \frac{\partial}{\partial \nu_{\epsilon}} \\
iL_{G1} &\equiv G_1 \frac{\partial}{\partial \nu_{\xi_1}} \\
iL_{\nu\xi_1} &\equiv -\left(\nu_{\xi_1} \nu_{\xi_2}\right) \frac{\partial}{\partial \nu_{\xi_1}} \\
iL_{G2} &\equiv G_2 \frac{\partial}{\partial \nu_{\xi_2}}.
\end{aligned}
$$

The Trotter expansion of the term iL_C is

$$
\begin{aligned}
e^{(iL_{CP}\Delta t/2)} &= e^{\left(iL_{G2}\Delta t/4 + iL_{\nu\xi_1}\Delta t/4\right)} e^{(iL_{G1}\Delta t/4)} e^{(iL_{G\epsilon}\Delta t/4 + iL_{\nu\epsilon}\Delta t/4)} \\
&\quad \times e^{(iL_{\xi}\Delta t/2)} e^{(iL_{C_v}\Delta t/2)} \\
&\quad \times e^{(iL_{\nu\epsilon}\Delta t/4 + iL_{G\epsilon}\Delta t/4)} e^{\left(iL_{G1}\Delta t/4 + iL_{\nu\xi_1}\Delta t/4\right)} e^{(iL_{G2}\Delta t/4)} \\
&= e^{\left(iL_{G_2}\Delta t/4\right)} \left[e^{\left(iL_{\nu\xi_1}\Delta t/8\right)} e^{(iL_{G1}\Delta t/4)} e^{\left(iL_{\nu\xi_1}\Delta t/8\right)} \right] \\
&\quad \times \left[e^{(iL_{\nu\epsilon}\Delta t/8)} e^{(iL_{G\epsilon}\Delta t/4)} e^{(iL_{\nu\epsilon}\Delta t/8)} \right] \\
&\quad \times e^{(iL_{\xi}\Delta t/2)} e^{(iL_{C_v}\Delta t/2)} \left[e^{(iL_{\nu\epsilon}\Delta t/8)} e^{(iL_{G\epsilon}\Delta t/4)} e^{(iL_{\nu\epsilon}\Delta t/8)} \right] \\
&\quad \times \left[e^{\left(iL_{\nu\xi_1}\Delta t/8\right)} e^{(iL_{G1}\Delta t/4)} e^{\left(iL_{\nu\xi_1}\Delta t/8\right)} \right] e^{(iL_{G2}\Delta t/4)}. \qquad \text{(E.2.15)}
\end{aligned}
$$

Similar to the NVT version the transformation rules of the various operators can be derived and translated into an algorithm. Such an algorithm is presented in ref. [85].

Appendix F

Saving CPU Time

The energy or force calculation is the most time-consuming part of almost all Molecular Dynamics and Monte Carlo simulations. If we consider a model system with pairwise additive interactions (as is done in many molecular simulations), we have to consider the contribution to the force on particle i, by all its neighbors. If we do not truncate the interactions, this implies that, for a system of N particles, we must evaluate $N(N-1)/2$ pair interactions. And even if we do truncate the potential, we still would have to compute all $N(N-1)/2$ pair distances to describe which pairs can interact. This implies that, if we use no tricks, the time needed for the evaluation of the energy scales as N^2. There exist efficient techniques for speeding up the evaluation of both short-range and long-range interactions in such a way that the computing time scales as $N^{3/2}$, rather than N^2. The techniques for the long-range interactions were discussed in Chapter 12.1; here, we discuss some of the techniques used for the short-range interactions. These techniques are:

1. Verlet list
2. Cell (or linked) list
3. Combination of Verlet and cell lists

F.1 Verlet List

If we simulate a large system and use a cutoff that is smaller than the simulation box, many particles do not contribute to the energy of a particle i. It is advantageous therefore to exclude the particles that do not interact from the expensive energy calculation. Verlet [13] developed a bookkeeping technique, commonly referred to as the Verlet list or neighbor list, which is illustrated in Figure F.1. In this method a second cutoff radius $r_v > r_c$ is in-

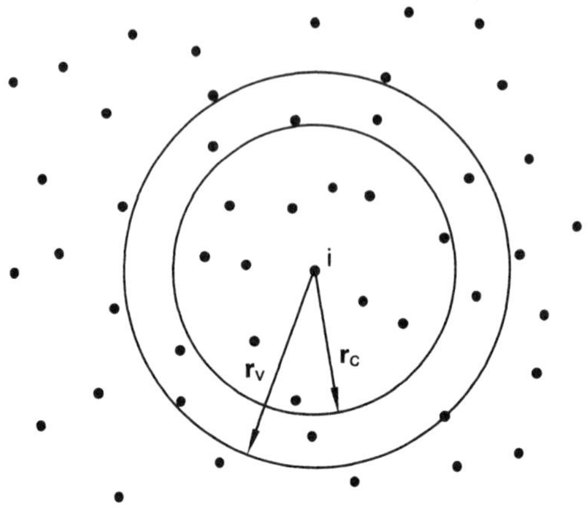

Figure F.1: The Verlet list: a particle i interacts with those particles within the cutoff radius r_c; the Verlet list contains all the particles within a sphere with radius $r_v > r_c$.

troduced, and before we calculate the interactions, a list is made (the Verlet list) of all particles within a radius r_v of particle i. In the subsequent calculation of the interactions, only those particles in this list have to be considered. Until now we have not saved any CPU time. We gain such time when we next calculate the interactions; if the maximum displacement of the particles is less than $r_v - r_c$, then we have to consider only the particles in the Verlet list of particle i. This is a calculation of order N. As soon as one of the particles is displaced more than $r_v - r_c$, we have to update the Verlet list. The latter operation is of order N^2, and although this step is not performed each time an interaction is calculated, it will dominate for a very large number of particles.

The Verlet list can be used for both Molecular Dynamics and Monte Carlo simulations. However, there are some small differences in the implementation. For example, in a Molecular Dynamics simulation, the force on all particles is calculated at the same time. It is sufficient therefore to have a Verlet list with half the number of particles for each particle as long as the interaction i-j is accounted for in either the list of particle i or that of j. In a Monte Carlo simulation each particle is considered separately, therefore it is convenient to have for each particle the complete list. Algorithm 33 shows the use of the Verlet list in a Monte Carlo simulation.

Bekker *et al.* have developed an elegant extension of the Verlet list for systems with periodic boundary conditions [530]. To calculate the force or potential energy of particle i one has to locate the nearest image of the particles in the Verlet list of particle j (see, Algorithm 34). Bekker *et al.* have

Algorithm 33 (Use of Verlet List in a Monte Carlo Move)

```
      SUBROUTINE mcmove_verlet               attempts to displace a particle
                                             using a Verlet list
      o=int(ranf()*npart)+1                  select a particle at random
      if (abs(x(o)-xv(o)).gt.(rv-rc)         check to make a new list
     +   /2) call new_vlist
      call en_vlist(o,x(o),eno)              energy old configuration
      xn=x(o)+(ranf()-0.5)*delx              random displacement
      if (abs(xn-xv(o)).gt.(rv-rc)/2)        check to make a new list
     +   call new_vlist
      call en_vlist(o,xn,enn)                energy new configuration
      arg=exp(-beta*(enn-eno))
      if (ranf().lt.arg)
     +   x(o)=xn                             accepted: replace x(o) by xn
      return
      end
```

Comments to this algorithm:

1. *The algorithm is based on Algorithm 2.*
2. *Subroutine* `newvlist` *makes the Verlet list (see Algorithm 34) and subroutine* `en_vlist` *calculates the energy of a particle at the given position using the Verlet list (see Algorithm 35).*

shown that this nearest image calculation in the inner loop of a MD or MC simulation can be avoided.

In a periodic system, the total force on particle i can be written as

$$\mathbf{F}_i = \sum_{j=1}^{N} \sum_{k=-13}^{13} {}' \mathbf{F}_{i(j.k)},$$

where the prime denotes that the summation is performed over the nearest image of particle j in the central box ($k = 0$) or in one of its 26 periodic images. Here, $(j.k)$ denotes the periodic image of particle j in box k. Box k is defined by the integer numbers n_x, n_y, n_z:

$$k = 9n_x + 3n_y + n_z$$

and

$$\mathbf{t}_k = n_x \mathbf{L}_x + n_y \mathbf{L}_y + n_z \mathbf{L}_z,$$

Algorithm 34 (Making a Verlet List)

```
SUBROUTINE new_vlist                        makes a new Verlet list
do i=1,npart                                initialize list
  nlist(i)=0
  xv(i)=x(i)                                store position of particles
enddo
do i=1,npart-1
  do j=i+1,npart
    xr=x(i)-x(j)
    if (xr.gt.hbox) then                    nearest image
      xr=xr-box
    else if (xr.lt.-hbox) then
      xr=xr+box
    endif
    if (abs(xr).lt.rv) then                 add to the lists
      nlist(i)=nlist(i)+1
      nlist(j)=nlist(j)+1
      list(i,nlist(i))=j
      list(j,nlist(j))=i
    endif
  enddo
enddo
return
end
```

Comments to this algorithm:

1. *Array* `list(i,itel)` *is the Verlet list of particle* `i`*, the total number of particles in the Verlet list of particle* `i` *is given by* `nlist(i)`*, and the array* `xv(i)` *contains the position of the particles at the moment the list is made (is used to see when a new list has to be made).*
2. *Note that in this algorithm we assume* all *particles are in the simulation box; hence* `x(i)` $\in$ *[0,box].*

where $\mathbf{t}_k$ is the translation vector of the central box to its periodic image k. A particle in the central box is denoted by $(i.0) = i$. Using this notation, we can write, for the interaction between particles i and j,

$$\mathbf{F}_{i(j.k)} = \mathbf{F}_{(i.-k)j} = -\mathbf{F}_{(j.k)i} = -\mathbf{F}_{j(i.-k)}.$$

Algorithm 35 (Calculating the Energy Using a Verlet List)

```
SUBROUTINE en_vlist(i,xi,en)      calculates energy using
                                  the Verlet list
en=0
do jj=1,nlist(i)                  loop over the particles in the list
  j=list(i,jj)                    next particle in the list
  en=en+enij(i,xi,j,x(j))
enddo
return
end
```

Comment to this algorithm:

1. *Array* `list(i,itel)` *and* `nlist` *are made in Algorithm 34 and* `enij` *gives the energy between particles* `i` *and* `j` *at the given positions.*

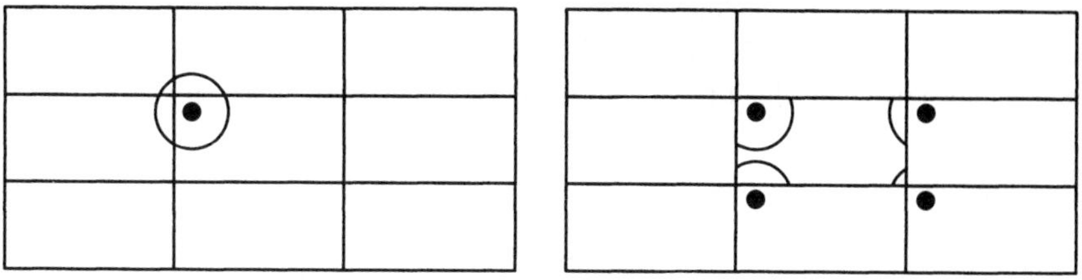

Figure F.2: Verlet lists: (left) conventional approach in which each particle has a Verlet list; (right) the approach of Bekker *et al.* in which each periodic image of a particle has its own Verlet list that contains only those particles in the central box.

We can write

$$\mathbf{F}_i = \sum_{j=1}^{N} \sum_{k=-13}^{13} {}'\mathbf{F}_{(i.k)j}.$$

The importance of this seemingly trivial result is that the summation is over all particles j in the *central* box with the nearest image of particle i. The difference between the two approaches is shown in Figure F.2.

This method is implemented using different Verlet lists for each periodic image of particle i. These lists contain only those particles that interact with particle i *and* are in the central box. If these lists are used, it is not necessary to

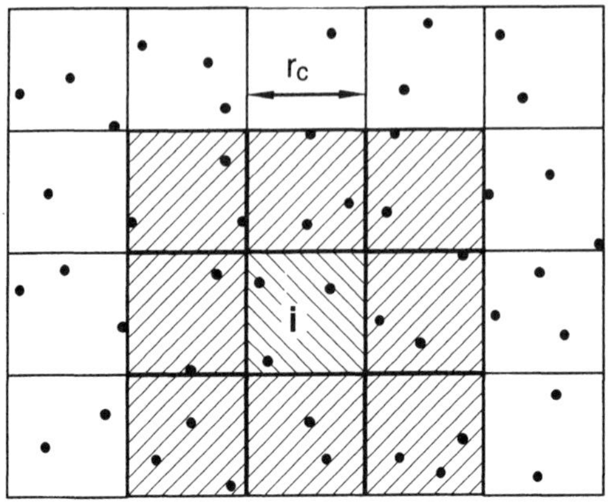

Figure F.3: The cell list: the simulation cell is divided into cells of size $r_c \times r_c$; a particle i interacts with those particles in the same cell or neighboring cells (in 2D there are 9 cells; and in 3D, 27 cells).

use the nearest image operation during the calculation of the force or energy. In [530] the use of these lists is shown to speed up an MD simulation by a factor 1.5. In addition, Bekker *et al.* have shown that a similar trick can be used to take the calculation of the virial (pressure) out of the inner loop.

F.2 Cell Lists

An algorithm that scales with N is the cell list or linked-list method [24]. The idea of the cell list is illustrated in Figure F.3. The simulation box is divided into cells with a size equal to or slightly larger than the cutoff radius r_c; each particle in a given cell interacts with only those particles in the same or neighboring cells. Since the allocation of a particle to a cell is an operation that scales with N and the total number of cells that needs to be considered for the calculation of the interaction is independent of the system size, the cell list method scales as N. Algorithm 36 shows how a cell list can be used in a Monte Carlo simulation.

F.3 Combining the Verlet and Cell Lists

It is instructive to compare the efficiency of the Verlet list and cell list in more detail. In the Verlet list the number of particles for which the distance needs to be calculated is in three dimensions, given by

$$n_v = \frac{4}{3}\pi\rho r_v^3;$$

Algorithm 36 (Use of Cell List in a Monte Carlo Move)

```
      SUBROUTINE mcmove_neigh          attempts to displace a particle
                                       using a cell list
      call newnlist(rc)                make the cell list
      o=int(ranf()*npart)+1            select a particle at random
      call en_nlist(o,x(o),eno)        calculate energy old configuration
      xn=x(o)+(ranf()-0.5)*delx        give particle random displacement
      call en_nlist(o,xn,enn)          calculate energy new configuration
      arg=exp(-beta*(enn-eno))
      if (ranf().lt.arg)
     +   x(o)=xn                       accepted: replace x(o) by xn
      return
      end
```

Comments to this algorithm:

1. *This algorithm is based on Algorithm 2.*
2. *Subroutine* `new_nlist` *makes the cell list (see Algorithm 37) and subroutine* `en_nlist` *calculates the energy of a particle at the given position using the cell list (see Algorithm 38). Note that it is possible, at the expense of some extra bookkeeping, to update the list once a move is accepted instead of making a new list every move.*

for the cell list the corresponding number is

$$n_l = 27\rho r_c^3.$$

If we use typical values for the parameters in these equations (Lennard-Jones potential with $r_c = 2.5\sigma$ and $r_v = 2.7\sigma$), we find that n_l is five times larger than n_v. As a consequence, in the Verlet scheme, the number of pair distances that needs to be calculated is 16 times less than in the cell list.

The observation that the Verlet scheme is more efficient in evaluating the interactions motivated Auerbach *et al.* [531] to use a combination of the two lists: use a cell list to construct a Verlet list. The use of the cell list removes the main disadvantage of the Verlet list for a large number of particles—scales as N^2—but keeps the advantage of an efficient energy calculation. An implementation of this method in a Monte Carlo simulation is shown in Algorithm 39.

Algorithm 37 (Making a Cell List)

```
SUBROUTINE new_nlist(rc)       makes a new cell list with cell size rc
                               using a linked-list algorithm
rn=box/int(box/rc)             determine size of cells rn >= rc
do icel=0,ncel-1
  hoc(icel)=0                  set head of chain to 0 for each cell
enddo
do i=1,npart                   loop over the particles
  icel=int(x(i)/rn)            determine cell number
  ll(i)=hoc(icel)              link list the head of chain of cell icel
  hoc(icel)=i                  make particle i the head of chain
enddo
return
end
```

Comment to this algorithm:

1. *This algorithm uses the linked-list method. To each cell a particle* `i` *is named head of chain and stored in the array* `hoc(icel)`. *To this particle the next particle in the cell (chain) is linked via the linked-list array* `ll(i)`. *If the value of the* `ll(i)` *is 0 no more particles are in the cell (chain). The desired (optimum) cell size is* `rc`*, and* `rn` *is the closest size that fits in the box.*

F.4 Efficiency

The first question that arises is when to use which method. This depends very strongly on the details of the systems. In any event, we always start with a scheme as simple as possible, hence no tricks at all. Although the algorithm scales as N^2, it is straightforward to implement and therefore the probability of programming errors is relatively small. In addition we should take into account how often the program will be used.

The use of the Verlet list becomes advantageous if the number of particles in the list is significantly less than the total number of particles; in three dimensions this means

$$n_v = \frac{4}{3}\pi r_v^3 \rho \ll N.$$

If we substitute some typical values for a Lennard-Jones potential ($r_v = 2.7\sigma$ and $\rho = 0.8\sigma^{-3}$), we find $n_v \approx 66$, which means that only if the number of particles in the box is more than 100 does it make sense to use a Verlet list.

To see when to use one of the other techniques, we have to analyze the algorithms in somewhat more detail. If we use no tricks, the amount of CPU

Algorithm 38 (Calculating the Energy Using a Cell List)

```
   SUBROUTINE ennlist(i,xi,en)     calculates energy using
                                   the cell list
   en=0
   icel=int(xi/rn)                 determine the cell number
   do ncel=1,neigh                 loop over the neighbor cells
     jcel=neigh(icel,ncel)         number of the neighbor
     j=hoc(jcel)                   head of chain of cell jcel
     do while (j.ne.0)
       if (i.ne.j)
+        en=en+enij(i,xi,j,x(j))
       j=ll(j)                     next particle in the list
     enddo
   enddo
   return
   end
```

Comment to this algorithm:

1. *Array* `ll(i)` *and* `hoc(icel)` *are constructed in Algorithm 37;* `enij` *is a function that gives the energy between particles* `i` *and* `j` *at the given positions.* `neigh(icel,ncel)` *gives the location of the* `ncel`*th neighbor of cell* `icel`.

time to calculate the total energy is given by

$$\tau = cN(N-1)/2.$$

The constant gives the required CPU time for an energy calculation between a pair of particles. If we use the Verlet list, the CPU time is

$$\tau_v = cn_vN + \frac{c_v}{n_u}N^2,$$

where the first term arises from the calculation of the interactions and the second term from the update of the Verlet list, which is done every n_u^{th} cycle.

The cell list scales with N and the CPU time can be split into two contributions: one that accounts for the calculation of the energy and the other for the making of the list,

$$\tau_l = cn_lN + c_lN.$$

If we use a combination of the two lists, the total CPU time becomes

$$\tau_c = cn_vN + \frac{c_l}{n_u}N.$$

Algorithm 39 (Combination of Verlet and Cell Lists)

```
   SUBROUTINE mcmove_clist             displace a particle
                                       using a combined list
   o=int(ranf()*npart)+1               select a particle at random
   if (abs(x(o)-xv(o)).gt.rv-rc)       check to make a new list
 +   call new_clist
   call en_vlist(o,x(o),eno)           energy old configuration
   xn=x(o)+(ranf()-0.5)*delx           random displacement
   if (abs(xn-xv(o)).gt.rv-rc)         check to make a new list
 +   call new_clist
   call en_vlist(o,xn,enn)             energy new configuration
   arg=exp(-beta*(enn-eno))
   if (ranf().lt.arg)
 +   x(o)=xn                           accepted: replace x(o) by xn
   return
   end
```

Comments to this algorithm:

1. *The algorithm is based on Algorithm 33.*
2. *Subroutine* `newclist` *makes the Verlet list using a cell list (see Algorithm 40) and subroutine* `en_vlist` *calculates the energy of a particle at the given position using the Verlet list (see Algorithm 35).*

The way to proceed is to perform some test simulations to estimate the various constants, and from the equations, it will become clear which technique is preferred. In Case Study 26, we have made such an estimate for a simulation of the Lennard-Jones fluid.

Case Study 26 (Comparison of Schemes for the Lennard-Jones Fluid)
It is instructive to make a detailed comparison of the various schemes to save CPU time for the Lennard-Jones fluid. We compare the following schemes:

1. Verlet list
2. Cell list
3. Combination of Verlet and cell lists
4. Simple N^2 algorithm

We have used the program of Case Study 1 as a starting point. At this point it is important to note that we have not tried to optimize the parameters (such

Algorithm 40 (Making a Verlet List Using a Cell List)

```
SUBROUTINE new_clist                          makes a new Verlet list
                                              using a cell list
call new_nlist(rv)                            make the cell lists
do i=1,npart                                  initialize list
  nlist(i)=0
  xv(i)=x(i)                                  store position of particles
enddo
do i=1,npart
 icel=int(x(i)/rn)                            determine cell number
 do ncel=1,neigh                              loop over the neighbor cells
   jcel=neigh(icel,ncel)                      number of the neighbor
   j=hoc(jcel)                                head of chain of cell jcel
   do while (j.ne.0)
     if (i.ne.j) then
      xr=x(i)-x(j)
      if (xr.gt.hbox) then                    nearest image
       xr=xr-box
      else if(xr.lt.-hbox)then
       xr=xr+box
      endif
      if (abs(xr).lt.rv) then                 add to the Verlet lists
       nlist(i)=nlist(i)+1
       nlist(j)=nlist(j)+1
       list(i,nlist(i))=j
       list(j,nlist(j))=i
      endif
     endif
    j=ll(j)                                   next particle in the cell list
   enddo
 enddo
enddo
return
end
```

Comments to this algorithm:

1. *Array* `list(i,itel)` *is the Verlet list of particle* `i`*, the number of particles in the Verlet list of particle* `i` *is given by* `nlist(i)`*, and the array* `xv(i)` *contains the position of the particles at the moment the list is made (is used to see when a new list has to be made). We assume that* all *particles are in the simulation box; hence* `x(i)` $\in$ *[0,*box*].*
2. *Subroutine* `new_nlist(rv,rn)` *makes a cell list (Algorithm 37). The desired cell size is* `rv` *and the actual cell size is* `rn`.

as the Verlet radius) for the various methods; we have simply taken some reasonable values.

For the Verlet list (and for the combination of Verlet and cell lists) it is important that the maximum displacement be smaller than twice the difference between the Verlet radius and cutoff radius. For the cutoff radius we have used $r_c = 2.5\sigma$, and for the Verlet radius $r_v = 3.0\sigma$. This limits the maximum displacement to $\Delta_x = 0.25\sigma$ and implies for the Lennard-Jones fluid that, if we want to use a optimum acceptance of 50%, we can use the Verlet method only for densities larger than $\rho > 0.6\sigma^{-3}$. For smaller densities, the optimum displacement is larger than 0.25. Note that this density dependence does not exist in a Molecular Dynamics simulation. In a Molecular Dynamics simulation, the maximum displacement is determined by the integration scheme and therefore is independent of density. This makes the Verlet method much more appropriate for a Molecular Dynamics simulation than for a Monte Carlo simulation. Only at high densities does it make sense to use the Verlet list.

The cell list method is advantageous only if the number of cells is larger than 3 in at least one direction. For the Lennard-Jones fluid this means that, if the number of particles is 400, the density should be lower than $\rho < 0.5\sigma^{-3}$. An important advantage of the cell list over the Verlet list is that this list can also be used for moves in which a particle is given a random position.

From these arguments it is clear that, if the number of particles is smaller than 200–500, the simple N^2 algorithm is the best choice. If the number of particles is significantly larger and the density is low, the cell list method is probably more efficient. At high density, all methods can be efficient and we have to make a detailed comparison.

To test these conclusions about the N dependence of the CPU time of the various methods, we have performed several simulations with a fixed number of Monte Carlo cycles. For the simple N^2 algorithm the CPU time per attempt is

$$\tau_{N^2} = cN,$$

where c is the CPU time required to calculate one interaction. This implies that the total amount of CPU time is independent of the density. For a calculation of the total energy, we have to do this calculation N times, which gives the scaling of N^2. Figure F.4 shows that indeed for the Lennard-Jones fluid, the τ_{N^2} increases linearly with the number of particles.

If we use the cell list, the CPU time will be

$$\tau_n = cV_l\rho + c_l p_l N,$$

where V_l is the total volume of the cells that contribute to the interaction (in three dimensions, $V_l = 27r_c^3$), c_n is the amount of CPU time required to

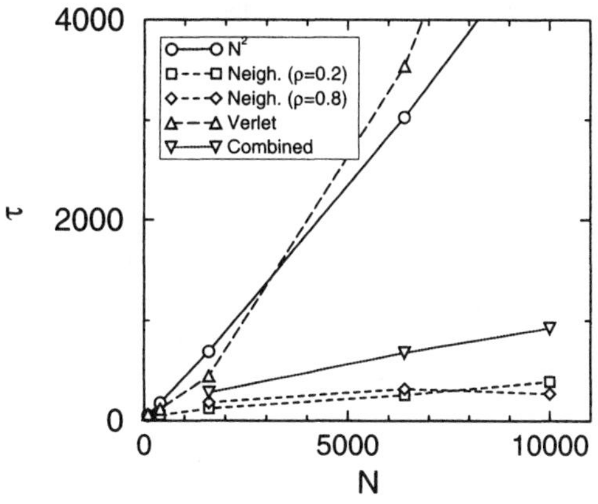

Figure F.4: Comparison of various schemes to calculate the energy: τ is in arbitrary units and N is the number of particles. As a test case the Lennard-Jones fluid is used. The temperature was $T^* = 2$ and per cycle the number of attempts to displace a particle was set to 100 for all systems. The lines serve to guide the eye.

make a cell list, and p_n is the probability that a new list has to be made. Figure F.4 shows that the use of a cell list reduces the CPU time for 10,000 particles with a factor 18. Interestingly, the CPU time does not increase with increasing density. We would expect an increase since the number of particles that contribute to the interaction of a particle i increases with density. However, the second contribution to τ_{Neigh} (p_n) is the probability that a new list has to be made, depends on the maximum displacement, which decreases when the density increases. Therefore, this last term will contribute less at higher densities.

For the Verlet scheme the CPU time is

$$\tau_v = cV_v\rho + c_v p_v N^2,$$

where V_v is the volume of the Verlet sphere (in three dimensions, $V_v = 4\pi r_v^3/3$), c_v is the amount of CPU time required to make the Verlet-list, and p_v is the probability that a new list has to be made. Figure F.4 shows that this scheme is not very efficient. The N^2 operation dominates the calculation. Note that we use a program in which a new list for all particles has to be made as soon as one of the particles has moved more than $(r_v - r_c)/2$; with some more bookkeeping it is possible to make a much more efficient program, in which a new list is made for only the particle that has moved out of the list.

The combination of the cell and Verlet lists removes the N^2 dependence of the simple Verlet algorithm. The CPU time is given by

$$\tau_c = cV_\nu \rho + c_\nu p_\nu c_n N.$$

Figure F.4 shows that indeed the N^2 dependence is removed, but the resulting scheme is not more efficient than the cell list alone.

This case study demonstrates that it is not simple to give a general recipe for which method to use. Depending on the conditions and number of particles, different algorithms are optimal. It is important to note that for a Molecular Dynamics simulation the conclusions may be different.

Appendix G

Reference States

G.1 Grand-Canonical Ensemble Simulation

In a grand-canonical ensemble simulation, we impose the temperature and chemical potential. Experimentally, however, usually the pressure rather than the chemical potential of the reservoir is fixed. To compare the experimental data with the simulation results it is necessary therefore to determine the pressure that corresponds to a given value of the chemical potential and temperature of our reservoir.

Preliminaries

The partition function of a system with N atoms in the N, V, T-ensemble is given by

$$Q(N, V, T) = \frac{V^N}{\Lambda^{3N} N!} \int ds^N \exp[-\beta \mathcal{U}(\mathbf{s}^N)], \tag{G.1.1}$$

where $\mathbf{s}^N$ are the scaled coordinates of the N particles. The free energy is related to the partition function via

$$F = -\frac{1}{\beta} \ln Q(N, V, T),$$

which gives us for the chemical potential

$$\mu = \frac{\partial F}{\partial N} = -\frac{1}{\beta} \ln[Q(N+1, V, T)/Q(N, V, T)]. \tag{G.1.2}$$

For a system consisting of N molecules with each molecule having M atoms, the partition function is

$$Q(N, M, V, T) = \frac{q(T)^N V^N}{N!} \prod_{i=1}^{M} \int ds_i^N \exp[-\beta \mathcal{U}(s_i^N)], \tag{G.1.3}$$

where $q(T)$ is the part of the partition function of a molecule that contains the integration over momenta (for an atom, $q(T)$ is simply Λ^{-3}) and the $\mathbf{s}^N$ are the Cartesian coordinates of atoms in the molecule. It should be stressed that, in writing equation (G.1.3), we are making the assumption that there are no "hard" constraints on these intramolecular coordinates. In the presence of hard constraints, the integral in equation (G.1.3) would contain a Jacobian (see section 15.1).

Ideal Gas

In the limit of zero density, any system will behave as an ideal gas. In this limit only the intramolecular interactions contribute to the total potential energy

$$\mathcal{U} \approx \sum_{i=1}^{N} \mathcal{U}^{\text{intra}}(i).$$

For a system consisting of noninteracting atoms, the partition function (G.1.1) reduces to

$$Q_{\text{IG}}(N, V, T) = \frac{V^N}{\Lambda^{3N} N!}. \tag{G.1.4}$$

We can write, for the chemical potential of such an ideal gas of atoms,

$$\mu_{\text{id.gas}} = \mu^0_{\text{id.gas}} + k_B T \ln \rho, \tag{G.1.5}$$

with the chemical potential of the reference state defined by

$$\mu^0_{\text{id.gas}} \equiv k_B T \ln \Lambda^3. \tag{G.1.6}$$

In case of gas of noninteracting molecules, the partition function (G.1.3) reduces to

$$Q_{\text{id.gas}}(N, M, V, T) = \frac{q(T)^N V^N}{N!} \left\{ \prod_{i=1}^{M} \int d\mathbf{s}_i \exp[-\beta \mathcal{U}^{\text{intra}}(\mathbf{s}_i)] \right\}^N. \tag{G.1.7}$$

Substitution into equation (G.1.2) yields, for the chemical potential,

$$\mu_{\text{id.gas}} = \mu^0_{\text{id.gas}} + k_B T \ln \rho, \tag{G.1.8}$$

where the reference chemical potential is defined as

$$\begin{aligned}\beta\mu^0_{\text{id.gas}} &\equiv -\ln q(T) + \beta\mu^0_{\text{intra}} \\ &= -\ln q(T) - \ln\left[\prod_{i=1}^{M}\int d\mathbf{s}_i \exp[-\beta\mathcal{U}^{\text{intra}}(\mathbf{s}_i)]\right]. \qquad \text{(G.1.9)}\end{aligned}$$

Note that $\mu^0_{\text{id.gas}}$ depends only on temperature. At any given temperature, it simply acts as a constant shift of the chemical potential that has no effect on the observable thermodynamic properties of the system.

Grand-Canonical Simulations

In a grand-canonical simulation, we use the following acceptance rules (see section 5.6.2). For the addition, we have

$$\begin{aligned}&\text{acc}(N \to N+1) \\ &= \min\left[1, \frac{Vq(T)\exp(-\beta\mu^0_{\text{intra}})}{(N+1)}\exp\{\beta[\mu^B - \mathcal{U}(N+1) + \mathcal{U}(N)]\}\right].\end{aligned}$$

For the removal of a particle, we have

$$\begin{aligned}&\text{acc}(N \to N-1) \\ &= \min\left[1, \frac{N}{q(T)\exp(-\beta\mu^0_{\text{intra}})V}\exp\{-\beta[\mu^B + \mathcal{U}(N-1) - \mathcal{U}(N)]\}\right].\end{aligned}$$

These equations are based on the idea that particles are exchanged with a reservoir containing the same molecules at the same chemical potential, the only difference being that, in the reservoir, the molecules do not interact. In practical cases (e.g., adsorption), this means that we have a dense phase in equilibrium with a dilute vapor. And, whereas the absolute chemical potential of the vapor is of little interest, the absolute pressure is clearly an important quantity. The pressure in the reservoir is related to the chemical potential through

$$\begin{aligned}\beta\mu^B &\equiv \beta\mu^0_{\text{id.gas}} + \ln(\rho) \\ &= \beta\mu^0_{\text{id.gas}} + \ln(\beta P_{\text{id.gas}}). \qquad \text{(G.1.10)}\end{aligned}$$

Substitution of this expression in the acceptance rules yields

$$\text{acc}(N \to N+1) = \min\left[1, \frac{V\beta P_{\text{id.gas}}}{(N+1)}\exp\{-\beta[\mathcal{U}(N+1) - \mathcal{U}(N)]\}\right] \qquad \text{(G.1.11)}$$

for the addition of a particle, and a similar expression for particle removal. In other words, if the experimental conditions are such that the system of interest is in equilibrium with a reservoir that behaves like an ideal gas, then only the pressure of this effectively ideal gas enters into the acceptance rules for trial moves. All information about the reference state drops out (as expected).

If the pressure in the reservoir is too high for the ideal gas law to hold, we have to use an equation of state to relate the chemical potential of the reservoir to its pressure:

$$\beta\mu_B = \beta\mu^0_{\text{id.gas}} + \ln(\beta P\phi), \tag{G.1.12}$$

where ϕ is the fugacity coefficient of the fluid in the reservoir. The fugacity coefficient can be computed directly from the equation of state of the vapor in the reservoir. It is important to note that this fugacity coefficient is a function of the temperature and pressure. In summary, for a nonideal gas, we should replace $P_{\text{id.gas}}$ in the acceptance rule (G.1.11) by $P\phi$.

Appendix H

Statistical Mechanics of the Gibbs "Ensemble"

The introduction of a new ensemble brings up the question of whether it is a "proper ensemble"; that is, does it yield the same results as the conventional ensembles? To prove it does, we use the partition function (8.2.1) as derived in section 8.2 to define a free energy. This free energy is used to show that, in the thermodynamic limit, the Gibbs ensemble and the canonical ensemble are equivalent. This proof gives considerable insight into why the method works. Before we proceed, we first list a few basic results for the free energy in the canonical ensemble.

H.1 Free Energy of the Gibbs Ensemble

H.1.1 Basic Definitions and Results for the Canonical Ensemble

Consider a system of N particles in a volume V and temperature T (canonical ensemble). The partition function is defined as (see Ruelle [532])

$$Q(N,V,T) \equiv \frac{1}{\Lambda^{3N} N!} \int_V d\mathbf{r}^N \ \exp\left[-\beta \mathcal{U}(N)\right]. \tag{H.1.1}$$

The free energy density is defined in the thermodynamic limit by

$$f(\rho) = \lim_{V\to\infty} f_V(\rho) \equiv \lim_{\substack{V\to\infty \\ N/V=\rho}} -\frac{1}{\beta V} \ln Q_{N,V,T},$$

where $\rho = N/V$ is the density of the system. For a finite number of particles we can write

$$Q(N, V, T) = \exp\{-\beta V [f(\rho) + o(V)]\}, \tag{H.1.2}$$

where $g(V) = o(V)$ means $g(V)/V$ approaches 0 as $V \to \infty$. With this free energy, we can derive some interesting properties of a canonical system in the thermodynamic limit.

For example, it can be shown that this free energy is a convex function of the density ρ [532]:

$$f(x\rho_1 + (1-x)\rho_2) \leq xf(\rho_1) + (1-x)f(\rho_2), \tag{H.1.3}$$

for every ρ_1, ρ_2, and x where $0 \leq x \leq 1$. The equality holds in the case of a first-order transition, if $\rho_g \leq \rho_1 \leq \rho_2 \leq \rho_l$, where ρ_g, ρ_l denote the density of coexisting gas and liquid phases, respectively.

Another interesting result, which plays a central role on the following pages, is the well-known saddle point theorem [533] (also called the *steepest descent method*). This theorem is based on the observation that, for a macroscopic system (N very large) in equilibrium, the probability that the free energy density deviates from its minimum value is extremely small. Therefore, when we calculate for such a system an ensemble average, we have to take into account only those contributions where the free energy has its minimum value. Assume that $Q(N, V, T)$ can be written as

$$Q(N, V, T) \equiv \int da_1, \cdots, da_m \, \exp\left[-\beta V \left(f_m(a_1, \cdots, a_m) + o(V)\right)\right],$$

where $a_1, \cdots, a_m$ are variables that characterize the thermodynamic state of the system. Furthermore, define

$$f(\rho) \equiv \min_{a_1, \cdots, a_m} f_m(a_1, \cdots, a_m)$$

and assume that $f_m(a_1, \cdots, a_m)$ and the term $o(V)$ satisfy a few technical conditions [533], which hold for most statistical mechanics systems. The saddle point theorem states that, in the thermodynamic limit, the free energy of the system is equal to this minimum value $f(\rho)$ or

$$\lim_{\substack{V \to \infty \\ N/V = \rho}} -\frac{1}{\beta V} \ln Q(N, V, T) = f(\rho). \tag{H.1.4}$$

Moreover, this saddle point theorem can also be used to calculate the ensemble average of a quantity A:

$$\begin{aligned} \langle A(a_1, \cdots, a_m) \rangle_V \equiv & \frac{1}{Q(N, V, T)} \int da_1, \cdots, da_m \\ & \times A(a_1, \cdots, a_m) \exp\{-\beta V [f_m(a_1, \cdots, a_m) + o(V)]\}. \end{aligned} \tag{H.1.5}$$

In the thermodynamic limit, this ensemble average again has contributions only from those configurations where $f_m(a_1, \cdots, a_m)$ has its minimum value. Let us define S as the collection of these minima:

$$S = \left\{ y_1, \cdots, y_m \middle| f_m(y_1, \cdots, y_m) = \min_{a_1, \cdots, a_m} f_m(a_1, \cdots, a_m) \right\}.$$

We now can state the saddle point theorem in a convenient form by introducing a function $G(a_1, \cdots, a_m) \geq 0$ with support on the surface S and normalization

$$\int_S da_1, \cdots, da_m \, G(a_1, \cdots, a_m) = 1,$$

such that, for an arbitrary function A,

$$\begin{aligned} \langle A(a_1, \cdots, a_m) \rangle &\equiv \lim_{V \to \infty} \langle A(a_1, \cdots, a_m) \rangle_V \\ &= \int_S da_1, \cdots, da_m \, G(a_1, \cdots, a_m) A(a_1, \cdots, a_m). \end{aligned} \tag{H.1.6}$$

H.1.2 The Free Energy Density in the Gibbs Ensemble

The Gibbs ensemble is introduced in section 8.2 as an N, V, T ensemble to which an additional degree of freedom is added: the system is divided into two subsystems that have *no* interaction with each other. We can rewrite the partition function of the canonical ensemble (H.1.1):

$$\begin{aligned} Q(N, V, T) &= \frac{1}{\Lambda^{3N} N!} \sum_{n_1=0}^{N} \binom{N}{n_1} \int_0^V dV_1 \int dr_1^{n_1} \int dr_2^{N-n_1} \exp\{-\beta \, [\mathcal{U}(n_1) \\ &\quad + \mathcal{U}(N - n_1) + \text{interactions between the two volumes}]\}. \end{aligned} \tag{H.1.7}$$

The difference between this equation and the partition function of the Gibbs ensemble (8.2.1) is that, in equation (H.1.1), we have interactions between the subsystems. In the case of short-range interactions, the last term in the exponent of equation (H.1.7) is proportional to a surface term. This already suggests that both ensembles should behave similarly in many respects. We work out these ideas more rigorously in the following pages.

In the usual way, we define, as a free energy in the Gibbs ensemble,

$$\bar{f}(\rho) \equiv \lim_{\substack{V \to \infty \\ N/V = \rho}} -\frac{1}{\beta V} \ln \bar{Q}_{N,V,T}. \tag{H.1.8}$$

In the partition function of the Gibbs ensemble (8.2.1), we can substitute equation (H.1.1):

$$\bar{Q}(N, V, T) = \sum_{n_1=0}^{N} \int_0^V dV_1 \, Q(n_1, V_1, T) Q(N - n_1, V - V_1, T).$$

Introducing $x = N_1/N$ and $y = V_1/V$, and assuming that the number of particles is very large, we can then write

$$\bar{Q}(N,V,T) = NV \int_0^1 dx \int_0^1 dy\, \bar{Q}_N(x,y),$$

where

$$\begin{aligned}\bar{Q}_N(x,y) &= Q_{xN,yV,T}\, Q_{(1-x)N,(1-y)V,T} \\ &= \exp\left\{-\beta V \left[y f\left(\frac{x}{y}\rho\right) + (1-y) f\left(\frac{1-x}{1-y}\rho\right) + o(V)\right]\right\}.\end{aligned}$$

Note that, in this equation, $f(\rho)$ is the free energy of a canonical system. So, we can apply the saddle point theorem of the previous section (H.1.4) to calculate the free energy density of the Gibbs ensemble $\bar{f}(\rho)$

$$\bar{f}(\rho) = \min_{\substack{0\leq x\leq 1\\ 0\leq y\leq 1}} \left[y f\left(\frac{x}{y}\rho\right) + (1-y) f\left(\frac{1-x}{1-y}\rho\right)\right] \equiv \min_{\substack{0\leq x\leq 1\\ 0\leq y\leq 1}} \bar{f}(x,y).$$

We now have to find the surface S on which the function $\bar{f}(x,y)$ reaches its minimum. For this, we can use that $f(\rho)$ is a convex function of the density (H.1.3). This gives, for $\bar{f}(x,y)$,

$$\bar{f}(x,y) \geq f\left(y\frac{x}{y}\rho + (1-y)\frac{1-x}{1-y}\rho\right) = f(\rho). \qquad \text{(H.1.9)}$$

We first consider the case where there is only one phase. For this case any combination of x and y that results in densities ρ_1 and ρ_2 in the subsystems different from ρ will give a higher free energy. So, the equality in equation (H.1.9) holds only if

$$\frac{x}{y}\rho = \frac{1-x}{1-y}\rho, \quad \text{or} \quad x = y.$$

Thus, when there is only one phase, the free energy of the Gibbs ensemble has its minimum value (in the thermodynamic limit) when both boxes have a density equal to the equilibrium density of the canonical ensemble. Therefore, the surface S is given by

$$S = \{(x,y)\,|x = y\}.$$

Second, we consider the case of a first-order phase transition. Let ρ be such that $\rho_l \leq \rho \leq \rho_g$, and let us choose x and y such that

$$\rho_g \leq \frac{x}{y}\rho \equiv \rho_3 \leq \rho_l \quad \text{and} \quad \rho_g \leq \frac{1-x}{1-y}\rho \equiv \rho_4 \leq \rho_l. \qquad \text{(H.1.10)}$$

For this case the equality in equation (H.1.3) holds, and we can write, for $\bar{f}(x, y)$,

$$\begin{aligned} \bar{f}(x, y) &= y f(\rho_3) + (1 - y) f(\rho_4) \\ &= f(y\rho_3 + (1 - y)\rho_4). \end{aligned} \tag{H.1.11}$$

Note that

$$(y\rho_3 + (1 - y)\rho_4) = \rho, \tag{H.1.12}$$

which gives

$$\bar{f}(x, y) = f(\rho). \tag{H.1.13}$$

It can be shown that, if x, y do not satisfy equation (H.1.10),

$$\bar{f}(x, y) > f(\rho).$$

Therefore, the surface S in the case of a first-order phase transition is given by

$$S = \left\{ (x, y) \middle| \rho_g \leq \frac{x}{y}\rho \leq \rho_l, \quad \rho_g \leq \frac{1 - x}{1 - y}\rho \leq \rho_l \right\}. \tag{H.1.14}$$

This result shows that, in the case of a first-order transition, the (bulk) free energy of the Gibbs ensemble has its minimum value (in the thermodynamic limit) for all values of x, y where there is vapor-liquid coexistence in *both* boxes.

Equations (H.1.9) and (H.1.13) show that, in the thermodynamic limit, the free energy of the Gibbs ensemble is equal to the free energy of the canonical ensemble. To calculate an ensemble average, it remains to determine the function $G(x, y)$ using equation (H.1.6).

In the case of a pure phase $G(x, y)$ needs to be of the form

$$G(x, y) = g(x)\,\delta(x - y). \tag{H.1.15}$$

It is shown in the Appendix of [148] that, for an ideal gas, $g(x) = 1$. We expect that the same holds for an interacting gas. Figure H.1 shows a probability plot in the x, y plane for a simulation of a finite system at high temperature. This figure shows that $x \approx y$.

In the case of two phases, we will show that the system will split up into a liquid phase, with density, ρ_l, in one box, and a vapor phase, with density, ρ_g, in the other box.

Until now we have ignored surface effects, which arise from the presence of a liquid-vapor interface in the boxes. When the density in one box is between the vapor and liquid density the system will form droplets of gas or liquid. The interfacial free energy associated with these droplets has (in the thermodynamic limit) a negligible contribution to the bulk free energy of the Gibbs ensemble. Nevertheless, this surface free energy is the driving force that causes the system to separate into a homogeneous liquid in one box and a homogeneous vapor phase in the other.

These surface effects are taken into account in the next significant term in the expression for the free energy (H.1.2), which is the term due to the surface tension. This gives, for the partition function,

$$Q(N,V,T) = \exp\{-\beta\,[Vf(\rho) + \gamma A + o(A)]\}, \tag{H.1.16}$$

where A denotes the area of the interface and γ denotes the interfacial tension. For three-dimensional systems, in general this area will be proportional to $V^{2/3}$. Using this form of the partition function for the Gibbs ensemble, equation (H.1.5) can be written as

$$\begin{aligned} &\langle A(x,y)\rangle_V \\ &= \frac{\iint dxdy\, A(x,y)\exp\left\{-\beta\left[Vf(x,y) + \gamma V^{2/3}a(x,y) + o(V^{2/3})\right]\right\}}{Q(N,V,T)}, \end{aligned} \tag{H.1.17}$$

where $a(x,y)$ is a function of the order of unity.

We know from the saddle point theorem that the most important contribution to the integrals comes from the region S, defined by equation (H.1.14). Thus,

$$\begin{aligned} &\langle A(x,y)\rangle_V \\ &\approx \frac{\iint_S dxdy\, A(x,y)\exp\left\{-\beta\left[Vf(x,y) + \gamma V^{2/3}a(x,y) + o(V^{2/3})\right]\right\}}{\iint_S dxdy\, \exp\left\{-\beta\left[Vf(x,y) + \gamma V^{2/3}a(x,y) + o(V^{2/3})\right]\right\}} \\ &= \frac{\iint_S dxdy\, A(x,y)\exp\left\{-\beta\left[\gamma V^{2/3}a(x,y) + o(V^{2/3})\right]\right\}}{\iint_S dxdy\, \exp\left\{-\beta\left[\gamma V^{2/3}a(x,y) + o(V^{2/3})\right]\right\}} \end{aligned} \tag{H.1.18}$$

and applying the saddle point theorem again

$$\langle A(x,y)\rangle_V \approx \frac{\iint_{S_A} dxdy\, A(x,y)\exp\left[-\beta\gamma V^{2/3}a(x,y) + o(V^{2/3})\right]}{\iint_{S_A} dxdy\, \exp\left[-\beta\gamma V^{2/3}a(x,y) + o(V^{2/3})\right]} \tag{H.1.19}$$

and

$$\lim_{V\to\infty} \langle A(x,y)\rangle_V = \iint_{S_A} dxdy\, G(x,y)\, A(x,y), \tag{H.1.20}$$

where the surface S_A is now given by

$$S_A = \left\{ (x,y) \,\middle|\, a(x,y) = \min_{\bar{x},\bar{y}} a(\bar{x},\bar{y}) \right\}. \tag{H.1.21}$$

In the infinite system it is easily seen that the area of the interface is 0, if box 1 contains only gas (liquid) and box 2 only liquid (gas). Therefore, the

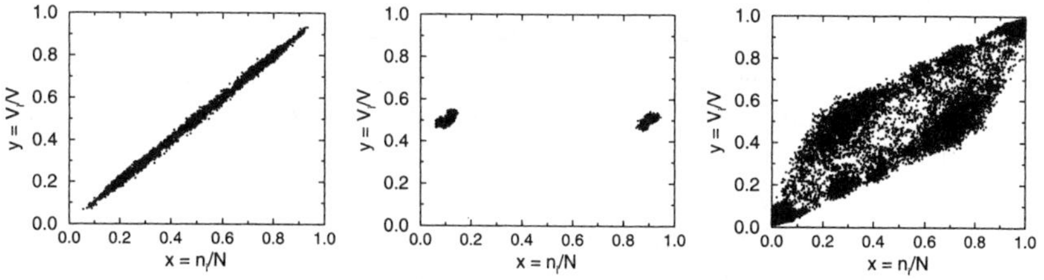

Figure H.1: Probability plot in the x, y plane ($x = n_1/N$, $y = V_1/V$ and $x = (N - n_1)/N$, $y = (V - V_1)/V$) for a Lennard-Jones fluid at various temperatures: (left) high temperature ($T = 10$), (middle) well below the critical temperature ($T = 1.15$), and (right) slightly below the critical temperature ($T = 1.30$).

surface S_A contains only two points, which correspond to the vapor and liquid densities:

$$S_A = \left\{ (x, y) \left| \frac{x}{y} = \rho_l \text{ and } \frac{1-x}{1-y} = \rho_g \text{ or } \frac{x}{y} = \rho_g \text{ and } \frac{1-x}{1-y} = \rho_l \right. \right\}. \tag{H.1.22}$$

It is straightforward to show that this surface gives, for $G(x, y)$,

$$\begin{aligned} G(x, y) &= \frac{1}{2} \delta \left(x - \frac{\rho_g}{\rho} \frac{\rho - \rho_g}{\rho_l - \rho_g} \right) \delta \left(y - \frac{\rho - \rho_g}{\rho_l - \rho_g} \right) \\ &+ \frac{1}{2} \delta \left(x - \frac{\rho_l}{\rho} \frac{\rho_l - \rho}{\rho_l - \rho_g} \right) \delta \left(y - \frac{\rho_l - \rho}{\rho_l - \rho_g} \right). \end{aligned} \tag{H.1.23}$$

We have shown more formally that the free energy density for the Gibbs ensemble, as defined by equation (H.1.8), becomes identical to the free energy density of the canonical ensemble. Furthermore, it is shown that, at high temperatures, $x = y$; that is, the densities in the two subsystems of the Gibbs ensemble are equal and equal to the density in the canonical ensemble (see Figure H.1).

In the case of a first-order phase transition, if surface terms would be unimportant, then x and y are restricted to the area defined by equation (H.1.14):

$$\rho_g \leq \frac{x}{y} \rho \equiv \rho_3 \leq \rho_l \quad \text{and} \quad \rho_g \leq \frac{1-x}{1-y} \rho \equiv \rho_4 \leq \rho_l. \tag{H.1.24}$$

If we take surface effects into account, this surface (equation (H.1.24)) reduces to two points in the x, y plane. The densities of these points correspond to the density of the gas or liquid phase in the canonical ensemble.

It is interesting to compare this with the results of an actual simulation of a finite system. In Figure H.1, the results are shown for a simulation at a temperature well below the critical point. Under such conditions, the surface reduces to two points. This should be compared to the results of a simulation close to the critical point (Figure H.1). Under such conditions the interfacial tension is very small and we see that the simulation samples the entire surface S. Note that due to the finite size of this system, fluctuations are also possible in which the density of a subsystem becomes greater or smaller than the density of the liquid or gas phase.

H.2 Chemical Potential in the Gibbs Ensemble

One of the steps in the Gibbs ensemble involves the insertion of a particle in one of the boxes. During this step, the energy of this particle has to be calculated (see section 8.3.4). Since this energy corresponds to the energy of a test particle, we can use the Widom insertion method [172] to calculate the chemical potential without additional costs [147]. At this point it is important to note that the Gibbs method requires no computation of the chemical potentials. However, to test whether the system under consideration has reached equilibrium or for comparison with other results, it is important to calculate the chemical potential of the individual phases correctly. The original Widom expression is valid only in the N, V, T ensemble and can be modified for applications in other ensembles (see section 7.2.1). Here we derive an expression for the chemical potential for the Gibbs ensemble. We restrict ourselves to temperatures sufficiently far below the critical temperature that the two boxes, after equilibration, do not change identity. For the more general case, we refer to [203].

If we rescale the coordinates of the particles with the box length, the partition function for the Gibbs ensemble (8.2.1) becomes

$$\begin{aligned}
\bar{Q}_{N,V,T} &\equiv \frac{1}{V\Lambda^{3N}N!}\sum_{n_1=0}^{N}\binom{N}{n_1}\int_0^V dV_1\, V_1^{n_1}(V-V_1)^{N-n_1} \\
&\quad\times \int d\mathbf{s}_1^{n_1}\exp\left[-\beta\mathcal{U}_1(n_1)\right]\int d\mathbf{s}_2^{N-n_1}\exp\left[-\beta\mathcal{U}_2(N-n_1)\right] \\
&= \sum_{n_1=0}^{N}\int_0^V dV_1\, V_1^{n_1}(V-V_1)^{N-n_1}Q_1(n_1,V_1)\,Q_2(N-n_1,V-V_1),
\end{aligned} \tag{H.2.1}$$

where $\mathbf{s} = \mathbf{r}/L$ is the scaled coordinates of a particle, L is the box length of the subsystem in which the particle is located, and $Q_i(n_i, V_i)$ is the partition of the canonical ensemble (see also section H.1.2).

The chemical potential of box 1 can be defined as

$$\begin{aligned}\mu_1 &\equiv -k_B T \ln \sum_{n_1=0}^{N} \int_0^V dV_1\, V_1^{n_1} (V - V_1)^{N-n_1} \\ &\quad \times \left[\frac{Q_1(n_1+1, V_1)}{Q_1(n_1, V_1)}\right] Q_2(N - n_1, V - V_1). \end{aligned} \tag{H.2.2}$$

For the ratio of the partition functions of box 1, we can write

$$\begin{aligned}\frac{Q_1(n_1+1, V_1)}{Q_1(n_1, V_1)} &= \frac{V_1}{(n_1+1)\Lambda^3} \frac{\int ds_1^{n_1+1} \exp\left[-\beta \mathcal{U}_1(n_1+1)\right]}{\int ds_1^{n_1} \exp\left[-\beta \mathcal{U}_1(n_1)\right]} \\ &= \frac{V_1}{(n_1+1)\Lambda^3} \frac{\int ds_1^{n_1} \exp\left[-\beta \Delta\mathcal{U}_1^+\right] \exp\left[-\beta \mathcal{U}_1(n_1)\right]}{\int ds_1^{n_1} \exp\left[-\beta \mathcal{U}_1(n_1)\right]}, \end{aligned} \tag{H.2.3}$$

in which we have used the notation

$$\mathcal{U}_1(n_1+1) = \Delta\mathcal{U}_1^+ + \mathcal{U}_1(n_1),$$

where $\Delta\mathcal{U}_1^+$ is the test particle energy of a (ghost) particle in box 1. We can write equation (H.2.2) as an ensemble average restricted to box 1:

$$\mu_1 = -k_B T \ln \frac{1}{\Lambda^3} \left\langle \frac{V_1}{n_1+1} \exp\left[-\beta \Delta\mathcal{U}_1^+\right] \right\rangle_{\text{Gibbs, box 1}}, \tag{H.2.4}$$

where $< \cdots >_{\text{Gibbs, box i}}$ denotes an ensemble average in the Gibbs ensemble restricted to box i (note that this ensemble average is well defined if the boxes do not change identity during a simulation [203]).

Appendix I

Overlapping Distribution for Polymers

Let us first consider how the basic idea behind the overlapping distribution method can be applied to the Rosenbluth insertion scheme. The simplest approach would be to consider the histogram of the potential energy change on addition or removal of a chain molecule (see section 7.2.3). However, for chain molecules, this approach differs from the original Shing-Gubbins approach in that it has little, if any, diagnostic value. For instance, if we consider the chemical potential of hard-core chain molecules, the distributions of $\Delta\mathcal{U}$ will always overlap (namely, at $\Delta\mathcal{U} = 0$), even in the regime where the method cannot be trusted. Here, we shall describe an overlapping distribution method based on histograms of Rosenbluth weights [305]. This method will prove to be a useful diagnostic tool.

Consider again a model with internal potential energy u_{int} and external potential energy u_{ext}. In what follows, we shall compare two systems. The first, denoted by 0, contains N chain molecules ($N \geq 0$). The second system, denoted by 1, contains $N+1$ chain molecules. In addition, both systems may contain a fixed number of other (solvent) molecules. Let us first consider system 1. Around every segment j of a particular chain molecule (say, i), we can generate $k-1$ trial directions according to an internal probability distribution given by equation (11.2.19). Note that the set does *not* include the actual orientation of segment j. We denote this set of trial orientations by

$$\{\gamma_{rest}(j)\} \equiv \prod_{j'=1}^{k-1} \{\gamma\}_{j'},$$

where the subscript rest indicates that this set excludes the actual segment j. The probability of generating this set of trial directions is given by $P_{rest}(j)$, given by equation (11.2.19). Having thus constructed an umbrella of trial directions around every segment $1 \leq j \leq \ell$, we can compute the Rosenbluth weight $\mathcal{W}_i$ of molecule i. Clearly, $\mathcal{W}_i$ depends on all coordinates of the remaining N molecules (for convenience, we assume that we are dealing with a neat liquid), on the position $\mathbf{r}_i$ and conformation Γ_i of molecule i, and on the ℓ sets of $k-1$ trial directions:

$$\{\boldsymbol{\Gamma}_{rest}\} \equiv \prod_{j=1}^{\ell} \{\gamma\}_{rest}(j).$$

We now define a quantity x through

$$x \equiv \ln \mathcal{W}_i(\mathbf{Q}^{N+1}, \{\boldsymbol{\Gamma}_{rest}\}),$$

where we use $\mathbf{Q}$ to denote the translational coordinates $\mathbf{r}$ and conformational coordinates Γ of a molecule. Next, consider the expression for the probability density of x, $p_1(x)$:

$$p_1(x) = \frac{\int d\mathbf{Q}^{N+1} d\{\boldsymbol{\Gamma}_{rest}\} \exp\left[-\beta\mathcal{U}(\mathbf{Q}^{N+1})\right] \prod_{j=1}^{\ell} P_{rest}(j)\delta(x - \ln \mathcal{W}_i)}{Z_{N+1}},$$

where

$$\begin{aligned} Z_{N+1} &= \int d\mathbf{Q}^{N+1} d\{\boldsymbol{\Gamma}_{rest}\} \exp\left[-\beta\mathcal{U}(\mathbf{Q}^{N+1})\right] \prod_{j=1}^{\ell} P_{rest}(j) \\ &= \int \cdots \int d\mathbf{Q}^{N+1} \exp\left[-\beta\mathcal{U}(\mathbf{Q}^{N+1})\right]. \end{aligned}$$

The second line of this equation follows from the fact that all $P_{int}(j)$ are normalized. We shall now try to relate $p_1(x)$ to an average in system 0 (i.e., the system containing only N chain molecules). To this end, we write $\mathcal{U}(\mathbf{Q}^{N+1})$ as $\mathcal{U}(\mathbf{Q}^N) = u_{ex}(\mathbf{Q}^N, \mathbf{Q}_i) + u_{int}(\mathbf{Q}_i)$. Second, we use the fact that

$$\exp\left[-\beta u_{int}(i)\right] = Z_{id} \times \prod_{j=1}^{\ell} P_{int}(j),$$

where

$$Z_{id} \equiv \int d\Gamma_i \prod_{j=1}^{\ell} \exp\left[-\beta u_{int}(j)\right].$$

Our expression for $p_1(x)$ now becomes

$$\begin{aligned} p_1(x) &= \frac{Z_{id}}{Z_{N+1}} \int d\mathbf{Q}^N d\mathbf{r}_i d\{\boldsymbol{\Gamma}_{trial}\} \exp\left[-\beta\mathcal{U}(\mathbf{Q}^N)\right] \\ &\quad \times \prod_{j=1}^{\ell} P_{trial}(j) \exp\left[-\beta u_{ex}(j)\right] \delta(x - \ln \mathcal{W}_i). \end{aligned}$$

We use the symbol $\{\Gamma_{\text{trial}}\}$ to denote the set of *all* trial segments, that is, the "umbrella" of trial directions around all segments of the chain molecule, *plus the segments themselves*. Next, every term $\exp(-\beta u_{\text{ex}}(j))$ is multiplied and divided by Z_j, defined as

$$Z_j \equiv \sum_{j'=1}^{k} \exp(-\beta u_{\text{ext}}(j')).$$

This allows us to write, for $p_1(x)$,

$$\begin{aligned} p_1(x) &= \frac{VZ_{\text{id}}}{Z_{N+1}} \\ &\times \sum_{\text{trials}} \int d\mathbf{Q}^N d\mathbf{s}_i d\{\Gamma_{\text{trial}}\} \exp\left[-\beta \mathcal{U}(\mathbf{Q}^N)\right] P_{\text{sel}}(\mathbf{Q}_i) \mathcal{W}_i \delta\left(x - \ln \mathcal{W}_i\right), \end{aligned}$$

where we have transformed from real coordinates $\mathbf{r}_i$ to scaled coordinates $\mathbf{s}_i$ by factoring out V, the volume of the system. Here, $P_{\text{sel}}(\mathbf{Q}_i)$ denotes the probability of selecting the actual conformation of the molecule from the given set of trial segments according to the rule given in equation (11.2.20). Finally, we multiply and divide by Z_N and employ the fact that the δ function ensures that $\mathcal{W}_i = \exp(x)$:

$$\begin{aligned} p_1(x) &= e^x \left(\frac{VZ_{\text{id}} Z_N}{Z_{N+1}} \right) \\ &\times \frac{\sum_{\text{trials}} \int d\mathbf{Q}^N d\mathbf{s}_i d\{\Gamma_{\text{trial}}\} \exp(-\beta \mathcal{U}(\mathbf{Q}^N)) P_{\text{sel}}(\mathbf{Q}_i) \delta(x - \ln \mathcal{W}_i)}{Z_N}. \end{aligned}$$

Finally, we obtain

$$p_1(x) = e^x \frac{VZ_{\text{id}} Z_N}{Z_{N+1}} p_0(x)$$

or

$$\ln p_1(x) = x + \beta \mu_{\text{ex}} + \ln p_0(x).$$

Hence, by constructing a histogram of $\ln w$ both in system 0 (with N chains) and in system 1 (with N + 1 chains), we can derive the excess chemical potential of the chain molecules by studying $\ln p_1(x) - \ln p_0(x)$. As in the original Bennett/Shing-Gubbins scheme [182–184], the method works only if there is a range of x values where we have good statistics on both $p_1(x)$ and $p_0(x)$. The advantage of this overlapping distribution scheme over the simple Rosenbluth particle insertion method is that, with the present method, sampling problems for long chains will manifest themselves as a breakdown of the overlap of p_0 and p_1. Figure I.1 shows an example of an application of this overlapping distribution method to hard-sphere polymers.

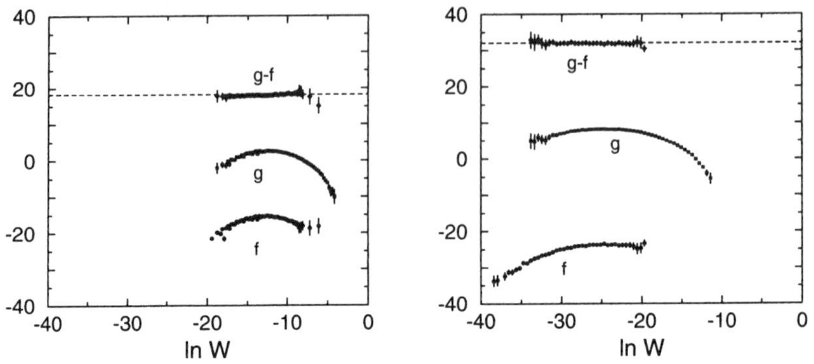

Figure I.1: The functions $f(\ln \mathcal{W}) \equiv p_0(\ln \mathcal{W}) + \frac{1}{2}\ln \mathcal{W}$, $g(\ln \mathcal{W}) \equiv p_1(\ln \mathcal{W}) - \frac{1}{2}\ln \mathcal{W}$ for fully flexible chains of hard sphere of length (left) $\ell = 8$ and (right) $\ell = 14$ in a hard-sphere fluid at density $\rho\sigma^3 = 0.4$. Note that the overlap between the distributions decreases as the chains become longer. The difference $g(\ln \mathcal{W}) - f(\ln \mathcal{W})$ is the overlapping distribution estimated for $\beta\mu^{ex}$. For the sake of comparison, we also show the value for $\beta\mu^{ex}$, obtained using the Rosenbluth test particle insertion method (dashed lines).

Appendix J

Some General Purpose Algorithms

This Appendix describes a few algorithms used in the main text.

Algorithm 41 (Selection of Trial Orientations)

```
SUBROUTINE select(w,sumw,n)        selects a trial position with
                                   prob. p(i) = w(i)/∑_j w(j)
ws=ranf()*sumw
cumw=w(1)
n=1
do while (cumw.lt.ws)
   n=n+1
   cumw=cumw+w(n)
enddo
return                             n is the selected trial position
end
```

Comments to this algorithm:

1. *This subroutine selects a trial position with probability*

$$p(i) = \frac{w(i)}{\sum_{j=1}^{k} w(j)} = \frac{w(i)}{\text{sumw}}.$$

2. *Note that for large values of* k *bisection [33] can be more efficient.*

Algorithm 42 (Random Vector on a Unit Sphere)

```
SUBROUTINE ranor(bx,by,bz)          generates a random vector
ransq=2.                            on a unit sphere
do while (ransq.ge.1)
   ran1=1.-2.*ranf()
   ran2=1.-2.*ranf()
   ransq=ran1*ran1+ran2*ran2
enddo
ranh=2.*sqrt(1.-ransq)
bx=ran1*ranh
by=ran2*ranh
bz=(1-2.*ransq)
return
end
```

Comment to this algorithm:

1. *The algorithm is based on [19] (p. 349).*

Algorithm 43 (Generate Bond Length with Harmonic Springs)

```
SUBROUTINE bondl(l)              generate the bond length
                                 assume harmonic springs.
sigma=sqrt(1./(beta*kv))         β = 1/kBT
a=(l0+3.*sigma)**2
ready=.false.
do while (.not.ready)
  call gauss(sigma,l0,l)         generate l with a Gaussian distr.
  if (ranf().le.l*l/a)           with mean l0 and
+    ready=.true.                standard deviation sigma
enddo                            correct not being truly Gaussian
return
end
```

Comment to this algorithm:

1. *The bond length has the following distribution:*

$$p(l) \propto \exp[-\beta 0.5 k_\nu (l - l_0)^2] \mathbf{dl} \propto l^2 \exp[-\beta 0.5 k_\nu (l - l_0)^2] dl.$$

This distribution is close to a Gaussian generated by `gauss` *(Algorithm 44) but not identical; we "correct" the Gaussian distribution for this, using the method described in [19] (p. 349).*

Algorithm 44 (Gaussian Distribution)

```
SUBROUTINE gauss(sigma,l0,l)        generate l Gaussian distributed
                                     with mean l_0 and standard
                                     deviation σ
r=2.
do while (r.ge.1.)
   v1=2.*ranf()-1.
   v2=2.*ranf()-1.
   r=v1*v1+v2*v2
enddo
l=v1*sqrt(-2.*log(r)/(r))            l has now zero mean
l=l0+sigma*l                         and unit variance
return
end
```

Comment to this algorithm:

1. *This algorithm is based on "numerical recipes" [33] (p. 203).*

Algorithm 45 (Generate Bond Angle)

```
 SUBROUTINE bonda(xn,b,i)            generate a bond angle pre-
                                     scribed by the internal inter-
 ready=.false.                       actions
 do while (.not.ready)
   call ranor(b)                     unit vector on a sphere
   dx1x2=xn(i-1)-xn(i-2)             vector r_21 = r_{i-1} - r_{i-2}
   dx1x2=dx1x2/|dx1x2|               normalize vector
   phi=acos(b • dx1x2)               calculate the bond angle φ
   ubb=ubb(phi)                      bond-bending energy
   if (ranf().lt.exp                 rejection test
+    (-beta*ubb))ready=.true.
 enddo
 return
 end
```

Comment to this algorithm:

1. *The algorithm uses the "acceptance-rejection" method (see [33]). The subroutine* `ranor` *generates a random vector on a unit sphere (Algorithm 42), the function* `ubb` *gives the bond-bending energy for the given angle.*

Algorithm 46 (Generate Bond and Torsion Angle)

`SUBROUTINE tors_bonda(xn,b,i)`	generate a vector on unit sphere with torsional angle and bond-bending angle prescribed by torsion and bond-bending potentials
`ready=.false.`	
`do while (.not.ready)`	
`  call ranor(b)`	generate unit vector on a sphere
`  dx1x2=xn(i-1)-xn(i-2)`	vector $\mathbf{r}_{21} = \mathbf{r}_{i-1} - \mathbf{r}_{i-2}$
`  dx1x2=dx1x2/|dx1x2|`	normalize vector
`  dx2x3=xn(i-2)-xn(i-3)`	vector $\mathbf{r}_{23} = \mathbf{r}_{i-2} - \mathbf{r}_{i-3}$
`  dx2x3=dx2x3/|dx2x3|`	normalize vector
`  phi=acos(b • dx1x2)`	calculate the bond angle ϕ
`  ubb=ubb(phi)`	bond-bending energy
`  xx1=b × dx1x2`	cross product: $\mathbf{b}$ and $\mathbf{r}_{21}$
`  xx2=dx1x2 × dx2x3`	cross product: $\mathbf{r}_{21}$ and $\mathbf{r}_{23}$
`  theta=acos(xx1 • xx2)`	
`  utors=utors(theta)`	determine torsion energy
`  usum=ubb+utors`	
`  if (ranf().lt.exp`	rejection test
`+   (-beta*usum))ready=.true.`	
`enddo`	
`return`	
`end`	

Comments to this algorithm:

1. *The algorithm uses the "acceptance-rejection" method (see [33]).*
2. *In the literature, the torsion angle is sometimes defined differently.*
3. *The subroutine* `ranor` *generates a random vector on a unit sphere (Algorithm 42), and the function* `ubb` *gives the bond-bending energy for the given angle ϕ, and the function* `utors` *gives the torsion for the angle θ.*

Appendix K

Small Research Projects

In this Appendix we list a few small research projects. These projects involve the development of your own program. A possible strategy is to use the source code of one of the Case Studies as the starting point. It is our experience that the completion of such a research project takes about two weeks, depending on your experience.

K.1 Adsorption in Porous Media

In this project we will investigate the adsorption behavior in porous media. As a model we use a slit-like pore. The interactions with the pore are given by

$$U(z) = \begin{cases} 0 & 0 < z < L \\ \infty & \text{otherwise} \end{cases}, \tag{K.1.1}$$

where L is the width of the slit. We will investigate the adsorption of methane, which we model with a Lennard-Jones potential. The starting point is Case Study 9, which uses a program to simulate the Lennard-Jones fluid in the grand-canonical ensemble. The project is to develop a program to simulate an adsorption isotherm of methane in the slit-like pore. Before starting to program you may want to think about the following points:

1. What is the geometry of the system and how should one apply the periodic boundary conditions?
2. What is the type of Lennard-Jones potential (truncated, truncated and shifted, with or without tail corrections)? And what are the parameters for modeling methane?
3. How is an adsorption isotherm defined thermodynamically?

With your program, try to answer the following questions:

1. Compute the density profile (density as function on the distance z between the plates) for $L = 1, 2, 5$, and 10 for $\rho = 0.7$ and $T = 2.0$. The total number of particles should be of the order of 100 to 500 for the widest slit.
2. Compute the excess chemical potential and chemical potential as a function of the distance between the plates for $L = 1, 2, 5$, and 10 for $\rho = 0.6$ and $T = 2.0$. Try to explain the differences.
3. Compute the adsorption isotherms for $L = 2$ and 5 for $T = 2.0$ and 0.8. Try to explain the results.

K.2 Transport Properties in Liquids

Molecular Dynamics simulations can be used to compute the transport properties of liquids. Examples of these transport properties are the (self-) diffusion coefficient and the viscosity. In Case Study 5 the results for the diffusion coefficient are shown. In this project we extend these results to mixtures of Lennard-Jones (LJ) molecules. Before starting to program you may want to think about the following points:

1. The generalization of the diffusion coefficient from a pure component to a mixture is not trivial. Try to find in the literature how one should define a diffusion coefficient of a mixture and how can one compute this in a simulation. See, for example, refs. [79,534].
2. What is the type of LJ potential (truncated, truncated and shifted, with or without tail corrections)? And what are the parameters for modeling argon and krypton?

With your program, try to answer the following questions:

1. Compute the pressure, viscosity, and diffusion coefficient of the LJ fluid at $T = 1.0, 1.5$, and 2.0 for $\rho = 0.7$.
2. Compute the diffusion coefficients D_{11}, D_{12}, and D_{22} for a mixture of 50% -50% LJ particles in which the components 1 and 2 have the same interactions ($\sigma_{12} = \sigma_{11} = \sigma_{22}$ and $\epsilon_{12} = \epsilon_{11} = \epsilon_{12}$) but carry a different color. This means that the particles are labeled. Experimentally this could be done by radioactive labeling.
3. Compute the pressure, viscosity, and diffusion coefficient of the LJ fluid at $T = 1.0, 1.5$, and 2.0 for $\rho = 0.7$. Compute the diffusion coefficients D_{11}, D_{12}, and D_{22} for a mixture of 50-50% LJ particles, but now for a system of 50% argon and 50% krypton (use the parameters of

argon to compute the reduced temperatures). Does the Einstein equation for the diffusivity hold for this system?

K.3 Diffusion in a Porous Media

The behavior of liquid in confined geometries is different from the behavior in the bulk liquid. In the project we investigate the diffusion coefficient in a cylindrical pore. The starting point is Case Study 5 in which the dynamic properties of a Lennard-Jones fluid are simulated. This Case Study can be used as a starting point for our study. The interactions with the walls are described with a repulsive potential:

$$U(r) = \begin{cases} \epsilon \left(\frac{\sigma}{L-r}\right)^{10} & 0 \leq r \leq L \\ \infty & r > L \end{cases}, \tag{K.3.1}$$

where L is a radius characterizing the size of the pore and the center of the cylinder is located at $r = 0$ and $\epsilon > 0$, $\sigma > 0$. Some questions that one should answer before one starts programming are the following:

1. Is the potential for the interactions with the walls appropriate for a Molecular Dynamics simulation?
2. What is the volume of the pore as a function of the parameter L?
3. What is the dimension of our problem? Do we have diffusion in 1, 2, or 3 dimensions?

In the first part of the project we study the diffusion in a smooth pore as defined by the above potential as a function of the pore diameter.

1. Compute the diffusion coefficient of a bulk Lennard-Jones liquid for $\rho = 0.6$ and $T = 2.0$ and 1.5. Since the program uses an NVE ensemble, it is not possible to simulate at exactly the requested temperature. However, one can ensure to be close to this temperature by an appropriate equilibration of the system (this is also the case for the following two questions) during the first part of the MD simulation.
2. Compute the density as a function of the distance from the center of the pore for $\rho = 0.6$ and $T = 2.0$ and 1.5 and $L = 5$ and 2. Interpret the results.
3. Compute the diffusion coefficient for $\rho = 0.6$ and $T = 2.0$ and $T = 1.5$ and $L = 5$, 2, and 1. Interpret the results. The interpretation is not trivial.
4. The above calculations have been performed using the NVT ensemble. This implies that there is no coupling with the atoms of the walls. In a

real system the walls are not smooth and can exchange heat with the adsorbed molecules. A possible way of modeling this is to assume that we have an Andersen thermostat in the boundary layer with the wall. Investigate how the results depend on the thickness of the boundary layer and the constant ν of the Andersen algorithm.

The next step is to model the corrugation caused by the atoms. This corrugation could be a term:

$$U(z, r) = A \sin^2(\pi z/\sigma_w) \exp\left[-\left(\frac{r-L}{L_0}\right)^2\right], \tag{K.3.2}$$

where z is the distance to the wall, and σ_w is a term characterizing the size of the atoms of the wall and A is the strength of the interaction. The exponential is added to ensure that the potential is localized close to the walls of the cylinder. Investigate the diffusion coefficient as a function of the parameters σ_w and A both in the NVE and in the Andersen thermostat cases.

K.4 Multiple-Time-Step Integrators

The time step in a Molecular Dynamics simulation strongly depends on the steepness of the potential energy surface. However, most potentials like the Lennard-Jones potential are steep at short distances. As short-range interactions can be computed very fast, it would be interesting to use a multiple-time-step integration algorithm, in which short-range (computationally cheap) interactions are computed every time step and in which long-range (computationally expensive) interactions are evaluated every n time steps ($n > 1$). Recently, there has been a considerable effort to construct time-reversible multiple-time-step algorithms [71,85].

1. Why is it important to use time-reversible integration schemes in MD?
2. Modify Case Study 4 in such a way that pairwise interactions are calculated using a Verlet neighbor list. For every particle, a list is made of neighboring particles within a distance of $r_{cut} + \Delta$. All lists have to be updated only when the displacement of a single particle is larger than $\Delta/2$. Hint: The algorithm on microfiche F.19 of the book of Allen and Tildesley [19,535] is a good starting point.
3. Investigate how the CPU time per time step depends on the size of Δ for various system sizes. Compare your results with Table 5.1 from ref. [19].
4. Modify the code in such a way that the NVE multiple-time-step algorithm of ref. [85] is used to integrate the equations of motion. You will

have to use separate neighbor lists for the short-range and the long-range parts of the potential.

5. Why does one have to use a switching function in this algorithm? Why is it a good idea to use a linear interpolation scheme to compute the switching function from ref. [85]?
6. Make a detailed comparison between this algorithm and the standard Leap-Frog integrator (with the use of a neighbor list) at the same energy drift.

K.5 Thermodynamic Integration

The difference in free energy between state A and state B can be calculated by thermodynamic integration:

$$F_A - F_B = \int_{\lambda=0}^{\lambda=1} d\lambda \left\langle \frac{\partial U(\lambda)}{\partial \lambda} \right\rangle_{\lambda}, \tag{K.5.1}$$

in which $\lambda = 1$ in state A and $\lambda = 0$ in state B. In order to calculate the excess chemical potential of a Lennard-Jones system, we might use the following modified potential [536]:

$$U(r, \lambda) = 4\epsilon \left(\lambda^5 \left(\frac{\sigma}{r}\right)^{12} - \lambda^3 \left(\frac{\sigma}{r}\right)^{6} \right). \tag{K.5.2}$$

Recall that the excess chemical potential is the difference in chemical potential between a real gas ($\lambda = 1$) and an ideal gas ($\lambda = 0$).

1. Make a plot of the modified LJ potential for various values of λ.
2. Show that

$$\left(\frac{\partial^2 F}{\partial \lambda^2} \right) < 0 \tag{K.5.3}$$

   when

$$U = U_0 \lambda + U_1 (1 - \lambda). \tag{K.5.4}$$

3. Derive equation (K.5.1).
4. Modify the code of Case Study 1 for this modified potential.
5. Perform the thermodynamic integration and compare your results with the conventional particle insertion method.
6. Calculate the chemical potential as a function of the density by scaling σ.

Appendix L

Hints for Programming

The official rules for writing a program are that it should involve three persons. One for the design, one for the implementation, and one for the testing of it. Most probably you are doing all steps on your own! The reason there is a need for three persons is that if you do not realize a certain aspect in the design phase, it is very likely that you do not think about this either while implementing or testing a program. The following tips may be useful in developing a program:

- First try to understand every line of the starting code. If you do not understand the starting point, it is impossible to make modifications.
- Try to develop a modification plan consisting of the following steps:
  1. Make a copy of the program in a new directory.
  2. Describe in words why and how a certain subroutine needs to be modified.
  3. Start the programming by writing comment lines describing what the modifications are and make the modifications according to your written instructions. If you find during the implementations that the original ideas were not correct, describe this in your notes. This is to ensure that you think before you do. This may sound obvious and it is, but as you may find out it is a very difficult rule to stick to.
  4. Do not start modifying the entire program. Try to do it in steps. Try to test the modifications as soon as possible. For example, if the program is a Gibbs ensemble simulation you probably have to modify the particle displacement routine, the volume change, and the particle exchange. However, there is no need to do this in all subroutines at the same time. First modify the particle displacement and make all the tests you can (for example, energy

conservation) without using the volume displacement and particle exchange subroutines, before you continue to the next subroutine.

5. The first time you run the modified code, you might want to use array-bound checking (usually, this option is not used by default). This will slow down the code, but it is very useful for detecting errors. It is also very useful to use code checkers like, for example, ftnchek [537].
6. Try to find limiting cases for which your program gives known results. For example, if you have written a program for a mixture the results should be identical to the pure component results if all interaction parameters are taken to be identical. Or does the program give in the ideal gas limit the same results as the ideal gas results that can be obtained from theory? This is, of course, not alway possible, but if it is possible it is a good test case.
7. For a Molecular Dynamics program, it is a good idea to check the conservation of the total energy. In most systems, the total impulse should also be conserved (usually zero).
8. In a Monte Carlo program, one has to calculate energy differences for each trial move. It is a good idea to write a subroutine that rigorously calculates the total energy of the system without using any sophisticated tricks. In this way, the sum of the initial energy and all energy differences of all accepted trial moves should equal the total energy; any difference should only be due to the limited accuracy of the computer.
9. Especially for lattice simulations, the result of a MC simulation might be dependent on the random number generator that is used. In principle, the perfect random number generator does not exsist, but some are less worse than others. It is always a good idea to do a MC simulation with two different random number generators. See, for example, ref. [538] for a discussion about this topic.

- If you see numbers try to understand them and check whether they are reasonable. Since we use reduced units most properties are in the range $[-1;1]$; this implies that if we find -2.4 it is probably all right, but if we find 2.4×10^{18} you should be suspicious.

- If you are writing a parallel code, we strongly recommend you make a detailed workplan before you start programming. A common mistake is that you forget to pass system variables (for example, temperature or masses of the particles) to all processors.

Bibliography

[1] W.W. Wood. Early history of computer simulation in statistical mechanics. In G. Ciccotti and W.G. Hoover, editors, *Molecular Dynamics Simulations of Statistical Mechanics Systems*, pages 2–14. Proceedings of the 97th Int. "Enrico Fermi" School of Physics, North Holland, Amsterdam, 1986.

[2] G. Ciccotti, D. Frenkel, and I.R. McDonald. *Simulation of Liquids and Solids: Molecular Dynamics and Monte Carlo Methods in Statistical Mechanics*. North-Holland, Amsterdam, 1987.

[3] J.A. Prins. Onze voorstelling omtrent de bouw van de stof. *Physica*, 8:257–268, 1928.

[4] O.K. Rice. On the statistical mechanics of liquids, and the gas of hard elastic spheres. *J. Chem. Phys.*, 12:1–18, 1944.

[5] J.D. Bernal. The Bakerian lecture, 1962: The structure of liquids. *Proc. R. Soc.*, 280:299–322, 1964.

[6] N. Metropolis, A.W. Rosenbluth, M.N. Rosenbluth, A.N. Teller, and E. Teller. Equation of state calculations by fast computing machines. *J. Chem. Phys.*, 21:1087–1092, 1953.

[7] N. Metropolis. The begining of the Monte Carlo method. *Los Alamos Science*, 12:125–130, 1987.

[8] E. Fermi, J.G. Pasta, and S.M. Ulam. Studies of non-linear problems. LASL Report LA-1940, 1955.

[9] B.J. Alder and T.E. Wainwright. Molecular dynamics by electronic computers. In I. Progigine, editor, *Proc. of the Int. Symp. on Statistical Mechanical Theory of Transport Processes (Brussels, 1956)*, pages 97–131. Interscience, Wiley, New York, 1958.

[10] J.B. Gibson, A.N. Goland, M. Milgram, and G.-H. Vineyard. Dynamics of radiation damage. *Phys. Rev.*, 120:1229–1253, 1960.

[11] G.-H. Vineyard. Autobiographical remarks of G.-H. Vineyard. In P.C. Gehlen, J.R. Beeler, and R.I. Jaffe, editors, *Interatomic Potentials and Simulation of Lattice Defects*, pages xiii–xvi. Plenum, New York, 1972.

[12] A. Rahman. Correlations in the motion of atoms in liquid argon. *Phys. Rev.*, 136:A405–A411, 1964.

[13] L. Verlet. Computer "experiments" on classical fluids. I. Thermodynamical properties of Lennard-Jones molecules. *Phys. Rev.*, 159:98–103, 1967.

[14] J.A. Barker and R.O. Watts. Structure of water: A Monte Carlo calculation. *Chem. Phys. Lett.*, 3:144–145, 1969.

[15] I.R. McDonald and K. Singer. Calculation of the thermodynamic properties of liquid argon from Lennard-Jones parameters by a Monte Carlo method. *Discuss. Faraday Soc.*, 43:40–49, 1967.

[16] P.N. Vorontsov-Vel'yaminov, A.M. El'yashevich, and A.K. Kron. Theoretical investigation of the thermodynamics properties of solutions of strong electrolytes by the Monte Carlo method. *Elektrokhimiya*, 2:708–716, 1966.

[17] B.J. Alder and T.E. Wainwright. Phase transition for a hard sphere system. *J. Chem. Phys.*, 27:1208–1209, 1957.

[18] W.W. Wood and J.D. Jacobson. Preliminary results from a recalculation of the Monte Carlo equation of state of hard-spheres. *J. Chem. Phys.*, 27:1207–1208, 1957.

[19] M.P. Allen and D.J. Tildesley. *Computer Simulation of Liquids*. Clarendon Press, Oxford, 1987.

[20] J.M. Haile. *Molecular Dynamics Simulations: Elementary Methods*. Wiley, New York, 1992.

[21] D.P Landau and K. Binder. *A Guide to Monte Carlo Simulation in Statistical Physics*. Cambridge University Press, Cambridge, 2000.

[22] D.C. Rapaport. *The Art of Molecular Dynamics Simulation*. Cambridge University Press, Cambridge, 1995.

[23] M.E.J. Newman and G.T. Barkema. *Monte Carlo Methods in Statistical Physics*. Oxford University Press, Oxford, 1999.

[24] R.W. Hockney and J.W. Eastwood. *Computer Simulations Using Particles*. McGraw-Hill, New York, 1981.

[25] W.G. Hoover. *Molecular Dynamics*. Springer, Berlin, 1986.

[26] W.G. Hoover. *Computational Physics Statistical Mechanics*. Elsevier Sci. Publ., Amsterdam, 1991.

[27] F.J. Vesely. *Computational Physics. An Introduction*. Plenum, New York, 1994.

[28] D.W. Heermann. *Computer Simulation Methods in Theoretical Physics*. Springer, Berlin, 1990.

[29] D.J. Evans and G.P. Morriss. *Statistical Mechanics of Non-Equilibrium Liquids*. Academic Press, London, 1990.

[30] S.E. Koonin. *Computational Physics*. Benjamin/Cummings, Menlo Park, Calif., 1986.

[31] H. Gould and J. Tobochnik. *Computer Simulation Methods, Vols. I and II*. Addison-Wesley, Reading, Mass., 1988.

[32] M.H. Kalos and P.A. Whitlock. *Monte Carlo Methods*. Wiley, New York, 1986.

[33] W.H. Press, B.P. Flannery, S.A. Teukolsky, and W.T. Vetterling. *Numerical Recipes: The Art of Scientific Computing*. Cambridge University Press, Cambridge, 1986.

[34] J.P. Valleau and S.G. Whittington. A guide to Monte Carlo simulations for statistical mechanics: 1. Highways. In B.J. Berne, editor, *Statistical Mechanics, Part A*, pages 137–168. Plenum, New York, 1977.

[35] J.P. Valleau and S.G. Whittington. A guide to Monte Carlo simulations for statistical mechanics: 2. Byways. In B.J. Berne, editor, *Statistical Mechanics, Part A*, pages 169–194. Plenum, New York, 1977.

[36] K. Binder. *Applications of the Monte Carlo Method in Statistical Physics*. Springer, Berlin, 1984.

[37] K. Binder. *The Monte Carlo Method in Condensed Matter Physics*. Springer, Berlin, 1992.

[38] O.G. Mouritsen. *Computer Studies of Phase Transitions and Critical Phenomena*. Springer, Berlin, 1984.

[39] G. Ciccotti and W.G. Hoover. *Molecular-Dynamics Simulations of Statistical-Mechanical Systems. Proceedings of the 97th International "Enrico Fermi" School of Physics*. North-Holland, Amsterdam, 1986.

[40] M. Meyer and V. Pontikis. *Proceedings of the NATO ASI on Computer Simulation in Materials Science*. Kluwer, Dordrecht, 1991.

[41] M.P. Allen and D.J. Tildesley. *Proceedings of the NATO ASI on Computer Simulation in Chemical Physics*. Kluwer, Dordrecht, 1993.

[42] M. Baus, L.F. Rull, and J.P. Ryckaert. *Observation, Prediction and Simulation of Phase Transitions*. Kluwer, Dordrecht, 1995.

[43] R. Balian. *From Microphysics to Macrophysics*. Springer, Berlin, 1991.

[44] L.E. Reichl. *A Modern Course in Statistical Physics*. University of Texas Press, Austin, 1980.

[45] H. Goldstein. *Classical Mechanics*. Addison-Wesley, Reading, 2nd edition, 1980.

[46] H.L. Anderson. Scientific uses of the MANIAC. *J. Stat. Phys.*, 43:731–748, 1986.

[47] N.G. van Kampen. *Stochastic Processes in Physics and Chemistry*. North-Holland, Amsterdam, 1981.

[48] K.A. Fichthorn and W.H. Weinberg. Theoretical foundations of dynamical Monte Carlo simulations. *J. Chem. Phys.*, 95:1090–1096, 1991.

[49] W.W. Wood and F.R. Parker. Monte Carlo equation of state of molecular interactions with Lennard-Jones potential. I. A supercritical isotherm at about twice the critical temperature. *J. Chem. Phys.*, 27:720–733, 1957.

[50] M.J. Mandel. On the properties of a periodic fluid. *J. Stat. Phys.*, 15:299–305, 1976.

[51] J.G. Powles. The liquid-vapour coexistence line for Lennard-Jones-type fluids. *Physica*, 126A:289–299, 1984.

[52] B. Smit and D. Frenkel. Vapour-liquid equilibria of the two dimensional Lennard-Jones fluid(s). *J. Chem. Phys.*, 94:5663–5668, 1991.

[53] B. Smit. Phase diagrams of Lennard-Jones fluids. *J. Chem. Phys.*, 96:8639–8640, 1992.

[54] V.I. Manousiouthakis and M.W. Deem. Strickt detiled balance is unnecessary in Monte Carlo simulation. *J. Chem. Phys.*, 110:2753–2756, 1999.

[55] W.G.T. Kranendonk and D. Frenkel. Simulation of the adhesive-hard-sphere model. *Mol. Phys.*, 64:403–424, 1988.

[56] R.D. Mountain and D. Thirumalai. Quantative measure of efficiency of Monte Carlo simulations. *Physica A*, 210:453–460, 1994.

[57] F.J. Vesely. Angular Monte Carlo integration using quaternion parameters: A spherical reference potential for CCl_4. *J. Comp. Phys.*, 47:291–296, 1982.

[58] M. Fixman. Classical statistical mechanics of constraints: A theorem and application to polymers. *Proc. Natl. Acad. Sci. USA*, 71:3050–3053, 1974.

[59] G. Ciccotti and J.P. Ryckaert. Molecular dynamics simulation of rigid molecules. *Comp. Phys. Rep.*, 4:345–392, 1986.

[60] L.R. Dodd, T.D. Boone, and D.N. Theodorou. A concerted rotation algorithm for atomistic Monte Carlo simulation of polymer melts and glasses. *Mol. Phys.*, 78:961–996, 1993.

[61] J.J. Nicolas, K.E. Gubbins, W.B. Streett, and D.J. Tildesley. Equation of state for the Lennard-Jones fluid. *Mol. Phys.*, 37:1429–1454, 1979.

[62] J.K. Johnson, J.A. Zollweg, and K.E. Gubbins. The Lennard-Jones equation of state revisited. *Mol. Phys.*, 78:591–618, 1993.

[63] Z.W. Salsburg and W.W. Wood. Equation of state of classical hard spheres at high density. *J. Chem. Phys.*, 37:798–804, 1962.

[64] P. Diaconis, S. Holmes, and R.M. Neal. Analysis of a non-reversible Markov chain sampler. Technical Report BU-1385-M, Biometrics Unit, Cornell University, 1997.

[65] M.A. Miller, L.M. Amon, and W.P. Reinhardt. Should one adjust the maximum step size in a Metropolis Monte Carlo simulation? *Chem. Phys. Lett.*, 331:278–284, 2000.

[66] G.D. Quinlan and S. Tremaine. On the reliability of gravitational n-body integrations. *Mon. Not. R. Astron. Soc.*, 259:505–518, 1992.

[67] R.E. Gillilan and K.R. Wilson. Shadowing, rare events, and rubber bands - A variational Verlet algorithm for molecular-dynamics. *J. Chem. Phys.*, 97:1757–1772, 1992.

[68] S. Toxvaerd. Hamiltonians for discrete dynamics. *Phys. Rev. E*, 50:2271–2274, 1994.

[69] W.C. Swope, H.C. Andersen, P.H. Berens, and K.R. Wilson. A computer simulation method for the calculation of equilibrium constants for the formation of physical clusters of molecules: Application to small water clusters. *J. Chem. Phys.*, 76:637–649, 1982.

[70] H.J.C. Berendsen and W.F. van Gunsteren. Practical algorithms for dynamics simulations. In G. Ciccotti and W. G. Hoover, editors, *Molecular Dynamics Simulations of Statistical Mechanics Systems*, pages 43–65. Proceedings of the 97th Int. "Enrico Fermi" School of Physics, North Holland, Amsterdam, 1986.

[71] M.E. Tuckerman, B.J. Berne, and G.J. Martyna. Reversible multiple time scale molecular dynamics. *J. Chem. Phys.*, 97:1990–2001, 1992.

[72] J.C. Sexton and D.H. Weingarten. Hamiltonian evolution for the hybrid Monte Carlo algorithm. *Nucl. Phys. B*, 380:665–677, 1992.

[73] H. Yoshida. Symplectic integrators for hamiltonian systems: Basic theory. In S. Ferraz-Mello, editor, *Chaos, Resonance and Collective Dynamical Phenomena in the Solar System*, pages 407–411. Kluwer, Dordrecht, 1992.

[74] H. Yoshida. Recent progress in the theory and application of symplectic integrators. *Celest. Mech. Dyn. Astron.*, 56:27–43, 1993.

[75] P. Saha and S. Tremaine. Symplectic integrators for solar system dynamics. *Astron. J.*, 104:1633–1640, 1992.

[76] R. Olender and R. Elber. Calculation of classical trajectories with a very large time step: Formalism and numerical examples. *J. Chem. Phys.*, 105:9299–9315, 1996.

[77] R. Elber, J. Meller, and R. Olender. Stochastic path approach to compute atomically detailed trajectories: Application to the folding of c peptide. *J. Phys. Chem. B*, 103:899–911, 1999.

[78] L. Onsager and S. Machlup. Fluctuations and irreversible processes. *Phys. Rev.*, 91:1505–1512, 1953.

[79] J.-P. Hansen and I.R. McDonald. *Theory of Simple Liquids*. Academic Press, London, 2nd edition, 1986.

[80] J.L. Lebowitz, J.K. Percus, and L. Verlet. Ensemble dependence of fluctuations with application to machine computations. *Phys. Rev.*, 153:250–254, 1967.

[81] J.-P. Hansen and L. Verlet. Phase transitions of the Lennard-Jones system. *Phys. Rev.*, 184:151–161, 1969.

[82] A.J.C. Ladd and L.V. Woodcock. Triple-point coexistence properties of the Lennard-Jones system. *Chem. Phys. Lett.*, 51:155–159, 1977.

[83] A.J.C. Ladd and L.V. Woodcock. Interfacial and co-existence properties of the Lennard-Jones system at the triple point. *Mol. Phys.*, 36:611–619, 1978.

[84] H. Flyvbjerg and H.G. Petersen. Error estimates on averages of correlated data. *J. Chem. Phys.*, 91:461–466, 1989.

[85] G.J. Martyna, M.E. Tuckerman, D.J. Tobias, and M.L. Klein. Explicit reversible integrators for extended systems dynamics. *Mol. Phys.*, 87:1117–1157, 1996.

[86] J. Naghizadeh and S.A. Rice. Kinetic theory of dense fluids. X. measurement and interpretation of self-diffusion in liquid ar, kr, xe and ch_4. *J. Chem. Phys.*, 36:2710–2720, 1962.

[87] J.J. Erpenbeck and W.W. Wood. Molecular dynamics techniques for hard-core systems. In B.J. Berne, editor, *Statistical Mechanics, Part B*, pages 1–40. Plenum, New York, 1976.

[88] W.W. Wood. Monte Carlo calculations for hard disks in the isothermal-isobaric ensemble. *J. Chem. Phys.*, 48:415–434, 1968.

[89] I.R. McDonald. NpT-ensemble Monte Carlo calculations for binary liquid mixtures. *Mol. Phys.*, 23:41–58, 1972.

[90] R. Najafabadi and S. Yip. Observation of finite-temperature strain transformation (f.c.c. $\leftrightarrow$ b.c.c.) in Monte Carlo simulation of iron. *Scripta Metall.*, 17:1199–1204, 1983.

[91] G.E. Norman and V.S Filinov. Investigation of phase transitions by a Monte-Carlo method. *High Temp. (USSR)*, 7:216–222, 1969.

[92] D.J. Adams. Chemical potential of hard-sphere fluids by Monte Carlo methods. *Mol. Phys.*, 28:1241–1252, 1974.

[93] M. Creutz. Microcanonical Monte Carlo simulation. *Phys. Rev. Lett.*, 50:1411–1414, 1983.

[94] A.Z. Panagiotopoulos. Direct determination of phase coexistence properties of fluids by Monte Carlo simulation in a new ensemble. *Mol. Phys.*, 61:813–826, 1987.

[95] P. Attard. On the density of volume states in the isobaric ensemble. *J. Chem. Phys.*, 103:9884–9885, 1995.

[96] G.J.M. Koper and H. Reiss. Length scale for the constant pressure ensemble: Application to small systems and relation to einstein fluctuation theory. *J. Chem. Phys.*, 100:422–432, 1996.

[97] R. Eppinga and D. Frenkel. Monte Carlo study of the isotropic and nematic phases of infinitely thin hard platelets. *Mol. Phys.*, 52:1303–1334, 1984.

[98] K. Binder. Finite size scaling analysis of Ising model block distribution functions. *Z. Phys. B*, 43:119–140, 1981.

[99] M. Rovere, D.W. Hermann, and K. Binder. Block density distribution function analyses of two-dimensional Lennard-Jones fluids. *Europhys. Lett.*, 6:585–590, 1988.

[100] N.B. Wilding and A.D. Bruce. Density fluctuations and field mixing in the critical fluid. *J. Phys.: Condens. Matter*, 4:3087–3108, 1992.

[101] M. Rovere, P. Nielaba, and K. Binder. Simulation studies of gas-liquid transitions in two dimensions via subsystem-block-density distribution analysis. Z. Phys., 90:215–228, 1993.

[102] M. Parrinello and A. Rahman. Crystal structure and pair potentials: A molecular-dynamics study. *Phys. Rev. Lett.*, 45:1196–1199, 1980.

[103] M. Parrinello and A. Rahman. Polymorphic transitions in single crystals: A new molecular dynamics method. *J. Appl. Phys.*, 52:7182–7190, 1981.

[104] H.C. Andersen. Molecular dynamics at constant pressure and/or temperature. *J. Chem. Phys.*, 72:2384–2393, 1980.

[105] D.C. Wallace. Thermodynamic theory of stressed crystals and higher-order elastic constants. In H. Ehrenreich, F. Seitz, and D. Turnbull, editors, *Solid State Physics: Advances in Research and Applications*, pages 301–404. Academic Press, New York, 1970.

[106] D.J. Adams. Grand canonical ensemble Monte Carlo for a Lennard-Jones fluid. *Mol. Phys.*, 29:307–311, 1975.

[107] D.J. Adams. Calculating the low temperature vapour line by Monte Carlo. *Mol. Phys.*, 32:647–657, 1976.

[108] D.J. Adams. Calculating the high-temperature vapour line by Monte Carlo. *Mol. Phys.*, 37:211–221, 1979.

[109] L.A. Rowley, D. Nicholson, and N.G. Parsonage. Monte Carlo grand canonical ensemble calculation in a gas-liquid transition region for 12-6 argon. *J. Comp. Phys.*, 17:401–414, 1975.

[110] J. Yao, R.A. Greenkorn, and K.C. Chao. Monte Carlo simulation of the grand canonical ensemble. *Mol. Phys.*, 46:587–594, 1982.

[111] M. Mezei. A cavity-biased (T,V,μ) Monte Carlo method for the computer simulation of fluids. *Mol. Phys.*, 40:901–906, 1980.

[112] J.P. Valleau and L.K. Cohen. Primitive model electrolytes. I. Grand canonical Monte Carlo computations. *J. Chem. Phys.*, 72:5935–5941, 1980.

[113] W. van Megen and I.K. Snook. The grand canonical ensemble Monte Carlo method applied to the electrical double layer. *J. Chem. Phys.*, 73:4656–4662, 1980.

[114] H.J.F. Stroud, E. Richards, P. Limcharoen, and N.G. Parsonage. Thermodynamic study of the Linde 5A + methane system. *J. Chem. Soc., Faraday Trans. I*, 72:942–954, 1976.

[115] C.R.A Catlow. *Modelling of Structure and Reactivity in Zeolites*. Academic Press, London, 1992.

[116] L.D. Gelb, K.E. Gubbins, R. Radhkrishnan, and M. Sliwinska-Bartkowiak. Phase separation in confined systems. *Rep. Prog. Phys.*, 62:1573–1659, 1999.

[117] J.L. Soto and A.L. Myers. Monte Carlo studies of adsorption in molecular sieves. *Mol. Phys.*, 42:971–983, 1981.

[118] G.B. Wood and J.S. Rowlinson. Computer simulations of fluids in zeolite X and Y. *J. Chem. Soc., Faraday Trans. 2*, 85:765–781, 1989.

[119] S.J. Goodbody, K. Watanabe, D. MacGowan, J.P.R.B. Walton, and N. Quirke. Molecular simulations of methane and butane in silicalite. *J. Chem. Soc., Faraday Trans.*, 87:1951–1958, 1991.

[120] R.Q. Snurr, R.L. June, A.T. Bell, and D.N. Theodorou. Molecular simulations of methane adsorption in silicalite. *Mol. Sim.*, 8:73–92, 1991.

[121] F. Karavias and A.L. Myers. Isosteric heat of multicomponent adsorption: Thermodynamics and computer simulations. *Langmuir*, 7:3118–3126, 1991.

[122] P.R. van Tassel, H.T. Davis, and A.V. McCormick. Open-system Monte Carlo simulations of Xe in NaA. *J. Chem. Phys.*, 98:8919–8929, 1993.

[123] M.W. Maddox and J.S. Rowlinson. Computer simulation of the adsorption of a fluid mixture in zeolite Y. *J. Chem. Soc. Faraday Trans.*, 89:3619–3621, 1993.

[124] B. Smit. Simulating the adsorption isotherms of methane, ethane, and propane in the zeolite silicalite. *J. Phys. Chem.*, 99:5597–5603, 1995.

[125] L.F. Rull, G. Jackson, and B. Smit. The condition of microscopic reversibility in the Gibbs-ensemble Monte Carlo simulations of phase equilibria. *Mol. Phys.*, 85:435–447, 1995.

[126] S. Nosé. A unified formulation of the constant temperature molecular dynamics method. *J. Chem. Phys.*, 81:511–519, 1984.

[127] W. Feller. *An Introduction to Probability Theory and Its Applications, Vol I.* Wiley, New York, 1957.

[128] W. Feller. *An Introduction to Probability Theory and Its Applications, Vol II.* Wiley, New York, 1966.

[129] H. Tanaka, K. Nakanishi, and N. Watanabe. Constant temperature molecular dynamics calculation on Lennard-Jones fluid and its application to water. *J. Chem. Phys.*, 78:2626–2634, 1983.

[130] S. Nosé. A molecular dynamics method for simulation in the canonical ensemble. *Mol. Phys.*, 52:255–268, 1984.

[131] W.G. Hoover. Canonical dynamics: Equilibrium phase-space distributions. *Phys. Rev. A*, 31:1695–1697, 1985.

[132] W.G. Hoover. Constant pressure equations of motion. *Phys. Rev. A*, 34:2499–2500, 1986.

[133] S. Nosé. An extension of the canonical ensemble molecular dynamics method. *Mol. Phys.*, 57:187–191, 1986.

[134] S. Toxvaerd and O.H. Olsen. Canonical molecular dynamics of molecules with internal degrees of freedom. *Ber. Bunsenges. Phys. Chem.*, 94:274–278, 1990.

[135] M.E. Tuckerman, Yi Liu, G. Ciccotti, and G.J. Martyna. Non-Hamiltonian molecular dynamics: Generalizing Hamilton phase space principles to non-Hamiltonian systems. *J. Chem. Phys.*, 116:1678, 2001.

[136] G.J. Martyna, M.L. Klein, and M.E. Tuckerman. Nosé-Hoover chains: The canonical ensemble via continuous dynamics. *J. Chem. Phys.*, 97:2635–2645, 1992.

[137] M.E. Tuckerman and G.J. Martyna. Understanding modern molecular dynamics: Techniques and applications. *J. Phys. Chem. B*, 104:159–178, 2000.

[138] G.J. Martyna, D.J. Tobias, and M.L. Klein. Constant-pressure molecular-dynamics algorithms. *J. Chem. Phys.*, 101:4177–4189, 1994.

[139] H.J.C. Berendsen, J.P.M. Postma, W.F. van Gunsteren, A. DiNola, and J.R. Haak. Molecular dynamics with coupling to an external bath. *J. Chem. Phys.*, 81:3684–3690, 1984.

[140] K. Binder. *Monte Carlo Methods in Statistical Physics*. Springer, Berlin, 1979.

[141] B.J. Alder and T.E. Wainwright. Phase transition in elastic disks. *Phys. Rev.*, 127:359–361, 1962.

[142] S. Toxvaerd and E. Praestgaard. Molecular dynamics calculation of the liquid structure up to a solid surface. *J. Chem. Phys.*, 67:5291–5295, 1977.

[143] J.N. Cape and L.V. Woodcock. Molecular dynamics calculation of phase coexistence properties: The soft-sphere melting transition. *Chem. Phys. Lett.*, 59:271–274, 1978.

[144] Y. Hiwatari, E. Stoll, and T. Schneider. Molecular-dynamics investigation of solid-liquid coexistence. *J. Chem. Phys.*, 68:3401–3404, 1978.

[145] A. Ueda, J. Takada, and Y. Hiwatari. Molecular dynamics studies of solid-liquid interface of soft-core model. *J. Phys. Soc. Jpn.*, 50:307–314, 1981.

[146] A.Z. Panagiotopoulos. Adsorption and capillary condensation of fluids in cylindrical pores by Monte Carlo simulation in the Gibbs ensemble. *Mol. Phys.*, 62:701, 1987.

[147] A.Z. Panagiotopoulos, N. Quirke, M.R. Stapleton, and D.J. Tildesley. Phase equilibria by simulations in the Gibbs ensemble: Alternative derivation, generalization and application to mixtures and membrane equilibria. *Mol. Phys.*, 63:527–545, 1988.

[148] B. Smit, Ph. de Smedt, and D. Frenkel. Computer simulations in the Gibbs ensemble. *Mol. Phys.*, 68:931–950, 1989.

[149] D.A. Kofke. Gibbs-Duhem integration: A new method for direct evaluation of phase coexistence by molecular simulations. *Mol. Phys.*, 78:1331–1336, 1993.

[150] D.A. Kofke. Direct evaluation of phase coexistence by molecular simulation via integration along the coexistence line. *J. Chem. Phys.*, 98:4149–4162, 1993.

[151] R. Agrawal and D.A. Kofke. Solid-fluid coexistence for inverse-power potentials. *Phys. Rev. Lett.*, 74:122–125, 1995.

[152] D.A. Kofke and P.T. Cummings. Quantitative comparison and optimization of methods for evaluating the chemical potential by molecular simulation. *Mol. Phys.*, 92:973–996, 1997.

[153] S.W. de Leeuw, B. Smit, and C.P. Williams. Molecular dynamics studies of polar/nonpolar fluid mixtures: I. mixtures of Lennard-Jones and Stockmayer fluids. *J. Chem. Phys.*, 93:2704–2719, 1990.

[154] J.G. Kirkwood. Statistical mechanics of fluid mixtures. *J. Chem. Phys.*, 3:300–313, 1935.

[155] E.J. Meijer, D. Frenkel, R.A. LeSar, and A.J.C. Ladd. Location of melting point at 300 K of nitrogen by Monte Carlo simulation. *J. Chem. Phys.*, 92:7570–7575, 1990.

[156] D. Frenkel. Free energy computations and first order phase transitions. In G. Ciccotti and W.G. Hoover, editors, *Molecular Dynamics Simulations of Statistical Mechanics Systems*, pages 151–188. Proceedings of the 97th Int. "Enrico Fermi" School of Physics, North Holland, Amsterdam, 1986.

[157] E.J. Meijer and D. Frenkel. Melting line of Yukawa system by computer simulation. *J. Chem. Phys.*, 94:2269–2271, 1991.

[158] D. Frenkel and B.M. Mulder. The hard ellipsoid-of-revolution fluid I. Monte Carlo simulations. *Mol. Phys.*, 55:1171–1192, 1985.

[159] A. Stroobants, H.N.W. Lekkerkerker, and D. Frenkel. Evidence for one-, two-, and three-dimensional order in a system of hard parallel spherocylinders. *Phys. Rev. A*, 36:2929–2945, 1987.

[160] D. Frenkel, H.N.W. Lekkerkerker, and A. Stroobants. Thermodynamic stability of a smectic phase in a system of hard rods. *Nature*, 332:822–823, 1988.

[161] J.A.C. Veerman and D. Frenkel. Phase diagram of a system of hard spherocylinders by computer simulation. *Phys. Rev. A*, 41:3237–3244, 1990.

[162] J.A.C. Veerman and D. Frenkel. Relative stability of columnar crystalline phases in a system of parallel hard spherocylinders. *Phys. Rev. A*, 43:4334–4343, 1991.

[163] J.A.C. Veerman and D. Frenkel. Phase behaviour of disklike hard-core mesogens. *Phys. Rev. A*, 45:5632–5648, 1992.

[164] W.G.T. Kranendonk and D. Frenkel. Computer simulation of solid-liquid coexistence in binary hard-sphere mixtures. *Mol. Phys.*, 72:679–697, 1991.

[165] W.G.T. Kranendonk and D. Frenkel. Free energy calculations for solid solutions by computer simulations. *Mol. Phys.*, 72:715–733, 1991.

[166] M.D. Eldridge, P.A. Madden, and D. Frenkel. Entropy-driven formation of a superlattice in a hard-sphere binary mixture. *Nature*, 365:35–37, 1993.

[167] M.D. Eldridge, P.A. Madden, and D. Frenkel. The stability of the AB_{13} crystal in a binary hard sphere system. *Mol. Phys.*, 79:105–120, 1993.

[168] M.D. Eldridge, P.A. Madden, and D. Frenkel. A computer investigation into the stability of the AB_2 superlattice in a binary hard sphere system. *Mol. Phys.*, 80:987–995, 1993.

[169] W.F. van Gunsteren, P.K. Weiner, and A. J. Wilkinson. *Computer Simulation of Biomolecular Systems: Theoretical and Experimental Applications.* Escom, Leiden, 1993.

[170] M. Watanabe and W.P. Reinhardt. Direct dynamical calculation of entropy and free energy by adiabatic switching. *Phys. Rev. Lett.*, 65:3301–3304, 1990.

[171] T.P. Straatsma, H.J.C. Berendsen, and J.P.M. Postma. Free energy of hydrophobic hydration: A molecular dynamics study of noble gases in water. *J. Chem. Phys.*, 85:6720–6727, 1986.

[172] B. Widom. Some topics in the theory of fluids. *J. Chem. Phys.*, 39:2802–2812, 1963.

[173] B. Widom. Structure of interfaces from uniformity of the chemical potential. *J. Stat. Phys.*, 19:563–574, 1978.

[174] K.S. Shing. Infinite-dilution activity coefficients from computer simulation. *Chem. Phys. Lett.*, 119:149–151, 1985.

[175] P. Sindzingre, G. Ciccotti, C. Massobrio, and D. Frenkel. Partial enthalpies and related quantities in mixtures from computer simulation. *Chem. Phys. Lett.*, 136:35–41, 1987.

[176] R. Lustig. Statistical mechanics in the classical molecular dynamics ensemble. I. Fundamentals. *J. Chem. Phys.*, 101:3048–3059, 1994.

[177] R. Lustig. Statistical thermodynamics in the classical molecular dynamics ensemble. II. Application to computer simulation. *J. Chem. Phys.*, 101:3060–3067, 1994.

[178] U. Heinbruch and J. Fischer. On the application of Widom's test particle method to homogeneous and inhomogeneous fluids. *Mol. Sim.*, 1:109–120, 1987.

[179] J.I. Siepmann, I.R. McDonald, and D. Frenkel. Finite-size corrections to the chemical potential. *J. Phys.: Condens. Matter*, 4:679–691, 1992.

[180] K. Shing and S.T. Chung. Computer simulation methods for the calculation of the solubility in supercritical extraction systems. *J. Phys. Chem.*, 91:1674–1681, 1987.

[181] P. Sindzingre, C. Massobrio, G. Ciccotti, and D. Frenkel. Calculation of partial enthalpies of an argon-krypton mixture by NPT molecular dynamics. *Chem. Phys.*, 129:213–224, 1989.

[182] K.S. Shing and K.E. Gubbins. Free energy and vapour-liquid equilibria for a quadrupolar Lennard-Jones fluid. *Mol. Phys.*, 46:1109–1128, 1982.

[183] K.S. Shing and K.E. Gubbins. The chemical potential in non-ideal liquid mixtures: Computer simulation and theory. *Mol. Phys.*, 49:1121–1138, 1983.

[184] C.H. Bennett. Efficient estimation of free energy differences from Monte Carlo data. *J. Comp. Phys*, 22:245–268, 1976.

[185] A.M. Ferrenberg and R.H. Swendsen. Optimized Monte Carlo data analysis. *Phys. Rev. Lett.*, 63:1195–1198, 1989.

[186] G.M. Torrie and J.P. Valleau. Nonphysical sampling distributions in Monte Carlo free-energy estimation: Umbrella sampling. *J. Comp. Phys.*, 23:187–199, 1977.

[187] D. Chandler. *An Introduction to Modern Statistical Mechanics*. Oxford University Press, New York, 1987.

[188] J.P. Valleau. Density-scaling: A new Monte Carlo technique in statistical mechanics. *J. Comp. Phys.*, 96:193–216, 1991.

[189] J.P. Valleau. Monte Carlo: Choosing which game to play. In M. Meyer and V. Pontikis, editors, *Proceedings of the NATO ASI on Computer Simulation in Materials Science*, pages 67–84. Kluwer, Dordrecht, 1991.

[190] J.P. Valleau. Density-scaling Monte Carlo study of subcritical Lennard-Jonesium. *J. Chem. Phys.*, 99:4718–4728, 1977.

[191] I.R. McDonald and K. Singer. Machine calculation of thermodynamic properties of a simple fluid at supercritical temperatures. *J. Chem. Phys.*, 47:4766–4772, 1967.

[192] I.R. McDonald and K. Singer. Examination of the adequacy of the 12-6 potential for liquid argon by means of Monte Carlo calculations. *J. Chem. Phys.*, 50:2308–2315, 1969.

[193] C. Jarzynski. Non-equilibrium equality for free energy differences. *Phys. Rev. Lett.*, 78:2690–2693, 1997.

[194] C. Jarzynski. Equilibrium free-energy differences from non-equilibrium measurements: A master-equation approach. *Phys. Rev. E*, 56:5018–5035, 1997.

[195] G.E. Crooks. Non-equilibrium measurements of free energy differences for microscopically reversible markovian systems. *J. Stat. Phys.*, 90:1480–1487, 1997.

[196] G.E. Crooks. Entropy production fluctuation theorem and the non-equilibrium work relation for free-energy differences. *Phys. Rev. E*, 60:2721–2726, 1999.

[197] G.E. Crooks. Path-ensemble averages in systems driven far from equilibrium. *Phys. Rev. E*, 61:2361–2366, 2000.

[198] D.A. Kofke and E.D. Glandt. Monte Carlo simulation of multicomponent equilibria in a semigrand canonical ensemble. *Mol. Phys.*, 64:1105–1131, 1988.

[199] P Tilwani and D. Wu. Direct simulation of phase coexistence in solids using the Gibbs ensemble: Configuration annealing Monte Carlo. Master's thesis, Department of Chemical Engineering, Colorado School of Mines, Golden, Colorado, USA, 1999.

[200] P.G. Bolhuis and D. Frenkel. Prediction of an expanded-to-condensed transition in colloidal crystals. *Phys. Rev. Lett.*, 72:2211–2214, 1994.

[201] A.Z. Panagiotopoulos. Direct determination of fluid phase equilibria by simulation in the Gibbs ensemble: A review. *Mol. Sim.*, 9:1–23, 1992.

[202] D. Frenkel. Monte Carlo simulations. In C.R.A. Catlow, editor, *Computer Modelling of Fluids, Polymers and Solids*. NATO ASI, Kluwer, Dordrecht, 1990.

[203] B. Smit and D. Frenkel. Calculation of the chemical potential in the Gibbs ensemble. *Mol. Phys.*, 68:951–958, 1989.

[204] A. Lofti, J. Vrabec, and J. Fischer. Vapour liquid equilibria of the Lennard-Jones fluid from the NPT plus test particle method. *Mol. Phys.*, 76:1319–1333, 1992.

[205] B. Smit. *Computer Simulation of Phase Coexistence: from Atoms to Surfactants*. Ph.D. thesis, Rijksuniversiteit Utrecht, The Netherlands, 1990.

[206] J.S. Rowlinson and F.L. Swinton. *Liquids and Liquid Mixtures*. 3rd edn. Butterworth, London, 1982.

[207] J.S. Rowlinson and B. Widom. *Molecular Theory of Capillarity*. Clarendon Press, Oxford, 1982.

[208] A.Z. Panagiotopoulos. Molecular simulations of phase coexistence: Finite-size effects and the determination of critical parameters for two- and three dimensional Lennard-Jones fluids. *Int. J. Thermophys.*, 15:1057–1072, 1994.

[209] M. Rovere, D.W. Hermann, and K. Binder. The gas-liquid transition of the two-dimensional Lennard-Jones fluid. *J. Phys.: Condens. Matter*, 2:7009–7032, 1990.

[210] K.K. Mon and K. Binder. Finite size effects for the simulation of phase coexistence in the Gibbs ensemble near the critical point. *J. Chem. Phys.*, 96:6989–6995, 1992.

[211] J.R. Recht and A.Z. Panagiotopoulos. Finite-size effects and approach to criticality in Gibbs ensemble simulations. *Mol. Phys.*, 80:843–852, 1993.

[212] E. de Miguel, E. Martin del Rio, and M.M. Telo da Gama. Liquid-liquid phase equilibria of symmetrical mixtures by simulation in the semi-grand canonical ensemble. *J. Chem. Phys.*, 103:6188–6196, 1995.

[213] A.Z. Panagiotopoulos. Molecular simulation of phase equilibria. In E. Kiran and J.M.H. Levelt Sengers, editors, *Supercritical Fluids: Fundamentals for Applications*, pages 411–437. NATO ASI, Kluwer, Dordrecht, 1994.

[214] A.Z. Panagiotopoulos. Gibbs ensemble techniques. In M. Bauss *et al*, editor, *Observation, Prediction, and Simulation of Phase Transitions in Complex Fluids*, pages 463–501. NATO ASI, Kluwer, Dordrecht, 1995.

[215] G.L. Deitrick, L.E. Scriven, and H.T. Davis. Efficient molecular simulation of chemical potentials. *J. Chem. Phys.*, 90:2370, 1989.

[216] M.R. Stapleton and A.Z. Panagiotopoulos. Application of the excluded volume map sampling to phase equilibrium calculations in the Gibbs ensemble. *J. Chem. Phys.*, 92:1285–1293, 1990.

[217] A.Z. Panagiotopoulos. Molecular simulation of phase equilibria: Simple, ionic and polymeric fluids. *Fluid Phase Equilibria*, 76:97–112, 1992.

[218] J.P. Valleau. The Coulombic phase transition: Density-scaling Monte Carlo. *J. Chem. Phys.*, 95:584–589, 1991.

[219] B. Smit, C.P. Williams, E.M. Hendriks, and S.W. de Leeuw. Vapour-liquid equilibria for Stockmayer fluids. *Mol. Phys.*, 68:765–769, 1989.

[220] M.E. van Leeuwen, B. Smit, and E.M. Hendriks. Vapour-liquid equilibria of stockmayer fluids: computer simulations and perturbation theory. *Mol. Phys.*, 78:271–283, 1993.

[221] P.G. de Gennes and P.G. Pincus. Pair correlations in a ferromagnetic colloid. *Phys. kondens. Materie*, 11:189–198, 1970.

[222] V.I. Kalikmanov. Statistical thermodynamics of ferrofluids. *Physica A*, 183:25–50, 1992.

[223] G.S. Rushbrooke, G. Stell, and J.S. Hoye. Theory of polar liquids I. Dipolar hard spheres. *Mol. Phys.*, 26:1199–1215, 1973.

[224] K.-C. Ng, J.P. Valleau, G.M. Torrie, and G.N. Patey. Liquid-vapour co-existence of dipolar hard spheres. *Mol. Phys.*, 38:781–788, 1979.

[225] J.-M. Caillol. Search of the gas-liquid transition of dipolar hard spheres. *J. Chem. Phys.*, 98:9835–9849, 1993.

[226] M.E. van Leeuwen and B. Smit. What makes a polar fluid a liquid? *Phys. Rev. Lett.*, 71:3991–3994, 1993.

[227] J.J. Weis and D. Levesque. Chain formation in low density dipolar hard spheres: A Monte Carlo study. *Phys. Rev. Lett.*, 71:2729–2732, 1993.

[228] P.J. Camp, J.C. Shelley, and G.N. Patey. Isotropic fluid phases of dipolar hard spheres. *Phys. Rev. Lett.*, 84:115–118, 2000.

[229] T. Tlusty and S.A. Safran. Defect-induced phase separation in dipolar fluids. *Science*, 290:1328–1331, 2000.

[230] M.E. van Leeuwen, C.J. Peter, J de Swaan Arons, and A.Z. Panagiotopoulos. Investigation of the transition to liquid-liquid immiscibility for Lennard-Jones (12,6) systems, using the Gibbs-ensemble molecular simulations. *Fluid Phase Equilibria*, 66:57–75, 1991.

[231] V.I. Harismiadis, N.K. Koutras, D.P. Tassios, and A.Z. Panagiotopoulos. How good is conformal solution theory for phase equilibrium predictions. *Fluid Phase Equilibria*, 65:1–18, 1991.

[232] A.Z. Panagiotopoulos. Exact calculations of fluid-phase equilibria by Monte Carlo simulations in a new ensemble. *Int. J. Thermophys.*, 10:447, 1989.

[233] J.G. Amar. Application of the Gibbs ensemble to the study of fluid-fluid phase equilibria in a binary mixture of symmetric non-additive hard-spheres. *Mol. Phys.*, 4:739–745, 1989.

[234] R.D. Mountain and A.H. Harvey. Computer simulation of fluid-fluid phase equilibria in mixtures of non-additive soft disks. *J. Chem. Phys.*, 94:2238–2243, 1991.

[235] K.S. Shing. Infinite-dilution activity coefficients of quadrupolar Lennard-Jones mixtures from computer simulation. *J. Chem. Phys.*, 85:4633–4637, 1986.

[236] W.G.T. Kranendonk and D. Frenkel. Thermodynamic properties of binary hard sphere mixtures. *Mol. Phys.*, 72:699–713, 1991.

[237] J.K. Johnson, A.Z. Panagiotopoulos, and K.E. Gubbins. Reactive canonical Monte Carlo: A new simulation technique for reacting or associating fluids. *Mol. Phys.*, 81:717–733, 1994.

[238] W.R. Smith and B. Triska. The reaction ensemble method for the computer simulation of chemical and phase equilibria. I. Theory and basic examples. *Macromol. Symp.*, 81:343–354, 1994.

[239] M.R. Stapleton, D.J. Tildesley, and N. Quirke. Phase equilibria in polydisperse fluids. *J. Chem. Phys.*, 92:4456–4467, 1990.

[240] R. Agrawal and D.A. Kofke. Thermodynamic and structural properties of model systems at solid-fluid coexistence: II. Melting and sublimation of the Lennard-Jones system. *Mol. Phys.*, 85:43–59, 1995.

[241] M.H.J. Hagen, E.J. Meijer, G.C.A.M. Mooij, D. Frenkel, and H.N.W. Lekkerkerker. Does C_{60} have a liquid phase? *Nature*, 365:425–426, 1993.

[242] E.J. Meijer and D. Frenkel. Colloids dispered in polymer solution. A computer simulation study. *J. Chem. Phys.*, 100:6873–6887, 1994.

[243] W.O. Haag. Catalysis by zeolites—science and technology. In J. Weitkamp, H.G. Karge, H. Pfeifer, and W. Hölderich, editors, *Zeolites and Related Microp-*

orous Materials: State of the Art 1994, volume 84, pages 1375–1394. Studies in Surface Science and Catalysis, Elsevier, Amsterdam, 1994.

[244] P.G. Bolhuis, M.H.J. Hagen, and D. Frenkel. Isostructural solid-solid transition in crystalline systems with short-ranged interactions. *Phys. Rev. E*, 50:4880–4890, 1994.

[245] M. Dijkstra. *The Effect of Entropy on the Stability and Structure of Complex Fluids.* Ph.D. thesis, Rijksuniversiteit Utrecht, The Netherlands, 1994.

[246] M. Dijkstra and D. Frenkel. Simulation study of a two-dimensional system of semiflexible polymers. *Phys. Rev. E*, 50:349–357, 1994.

[247] P.G. Bolhuis and D.A. Kofke. Numerical study of freezing in polydisperse colloidal suspensions. *J. Phys.: Condens. Matt.*, 8:9627–9631, 1996.

[248] P.G. Bolhuis and D. Frenkel. Tracing the phase boundaries of hard spherocylinders. *J. Chem. Phys.*, 106:666–687, 1997.

[249] E.J. Meijer and F.E. Azhar. Novel procedure to determine coexistence lines by computer simulation. Application to hard-core Yukawa model for charge-stabilized colloids. *J. Chem. Phys.*, 106:4678–4683, 1997.

[250] F.A. Escobedo and J.J. de Pablo. Pseudo-ensemble simulations and Gibbs-Duhem integrations for polymers. *J. Chem. Phys.*, 106:2911–2923, 1997.

[251] F.A. Escobedo. Tracing coexistence lines in multicomponent fluid mixtures by molecular simulation. *J. Chem. Phys.*, 110:11999–12010, 1999.

[252] W.G. Hoover and F.H. Ree. Melting transition and communal entropy for hard spheres. *J. Chem. Phys.*, 49:3609–3617, 1968.

[253] G.S. Stringfellow, H.E. DeWitt, and W.L. Slattery. Equation of state of the one-component plasma derived from precision Monte Carlo calculations. *Phys. Rev. A*, 41:1105–1111, 1990.

[254] W.G. Hoover, S.G. Gray, and K.W. Johnson. Thermodynamic properties of the fluid and solid phases for the inverse power potentials. *J. Chem. Phys.*, 55:1228–1136, 1971.

[255] B.B. Laird and A.D.J. Haymet. Phase diagram for the inverse sixth power potential system from molecular dynamics simulations. *Mol. Phys.*, 75:71–80, 1992.

[256] W.G. Hoover, M. Ross, K.W. Johnson, D. Henderson, J.A. Barker, and B.C. Brown. Soft-sphere equation of state. *J. Chem. Phys.*, 52:4931–4941, 1970.

[257] J.-P. Hansen. Phase transition of the Lennard-Jones system. II. High-temperature limit. *Phys. Rev. A*, 2:221–230, 1970.

[258] H. Ogura, H. Matsuda, T. Ogawa, N. Ogita, and A. Ueda. Computer simulations for the melting curve maximum phenomenon. *Prog. Theoret. Phys.*, 58:419–433, 1977.

[259] D. Frenkel and A.J.C. Ladd. New Monte Carlo method to compute the free energy of arbitrary solids. Application to the fcc and hcp phases of hard spheres. *J. Chem. Phys.*, 81:3188–3194, 1984.

[260] P.G. Bolhuis and D.A. Kofke. Monte Carlo study of freezing of polydisperse hard spheres. *Phys. Rev. E*, 54:634–643, 1996.

[261] D.A. Kofke and P.G. Bolhuis. Freezing of polydisperse hard spheres. *Phys. Rev. E*, 59:618–622, 1999.

[262] P.A. Monson and D.A. Kofke. Solid-fluid equilibrium: Insights from simple molecular models. In I. Prigogine and S.A. Rice, editors, *Advances in Chemical Physics*, volume 115, pages 113–179. Wiley, New York, 2000.

[263] W.G. Hoover and F.H. Ree. Use of computer experiments to locate the melting transition and calculate the entropy in the solid phase. *J. Chem. Phys.*, 47:4873–4878, 1967.

[264] E.J. Meijer. *Computer Simulations of Molecular Solids and Colloidal Dispersions*. Ph.D. thesis, Rijksuniversiteit Utrecht, The Netherlands, 1993.

[265] D. Frenkel. Stability of the high-pressure body-centered-cubic phase of helium. *Phys. Rev. Lett.*, 56:858–860, 1986.

[266] J.Q. Broughton and G.H. Gilmer. Molecular dynamics investigation of the crystal-fluid interface. I. Bulk properties. *J. Chem. Phys.*, 79:5095–5104, 1983.

[267] J.A. Schouten. Phase equilibria in binary systems at very high pressures. *Phys. Rep.*, 172:33–92, 1989.

[268] J.P. Ryckaert and G. Ciccotti. Introduction of Andersen's demon in the molecular dynamics of systems with constraints. *J. Chem. Phys.*, 78:7368–7374, 1983.

[269] W.G. Hoover. Entropy for small classical crystals. *J. Chem. Phys.*, 49:1981–1982, 1968.

[270] J.M. Polson, E. Trizac, S. Pronk, and D. Frenkel. Finite-size corrections to the free energies of crystalline solids. *J. Chem. Phys.*, 112:5339–5342, 2000.

[271] R.J. Speedy. Pressure of the metastable hard-sphere fluid. *J. Phys.: Condensed Matter*, 9:8591–8599, 1997.

[272] R.J. Speedy. Pressure and entropy of hard-sphere crystals. *J. Phys.: Condensed Matter*, 10:4387–4391, 1998.

[273] B.J. Alder and T.E. Wainwright. Studies in molecular dynamics. II. Behavior of a small number of elastic spheres. *J. Chem. Phys.*, 33:1439–1451, 1960.

[274] B. Smit and D. Frenkel. Explicit expression for finite size corrections to the chemical potential. *J. Phys.: Condens. Matt.*, 1:8659–8665, 1989.

[275] M. Abramowitz and I. Stegun. *Handbook of Mathematical Functions*. Dover, New York, 1970.

[276] L.V. Woodcock. Entropy difference between the face-centred cubic and hexagonal close-packed crystal structures. *Nature*, 385:141–143, 1997.

[277] P.G. Bolhuis, D. Frenkel, S.-C. Mau, and D.A. Huse. Entropy differences between crystal phases. *Nature*, 388:235–237, 1997.

[278] A.D. Bruce, N.B. Wilding, and G.J. Ackland. Free energy of crystalline solids: A lattice-switch Monte Carlo method. *Phys. Rev. Lett.*, 79:3002–3005, 1997.

[279] B.A. Berg and T. Neuhaus. Multicanonical ensemble: A new approach to simulate first-order phase transitions. *Phys. Rev. Lett.*, 68:9–12, 1992.

[280] S.-C. Mau and D.A. Huse. Stacking entropy of hard-sphere crystals. *Phys. Rev. E*, 59:4396–4401, 1999.

[281] C.H. Bennett and B.J. Alder. Studies in molecular dynamics. IX. Vacancies in hard sphere crystals. *J. Chem. Phys.*, 54:4796–4808, 1970.

[282] R.K. Bowles and R.J. Speedy. Cavities in the hard-sphere crystal and fluid. *Mol. Phys.*, 83:113–125, 1994.

[283] W.C. Swope and H.C. Andersen. Thermodynamics, statistical mechanics and computer simulation of crystals with vacancies and interstitials. *Phys. Rev. A*, 46:4539–4548, 1992.

[284] R.J. Speedy and H. Reiss. Cavities in the hard-sphere fluid and crystal and the equation of state. *Mol. Phys.*, 72:999–1014, 1991.

[285] R.J. Speedy and H. Reiss. A computer simulation study of cavities in the hard-disc fluid and crystal. *Mol. Phys.*, 72:1015–1033, 1991.

[286] M.A. Bates and D. Frenkel. Influence of vacancies on the melting transition of hard disks in two dimensions. *Phys. Rev. E*, 61:5223–5227, 2000.

[287] S. Pronk and D. Frenkel. Vacancies and interstitials in hard sphere crystals. *J. Phys. Chem. B*, 105:6722–6727, 2001.

[288] M. Müller and W. Paul. Measuring the chemical potential of polymer solutions and melts in computer simulations. *J. Chem. Phys.*, 100:719–724, 1994.

[289] S.K. Kumar, I. Szleifer, and A.Z. Panagiotopoulos. Determination of the chemical potentials of polymeric systems from Monte Carlo simulations. *Phys. Rev. Lett.*, 66:2935–2938, 1991.

[290] S.K. Kumar, I. Szleifer, and A.Z. Panagiotopoulos. Determination of the chemical potentials of polymeric systems from Monte Carlo simulations. *Phys. Rev. Lett.*, 68:3658, 1992.

[291] K.K. Mon and R.B. Griffiths. Chemical potential by gradual insertion of a particle in Monte Carlo simulation. *Phys. Rev. A*, 31:956–959, 1985.

[292] B. Smit, G.C.A.M. Mooij, and D. Frenkel. Comment on "determination of the chemical potential of polymeric systems from Monte Carlo simulations". *Phys. Rev. Lett.*, 68:3657, 1992.

[293] J. Harris and S.A. Rice. A lattice model of a supported monolayer of amphiphilic molecules: Monte Carlo simulations. *J. Chem. Phys.*, 88:1298–1306, 1988.

[294] J.I. Siepmann. A method for the direct calculation of chemical potentials for dense chain systems. *Mol. Phys.*, 70:1145–1158, 1990.

[295] M.N. Rosenbluth and A.W. Rosenbluth. Monte Carlo simulations of the average extension of molecular chains. *J. Chem. Phys.*, 23:356–359, 1955.

[296] D. Frenkel and B. Smit. Unexpected length dependence of the solubility of chain molecules. *Mol. Phys.*, 75:983–988, 1992.

[297] D. Frenkel, G.C.A.M. Mooij, and B. Smit. Novel scheme to study structural and thermal properties of continuously deformable molecules. *J. Phys.: Condens. Matter*, 4:3053–3076, 1992.

[298] J.J. de Pablo, M. Laso, and U.W. Suter. Estimation of the chemical potential of chain molecules by simulation. *J. Chem. Phys.*, 96:6157–6162, 1992.

[299] K. Kremer and K. Binder. Monte Carlo simulations of lattice models for macromolecules. *Comp. Phys. Rep.*, 7:259–310, 1988.

[300] J. Batoulis and K. Kremer. Statistical properties of biased sampling methods for long polymer chains. *J. Phys. A: Math. Gen.*, 21:127–146, 1988.

[301] E.J. Maginn, A.T. Bell, and D.N. Theodorou. Sorption thermodynamics, siting and conformation of long n-alkanes in silicalite as predicted by

configurational-bias Monte Carlo integration. *J. Phys. Chem.*, 99:2057–2079, 1995.

[302] B. Smit and J.I. Siepmann. Simulating the adsorption of alkanes in zeolites. *Science*, 264:1118–1120, 1994.

[303] B. Smit and J.I. Siepmann. Computer simulations of the energetics and siting of n-alkanes in zeolites. *J. Phys. Chem.*, 98:8442–8452, 1994.

[304] B. Smit. Grand-canonical Monte Carlo simulations of chain molecules: Adsorption isotherms of alkanes in zeolites. *Mol. Phys.*, 85:153–172, 1995.

[305] G.C.A.M. Mooij and D. Frenkel. The overlapping distribution method to compute chemical potentials of chain molecules. *J. Phys.: Condens. Matter*, 6:3879–3888, 1994.

[306] P. Grassberger. Monte Carlo simulations of 3d self-avoiding walks. *J. Phys. A: Math. Gen.*, 26:2769–2776, 1993.

[307] P. Grassberger and R. Hegger. Monte Carlo simulations of off-lattice polymers. *J. Phys.: Condens. Matt.*, 7:3089–3097, 1993.

[308] P. Grassberger and R. Hegger. Monte Carlo simulations of off-lattice polymers (corrections and additions). *J. Phys.: Condens. Matter*, 7:3089–3097, 1995.

[309] T. Garel and H. Orland. Guided replication of random chains: a new Monte Carlo method. *J. Phys. A: Math. Gen.*, 23:L621–L626, 1990.

[310] D. Frenkel. Numerical techniques to study complex liquids. In M. Baus *et al.*, editor, *Observation, Prediction and Simulation of Phase Transitions in Complex Fluids*, pages 357–419. NATO ASI, Kluwer, Dordrecht, 1995.

[311] P. Grassberger. Pruned-enriched rosenbluth method: Simulations of theta polymers of chain length up to 1,000,000. *Phys. Rev. E*, 56:3682–3693, 1997.

[312] H. Frauenkron, U. Bastolla, E. Gerstner, P. Grassberger, and W. Nadler. New Monte Carlo algorithm for protein folding. *Phys. Rev. Lett.*, 80:3149–3152, 1998.

[313] H. Meirovitch. Scanning method as an unbiased simulation technique and its application to the study of self-attracting random walks. *Phys. Rev. A*, 32:3699–3708, 1985.

[314] P.J. Steinbach and B.R. Brooks. New spherical-cutoff methods for long-range forces in macromolecular simulation. *J. Comp. Chem.*, 15:667–683, 1994.

[315] J.W. Eastwood and R.W. Hockney. Shaping the force law in two-dimensional particle mesh models. *J. Comp. Phys.*, 16:342–359, 1974.

[316] L. Greengard and V. Rokhlin. A fast algorithm for particle simulations. *J. Comp. Phys.*, 73:325–348, 1987.

[317] T.A. Darden, D. York, and L. Pedersen. Particle mesh Ewald: An N log(N) method for Ewald sums in large systems. *J. Chem. Phys.*, 98:10089–10092, 1993.

[318] P.P. Ewald. Die Berechnung optischer und elektrostatischer Gitterpotentiale. *Ann. Phys.*, 64:253–287, 1921.

[319] S.W. de Leeuw, J.W. Perram, and E.R. Smith. Simulation of electrostatic systems in periodic boundary conditions. I. Lattice sums and dielectric constants. *Proc. R. Soc. London A*, 373:27–56, 1980.

[320] S.W. de Leeuw, J.W. Perram, and E.R. Smith. Simulation of electrostatic systems in periodic boundary conditions. II. Equivalence of boundary conditions. *Proc. R. Soc. London A*, 373:56–66, 1980.

[321] S.W. de Leeuw, J.W. Perram, and E.R. Smith. Simulation of electrostatic systems in periodic boundary conditions. III. Further theory and applications. *Proc. R. Soc. London A*, 388:177–193, 1983.

[322] J.-P. Hansen. Molecular-dynamics simulations of Coulomb systems in two and three dimensions. In G. Ciccotti and W.G. Hoover, editors, *Molecular Dynamics Simulations of Statistical Mechanics Systems*, pages 89–129. Proceedings of the 97th Int. "Enrico Fermi" School of Physics, North Holland, Amsterdam, 1986.

[323] M.P. Tosi. Cohesion of ionic solids in the Born model. In F. Seitz and D. Turnbull, editors, *Solid State Physics: Advances in Research and Applications*, pages 1–120. Academic Press, New York, 1964.

[324] J.-M. Caillol and D. Levesque. Numerical simulations of homogeneous and inhomogeneous ionic systems: An efficient alternative to the Ewald method. *J. Chem. Phys.*, 94:597–607, 1991.

[325] M. Neumann. Dipole moment fluctuation formulas in computer simulations of polar systems. *Mol. Phys.*, 50:841–858, 1983.

[326] J.A. Kolafa and J.W. Perram. Cutoff errors in the Ewald summation formulae for point charge systems. *Mol. Sim.*, 9:351–368, 1992.

[327] H.G. Petersen. Accuracy and efficiency of the particle mesh Ewald method. *J. Chem. Phys.*, 103:3668–3679, 1995.

[328] K. Esselink. A comparison of algorithms for long-range interactions. *Comp. Phys. Comm.*, 87:375–395, 1995.

[329] A.W. Appel. An efficient program for many-body simulation. *SIAM J. Sci. Stat. Comput.*, 6:85–103, 1985.

[330] K. Esselink. The order of Appel's algorithm. *Inf. Proc. Lett.*, 41:141–147, 1992.

[331] J. Barnes and P. Hut. A hierarchical $\mathcal{O}(N \log N)$ force-calculation algorithm. *Nature*, 324:446–449, 1986.

[332] K.E. Schmidt and M.A. Lee. Implementing the fast multipole method in three dimensions. *J. Stat. Phys.*, 63:1223–1235, 1991.

[333] H.Q. Ding, N. Karasawa, and W.A. Goddard III. Atomic level simulations on a million particles: The cell multipole method for Coulombic and London nonbond interactions. *J. Chem. Phys.*, 97:4309–4315, 1992.

[334] M. Deserno and C Holm. How to mesh up Ewald sums. I. A theoretical and numerical comparison of various particle mesh routines. *J. Chem. Phys.*, 109:7678–7693, 1998.

[335] U. Essmann, L. Perera, M.L. Berkowitz, T.A. Darden, H. Lee, and L. Pedersen. A smooth particle mesh Ewald method. *J. Chem. Phys.*, 103:8577–8593, 1995.

[336] E.L. Pollock and J. Glosli. Comments on P^3M, FMM, and the Ewald method for large periodic Coulombic systems. *Comp. Phys. Comm.*, 95:93–110, 1996.

[337] C. Sagui and T.A. Darden. Molecular dynamics simulations of biomolecules: Long-range electrostatic effects. *Annu. Rev. Biophys. Biomol. Struc.*, 28:155–179, 1999.

[338] J.V.L. Beckers, C.P. Lowe, and S.W. de Leeuw. An iterative PPPM method for simulating Coulombic systems on distributed memory parallel computers. *Mol. Sim.*, 20:369–383, 1998.

[339] M. Deserno and C Holm. How to mesh up Ewald sums. II. An accurate error estimate for the particle-particle-particle-mesh algorithm. *J. Chem. Phys.*, 109:7694–7701, 1998.

[340] H.Q. Ding, N. Karasawa, and W.A. Goddard III. The reduced cell multipole method for coulombic interactions in periodic systems with million-atom unit cells. *Chem. Phys. Lett.*, 196:6–10, 1992.

[341] B.A. Luty, I.G. Tironi, and W.F. van Gunsteren. Lattice-sum methods for calculating electrostatic interactions. *J. Chem. Phys.*, 103:3014–3021, 1995.

[342] F. Figueirido, R.M. Levy, R. Zhou, and B.J. Berne. Large scale simulation of macromolecules in solution: Combining the periodic fast multipole method with multiple time step integrators. *J. Chem. Phys.*, 106:9835–9849, 1997.

[343] D.M. Heyes, M. Barber, and J.H.R. Clarke. Molecular dynamics computer simulation of surface properties of crystalline potassium chloride. *J. Chem. Soc. Faraday Trans. II*, 73:1485–1496, 1977.

[344] A. Grzybowski, E. Gwozdz, and A. Brodka. Ewald summation of electrostatic interactions in molecular dynamics of a three-dimensional system with periodicity in two directions. *Phys. Rev. B*, 61:6706–6712, 2000.

[345] A.H. Widmann and D.B. Adolf. A comparison of Ewald summation techniques for planar surfaces. *Comput. Phys. Commun.*, 107:167–186, 1997.

[346] S.W. de Leeuw and J.W. Perram. Electrostatic lattice sums for semi-infinite lattices. *Mol. Phys.*, 37:1313–1322, 1979.

[347] E.R. Smith. Electrostatic potentials for simulations of thin layers. *Mol. Phys.*, 65:1089–1104, 1988.

[348] E. Spohr. Effect of boundary conditions and system size on the interfacial properties of water and aqueous solutions. *J. Chem. Phys.*, 107:6342–6348, 1994.

[349] J. Hautman and M.L. Klein. An Ewald summation method for planar surfaces and interfaces. *Mol. Phys.*, 75:379–395, 1992.

[350] I.C. Yeh and M.L. Berkowitz. Ewald summation for systems with slab geometry. *J. Chem. Phys.*, 111:3155–3162, 1999.

[351] E.R. Smith. Electrostatic energy in ionic crystals. *Proc. R. Soc. London A*, 375:475–505, 1981.

[352] P.S. Crozier, R.L. Rowley, E. Spohr, and D. Henderson. Comparison of charged sheets and corrected 3D Ewald calculations of long-range forces in slab geometry electrolyte systems with solvent molecules. *J. Chem. Phys.*, 112:9253–9257, 2000.

[353] R.F. Cracknell, D. Nicholson, N.G. Parsonage, and H. Evans. Rotational insertion bias: A novel method for simulating dense phases of structured particles, with particular application to water. *Mol. Phys.*, 71:931–943, 1990.

[354] J.I. Siepmann and D. Frenkel. Configurational-bias Monte Carlo: A new sampling scheme for flexible chains. *Mol. Phys.*, 75:59–70, 1992.

[355] J.J. de Pablo, M. Laso, and U.W. Suter. Simulation of polyethylene above and below the melting point. *J. Chem. Phys.*, 96:2395–2403, 1992.

[356] G.C.A.M. Mooij, D. Frenkel, and B. Smit. Direct simulation of phase equilibria of chain molecules. *J. Phys.: Condens. Matter*, 4:L255–L259, 1992.

[357] G.C.A.M. Mooij and D. Frenkel. A systematic optimization scheme for configurational bias Monte Carlo. *Mol. Sim.*, 17:41–55, 1996.

[358] B. Chen and J.I. Siepmann. Transferable potentials for phase equilibria. 3. Explicit-hydrogen description of normal alkanes. *J. Phys. Chem. B*, 103:5370–5379, 1999.

[359] M.D. Macedonia and E.J. Maginn. A biased grand canonical Monte Carlo method for simulating adsorption using all-atom and branched united atom models. *Mol. Phys.*, 96:1375–1390, 1999.

[360] T.J.H. Vlugt. *Adsorption and Diffusion in Zeolites: A Computational Study*. Ph.D. thesis, University of Amsterdam, 2000.

[361] T.J.H. Vlugt, R. Krishna, and B. Smit. Molecular simulations of adsorption isotherms for linear and branched alkanes and their mixtures in silicalite. *J. Phys. Chem. B*, 103:1102–1118, 1999.

[362] M. Dijkstra. Confined thin films of linear and branched alkanes. *J. Chem. Phys.*, 107:3277–3288, 1997.

[363] M.G. Martin and J.I. Siepmann. Novel configurational-bias Monte Carlo method for branched molecules. Transferable potentials for phase equilibria. 2. United-atom description of branched alkanes. *J. Phys. Chem. B*, 103:4508–4517, 1999.

[364] D. Chandler and P.G. Wolynes. Exploiting the isomorphism between quantum theory and classical statistical mechanics of polyatomic fluids. *J. Chem. Phys.*, 74:4078–4095, 1981.

[365] V.G. Mavrantzas, T.D. Boone, E. Zervopoulou, and D.N. Theodorou. End-bridging Monte Carlo: A fast algorithm for atomistic simulation of condensed phases of long polymer chains. *Macromolecules*, 32:5072–5096, 1999.

[366] M. Dijkstra and D. Frenkel. Evidence for entropy-driven demixing in hard-core fluids. *Phys. Rev. Lett.*, 72:298–300, 1994.

[367] M. Dijkstra, D. Frenkel, and J.-P. Hansen. Phase separation in binary hard-core mixtures. *J. Chem. Phys.*, 101:3179–3189, 1994.

[368] H. Yamakawa. *Modern Theory of Polymer Solutions*. Harper and Row, New York, 1971.

[369] M. Vendruscolo. Modified configurational bias Monte Carlo method for simulation of polymer systems. *J. Chem. Phys.*, 106:2970–2976, 1996.

[370] C.D. Wick and J.I. Siepmann. Self-adapting fixed-end-point configurational-bias Monte Carlo method for the regrowth of interior segments of chain molecules with strong intramolecular interactions. *Macromolecules*, 33:7207–7218, 2000.

[371] Z. Chen and F.A. Escobedo. A configurational-bias approach for the simulation of inner sections of linear and cyclic molecules. *J. Chem. Phys.*, 113:11382–11392, 2000.

[372] P.V.K. Pant and D.N. Theodorou. Variable connectivity method for the atomistic Monte Carlo simulation of polydisperse polymer melts. *Macromolecules*, 28:7224–7234, 1995.

[373] M.G. Wu and M.W. Deem. Analytical rebridging Monte Carlo: Application to cis/trans isomerization in proline-containing, cyclic peptites methods for cyclic peptides. *J. Chem. Phys.*, 111:6625–6632, 1999.

[374] M.G. Wu and M.W. Deem. Efficient Monte Carlo methods for cyclic peptides. *Mol. Phys.*, 97:559–580, 1999.

[375] T. Biben. *Structure et stabilité des fluides à deux composants: des fluides atomiques aux suspensions colloidales.* Ph.D. thesis, Université Claude Bernard, Lyon, 1993.

[376] T. Biben, P. Bladon, and D. Frenkel. Depletion effects in binary hard-sphere fluids. *J. Phys.: Condens. Matt.*, 8:10799–10821, 1996.

[377] P.G. Bolhuis and D. Frenkel. Numerical study of the phase diagram of a mixture of spherical and rodlike colloids. *J. Chem. Phys.*, 101:9869–9875, 1995.

[378] J.C. Shelley and G.N. Patey. A configurational bias Monte Carlo method for ionic solutions. *J. Chem. Phys.*, 100:8265–8270, 1994.

[379] K. Esselink, L.D.J.C. Loyens, and B. Smit. Parallel Monte Carlo simulations. *Phys. Rev. E*, 51:1560–1568, 1995.

[380] L.D.J.C. Loyens, B. Smit, and K. Esselink. Parallel Gibbs-ensemble simulations. *Mol. Phys.*, 86:171–183, 1995.

[381] K. Esselink, P.A.J. Hilbers, S. Karaborni, J.I. Siepmann, and B. Smit. Simulating complex fluids. *Mol. Sim.*, 14:259–274, 1995.

[382] T.J.H. Vlugt, M.G. Martin, B. Smit, J.I. Siepmann, and R. Krishna. Improving the efficiency of the cbmc algorithm. *Mol. Phys.*, 94:727–733, 1998.

[383] B. Smit and T.L.M. Maesen. Commensurate "freezing" of alkanes in the channels of a zeolite. *Nature*, 374:42–44, 1995.

[384] R. Evans. Microscopic theories of simple fluids and their interfaces. In J. Charvolin, J.F. Joanny, and J. Zinn-Justin, editors, *Liquides aux Interfaces/Liquids at interfaces*, pages 1–98. Les Houches, Session XLVIII, 1988, North Holland, Amsterdam, 1990.

[385] S.J. Gregg and K.S.W. Sing. *Adsorption, Surface Area and Porosity.* Academic Press, London, 1982.

[386] J.I. Siepmann, S. Karaborni, and B. Smit. Simulating the critical properties of complex fluids. *Nature*, 365:330–332, 1993.

[387] W.J.M. van Well, J.P. Wolthuizen, B. Smit, J.H.C. van Hooff, and R.A. van Santen. Commensurate freezing of n-alkanes in silicalite. *Angew. Chem. (Int. Ed.)*, 34:2543–2544, 1995.

[388] R. Krishna, B. Smit, and T.J.H. Vlugt. Sorption-induced diffusion-selective separation of hydrocarbon isomers using silicalite. *J. Phys. Chem. A*, 102:7727–7730, 1998.

[389] G.C.A.M. Mooij. *Novel Simulation Techniques for the Study of Polymer Phase Equilibria.* Ph.D. thesis, Rijksuniversiteit Utrecht, The Netherlands, 1993.

[390] B. Smit, S. Karaborni, and J.I. Siepmann. Computer simulations of vapour-liquid phase equilibria of n-alkanes. *J. Chem. Phys.*, 102:2126–2140, 1995. Erratum: *J. Chem. Phys.* 109:352, 1998.

[391] M. Laso, J.J. de Pablo, and U.W. Suter. Simulation of phase equilibria for chain molecules. *J. Chem. Phys.*, 97:2817–2819, 1992.

[392] J.I. Siepmann, S. Karaborni, and B. Smit. Vapor-liquid equilibria of model alkanes. *J. Am. Chem. Soc.*, 115:6454–6455, 1993.

[393] C. Tsonopoulos. Critical constant of normal alkanes from methane to polyethylene. *AIChE Journal*, 33:2080–2083, 1987.

[394] W.L. Jorgensen, J.D. Madura, and C.J. Swenson. Optimized intermolecular potential function for liquid hydrocarbons. *J. Am. Chem. Soc.*, 106:6638–6646, 1984.

[395] S. Toxvaerd. Molecular dynamics calculation of the equation of state of alkanes. *J. Chem. Phys.*, 93:4290–4295, 1990.

[396] C. Tsonopoulos and Z. Tan. Critical constant of normal alkanes from methane to polyethylene II. Application of the Flory theory. *Fluid Phase Equilibria*, 83:127–138, 1993.

[397] M.J. Anselme, M. Gude, and A.S. Teja. The critical temperatures and densities of the n-alkanes from pentane to octadecane. *Fluid Phase Equilibria*, 57:317–326, 1990.

[398] Y.-J. Sheng, A.Z. Panagiotopoulos, S.K. Kumar, and I. Szleifer. Monte Carlo calculation of phase equilibria for a bead-spring polymeric model. *Macromolecules*, 27:400–406, 1994.

[399] F.A. Escobedo and J.J. de Pablo. Simulation and prediction of vapour-liquid equilibria for chain molecules. *Mol. Phys.*, 87:347–366, 1996.

[400] M. Mondello and G.S. Grest. Molecular dynamics of linear and branched alkanes. *J. Chem. Phys.*, 103:7156–7165, 1995.

[401] M. Mondello, G.S. Grest, E.B. Webb III, and P. Peczak. Dynamics of n-alkanes: Comparison to Rouse model. *J. Chem. Phys.*, 109:798–805, 1998.

[402] J.D. Moore, S.T. Cui, H.D. Cochran, and P.T. Cummings. Rheology of lubricant basestocks: A molecular dynamics study of c-30 isomers. *J. Chem. Phys.*, 113:8833–8840, 2000.

[403] C. McCabe, S.T. Cui, P.T. Cummings, P.A. Gordon, and R.B. Saeger. Examining the rheology of 9-octylheptadecane to giga-pascal pressures. *J. Chem. Phys.*, 114:1887–1891, 2001.

[404] M.G. Martin and J.I. Siepmann. Transferable potentials for phase equilibria (trappe): I. united-atom description of n-alkanes. *J. Phys. Chem. B*, 102:2569–2577, 1998.

[405] S.K. Nath, F.A. Escobedo, and J.J. de Pablo. On the simulation of vapor-liquid equilibria for alkanes. *J. Chem. Phys.*, 108:9905–9911, 1998.

[406] C.D. Wick, M.G. Martin, and J.I. Siepmann. Transferable potentials for phase equilibria. 4. United-atom description of linear and branched alkenes and alkylbenzenes. *J. Phys. Chem. B*, 104:8008–8016, 2000.

[407] S.K. Nath, B.J. Banaszak, and J.J. de Pablo. A new ninted atom force field for α-olefins. *J. Chem. Phys.*, 1114:3612–3161, 2001.

[408] M.E. van Leeuwen and B. Smit. Molecular simulations of the vapour-liquid coexistence curve of methanol. *J. Phys. Chem.*, 99:1831–1833, 1995.

[409] B. Chen, J.J. Potoff, and J.I. Siepmann. Monte Carlo calculations for alcohols and their mixtures with alkanes. transferable potentials for phase equilibria. 5.

United-atom description of primary, secondary, and tertiary alcohols. *J. Phys. Chem. B*, 105:3093–3104, 2001.

[410] S. Consta, N.B. Wilding, D. Frenkel, and Z. Alexandrowicz. Recoil growth: An efficient simulation method for multi-polymer systems. *J. Chem. Phys.*, 110:3220–3228, 1999.

[411] S. Consta, T.J.H. Vlugt, J. Wichers Hoeth, B. Smit, and D. Frenkel. Recoil growth algorithm for chain molecules with continuous interactions. *Mol. Phys.*, 97:1243–1254, 1999.

[412] H. Meirovitch. Statistical properties of the scanning simulation method for polymer-chains. *J. Chem. Phys.*, 89:2514–2522, 1988.

[413] Z. Alexandrowicz and N.B. Wilding. Simulation of polymers with rebound selection. *J. Chem. Phys.*, 109:5622–5626, 1998.

[414] M. Falcioni and M.W. Deem. A biased Monte Carlo scheme for zeolite structure solution. *J. Chem. Phys.*, 110:1754–1766, 1999.

[415] A.P. Lyubartsev, A.A. Martsinovski, S.V. Shevkunov, and P.N. Vorontsov-Vel'yaminov. New approach to Monte Carlo calculation of the free energy: Method of expanded ensembles. *J. Chem. Phys.*, 96:1776–1783, 1992.

[416] E. Marinari and G. Parisi. Simulated tempering: A new Monte Carlo scheme. *Europhys. Lett.*, 19:451–458, 1992.

[417] C.J. Geyer and E.A. Thompson. Annealing markov chain Monte Carlo with applications to the ancestral inference. *J. Am. Stat. Soc.*, 90:909–920, 1995.

[418] S. Kirkpatrick, C.D. Gelatt Jr., and M.P. Vecchi. Optimization by simulated annealing. *Science*, 220:671–680, 1983.

[419] C.J. Geyer. Markov chain Monte Carlo maximum likelihood. In *Computing Science and Statistics*, pages 156–163. Proceedings of the 23rd Symposium on the Interface, 1991.

[420] D.D. Frantz, D.L. Freeman, and J.D. Doll. Reducing quasi-ergodic behaviour in Monte Carlo simulations by J-walking: Application to atomic clusters. *J. Chem. Phys.*, 93:2769–2784, 1990.

[421] I. Nezbeda and J.A. Kolafa. A new version of the insertion particle method for determining the chemical potential by Monte Carlo simulation. *Mol. Sim.*, 5:391–403, 1991.

[422] K. Shing and A.Z. Azadipour. A new simulation method for the grand canonical ensemble. *Chem. Phys. Lett.*, 190:386–390, 1992.

[423] L.F. Vega, K.S. Shing, and L.F. Rull. A new algorithm for molecular dynamics simulations in the grand canonical ensemble. *Mol. Phys.*, 82:439–453, 1994.

[424] Q.L. Yan and J.J. de Pablo. Hyper-parallel tempering Monte Carlo: Application to the Lennard-Jones fluid and the restricted primitive model. *J. Chem. Phys.*, 111:9509–9516, 1999.

[425] Q.L. Yan and J.J. de Pablo. Hyperparallel tempering Monte Carlo simulation of polymeric systems. *J. Chem. Phys.*, 113:1276–1282, 2000.

[426] A. Bunker and B. Dünweg. Parallel excluded volume tempering for polymer melts. *Phys. Rev. E*, 63:art. no. 010902, 2001.

[427] S. Auer and D. Frenkel. Prediction of absolute crystal-nucleation rate in hard-sphere colloids. *Nature*, 409:1020–1023, 2001.

[428] N.B. Wilding. Critical-point and coexistence-curve properties of the Lennard-Jones fluid: A finite-size scaling study. *J. Phys.: Condens. Matter*, 4:3087–3108, 1992.

[429] A.M. Ferrenberg and R.H. Swendsen. New Monte Carlo technique for studying phase transitions. *Phys. Rev. Lett.*, 61:2635–2638, 1988.

[430] S. Duane, A. Kennedy, B.J. Pendleton, and D. Roweth. Hybrid Monte Carlo. *Phys. Lett. B.*, 195:216–222, 1987.

[431] B. Mehlig, D.W. Heermann, and B.M. Forrest. Exact Langevin algorithms. *Mol. Phys.*, 76:1347–1357, 1992.

[432] B. Mehlig, D.W. Heermann, and B.M. Forrest. Hybrid Monte Carlo method for condensed matter systems. *Phys. Rev. B*, 45:679–685, 1992.

[433] B.M. Forrest and U.W. Suter. Generalized coordinate Hybrid Monte Carlo. *Mol. Phys.*, 82:393–410, 1994.

[434] G.M. Crippen. Conformational analysis by energy embedding. *J. Comp. Chem.*, 3:471–476, 1982.

[435] R.H. Swendsen and J.-S. Wang. Nonuniversal critical dynamics in Monte Carlo simulations. *Phys. Rev. Lett.*, 58:86–88, 1987.

[436] D. Wu, D. Chandler, and B. Smit. Electrostatic analogy for surfactant assemblies. *J. Phys. Chem.*, 96:4077–4083, 1992.

[437] B. Smit, K. Esselink, P.A.J. Hilbers, N.M. van Os, and I. Szleifer. Computer simulations of surfactant self-assembly. *Langmuir*, 9:9–11, 1993.

[438] F.H. Stillinger. Variational model for micelle structure. *J. Chem. Phys.*, 78:4654–4661, 1983.

[439] G. Orkoulas and A.Z. Panagiotopoulos. Chemical potentials in ionic systems from Monte Carlo simulations with distance-biased test particle insertion. *Fluid Phase Equilibria*, 93:223–231, 1993.

[440] G. Orkoulas and A.Z. Panagiotopoulos. Free energy and phase equilibria for the restricted primitive model of ionic fluids from Monte Carlo simulations. *J. Chem. Phys.*, 101:1452–1459, 1994.

[441] D. Frenkel. Advanced Monte Carlo techniques. In M.P. Allen and D.J. Tildesley, editors, *Computer Simulation in Chemical Physics*, pages 93–152. NATO ASI, Kluwer, Dordrecht, 1993.

[442] R. Car and M. Parrinello. Unified approach for molecular dynamics and density-functional theory. *Phys. Rev. Lett.*, 55:2471–2474, 1985.

[443] J.P. Ryckaert, G. Ciccotti, and H.J.C. Berendsen. Numerical integration of the Cartesian equations of motion of a system with constraints: Molecular dynamics of n-alkanes. *J. Comp. Phys.*, 23:237–341, 1977.

[444] S.W. de Leeuw, J.W. Perram, and H.G. Petersen. Hamilton's equations for constrained dynamical systems. *J. Stat. Phys.*, 61:1203–1222, 1990.

[445] G. Ciccotti. Molecular dynamics simulations of nonequilibrium phenomena and rare dynamical events. In M. Meyer and V. Pontikis, editors, *Proceedings of the NATO ASI on Computer Simulation in Materials Science*, pages 119–137. Kluwer, Dordrecht, 1991.

[446] G. Galli and A. Pasquarello. First-principle molecular dynamics. In M.P. Allen and D.J. Tildesley, editors, *Computer Simulation in Chemical Physics*, pages 261–313. NATO ASI, Kluwer, Dordrecht, 1993.

[447] D.K. Remler and P.A. Madden. Molecular dynamics without effective potentials via the Carr-Parrinello approach. *Mol. Phys.*, 70:921–966, 1990.

[448] H. Löwen, P.A. Madden, and J.-P. Hansen. Ab initio description of counterion screening in colloidal suspensions. *Phys. Rev. Lett.*, 68:1081–1084, 1992.

[449] H. Löwen, P.A. Madden, and J.-P. Hansen. Nonlinear counterion screening in colloidal suspensions. *J. Chem. Phys.*, 98:3275–3289, 1993.

[450] C.G. Gray and K.E. Gubbins. *Theory of Molecular Fluids. 1. Fundamentals*. Clarendon, Oxford, 1984.

[451] M-L. Saboungi, A. Rahman, J.W. Halley, and M. Blander. Molecular dynamics studies of complexing in binary molten salts with polarizable anions: MAX_4. *J. Chem. Phys.*, 88:5818–5823, 1988.

[452] M. Sprik and M.L. Klein. A polarizable model for water using distributed charge sites. *J. Chem. Phys.*, 89:7556–7560, 1988.

[453] M. Wilson and P.A. Madden. Polarization effects in ionic systems from first principles. *J. Phys.: Condens. Matter*, 5:2687–2706, 1993.

[454] M. Sprik. Computer simulation of the dynamics of induced polarization fluctuations in water. *J. Chem. Phys.*, 95:2283–2291, 1991.

[455] P. Procacci and M. Marchi. Taming the Ewald sum in molecular dynamics simulations of solvated proteins via a multiple time step algorithm. *J. Chem. Phys.*, 104:3003–3012, 1996.

[456] P. Procacci, M. Marchi, and G.J. Martyna. Electrostatic calculations and multiple time scales in molecular dynamics simulation of flexible molecular systems. *J. Chem. Phys.*, 108:8799–8803, 1998.

[457] C.H. Bennett. Exact defect calculations in model substances. In A.S. Nowick and J.J. Burton, editors, *Diffusion in Solids: Recent Developments*, pages 73–113. Academic Press, New York, 1975.

[458] D. Chandler. Statistical mechanics of isomerization dynamics in liquids and the transition state approximation. *J. Chem. Phys.*, 68:2959–2970, 1978.

[459] B.J. Berne, G. Ciccotti, and D.F. Coker. *Classical and Quantum Dynamics in Condensed Phase Simulations*. World Scientific, Singapore, 1998.

[460] M.J. Ruiz-Montero, D. Frenkel, and J.J. Brey. Efficient schemes to compute diffusive barrier crossing rates. *Mol. Phys.*, 90:925–941, 1997.

[461] W.H. Miller. Importance of nonseparability in quantum mechanical transition-state theory. *Acc. Chem. Res.*, 9:306–312, 1976.

[462] M.A. Wilson and D. Chandler. Molecular dynamics study of the cyclohexane interconversion. *Chem. Phys.*, 149:11–20, 1990.

[463] M.H. Müser and G. Ciccotti. Two-dimensional orientational motion as a multichannel reaction. *J. Chem. Phys.*, 103:4273–4278, 1995.

[464] H.A. Kramer. Brownian motion in a field of force and the diffusion model of chemical reactions. *Physica*, 7:284–304, 1940.

[465] P.G. Bolhuis, C. Dellago, and D. Chandler. Sampling ensembles of deterministic transition pathways. *Faraday Discuss.*, 110:421–436, 1998.

[466] D. Chandler. Finding transition pathways: Throwing ropes of rough mountain passes, in the dark. In B.J. Berne, G. Ciccotti, and D.F. Coker, editors, *Classical and Quantum Dynamics in Condensed Phase Simulations*, International School Enrico Fermi, pages 51–66. Italian Physical Society, World Scientific, Singapore, 1998.

[467] C. Dellago, P.G. Bolhuis, F.S. Csajka, and D. Chandler. Transition path sampling and the calculation of rate constants. *J. Chem. Phys.*, 108:1964–1977, 1998.

[468] C. Dellago, P.G. Bolhuis, and D. Chandler. On the calcualtion of reaction rate constants in the transition path ensemble. *J. Chem. Phys.*, 110:6617–6625, 1999.

[469] L.R Pratt. A statistical method for identifying transition states in high dimensional problems. *J. Chem. Phys.*, 9:5045–5048, 1986.

[470] T.J.H. Vlugt and B. Smit. On the efficient sampling of pathways in the transition path ensemble. *Phys. Chem. Comm.*, 2:3–7, 2001.

[471] P.L. Geissler, C. Dellago, and D. Chandler. Kinetic pathways of ion pair dissociation in water. *J. Phys. Chem. B*, 103:3706–3710, 1999.

[472] G. Mills, H. Jonsson, and G.K. Schenter. Reversible work transition-state theory—Application to dissociative adsorption of hydrogen. *Surface Science*, 324:305–337, 1995.

[473] G. Henkelman and H. Jonsson. A dimer method for finding saddle points on high dimensional potential surfaces using only first derivatives. *J. Chem. Phys.*, 111:7010–7022, 1999.

[474] G. Henkelman, B.P. Uberuaga, and H. Jonsson. A climbing image nudged elastic band method for finding saddle points and minimum energy paths. *J. Chem. Phys.*, 113:9901–9904, 2000.

[475] G. Henkelman and H. Jonsson. Improved tangent estimate in the nudged elastic band method for finding minimum energy paths and saddle points. *J. Chem. Phys.*, 113:9978–9985, 2000.

[476] G.T. Barkema and N. Mousseau. Event-based relaxation of continuous disordered systems. *Phys. Rev. Lett.*, 77:4358–4361, 1996.

[477] A.F. Voter. Hyperdynamics: Accelerated molecular dynamics of infrequent events. *Phys. Rev. Lett.*, 78:3908–3911, 1997.

[478] M.R. Sorensen and A.F. Voter. Temperature-accelerated dynamics for simulation of infrequent events. *J. Chem. Phys.*, 112:9599–9606, 2000.

[479] A.F. Voter. Parallel replica method for dynamics of infrequent events. *Phys. Rev. B*, 57:R13985–R13988, 1998.

[480] P.J. Hoogerbrugge and J.M.V.A. Koelman. Simulating microscopic hydrodynamics phenomena with dissipative particle dynamics. *Europhys. Lett.*, 19:155–160, 1992.

[481] J.M.V.A. Koelman and P.J. Hoogerbrugge. Dynamic simulation of hard-sphere suspensions under steady shear. *Europhys. Lett.*, 21:363–368, 1993.

[482] A.J.C. Ladd. Short-time motion of colloidal particles - numerical-simulation via a fluctuating lattice-boltzmann equation. *Phys. Rev. Lett.*, 70:1339–1342, 1993.

[483] A. Malevanets and R. Kapral. Solute molecular dynamics in a mesoscale solvent. *J. Chem. Phys.*, 112:7260–7269, 2000.

[484] I. Pagonabarraga and D. Frenkel. Dissipative particle dynamics for interacting systems. *J. Chem. Phys.*, 115:5015–5026, 2001.

[485] R.D. Groot and P.B. Warren. Dissipative particle dynamics: bridging the gap between atomistic and mesoscopic simulation. *J. Chem. Phys.*, 107:4423–4435, 1997.

[486] C.P. Lowe. An alternative approach to dissipative particle dynamics. *Europhys. Lett.*, 47:145–151, 1999.

[487] P. Español and P.B. Warren. Statistical mechanics of dissipative particle dynamics. *Europhys. Lett.*, 30:191–196, 1995.

[488] P. Español. Hydrodynamics from dissipative particle dynamics. *Phys. Rev. E*, 52:1734–1742, 1995.

[489] C.A. Marsh, G. Backx, and M.H. Ernst. Fokker-Planck-Boltzmann equation for dissipative particle dynamics. *Europhys. Lett.*, 38:411–415, 1997.

[490] C.A. Marsh, G. Backx, and M.H. Ernst. Static and dynamic properties of dissipative particle dynamics. *Phys. Rev. E*, 56:1676–1691, 1997.

[491] A.J. Masters and P.B. Warren. Kinetic theory for dissipative particle dynamics: The importance of collisions. *Europhys. Lett.*, 48:1–7, 1999.

[492] C.A. Marsh and J.M. Yeomans. Dissipative particle dynamics: The equilibrium for finite time steps. *Europhys. Lett.*, 37:511–516, 1997.

[493] I. Pagonabarraga, M.H.J. Hagen, and D. Frenkel. Self-consistent dissipative particle dynamics. *Europhys. Lett.*, 42:377–382, 1998.

[494] S.M. Willemsen, T.J.H. Vlugt, H.C.J. Hoefsloot, and B. Smit. Combining dissipative particle dynamics and Monte Carlo techniques. *J. Comp. Phys.*, 147:507–517, 1998.

[495] J.G. Kirkwood and F.P. Buff. . *J. Chem. Phys.*, 17:338, 1949.

[496] J.H. Irving and J.G. Kirkwood. . *J. Chem. Phys.*, 18:817, 1950.

[497] J.P.R.B. Walton, D.J. Tildesley, and J.S. Rowlinson. The pressure tensor at the planar surface of a liquid. *Mol. Phys.*, 48:1357–1368, 1983.

[498] M.J.P. Nijmeijer, A.F. Bakker, C. Bruin, and J.H. Sikkenk. A molecular dynamics simulation of the Lennard-Jones liquid-vapour interface. *J. Chem. Phys.*, 89:3789–3792, 1988.

[499] J. Bonet Avalos and A.D. Mackie. Dissipative particle dynamics with energy conservation. *Europhys. Lett.*, 40:141–146, 1997.

[500] P. Español. Dissipative particle dynamics with energy conservation. *Europhys. Lett.*, 40:631–636, 1997.

[501] J. Bonet Avalos and A.D. Mackie. Dynamic and transport properties of dissipative particle dynamics with energy conservation. *J. Chem. Phys.*, 111:5267–5276, 1997.

[502] A.D. Mackie, J. Bonet Avalos, and V. Navas. Dissipative particle dynamics with energy conservation: Modelling of heat flow. *Phys. Chem. Chem. Phys.*, 1:2039–2049, 1999.

[503] S.M. Willemsen, H.C.J. Hoefsloot, D.C. Visser, P.J. Hamersma, and P.D. Iedema. Modelling phase change with disspative particle dynamics using a consistent boundary condition. *J. Comp. Phys.*, 162:385–394, 2000.

[504] M. Ripoll, P. Español, and M.H. Ernst. Dissipative particle dynamics with energy conservation: Heat conduction. *Int. J. Mod. Phys. C*, 9:1329–1338, 1998.

[505] R.D. Groot, T.J. Madden, and D.J. Tildesley. On the role of hydrodynamic interactions in block copolymer microphase separation. *J. Chem. Phys.*, 110:9739–9749, 1999.

[506] G.R McNamara and G. Zanetti. Use of the Boltzmann-equation to simulate lattice-gas automata. *Phys. Rev. Lett.*, 61:2332–2335, 1988.

[507] R. Benzi, S. Succi, and M. Vergassola. The lattice Boltzmann-equation—Theory and applications. *Phys. Rep.*, 222:145–197, 1992.

[508] S. Chen and G.D. Doolen. Lattice Boltzmann method for fluid flows. *Ann. Rev. Fluid Mech.*, 30:329–364, 1998.

[509] U. Frisch, B. Hasslacher, and Y. Pomeau. Lattice-gas automata for the Navier-Stokes equation. *Phys. Rev. Lett.*, 56:1505–1508, 1986.

[510] A.J.C. Ladd. Numerical simulations of particulate suspensions via a discretized Boltzmann-equation. 1. Theoretical foundation. *J. Fluid Mech.*, 271:285–309, 1994.

[511] M.H.J. Hagen, C.P. Lowe, and D. Frenkel. Non-Boltzmann behavior from the Boltzmann equation. *Phys. Rev. E*, 51:4287–4291, 1995.

[512] M.R. Swift, W.M. Osborn, and J.M. Yeomans. Lattice Boltzmann simulation of nonideal fluids. *Phys. Rev. Lett.*, 75:830–833, 1995.

[513] G.A. Bird. *Molecular Gas Dynamics and the Direct Simulation of Gas Flows*. Clarendon, Oxford, 1994.

[514] F.J. Alexander, A.L. Garcia, and B.J. Alder. A consistent Boltzmann algorithm. *Phys. Rev. Lett.*, 74:5212–5215, 1995.

[515] G.A. Bird. Recent advances and current challenges for DSMC. *Comput. Math. Appl.*, 35:1–14, 1998.

[516] F.J. Alexander and A.L. Garcia. The direct simulation Monte Carlo method. *Comp. Phys.*, 11:588–593, 1997.

[517] R.P. Feynman, R.B. Leighton, and M. Sands. *The Feynmann Lectures on Physics*. Addison-Wesley, Reading, Mass., 1965.

[518] J.D. Meiss. Symplectic maps, variational principles, and transport. *Rev. Mod. Phys.*, 64:795–848, 1992.

[519] S.K. Gray, D.W. Noid, and B.Q. Sumpter. Symplectic integrators for large-scale molecular-dynamics simulations - A comparison of several explicit methods. *J. Chem. Phys.*, 101:4062–4072, 1994.

[520] M.E. Tuckerman, C.J. Mundy, and G.J. Martyna. On the classical statistical mechanics of non-Hamiltonian systems. *Europhys. Lett.*, 45:149–155, 2000.

[521] K. Cho, J.D. Joannopoulos, and L. Kleinman. Constant temperature molecular dynamics with momentum conservation. *Phys. Rev. E*, 47:3145–3151, 1993.

[522] D. Frenkel and A.J.C. Ladd. Elastic constants of hard-sphere crystals. *Phys. Rev. Lett.*, 59:1169, 1987.

[523] D.R. Squire, A.C. Holt, and W.G. Hoover. Isothermal elastic constants for argon. Theory and Monte Carlo simulations. *Physica A*, 42:388–397, 1969.

[524] M. Sprik, R.W. Impey, and M.L. Klein. Second-order elastic constants for the Lennard-Jones solid. *Phys. Rev. B*, 29:4368–4374, 1984.

[525] O. Farago and Y. Kantor. Fluctuation formalism for elastic constants in hard-spheres-and-tethers systems. *Phys. Rev. E*, 61:2478–2489, 2000.

[526] S. Sengupta, P. Nielaba, and K. Binder. Elastic moduli, dislocation core energy, and melting of hard disks in two dimensions. *Phys. Rev. E*, 61:6294–6301, 2000.

[527] R. Zwanzig and N.K. Ailawadi. Statistical error due to finite averaging in computer experiments. *Phys. Rev.*, 182:280–283, 1969.

[528] D. Frenkel. Intermolecular spectroscopy and computer simulations. In J. van Kranendonk, editor, *Intermolecular Spectroscopy and Dynamical Properties of Dense Systems*, International School of Physics "Enrico Fermi", pages 156–201. Italian Physical Society, North Holland, Amsterdam, 1980.

[529] G. Jacucci and A. Rahman. Comparing the efficiency of Metropolis Monte Carlo and molecular-dynamics methods for configuration space sampling. *Nuovo Cimento*, D4:341–356, 1984.

[530] H. Bekker, E.J. Dijkstra, M.K.R. Renardus, and H.J.C. Berendsen. An efficient, box shape independent non-bonded force and virial algorithm for molecular dynamics. *Mol. Sim.*, 14:137–151, 1995.

[531] D.J. Auerbach, W. Paul, C. Lutz, A.F. Bakker, W.E. Rudge, and F.F. Abraham. A special purpose parallel computer for molecular dynamics: motivation, design, implementation, and application. *J. Phys. Chem.*, 91:4881–4890, 1987.

[532] D. Ruelle. *Statistical Mechanics: Rigorous Results*. Benjamin, Reading, Mass., 1969.

[533] R.B. Dingle. *Asymptotic Expansions, Their Derivation and Interpretation*. Academic Press, New York, 1973.

[534] I.M.J.J. van de Ven-Lucassen, T.J.H. Vlugt, A.J.J. van der Zanden, and P.J.A.M. Kerkhof. Using molecular dynamics to obtain Maxwell-Stefan diffusion coefficients in liquid systems. *Mol. Phys.*, 94:495–503, 1998.

[535] http://www.dl.ac.uk/CCP/CCP5/main.html.

[536] G. Ilario, I.G. Tironi, and W.F. van Gunsteren. A molecular dynamics study of chloroform. *Mol. Phys.*, 83:381–403, 1994.

[537] http://www.dsm.fordham.edu/ ftnchek/.

[538] P. Hellekalek. Good random number generators are (not so) easy to find. *Math. Comput. Sim.*, 46:485–505, 1998.

Author Index

Index